A Handbook on

Instrumentation Engineering

Contains well illustrated formulae & key theory concepts

for

GATE, PSUs

& OTHER COMPETITIVE EXAMS

MADE EASY Publications

MADE EASY Publications

Corporate Office: 44-A/4, Kalu Sarai (Near Hauz Khas Metro Station), New Delhi-110016
E-mail: infomep@madeeasy.in
Contact: 011-45124660, 0-8860378007
Visit us at: www.madeeasypublications.org

A Handbook on Instrumentation Engineering

First Edition: 2016
Reprint: 2018
Reprint: 2020
Reprint: 2022

Director's Message

B. Singh (Ex. IES)

During the current age of international competition in Science and Technology, the Indian participation through skilled technical professionals have been challenging to the world. Constant efforts and desire to achieve top positions are still required.

I feel every candidate has ability to succeed but competitive environment and quality guidance is required to achieve high level goals. At MADE EASY, we help you to discover your hidden talent and success quotient to achieve your ultimate goals. In my opinion GATE & PSU's exams are tool to enter in to main stream of Nation serving. The real application of knowledge and talent starts, after you enter in to the working system. Here in MADE EASY you are also trained to become winner in your life and achieve job satisfaction.

MADE EASY alumni have shared their winning stories of success and expressed their gratitude towards quality guidance of MADE EASY. Our students have not only secured All India First Ranks in ESE, GATE and PSU entrance examinations but also secured top positions in their career profiles. Now, I invite you to become alumni of MADE EASY to explore and achieve ultimate goal of your life. I promise to provide you quality guidance with competitive environment which is far advanced and ahead than the reach of other institutions. You will get the guidance, support and inspiration that you need to reach the peak of your career.

I have true desire to serve Society and Nation by way of making easy path of the education for the people of India.

After a long experience of teaching in Instrumentation Engineering over the period of time MADE EASY team realised that there is a need of good *Handbook* which can provide the crux of Instrumentation Engineering in a concise form to the student to brush up the formulae and important concepts required for GATE, PSUs and other competitive examinations. This *handbook* contains all the formulae and important theoretical aspects of Instrumentation Engineering. It provides much needed revision aid and study guidance before examinations.

B. Singh (Ex. IES)
CMD, MADE EASY Group

A Handbook on

Instrumentation Engineering

CONTENTS

OOOO

A Handbook on
Instrumentation Engineering

1

Engineering Mathematics

CONTENTS

Linear Algebra

1

MATRIX

Principal Diagonal: In a square matrix all elements a_{ij} for which $i = j$ are elements of principal diagonal.

Matrices

1. **Upper Triangular matrix:** A square matrix in which all the elements below the principle diagonal are zero.
2. **Lower Triangular Matrix:** A square matrix in which all the elements above the principle diagonal are zero.
3. **Diagonal Matrix:** A square matrix in which all the elements other than the elements of principle diagonal are zero.
4. **Scalar Matrix:** *A* diagonal matrix with all elements of principle diagonal being same.
5. **Idempotent Matrix:** '*A*' is square matrix i.e. $A^2 = A$.
6. **Involutary Matrix:** '*A*' is square matrix i.e. $A^2 = I$.
7. **Nilpotent Matrix:** '*A*' is square matrix i.e. $A^m = 0$ where m is the least positive integer and m is also called as Index of class of Nilpotent matrix *A*.
8. **Transpose Matrix:** A^T is transpose matrix of matrix *A*. A^T can be obtained by switching the rows as columns and columns as rows of *A*.
9. **Symmetric Matrix:** '*A*' is a square matrix i.e. $A^T = A$.
10. **Skew-Symmetric Matrix:** '*A*' is a square matrix i.e. $A^T = -A$.
11. **Orthogonal Matrix:** '*A*' is a orthogonal matrix i.e. $A^T = A^{-1}$ or $AA^T = I = A^TA$.
12. **Conjugate Matrix of *A*** ($\bar{A}$) or ($\sim A$) : '*A*' is any matrix, by replacing the elements by corresponding conjugate complex numbers the matrix obtained is conjugate of '*A*'.

Example:

$$A = \begin{bmatrix} 2+3i & 4+7i & 5 \\ 2i & 3 & 9-i \end{bmatrix} \Rightarrow \bar{A} = \begin{bmatrix} 2-3i & 4-7i & 5 \\ -2i & 3 & 9+i \end{bmatrix}$$

13. **Transpose Conjugate Matrix (A^θ) or (A^*):** $(\bar{A})^T$.

14. **Hermitian Matrix:** 'A' is a square matrix i.e. $A^\theta = A$.

    All diagonal elements of hermitian matrix are real number and all off-diagonal elements above and below the principle diagonal must be conjugate of each other. i.e. $a_{ij} = \overline{a_{ji}}$.

    Example : $\begin{bmatrix} 2 & 3-4i \\ 3+4i & 5 \end{bmatrix}$

15. **Skew-Hermitian Matrix:** 'A' is a square matrix i.e. $A^\theta = -A$.

    All diagonal elements of Skew-Hermitian matrix are purely imaginary or zero and all off-diagonal elements above and below the principle diagonal must be conjugate of each other with opposite sign. i.e. $a_{ij} = -\overline{a_{ji}}$.

    Example : $\begin{bmatrix} 2i & 3-4i \\ -3-4i & 5i \end{bmatrix}$

16. **Unitary Matrix:** 'A' is a square matrix i.e. $A^\theta = A^{-1}$ or $A\,A^\theta = I = A^\theta\,A$.

17. **Boolean Matrix:** Any matrix with only elements '0' or '1'

18. **Sparse Matrix:** A matrix 'A' in which more number of elements are zeros.

19. **Singular and Non-singular Matrix:** A square matrix 'A' is singular if $|A| = 0$, and non singular if $|A| \neq 0$. Only non-singular matrices have inverse.

20. **Adjoint Matrix:** Transpose of cofactors matrix. i.e. Adj (A) = (Cof (A))^t

Properties of Matrices

- $A + B = B + A$ (Commutative)
- $(A + B) + C = A + (B + C)$ (Associative)
- $AB \neq BA$ (Not commutative)
- $(AB)\,C = A(BC)$ (Associative)
- $A(B + C) = AB + AC$ (Distributive)
- $A = \frac{1}{2}(A + A^T) + \frac{1}{2}(A - A^T)$
- $A(Adj\,A) = (Adj\,A)\,A = |A|\,I_n$
- $Adj\,(AB) = (Adj\,B).\,(Adj\,A)$
- $A^{-1} = \frac{Adj\,A}{|A|}; |A| \neq 0$

- $(A^{-1})^{-1} = A$ and $(A^{-1})^T = (A^T)^{-1}$
- $(AB)^{-1} = B^{-1} A^{-1}$.
- If A is a square matrix of order n then $|Adj\,A| = (\det A)^{n-1} = |A|^{n-1}$ and $|Adj\,(Adj\,A)| = |A|^{(n-1)^2}$
- If $|A| \neq 0$ then $|A^{-1}| = \frac{1}{|A|}$
- If $A_{n \times n}$ matrix then $|KA| = K^n|A|$
- If A is square matrix then
  (i) $A + A^T$ is always symmetric
  (ii) $A - A^T$ always skew-symmetric
- If A and B are symmetric then
  (i) $A + B$ is also symmetric.
  (ii) $A - B$ is also symmetric.
  (iii) $AB + BA$ is symmetric
  (iv) $AB - BA$ is skew-symmetric
  (v) A^n and B^n are symmetric
- If A and B are skew-symmetric then,
  (i) $A + B$ is also skew-symmetric.
  (ii) $A - B$ is also skew-symmetric.
  (i) A^n and B^n are symmetric, if 'n' is even
  (ii) A^n and B^n are skew-symmetric, if 'n' is odd
- The determinant of orthogonal matrix and unitary matrix A has absolute value '1'.
- If $A_{m \times n}$ and B_{nxp} then product of AB requires
  (i) mnp multiplications
  (ii) $m(n-1)p$ additions
  (iii) for each entry, n multiplications and $(n-1)$ additions.
- $(A^T)^T = A$, $(kA)^T = k(A^T)$
- $(A + B)^T = A^T + B^T$, $(AB)^T = B^T A^T$
- **Rank of Matrix ($r(A)$):** It is the order of its largest non-vanishing (non-zero) minor of the matrix.
- Rank is equal to the number of linearly independent rows or columns in the matrix.

- The system of linear equation $AX = B$ has a solution (consistent) iff rank of A = Rank of $(A|B)$.
- The system $AX = B$ has
  (*i*) A unique solution iff Rank (A) = Rank $(A|B)$ = Number of variables.
  (*ii*) Infinitely many solutions $\Leftrightarrow$ Rank (A) = Rank $(A|B)$ < number of variables.
  (*iii*) No solution if Rank $(A) \neq$ Rank $(A|B)$ i.e. Rank (A) < Rank $(A|B)$.
- The system $AX = 0$ has
  (*i*) Unique solution (zero solution or trivial solution) if Rank (A) = number of variables.
  (*ii*) Infinitely many number of solutions (non-trivial solutions) if Rank (A) < number of variables.
- If Rank $(A) = r$, and number of variables = n then, the number of linearly independent infinite solutions of $AX = 0$ is $(n - r)$.
- In the system of homogenous linear equation $AX = 0$
  (*i*) If A is singular then the system possesses non-trivial solution (i.e. infinite solution).
  (*ii*) If A is non-singular then the system possesses trivial (zero) solution (i.e. unique solution).
- Rank of a diagonal matrix = Number of non-zero elements in diagonal.
- If A and B are two matrices
  (*i*) $r(A + B) \leq r(A) + r(B)$
  (*ii*) $r(A - B) \geq r(A) - r(B)$
  (*iii*) $r(AB) \leq \min\{r(A), r(B)\}$
- If a matrix A has rank 'R', then A contains 'R' linearly independent vectors (row/column).
- The system of homogeneous linear equations such that number of unknowns (or variables) exceeds the number of equations necessarily possesses a non-zero solution.

Eigen Value

Let 'A' be a square matrix of order n and λ be a scalar then $|A - \lambda I| = 0$ is the characteristic equation of A. The roots of characteristic equation are called eigen values/lantent roots/Characteristic roots.

- The set of eigen values of matrix is called "spectrum of matrix".
- A matrix of order n will have n latent roots not necessarily distinct.

Eigen Vector

Corresponding to each eigen value λ, there exists a non-zero solution X such that $(A - \lambda I)X = 0$ then X is eigen vector/latent/vector/characteristic vector of A.

Properties of Eigen Values

- Sum of eigen values of a matrix = sum of elements of principal diagonal (trace).
$$\Sigma\lambda_i = \lambda_1 + \lambda_2 + \lambda_3 + \ldots + \lambda_n = \text{Trace of } A$$
- Product of eigen values = Determinant of matrix.
$$\Pi\lambda_i = \lambda_1 \cdot \lambda_2 \cdot \ldots \lambda_n = |A|$$
- If λ is eigen value of A then $\frac{1}{\lambda}$ is eigen value of A^{-1}. (provided $\lambda \neq 0$ i.e. A is non-singular).
- Eigen values of A and A^T are same
- If λ is eigen value of orthogonal matrix then $\frac{1}{\lambda}$ is also its eigen value $[\because \; A^T = A^{-1}]$
- If $\lambda_1, \lambda_2, \ldots, \lambda_n$ are eigen values of matrix 'A', then

(*i*) $\lambda_1^m, \lambda_2^m, \ldots, \lambda_n^m$ are eigen values of matrix A^m.

(*ii*) $\lambda_1 + K, \lambda_2 + K, \ldots, \lambda_n + K$ are eigen values of $A + KI$

(*iii*) $(\lambda_1 - K)^2, (\lambda_2 - K)^2, \ldots, (\lambda_n - K)^2$ are eigen values of $(A - KI)^2$

(*iv*) $K\lambda_1, K\lambda_2, \ldots, K\lambda_n$ are eigen values of KA.

- The eigen values of symmetric matrix are real.
- The eigen values of skew-symmetric matrix are either purely imaginary or zero.
- The modulus of the eigen values of orthogonal and unitary matrices = 1.
- If a matrix is either lower or upper triangular or diagonal then the principal diagonal elements themselves are the eigen values.
- Zero is eigen value of a matrix iff the matrix is singular.

■■■■

Differential Equations

2

Differential Equation

Order of differential equation : It is the order of the highest derivative appearing in it.

Degree of Differential Equation : It is the degree of the highest derivative occurring in it, after expressing the equation free from radicals and fractions as far as derivatives are concerned.

Differential Equations of First Order First Degree

Equations of first order and first degree can be expressed in the form $y' = f(x, y)$. Following are the different ways of solving equations of first order and first degree:

1. Variable Separable : $f(x)dx + g(y)dy = 0$

   $\int f(x)dx + \int g(y)dy = c$ is the solution.

2. Homogenous Equation : $\frac{dy}{dx} = \frac{f(x,y)}{g(x,y)}$

   - To solve a homogeneous equation, substitute $y = Vx$

   $$\frac{dy}{dx} = V + x\frac{dV}{dx}$$

   - Separate the variable V and x and integrate.

   Equations Reducible to Homogenous Equation

   The differential equation: $\frac{dy}{dx} = \frac{ax+by+x'}{a'x+b'y+c'}$

   This is non-homogeneous but can be converted to homogeneous equation.

   Case I : If $\frac{a}{a'} \neq \frac{b}{b'}$

   Substitude $x = X + h$, $y = Y + k$(h and K are constants)

   Solve for h and k, we get

   $$ah + bk + c = 0$$
   $$a'h + b'k + c' = 0$$

   $$\frac{dY}{dX} = \frac{aX + bY}{a'X + b'Y}$$

Case II : If $\frac{a}{a'} = \frac{b}{b'}$

$$\frac{a}{a'} = \frac{b}{b'} = \frac{1}{m} \text{ (say)}$$

$$\frac{dy}{dx} = \frac{(ax+by)+C}{m(ax+by)+C'}$$

Substitute $ax + by = t$, so that

$$\frac{dt}{dx} = \frac{b(t+C)}{mt+C^1} + a$$

Solve by variable separable method

3. **Linear Equations :** The standard form of a linear equation of first order

$\frac{dy}{dx} + P(x)y = Q(x)$, where P and Q are functions of x.

Second order linear equation $\frac{d^2y}{dx^2} + P(x)\frac{dy}{dx} + Q(x)y = R(x)$

It is commonly known as "Leibnitz's linear equations"

Integrating factor, $\text{I.F.} = e^{\int P dx}$

$$ye^{\int Pdx} = \int Q\cdot(I.F.) + dx + C \Rightarrow y(I.F.) = \int Q\cdot(I.F.)dx + C$$

Note: ....

- **The degree of every linear differential equation is always one but if the degree of the differential equation is one, then it need not be linear.**

....

For Ex. : $\frac{d^4y}{dx^4} + 3x^2\left(\frac{dy}{dx}\right)^3 + y^{2003} = 0$

Bernoulli's Equation

$$\frac{dy}{dx} + Py = Qy^n \quad \text{where } P \,\&\, Q \text{ are functions of } x \text{ only}$$

Divide by y^n

$$y^{-n}\frac{dy}{dx} + Py^{1-n} = Q$$

Substituting, $y^{1-n} = z$, we get

$$\frac{dz}{dx} + (1-n)Pz = Q(1-n)]$$

This is a linear equation and can be solved easily.

4. **Exact Differential Equations :** The equations are given as
$M(x, y)dx + N(x, y)dy = 0$
The necessary and sufficient condition for the differential equations $Mdx + Ndy = 0$ to be exact is $\frac{\partial M}{\partial y} = \frac{\partial N}{\partial x}$.

Solution of exact differential equation

$$\int_{y \text{ is constant}} M\,dx + \int(\text{terms of } N \text{ not containing } x)dy = C$$

Equation Reducible to Exact Equation

Integrating Factor : Sometimes an equation which is not exact may become so on multiplication by some function known as Integrating factgor ($I.F.$).

Case I : Finding by inspection

1. $x dy + y dx = d(x\,y)$
2. $\frac{xdy - ydx}{xy} = d\left[\log\left(\frac{y}{x}\right)\right]$
3. $\frac{xdy - ydx}{x^2 + y^2} = d\left[\tan^{-1}\left(\frac{y}{x}\right)\right]$
4. $\frac{xdy - ydx}{x^2} = d\left(\frac{y}{x}\right)$
5. $\frac{xdy - ydx}{y^2} = -d\left(\frac{y}{x}\right)$
6. $\frac{xdy - ydx}{x^2 - y^2} = d\left[\frac{1}{2}\log\left(\frac{x+y}{x-y}\right)\right]$

Case II : When $Mdx + Ndy = 0$ is homogenous in x and y and $Mx + Ny \neq 0$, then $I.F. = \frac{1}{Mx + Ny}$.

Case III : If the equation $f_1(x, y)ydx + f_2(x, y)ydy = 0$ and $Mx - Ny \neq 0$, then $I.F. = \frac{1}{Mx - Ny}$.

Case IV : If the $Mdx + Ndy = 0$ and $\frac{1}{N}\left(\frac{\partial M}{\partial y} - \frac{\partial n}{\partial x}\right) = f(x)$, then $I.F. = e^{\int f(x)dx}$.

Case V : If the equation $Mdx + Ndy = 0$ and $\frac{1}{M}\left(\frac{\partial N}{\partial x} - \frac{\partial M}{\partial y}\right) = f(y)$, then $I.F. = e^{\int f(y)dy}$. .

Linear Differential Equation with Constant Coefficients

$$\frac{d^n y}{dx^n} + \frac{d^{n-1} y}{dx^{n-1}} + \ldots + k_n y = X$$

The equation can be written as $(D^n + k_1 D^{n-1} + \ldots + k_n)\, y = X$ (where $D = d/dx$)

$f(D)y + X$; $f(D) = 0$ is called auxiliary equation.

Rules for Finding Complimentary Function

Case I : If all the roots of AE are real and different $(D - m_1)(D - m_2) \ldots (D - m_n)y = 0$

So, the solution is $y = C_1^{m_1 x} + C_2^{m_2 x} + \ldots + C_n^{m_n x}$

Case II : If two roots are equal, i.e. $m_1 = m_2$, then $y = (C_1 + C_2 x)\, e^{m_1 x}$

Similarly, if $m_1 = m_2 = m_3$, $y = (C_1 + C_2 x + C_3 x^2)e^{m_1 x}$

Case III : If one pair of roots are imaginary, i.e. $m_1 = \alpha + i\beta$, $m_2 = \alpha - i\beta$, then $y = e^{\alpha x}(C_1 \cos\beta x + C_2 \sin\beta x)$

Case IV : If two pairs of root are imaginary, i.e. repeated imaginary root, then $y = e^{\alpha x}[(C_1 x + C_2)\cos\beta x + (C_3 x + C_4)\sin\beta x]$

Rules for Finding Particualr Integral

$$P.I. = \frac{1}{D^n + k_1 D^{n-1} + \ldots + K_n} X = \frac{1}{f(D)} X$$

Case I :

When $X = e^{ax}$

$P.I. = \frac{1}{f(D)} e^{ax}$ put $D = a$ $[f(a) \neq 0]$

$P.I. = x \frac{1}{f'(D)} e^{ax}$ put $D = a$ $[f'(a) \neq 0, f(a) = 0]$

$P.I. = x^2 \frac{1}{f''(D)} e^{ax}$ put $D = a$ $[f'(a) = 0, f'(a) = 0, f''(a) \neq 0]$

Case II :

When $X = \sin(ax + b)$ or $\cos(ax + b)$

$P.I. = \frac{1}{\phi(D^2)} \sin(ax + b)$ put $D^2 = -a^2[\phi(-a^2) \neq 0]$

$$= x\frac{1}{\phi'(D^2)}\sin(ax+b) \quad \text{put } D^2 = -a^2[\phi'(-a^2) \neq 0,\ \phi(-a^2) = 0]$$

$$= x\frac{1}{\phi''(D^2)}\sin(ax+b) \quad \text{put } D^2 = -a^2[\phi''(-a^2) \neq 0,\ \phi'(-a^2) = 0,\ \phi(-a^2) = 0]$$

Case III :

When $X = x^m$, m being positive integer

$$P.I. = \frac{1}{f(D)}x^m = [f(D)]^{-1}x^m$$

$$= f(D)\left[1+\frac{1}{f(d)}\right]^{-1}x^m = f(D)[1 - f(D) + f^2(D) - f^3(D) + ...]x^m$$

Case IV :

When $X = e^{ax}V$ where V is function of x

$$P.I. = \frac{1}{f(D)}e^{ax}V$$

$$= e^{ax}\frac{1}{f(D+a)}V \text{ then evaluate } \frac{1}{f(D+a)}V \text{ as in Case I, II, III.}$$

Case V :

When $X = xV(x)$

$$P.I. = \frac{1}{f(D)}xV(x) = \left[x - \frac{f'(D)}{f(D)}\right]\frac{1}{f(D)}V(x)$$

Case VI :

When X is any other function of x

$$P.I. = \frac{1}{f(D)}X$$

Factorize $f(D) = (D - m_1)(D - m_2) \ldots (D - m_n)$ and resolve $1/f(D)$ into partial fractions and then apply, $\frac{1}{D-a}X = e^{ax}\int Xe^{-ax}\,dx$ on each terms

Complete Solution : y = C.F. + P.I.

Cauchy-Euler Equation (Homogenous Linear Equation)

$$x^n\frac{d^n y}{dx^n} + k_1 x^{n-1}\frac{d^{n-1}y}{dx^{n-1}} + \ldots + k_{n-1}\frac{dy}{dx} + k_n y = X$$

Substitute $x = e^t$

$$x\frac{dy}{dx} = Dy$$

$$x^2\frac{d^2y}{dx^2} = D(D-1)y$$

$$x^3\frac{d^3y}{dx^3} = D(D-1)(D-2)y$$

After substituting these differentials, the Cauchy-Euler equation results in a linear equation with constant coefficients.

Legendre's Linear Equation

$$(ax+b)^n\frac{d^ny}{dx^n} + k_1(ax+b)^{n-1}\frac{d^{n-1}y}{dx^{n-1}} + ...k_ny = X$$

$$ax + b = e^t$$

$$\Rightarrow \quad t = ln(ax+b)$$

$$(ax+b)\frac{dy}{dx} = aDy$$

$$(ax+b)^2\frac{d^2y}{dx^2} = a^2D(D-1)y$$

$$(ax+b)^3\frac{d^3y}{dx^3} = a^3D(D-1)(D-2)y$$

After substituting these differentials, the Legendre's equation results in a linear equation with constant coefficients.

Partial Differential Equation

$$z = f(x, y)$$

$$\frac{\partial z}{\partial x} = p, \frac{\partial z}{\partial y} = q, \frac{\partial^2 z}{\partial x^2} = r, \frac{\partial^2 z}{\partial x \partial y} = s, \frac{\partial^2 z}{\partial y^2} = t$$

Homogenous Linear Equation with Constant Coefficients

$\frac{\partial^n z}{\partial x^n} + k_1\frac{\partial^n z}{\partial x^{n-1}dy} + ... + k_n\frac{\partial^n z}{\partial y^n} = f(x,y) \rightarrow$ this is called homogenous because all terms containing derivative is of same order.

$D^n + k_1D^{n-1}D' + ... + k_nD'^n = f(x, y) \left(\text{where } D = \frac{\partial}{\partial x} \text{ and } D' = \frac{\partial}{\partial y}\right)$

$f(D, D') = f(x, y)$

Step 1 : Find the C.F.

1. Write A.E.

$$m^n + k_1 m^{n-1} + ...k_n = 0 \text{ where } m = \frac{D}{D'}. \text{ Roots are } m_1, m_2, ..., m_n.$$

$$m^n + k_1 m^{n-1} + k_n = 0, \text{ where } m = \frac{D}{D'}. \text{ Roots are } m_1, m_2, ..., m_n.$$

2. $CF = f_1(y + m_1x) + f_2(y + m_2x) + ...$ m_1, m_2 are distinct
   $CF = f_1(y + m_1x) + xf_2(y + m_2x) + f_3(y + m_3x) ...$ m_1, m_2, m_3 two equal roots.
   $CF = f_1(y + m_1x) + xf_2(y + m_2x) + x^2f_3(y + m_3x)...$ m_1, m_2, m_3 three equal roots.

Step 2 : Finding *P.I.*

$$P.I. = \frac{1}{f(D, D')} f(x, y)$$

1. When $F(ax + by) = e^{ax + by}$, put $[D = a, D' = b]$
2. When $F(x, y) = \sin(mx + ny)$, put $\left[D^2 = -m^2, DD' = -mn, D'^2 = -n^2\right]$
3. When $F(x, y) = x^m y^n$, $P.I. = \frac{1}{f(D, D')} x^m y^n = [f(D, D')]^{-1} x^m y^n$.
4. When $F(x, y)$ is any function of x and y. $\frac{1}{f(D, D')} f(x, y)$, resolve $\frac{1}{f(D, D')}$ into partial fractions considering $f(D, D')$ as a function of D alone and operate each partial fraction of $f(x, y)$ remembering that $\frac{1}{(D - mD')} f(x, y) = \int f(x, c - mx)dx$ where c is replaced by $y + mx$ after integration.

■■■■

Probability & Statistics

3

MEAN, MEDIAN AND MODE

- $$\text{Mean } (\bar{X}) = \frac{\sum_{i=1}^{n} x_i}{n} = \frac{\sum_{i=1}^{n} f_i x_i}{\sum f_i}$$

- $$\text{Median} = \frac{x_{n/2} + \left(x_{n/2+1}\right)}{2};\ n \text{ is even}$$
  $$= x_{n+1/2};\ n \text{ is odd}$$
  $$= L_{med} + \left(\frac{N/2 - F}{f}\right) \times w$$

where L_{med} = lower limit of median class
N = Σf_i
F = Cumulative frequency upto the median class (cumulative frequency of the preceding class)
f = Frequency of the median class

Mode : Value of 'x' corresponding to maximum frequency.

$$L_{mode} + \frac{f_m - f_1}{(2f_m - f_1 - f_2)} \times h$$

where L_{mode} = lower limit of modal class
f_m = frequency of modal class
f_1 = preceding frequency of modal class
f_2 = Following frequency of modal class

Note:

- Mode = 3 Median – 2 Mean [for Asymmetric distribution]
- Mean = Mode = Median [for Symmetric distribution]

AXIOMS OF PROBABILITY

Let A and B be two events. Then

1. $P(\bar{A}) = 1 - P(A)$
2. $P(\phi) = 0$; ϕ is the empty set
3. $P(A - B) = P(A) - P(A \cap B)$
4. $P(A - B) = P(A \cap \bar{B})$
5. $P(A \cup B) = P(A) + P(B) - P(A \cap B)$
6. $P(A \cup B) = P(A) + P(B)$; mutually exclusive events.
7. $P(A \cap B) = P(A) \cdot P(B/A) = P(B) \cdot P(A/B)$
8. $P(A \cap B) = P(A) \cdot P(B)$; independent events.
9. $P(A \cap B) = \phi$; mutually exclusive events.
10. $P(S) = 1$; S is sample space.
11. $\max(0, P(A) + P(B) - 1) \le P(A \cap B) \le \min(P(A), P(B))$
12. $\max(P(A), P(B)) \le P(A \cup B) \le \min(1, P(A) + P(B))$
13. $\max(0, P(A_1) + P(A_2) \ldots + P(A_n) - (n-1)) \le P(A_1 \cap A_2 \ldots \cap A_n)$
    $\le \min(P(A_1), P(A_2) \ldots, P(A_n))$
14. $\max(P(A_1), P(A_2) \ldots, P(A_n)) \le P(A_1 \cup A_2 \ldots \cup A_n)$
    $\le \min(1, P(A_1) + P(A_2) \ldots + P(A_n))$
15. $P(A|B) = \dfrac{P(A \quad B)}{P(B)}$
16. $P(A|B) = P(A)$; independent events.
17. $P(E_1 \cap E_2 \cap \ldots \cap E_n) = P(E_1) \cdot P(E_2) \ldots P(E_n)$; independent events.
18. Rule of total probability: $P(X) = \sum_{i=1}^{n} P(E_i)\, P(X|E_i)$
19. Baye's Theorem:

$$P(E_1|X) = \frac{P(E_1) \cdot P(X|E_1)}{P(E_1) \cdot P(X|E_1) + P(E_2) \cdot P(X|E_2) + P(E_3) \cdot P(X|E_3)}$$

In general,

$$P(E_i|X) = \frac{P(E_i) \cdot P(X|E_i)}{\sum_{j=1}^{n} P(E_j) \cdot P(X|E_j)}$$

RANDOM VARIABLE (STOCHASTIC VARIABLE)

Random variable assigns a real number to each possible outcome.

Let X be a discrete random variable, then

1. $F(x) = P(X \leq x)$ is called distribution function $\sum_{i=0}^{n} P(i)$ of X.
2. Mean or Expectation of $X = \mu = E(X) = \sum_{i=1}^{n} x_i\, P(x_i)$
3. Variance of $X = \sigma^2 = E(X^2) - [E(X)]^2 = \sum_{i=1}^{n} (x_i - \mu)^2 P(x_i)$
4. Standard deviation of $X = \sigma = \sqrt{\text{Variance}}$
5. $\sum_{i=1}^{n} P(x_i) = 1$

Types of Random Variables

1. **Discrete Random Variable:** "Finite set of values" or "Countably infinite".
2. **Continous Random Variable (Non-discrete):** "Infinite number of uncountable values".

Discrete Distributions

1. **Binomial Distribution:** The probability that the event will happen exactly r times in n trials i.e. r successes and $n-r$ failures will occur.

   $P(X = r) = P(r) = \Sigma\, {}^nC_r p^r q^{n-r}$

   Mean $= E(x) = np$

   Variance $(\sigma^2) = V(x) = npq = np(1 - p)$

   $S.D$ $(\sigma) = \sqrt{npq} = \sqrt{np(1-p)}$

   Where $r = 0, 1, \ldots, n$, $q = 1 - p$, n = fixed number of trials, p = probability of success

2. **Poisson Distribution:**

$$P(X = x) = \sum \frac{e^{-\lambda} \cdot \lambda^x}{x!};\ x = 0, 1, 2, \ldots, \infty$$

Where X = Discrete random variable

λ = Parameter of distribution (positive constant)

- Mean (μ) = Variance $(\sigma^2) = \lambda$
- $S.D = \sqrt{\lambda}$

  Poisson distribution is a limiting case of binomial distribution as $n \to \infty$ and $p \to 0$.

CONTINUOUS DISTRIBUTION

Let X be continous random variable. Then

(*i*) **Density functions:**

- $P(X \leq a) = \int_{-\infty}^{a} f(x) \cdot dx$

- $P(a \leq X \leq b) = \int_{a}^{b} f(x) \cdot dx$

(*ii*) **Mean** $= E(x) = \int_{-\infty}^{\infty} x \cdot f(x) \cdot dx$

(*iii*) **Variance of** $X = V(X) = \int_{-\infty}^{\infty} [x - E(x)]^2 \cdot f(x)\, dx = E(x^2) - (E(x))^2$

$$= \int_{-\infty}^{\infty} x^2 f(x) dx - \left[\int_{-\infty}^{\infty} x f(x) dx\right]^2$$

(*iv*) $\int_{-\infty}^{\infty} f(x)\, dx = 1$

1. Uniform Distribution (Rectangular Distribution)

(*i*) **Density function:**

$$f(x) = \frac{1}{b-a}; \quad a \leq x \leq b$$
$$= 0\text{ ; otherwise}$$

(*ii*) **Cumulative function:**

$$P(X \leq x) = \int_{-\infty}^{x} f(x) \cdot dx = \begin{cases} 0 & ; \quad \text{if } x \leq a \\ \dfrac{x-a}{b-a} & ; \quad \text{if } a \leq x \leq b \\ 1 & ; \quad \text{if } x > b \end{cases}$$

(*iii*) **Mean** (μ) $= (a + b)/2 = E(X)$

(*iv*) **Variance** (σ^2) $= (b - a)^2/12$

2. Normal Distribution

(*i*) Density function $f(x) = \dfrac{1}{\sigma\sqrt{2\pi}} \cdot e^{-\frac{1}{2}\left(\frac{x-\mu}{\sigma}\right)^2}$; $-\infty \le x \le \infty, \sigma > 0, -\infty < \mu < \infty$

(*ii*) Normal distribution is symmetrical

(*iii*) Mean = μ; Variance = σ^2

(*iv*) $f(x) \ge 0$ for all x

(*v*) $\int_{-\infty}^{\infty} f(x) \cdot dx = 1$

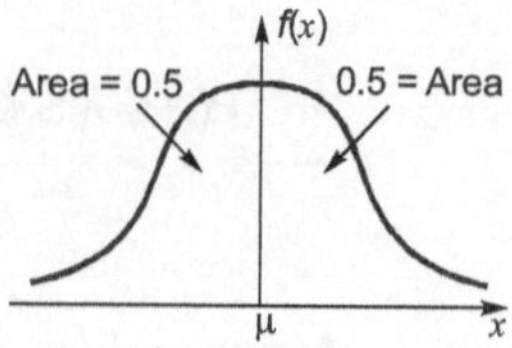

(*vi*) $P(Z) = \dfrac{1}{\sqrt{2\pi}} e^{-\frac{Z^2}{2}}$; $-\infty \le Z \le \infty$

and $Z = \dfrac{x-\mu}{\sigma}$

$Z = \dfrac{x - np}{\sqrt{npq}}$ (when approximating binomial by normal)

Z = Standard normal variate

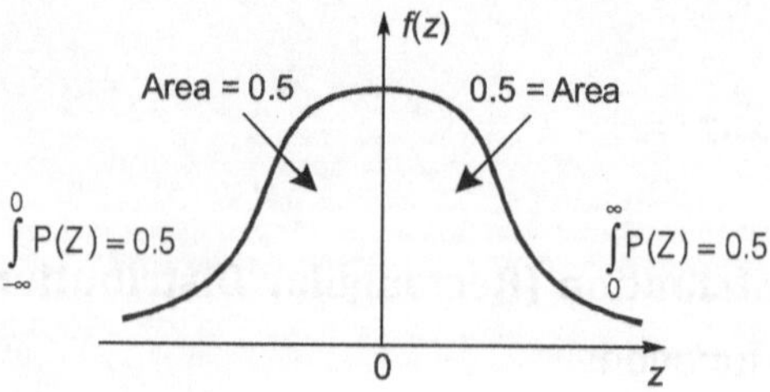

3. Exponential Distribution

(*i*) Density function:

$$f(x) = \lambda \cdot e^{-\lambda x} \quad ; x > 0$$
$$= 0 \quad ; \text{ Otherwise}$$

(*i*) Mean (μ) = $\dfrac{1}{\lambda} = S.D(\sigma)$

(*i*) Variance (σ^2) = $\dfrac{1}{\lambda^2}$

■■■■

Numerical Methods

4

Solution of Algebraic and Transcendental Equation

Consider an equation $f(x) = 0$.

Bisection Method

This method finds the root between points 'a' and 'b'. If $f(x)$ is continuous between a and b and $f(a)$ and $f(b)$ are of opposite signs then their is a root between a and b (Intermediate Value Theorem). First approximation to the root is $x_1 = (a + b)/2$. If $f(x_1) = 0$, then x_1 is the root of $f(x) = 0$ otherwise root lies between a and x_1 or x_1 and b, Similarly, x_2 and x_3 ..... are determined.

- Simplest iterative method.
- Bisection method always converges, but often slowly.
- This method can't be used for finding the complex roots.
- Rate of convergence is linear.

Newton-Raphson Method (or Successive Substitution Method or Tangent Method)

$$x_{n+1} = x_n - \frac{f(x_n)}{f'(x_n)}$$

- This method is commonly used for its simplicity and greater speed.
- Here $f(x)$ is assumed to have continuous derivative $f'(x)$.
- This method fails if $f'(x) = 0$.
- It has second order of convergence or quadratic convergence i.e. the subsequent error at each step is proportional to the square of the error at previous step.
- Sensitive to starting value, i.e., the Newton's method converges provided the initial approximation is chosen sufficiently close to the root.
- Rate of convergence is quadratic.

Secant Method

$$x_{n+1} = x_n - \frac{x_n - x_{n-1}}{f(x_n) - f(x_{n-1})} f(x_n)$$

- Convergence is not guaranteed.
- It converges, convergence super.

Regula Falsi Method or (Method of False Position)

- Regula falsi method always converges.
- However, it converges slowly.
- If converges, order of convergence is between 1 and 2 (closer to 1).
- It is superior to bisection method.

Given $f(x) = 0$

Select x_0 and x_1 such that $f(x_0)\, f(x_1) < 0$ (*i.e.* opposite sign)

$$x_2 = x_0 - \frac{x_1 - x_0}{f(x_1) - f(x_0)} f(x_0) = \frac{x_0 f(x_1) - x_1 f(x_0)}{f(x_1) - f(x_0)}$$

Check if $f(x_0)\, f(x_2) < 0$ $f(x_1)\, f(x_2) < 0$

Computer x_3 ....

which is an approximation to the root.

Solution of Linear System of Equations

Gauss Elimination Method

Here equations are converted into "upper triangular matrix" form, then solved by "back substitution" method.

Consider

$$\begin{aligned} a_1x + b_1y + c_1z &= d_1 \\ a_2x + b_2y + c_2z &= d_2 \\ a_3x + b_3y + c_3z &= d_3 \end{aligned}$$

Step 1 : To eliminate x from second and third equation (we do this by subtracting suitable multiple of first equation from second and third equation)

$$\begin{aligned} a_1x + b_1y + c_1z &= d_1 && \text{pivotal equation, } a_1 \text{ pivot point} \\ b_{2'}y + c_{2'}z &= d_{2'} \\ b_{3'}y + c_{3'}z &= d_{3'} \end{aligned}$$

Step 2 : Eliminate y from third equation

$$\begin{aligned} a_1x + b_1y + c_1z &= d_{1'} \\ b_{2'}y + c_{2'}z &= d_{2'} && \text{pivotal equation, } b_{2'} \text{ pivot point} \\ c_{3''}z &= d_{3''} \end{aligned}$$

Step 3 : The value of x, y and z can be found by back substitution.

Note: ....

- Number of operations : $N = \frac{n^3}{3} + n^2 - \frac{n}{3}$

....

Gauss - Jordon Method

- Used to find inverse of the matrix and solving linear equations.
- Here back substitution is avoided by additional computations that reduce the matrix to "diagonal form", instead to triangular form in Gauss elimination method.
- Number of operations is more than Gauss elimination as the effort of back substitution is saved at the cost of additional computation.

  Step 1 : Eliminate x from 2nd and 3rd.

  Step 2 : Eliminate y from 1st and 3rd.

  Step 3 : Eliminate z from 1st and 2nd.

LU Decomposition

- It is a modification of the Gauss elimination method.
- Also used for finding the inverse of the matrix.

$$[A]_{n\times n} = [L]_{n\times n}[U]_{n\times n}$$

$$\begin{pmatrix} a_{11} & a_{12} & a_{13} \\ a_{21} & b_{22} & c_{23} \\ a_{31} & b_{32} & c_{33} \end{pmatrix} = \begin{pmatrix} 1 & 0 & 0 \\ L_{21} & 1 & 0 \\ L_{31} & L_{32} & 1 \end{pmatrix}\begin{pmatrix} U_{11} & U_{12} & U_{13} \\ 0 & U_{22} & U_{23} \\ 0 & 0 & U_{31} \end{pmatrix}$$

$Ax = LUX = b$ can be written as

(a) $LY = b$, where (b) $UX = Y$

Solve for Y from (a) then solve for X from (b). This method is known as Doolittle's method.

- Similar methods are Crout's and Cholesky's methods.

Iterative Method

1. Jacobi Iteration Method

$$\begin{aligned} a_1x + b_1y + c_1z &= d_1 \\ a_2x + b_2y + c_2z &= d_2 \\ a_3x + b_3y + c_3z &= d_3 \end{aligned}$$

If a_1, b_2, c_3 are large compared to other coefficients, then solving these for x, y, z respectively.

$$\begin{aligned} x &= k_1 - l_1y - m_1z \\ y &= k_2 - l_2x - m_2z \\ z &= k_3 - l_3x - m_3y \end{aligned}$$

Let us start with initial approximation x_0, y_0, z_0

$$\begin{aligned} x_1 &= k_1 - l_1y_0 - m_1z_0 \\ y_1 &= k_2 - l_2y_0 - m_2z_0 \\ z_1 &= k_3 - l_3y_0 - m_3z_0 \end{aligned}$$

Note: ....

- No component of $x^{(k)}$ is used in computation unless $y^{(k)}$ and $z^{(k)}$ are computed.

....

The process is repeated till the difference between two consecutive approximatims is negligible.

In generalized form :

$$x^{(k+1)} = k_1 - l_1 y^{(k)} - m_1 z^{(k)}$$
$$y^{(k+1)} = k_2 - l_2 y^{(k)} - m_2 z^{(k)}$$
$$z^{(k+1)} = k_3 - l_3 x^{(k)} - m_3 y^{(k)}$$

2. Gauss-Siedel Iteration Method

Modification of the Jacobi's iteration method.

Start with $(x_0, y_0, z_0) = (0, 0, 0)$ or anything (no specific condition).

In first equation, put $y = y_0$ and $z = z_0$ which will give x_1.

In second equation, put $y = y_0$ and $z = z_0$ which will give y_1.

In third equation, put $x = x_1$ and $y = y_1$ which will give z_1.

Note: ....

- To compute any variable use the latest available value.

....

In generalized form :

$$x^{(k+1)} = k_1 - l_1 y^{(k)} - m_1 z^{(k)}$$
$$y^{(k+1)} = k_2 - l_2 y^{(k+1)} - m_2 z^{(k)}$$
$$z^{(k+1)} = k_3 - l_3 y^{(k+1)} - m_3 z^{(k+1)}$$

Numerical Integration

Trapezoidal Formula

Step size, $$h = \frac{b-a}{n}$$

$$\int_{x_0}^{x_0+nh} f(x)dx = \frac{h}{2}\{(\text{first term} + \text{last term}) + 2\,(\text{remaining terms})\}$$

Error = Exact - approx

The error in approximating an integral using Trapezoidal rule is bounded by

$$-\frac{h^2}{12}(b-a)\max_{\xi\in[a,b]}|f''(\xi)|$$

Simpson's One-Third Rule (Simpson's Rule)

$$\int_{x_0}^{x_0+nh} f(x)dx = \frac{h}{3}$$ {(first term + last term) + 4 (all odd terms) + 2 (all even terms)}

The error in approximating an integral using Simpson's one-third rule is

$$-\frac{h^4}{180}(b-a)\max_{\xi\in[a,b]}\left|f^{(4)}(\xi)\right|$$

Simpson's Three Eighth Rule

$$\int_{x_0}^{x_0+nh} f(x)dx = \frac{3h}{8}$$ {(first term + last term) + 2 (all multiple of 3 terms) + 3 (all remaining terms)}

The error in approximating an integral using Simpson's three eight rule is

$$-\frac{(b-a)^5}{6480}\max_{\xi\in[a,b]}\left|f^{(4)}(\xi)\right|$$

Solving Differential Equation

1. **Euler Method** (for first order differential equation):
   Given equation is $y' = f(x, y)$, $y(x_0) = y_0$
   Solution is given by $Y_{n+1} = y_n + hf(x_n, y_n)$
2. **Runge-Kutta Method :**
   Used for finding the y at a particular x without solving the first order differential equation $\frac{dy}{dx} = f(x, y)$

$$K_1 = hf(x_0, y_0)$$

$$K_2 = hf\left(x_0 + \frac{h}{2}, y_0 + \frac{k_1}{2}\right)$$

$$K_3 = hf\left(x_0 + \frac{h}{2}, y_0 + \frac{k_2}{2}\right)$$

$$K_4 = hf(x_0 + h, y_0 + k_3)$$

$$K = \frac{1}{6}(k_1 + 2k_2 + k_3 + k_4)$$

$$Y(x_0 + h) = y_0 + K$$

■■■■

Calculus

5

LIMIT, CONTINUITY AND DIFFERENTIABILITY

Limit Existence at a Point

- Left Limit (LHL): $\lim_{x \to a^-} f(x) = \lim_{h \to 0} f(a-h)$

  Right Limit (RHL): $\lim_{x \to a^+} f(x) = \lim_{h \to 0} f(a+h)$

  $\lim_{x \to a} f(x)$ exists, iff both LHL and RHL exists.
- Indeterminate forms: $0/0$, ∞/∞, $0 \times \infty$, $\infty - \infty$, 0^0, 1^∞, 0^∞
- **L Hospital's Rule:** If $\lim_{x \to a} [f(x)/g(x)]$ is of the form either $0/0$ or ∞/∞

  then $\lim_{x \to a} [f(x)/g(x)] = \lim_{x \to a} [f'(x)/g'(x)]$

Continuity

A function $f(x)$ is said to be continuous at $x = a$ if :

$$\lim_{x \to a^+} f(x) = \lim_{x \to a^-} f(x) = f(a)$$

- If $f(x)$ and $g(x)$ is continuous function then following are also continuous.
  (*i*) $f(x) + g(x)$
  (*ii*) $f(x) \cdot g(x)$
  (*iii*) $f(x) - g(x)$
  (*iv*) $f(x)/g(x)$ $[g(x) \neq 0]$
- Polynomial, Exponential, Sin and Cos functions are continuous everywhere.

Differentiability

A function $f(x)$ is said to be differentiable at $x = a$ if it is continuous at $x = a$ and if left derivative $= f'(a) =$ right derivative

i.e. $$\lim_{h \to 0^-} \frac{f(a) - f(a-h)}{h} = f'(a) = \lim_{h \to 0^+} \frac{f(a+h) - f(a)}{h}$$

- If $f(x)$ is differentiable in (a, b), then it is always continuous at (a, b), but converse need not be true.

MEAN VALUE THEOREMS

Rolle's Mean Value Theorem

If

(*i*) '$f(x)$' is continuous in $[a, b]$

(*ii*) '$f(x)$' is differentiable in (a, b)

(*iii*) $f(a) = f(b)$

then $\exists\, c \in (a, b)$ such that $f'(c) = 0$

Lagrange's Mean Value Theorem

If

(*i*) '$f(x)$' is continuous in $[a, b]$

(*ii*) '$f(x)$' is differentiable in (a, b)

then $\exists\, c \in (a, b)$ such that $f'(c) = \dfrac{f(b) - f(a)}{b - a}$

Cauchy's Mean Value Theorem

Let $f(x)$ and $g(x)$ are two functions if

(*i*) $f(x)$ and $g(x)$ are continuous in $[a, b]$

(*ii*) $f(x)$ and $g(x)$ are differentiable in (a, b)

(*iii*) $g'(x) \neq 0 \;\; \forall x \in (a, b)$

then $\exists\, c \in (a, b)$ such that $\dfrac{f'(c)}{g'(c)} = \dfrac{f(b) - f(a)}{g(b) - g(a)}$

Derivatives

- $\dfrac{d}{dx}(\sin x) = \cos x$
- $\dfrac{d}{dx}(\cos x) = -\sin x$
- $\dfrac{d}{dx}(\tan x) = \sec^2 x$
- $\dfrac{d}{dx}(\cot x) = -\operatorname{cosec}^2 x$
- $\dfrac{d}{dx}(\sec x) = \sec x \cdot \tan x$
- $\dfrac{d}{dx}(\operatorname{cosec} x) = \operatorname{cosec} x \cot x$
- $\dfrac{d}{dx}(\sin^{-1} x) = \dfrac{1}{\sqrt{1-x^2}}$
- $\dfrac{d}{dx}(\cos^{-1} x) = \dfrac{-1}{\sqrt{1-x^2}}$
- $\dfrac{d}{dx}(\tan^{-1} x) = \dfrac{1}{1+x^2}$
- $\dfrac{d}{dx}(\cot^{-1} x) = \dfrac{-1}{1+x^2}$

- $\frac{d}{dx}(\sec^{-1} x) = \frac{1}{x\sqrt{x^2 - 1}}$
- $\frac{d}{dx}(\text{cosec}^{-1} x) = \frac{-1}{x\sqrt{x^2 - 1}}$
- $\frac{d}{dx}(\sinh x) = \cosh x$
- $\frac{d}{dx}(\cosh x) = \sinh x$
- $\frac{d}{dx}(e^x) = e^x$
- $\frac{d}{dx}(a^x) = a^x \cdot \log a$
- $\frac{d}{dx}[f(x)]^n = n \cdot [f(x)]^{n-1}$
- $\frac{d}{dx}(ax + b)^n = n \cdot (ax + b)^{n-1} . a$
- $\frac{du}{dx} = \frac{du}{dy}\frac{dy}{dx}$ (chain rule)
- $\frac{d}{dx}(uv) = u \cdot \frac{dv}{dx} + v \cdot \frac{du}{dx}$
- $\frac{d}{dx}\left(\frac{u}{v}\right) = \frac{v . \frac{du}{dx} - u \cdot \frac{dv}{dx}}{v^2}$
- $\sinh x = \frac{e^x - e^{-x}}{2}$
- $\cosh x = \frac{e^x + e^{-x}}{2}$
- $\frac{d}{dx}(\log_e x) = \frac{1}{x}$

Integrals

- $\int x^n \cdot dx = \frac{x^{n+1}}{n+1}; \ (n \neq -1)$
- $\int \frac{1}{x} \cdot dx = \log_e x$
- $\int e^x \cdot dx = e^x$
- $\int a^x = \frac{a^x}{\log a}$
- $\int \frac{f'(x)}{f(x)} \cdot dx = \log [f(x)]$
- $\int [f(x)]^n \cdot f'(x) \cdot dx = \frac{[f(x)]^{n+1}}{n+1}$
- $\int \frac{1}{a^2 + x^2} dx = \frac{1}{a} \cdot \tan^{-1} \frac{x}{a}$
- $\int \frac{1}{a^2 - x^2} dx = \frac{1}{2a} \cdot \log\left[\frac{a+x}{a-x}\right]$
- $\int \frac{1}{x^2 - a^2} dx = \frac{1}{2a} \cdot \log\left[\frac{x-a}{x+a}\right]$
- $\int \frac{dx}{\sqrt{a^2 - x^2}} = \sin^{-1} \frac{x}{a}$
- $\int \frac{dx}{\sqrt{a^2 + x^2}} = \sinh^{-1} \frac{x}{a}$
- $\int \frac{dx}{\sqrt{x^2 - a^2}} = \cosh^{-1} \frac{x}{a}$
- $\int \sinh x = \cosh x$
- $\int \cosh x = \sinh x$

- $\int \sin x \cdot dx = -\cos x$
- $\int \cos x \cdot dx = \sin x$
- $\int \tan x \cdot dx = \log \cos x = \log \sec x$
- $\int \cot x = \log \sin x$
- $\int \sec x \cdot dx = \log (\sec x + \tan x)$
- $\int \text{cosec}\, x \cdot dx = \log (\text{cosec}\, x - \cot x)$
- $\int \sec^2 x \cdot dx = \tan x$
- $\int \text{cosec}^2 x \cdot dx = -\cot x$
- $\int e^{ax} \sin bx\, dx = \dfrac{e^{ax}}{a^2+b^2} (a \sin bx - b \cos bx)$
- $\int e^{ax} \cos bx\, dx = \dfrac{e^{ax}}{a^2+b^2} (a \cos bx + b \sin bx)$
- $\int \sqrt{a^2 - x^2} \cdot dx = \dfrac{x}{2}\sqrt{a^2 - x^2} + \dfrac{a^2}{2} \sin^{-1}\dfrac{x}{a}$
- $\int \sqrt{a^2 - x^2} \cdot dx = \dfrac{x}{2}\sqrt{a^2 + x^2} + \dfrac{a^2}{2} \sinh^{-1}\dfrac{x}{a}$
- $\int \sqrt{x^2 - a^2} \cdot dx = \dfrac{x}{2}\sqrt{x^2 - a^2} - \dfrac{a^2}{2} \cosh^{-1}\dfrac{x}{a}$

Definite Integrals

Theorem 1:

If $f(x)$ is continous in $[a, b]$ and $F(x)$ is any antiderivative of $f(x)$ in $[a, b]$ then

$$\int_a^b f(x) dx = F(b) - F(a)$$

Theorem 2:

If $f(x)$ is continous on $[a, b]$ then $F(x) = \int_a^b f(t) \cdot dt$ is differentiable at every point of x in $[a, b]$ and $\dfrac{dF}{dx} = \dfrac{d}{dx}\int_a^b f(t) \cdot dt = f(x)$.

i.e., If $f(x)$ is continous on $[a, b]$ then $\exists$ a function $F(x)$ whose derivative on $[a, b]$ is $f(x)$.

Theorem 3:

If $f(x)$ is continous on $[a, b]$ and $u(x)$ and $v(x)$ are differentiable functions of x whose values lie in $[a, b]$ then $\frac{d}{dx}\int_{u(x)}^{v(x)} f(t)\cdot dt = f(v(x))\frac{dv}{dx} - f(u(x))\cdot\frac{du}{dx}$

Properties of Definite Integrals

1. $\int_a^a f(x)\cdot dx = 0$

2. $\int_a^b f(x)\cdot dx = \int_b^a f(z)\cdot dz$

3. $\int_a^b f(x)\cdot dx = -\int_b^a f(x)\cdot dx$

4. $\int_a^b k\cdot dx = k(b-a)$

5. $\int_a^b f(x)\cdot dx = \int_a^c f(x)\cdot dx + \int_c^b f(x)\cdot dx$

6. $\int_0^a f(x)\cdot dx = \int_0^a f(a-x)\cdot dx$

7. $\int_{-a}^a f(x)\cdot dx = 2\int_0^a f(x)\cdot dx$; if $f(x)$ is even function $f(-x) = f(x)$

   $= 0$; if $f(x)$ is odd function $f(-x) = -f(x)$

8. $\int_a^b f(x)\cdot dx = \int_a^b f(a+b-x)\cdot dx$

9. $\int_0^{2a} f(x)\cdot dx = \int_0^a f(x)\cdot dx + \int_0^a f(2a-x)\cdot dx$

10. $\int_0^{2a} f(x)\cdot dx = 2\int_0^a f(x)\cdot dx$; if $f(2a-x) = f(x)$

    $= 0$; If $f(2a-x) = -f(x)$

11. $\int_0^{na} f(x)\cdot dx = n\cdot\int_0^{a} f(x)\cdot dx$; if $f(x+a)=f(x)$

[$f(x)$ is periodic function with period 'a']

12. $\int_0^{a} x\cdot f(x)\cdot dx = \frac{a}{2}\int_0^{a} f(x)\cdot dx$; if $f(a-x)=f(x)$

13. **Mean value theorem for definite integrals:** Let f be continuous in (a, b) then $\exists c \in (a, b)$ such that $f(c) = \frac{\int_a^b f(x)dx}{(b-a)}$

$f(c)$ is the average value of the function in the interval (a, b).

14. $\int_a^b f(x)dx \le \int_a^b g(x)dx$ If $f(x) \le g(x)$ on (a, b)

15. $\int_0^{\pi/2} \sin^n x\cdot dx = \int_0^{\pi/2} \cos^n x\cdot dx = \frac{(n-1)\cdot(n-3)\cdot(n-5)....}{n\cdot(n-2)(n-4)....}\times K$

Where, $K = 1$; if n is odd and $K = \pi/2$; if n is even

Improper Integrals

1. $\int_a^b f(x)\cdot dx$ is "improper integral";

   (*i*) If $a = -\infty$ or $b = \infty$ or both
   (*ii*) If $f(x) = \infty$ for one or more values of x in $[a, b]$.

2. $\int_a^b f(x)\cdot dx$ is "convergent", if the value of integral is "finite".

3. $\int_a^b f(x)\cdot dx$ is "divergent", if the value of integral is "infinite".

4. If $0 \le f(x) \le g(x)$; $\forall x$ and $\int_a^{\infty} g(x)\cdot dx$ converges then $\int_a^{\infty} f(x)$ also converges.

5. If $0 \le g(x) \le f(x);\ \forall x$ and $\int_a^\infty g(x)\cdot dx$ diverges then $\int_a^\infty f(x)$ also diverges.

6. If $f(x)$ and $g(x)$ are two functions, such that $\lim_{n\to\infty} \frac{f(x)}{g(x)} = k$ where k = finite and $k \ne 0$ then

$\int_a^\infty f(x)\cdot dx$ and $\int_a^\infty g(x)\cdot dx$ "converge" or "diverge" together.

7. $\int_a^\infty e^{-Px}\cdot dx$ and $\int_{-\infty}^b e^{Px}\cdot dx$ converges for any constant $P > 0$ and diverges for $P \le 0$.

8. $\int_a^b \frac{dx}{(b-x)^P}$ is convergent iff $P < 1$.

9. $\int_a^b \frac{dx}{(x-a)^P}$ is convergent iff $P < 1$.

10. $\int_1^\infty \frac{dx}{x^P}$ converges if $P > 1$ and diverges if $P \le 1$.

Total Derivatives

If $u = f(x, y)$ where x and y are functions of t, then the "total derivative" of u with respect to t is :

$$\frac{du}{dt} = \frac{\partial u}{\partial x}\left(\frac{dx}{dt}\right) + \frac{\partial u}{\partial y}\left(\frac{dy}{dt}\right)$$

By putting $t = x$ in above equation, we get

$\frac{du}{dx} = \frac{\partial u}{\partial x} + \frac{\partial u}{\partial y}\cdot\frac{dy}{dx}$; where x and y are connected by some relation.

Partial Derivatives

1. If $u = f(x, y)$, then $\frac{\partial u}{\partial x} = \lim_{h\to 0} \frac{[f(x+h, y) - f(x, y)]}{h}$ and

$\frac{\partial u}{\partial y} = \lim_{k\to 0} \frac{[f(x, y+k) - f(x, y)]}{k}$.

2. If $u = f(x, y)$, where $x = g(r, s)$ and $y = h(r, s)$ then

$$\frac{\partial u}{\partial r} = \frac{\partial u}{\partial x} \cdot \frac{\partial x}{\partial r} + \frac{\partial u}{\partial y} \cdot \frac{\partial y}{\partial r}$$

$$\frac{\partial u}{\partial s} = \frac{\partial u}{\partial x} \cdot \frac{\partial x}{\partial s} + \frac{\partial u}{\partial y} \cdot \frac{\partial y}{\partial s}$$

3. If $u = f(x, y)$ is a homogenous function of degree 'n' then

$$x \cdot u_x + y \cdot u_y = n \cdot u; \; x^2 u_{xx} + 2xy \cdot u_{xy} + y^2 u_{yy} = n(n-1)u$$

4. If $u = f(x, y, z)$ is a homogenous function of degree 'n', then

$$x \cdot u_x + y \cdot u_y + z \cdot u_z = n \cdot u$$

5. If $u = f(x, y)$ is not homogenous function, but $F(u)$ is a homogenous function of degree 'n', then

$$x \cdot u_x + y \cdot u_y = n\left[\frac{F(u)}{F'(u)}\right] = G(u)$$

$$x^2 \cdot u_{xx} + 2xy \cdot u_{xy} + y^2 \cdot u_{yy} = G(u)\,[G'(u) - 1]$$

Maximum and Minima [Extremum]

1. A necessary condition for $f(a)$ to be an extreme value of $f(x)$ is $f'(a) = 0$.
2. The vanishing of $f'(a) = 0$ is only a necessary but not a sufficient condition for $f(a)$ to be an extreme value of $f(x)$.
3. **Stationary Values:** A function $f(x)$ is said to be stationary for $x = C$ and $f(C)$ is a stationary value of $f(x)$ if $f'(C) = 0$.
4. To find points of maximum, points of minimum and saddle points the following steps are required:

   (*i*) Put $f'(x) = 0$ and find stationary points (x_i)

   (*ii*) If $f''(x_i) > 0$ then $f(x)$ is minimum at $x_i \; \forall x_i$.

   (*iii*) If $f''(x_i) < 0$ then $f(x)$ is maximum at $x_i \; \forall x_i$.

   (*iv*) $f'(x) = 0$ and $f''(x) = 0$ and so on

   (1) Continue finding derivatives at x_i untill the first non-zero derivative is found.

   (2) If the first non-zero derivative is obtained at odd derivative then x_i is a point of inflection or saddle point. If the first non-zero derivative is obtained at even derivative then x_i is either a point of maximum or a point of minimum depending on whether the derivative is negative or positive respectively.

6. **Sufficient condition for extrema:**

   Theorem: $f(C)$ is an extremum of $f(x)$ iff $f'(x)$ changes sign as x passes through C.

   Case (i): If $f'(x) = 0$ and $f'(x)$ changes sign from positive to –ve as 'x' passes through C then $f(C)$ is maximum.

   Case (ii): If $f'(x) = 0$ and $f(x)$ changes sign from –ve to +ve as x passes through C then $f(C)$ is minimum.

   Case (iii): If $f'(x) = 0$ and $f(x)$ does not changes sign as x passes through C then $f(C)$ is not extremum but rather a point of inflection or saddle point.

7. For 2-variable functions:

   (*i*) Put $f_x = 0$ and $f_y = 0$; find stationary points by solving simultaneously.

   (*ii*) If $f_{xx} f_{yy} > f_{xy}^2$ and

   (a) If $f_{xx} > 0$ then $f(x)$ is "minimum" at a

   (b) If $f_{xx} < 0$ then $f(x)$ is "maximum" at a

   (*iii*) If $f_{xx} f_{yy} - f_{xy}^2 < 0$ then $x = a$ is a point of inflection or saddle point.

■■■

Vector Calculus

6

Vector Calculus

Scalar Point Function

If corresponding to each point P of region R, there is a corresponding scalar $\phi(P)$ is said to be a scalar point function for the region R.

$$\phi(P) = \phi(x, y, z)$$

Vector Point Function

If corresponding to each point P of region R, there corresponds a vector defined by $F(P)$, then F is called a vector point function for region R.

$$F(P) = F(x, y, z) = f_1(x,y,z)\hat{i} + f_2(x,y,z)\hat{j} + f_3(x,y,z)\hat{k}$$

Vector Differential Operator or Del Operator

$$\Delta = \left(\hat{i}\frac{\partial}{\partial x} + \hat{j}\frac{\partial}{\partial y} + \hat{k}\frac{\partial}{\partial z}\right)$$

Directional Derivative

The directional derivative of f in a direction $\vec{N}$ is the resolved part of ∇f in direction $\vec{N}$.

$$\nabla f.\vec{N} = |\nabla f|\cos\alpha$$

where $\vec{N}$ is a unit vector in a particular direction.

Direction cosine: $l^2 + m^2 + n^2 = 1$,

where $\quad l = \cos\alpha,\ m = \cos\beta,\ n = \cos\gamma$

Gradient

The vector function ∇f is defined as the gradent of the scalar point function $f(x, y, z)$ and written as grad f.

$$\text{Grad} f = \nabla f = \hat{i}\frac{\partial f}{\partial x} + \hat{j}\frac{\partial f}{\partial y} + \hat{k}\frac{\partial f}{\partial z}$$

- ∇f is vector function.
- If $f(x, y, z) = 0$ is any surface, then ∇f is a vector normal to the surface f and has a magnitude equal to rate of change of f along this normal.
- Direction derivative of $f(x, y, z)$ is maximum along ∇f and magnitude of this maximum is $|\nabla f|$.

Divergence

The divergence of a continuously differentiable vector point function F is denoted by div. F and is defined by the equation div. $F = \nabla.F$.

$$F = f\hat{i} + \phi\hat{j} + \psi\hat{k}$$

$$\text{div.} \quad F = \nabla.F = \left(\hat{i}\frac{\partial}{\partial x} + \hat{j}\frac{\partial}{\partial y} + \hat{k}\frac{\partial}{\partial z}\right)\cdot\left(f\hat{i} + \Phi\hat{j} + \psi\hat{k}\right) = \frac{\partial f}{\partial x} + \frac{\partial \Phi}{\partial y} + \frac{\partial y}{\partial z}$$

- $\nabla \cdot f$ is scalar.
- $\nabla \cdot \nabla = \nabla^2$ is Laplacian operator.

Curl

The curl of a continuously differentiable vector point function F is denoted by curl F and is defined by the equation

$$\text{Curl } F = \nabla \times F = \begin{vmatrix} \hat{i} & \hat{j} & \hat{k} \\ \frac{\partial}{\partial x} & \frac{\partial}{\partial y} & \frac{\partial}{\partial z} \\ f & \Phi & \psi \end{vmatrix}$$

$\nabla \times F$ is a vector function.

Solenoidal Vector Function

If $\nabla \cdot F = 0$, then A is called as solenoidal vector function.

Irrotational Vector Function

If $\nabla \cdot A = 0$, then A is said to be irrotational otherwise rotational.

Del Applied Twice to Point Functions

1. div. grad $f = \nabla^2 f = \frac{\partial^2 f}{\partial x^2} + \frac{\partial^2 f}{\partial y^2} + \frac{\partial^2 f}{\partial z^2}$ ... this is Laplace equation.
2. curl grad $f = \nabla \times \nabla f = 0$
3. div. curl $F = \nabla \cdot \nabla \times F = 0$
4. curl curl $F = \nabla \times (\nabla \times F) = \nabla(\nabla \cdot F) - \nabla^2 \cdot F$
5. grad div $F = \nabla(\nabla \cdot F) = \nabla \times (\nabla \times F) + \nabla^2 \cdot F$

Vector Identities

- $\nabla(f + g) = \nabla f + \nabla g$
- $\nabla.(F + G) = \nabla.G + \nabla.G$
- $\nabla \times (F + G) = \nabla \times G + \nabla \times G$
- $\nabla(fg) = f\nabla g + g\nabla f$
- $\nabla.(fg) = \nabla f.G + f\nabla.G$
- $\nabla \times (fG) = \nabla f \times G + f\nabla \times G$
- $\nabla(F \cdot G) = F \times (\nabla \times G) + G \times (\nabla \times F)$
- $\nabla.(F \times G) = G \cdot (\nabla \times F) - F \cdot (\nabla \times G)$
- $\nabla \times (F \times G) = F(\nabla.G) - G(\nabla \times F)$

Note:

- $\nabla(f/g) = (g\nabla f - f\nabla g)/g^2$
- $(F \times G)' = F' \times G + F \times G'$
- $(F \cdot G)' = F' \cdot G + F \cdot G'$
- $\nabla^2(fg) = g\nabla^2 f + 2\nabla f \cdot \nabla g + f\nabla^2 g$

Vector Product

1. Dot product of $A \times B$ with C is called scalar triple product and denoted as $[ABC]$.

   Rule : For evaluating the scalar triplet product

   (i) Independent the position of dot and cross

   (ii) Dependent on the cyclic order of the vector

$$[ABC] = A \times B \cdot C = A \cdot B \times C$$
$$= B \times C \cdot A = B \cdot C \times A$$
$$= C \times A \cdot B = C \cdot A \times A \cdot B$$
$$A \times B \cdot C = -(B \times A \cdot C)$$

2. $(\vec{A}+\vec{B})\times\vec{C}$ = (extreme × adjacent) × Outer

   = (Outer. extreme)adjacent – (Outer. adjacent)extreme

   - $(\vec{A}+\vec{B})\times\vec{C} = (\vec{C}\cdot\vec{A})\vec{B}-(\vec{C}\cdot\vec{B})\vec{A}$
   - $\vec{A}\times(\vec{B}\times\vec{C}) = (\vec{A}\cdot\vec{C})\vec{B}-(\vec{A}\cdot\vec{B})\vec{C}$
   - $(\vec{A}\times\vec{B})\times\vec{C} = \vec{C}\times(\vec{B}\times\vec{C})$

Line Integral, Surface Integral and Volume Integral

- Line integral: $\int_C F(R)dR$

  If $F(R) = \hat{i}f(x,y,z)+\hat{j}\,\phi(x,y,z)+\psi(x,y,z)$

  $dR = \hat{i}dx+\hat{j}dy+\hat{k}dz$

  $\int_C F(R)dR = \int_C (fdx+\phi dy+\psi dz)$

- Surface integral: $\int_S \vec{F}\cdot\vec{ds}$ or $\int_S \vec{F}\cdot\vec{N}\,ds$, where N is unit outward normal to surface.

- Volume integral: $\int_V F\cdot dv$

  If $F(R) = f(x, y, z)\hat{i}+\phi(x,y,z)\hat{j}+\psi(x,y,z)\hat{k}$ and $\delta v = \delta x\delta y\delta z$, then

$$\int F\,dv = \hat{i}\iiint f\,dx\,dydz+\hat{j}\iiint \phi dx\,dydz+\hat{k}\iiint \psi dx\,dydz$$

Green's Theorem

If R is a closed region in the xy plane bounded by a simple closed curve c and if P and Q are continuous functions of x and y having continuous derivative in R, then according to Green's theorem

$$\oint_C (Pdx + Qdy) = \int_R\int \left(\frac{\partial Q}{\partial x} - \frac{\partial P}{\partial y}\right) dxdy$$

Stoke's Theorem

If F is continuously differentiable vector function in R, then $\oint F.dr = \int \nabla \times F.Nds$.

Gauss Divergence Theorem

The normal surface integral of a vector point function F which is continuously differentiable over the boundary of a closed region is equal to

$$\int_S F.N.ds = \int_V div\ Fdv$$

■■■■

Complex Numbers

7

COMPLEX VARIABLES

$z = x + iy$ is a complex number, where x and y are real numbers called as real and imaginary parts of z.

Modulus or absolute value $= |z| = \sqrt{x^2 + y^2}$; argument of $Z = u = \arg(z)$ $= \tan^{-1}(y/x)$

Function of a Complex Variable

It is a rule by means of which it is possible to find one or more complex numbers 'w' for every value of 'z' in a certain domain D, than $w = f(z)$, where $z = x + iy$, $w = f(z) = u(x, y) + iv(x, y)$.

Continuity of $f(z)$

- A function $w = f(z)$ is said to be continuous at $z = z_0$ if $\lim_{z \to z_0} f(z) = f(z_0)$.
- Further $f(z)$ is said to be continuous in any region R of the z-plane, if it is continuous at every point of that region.
- Also is $w = f(z) = u(x, y) + iv(x, y)$ is continuous at $z = z_0$, then $u(x, y)$ and $v(x, y)$ are also continuous at $x = x_0$ and $y = y_0$.

Theorem of Differentiability

The necessary and sufficient conditions for the derivative of the function $f(z)$ to exist for all values of z in a region R.

(i) $\frac{\partial u}{\partial x}, \frac{\partial u}{\partial y}, \frac{\partial y}{\partial x}, \frac{\partial v}{\partial y}$ are continuous functions of x and y in R.

(ii) $\frac{\partial u}{\partial x} = \frac{\partial v}{\partial y}, \frac{\partial u}{\partial y} = -\frac{\partial v}{\partial x}$ → Cauchy-Riemann equations (CR Equations).

Analytic Functions (or Regular Function)

- A single-valued function which is defined and differentiable at each point of a domain D is said to be analytic in that domain.
- A point at which an analytic function ceases to posseses a derivative is called singular point.

- Thus, if u and v are real single-valued functions of x and y such that $\frac{\partial u}{\partial x}, \frac{\partial u}{\partial y}, \frac{\partial v}{\partial y}, \frac{\partial v}{\partial y}$ are continuous throughout a region R, then C-R equations $\frac{\partial u}{\partial x} = \frac{\partial v}{\partial y}, \frac{\partial u}{\partial y} = -\frac{\partial v}{\partial x}$ are both "necessary and sufficient" condition for the function $f(z) = u + iv$ to be analytic in R.
- Real and imaginary parts, i.e. u, v of the function are called conjugate function.
- An analytic function posseses derivatives of all order an these are themselves analytic.

Harmonic Functions

If $f(z) = u + iv$ is an analytic function in some region of the z-plane then the $C - R$ equations are satisfied.

$$\frac{\partial u}{\partial x} = \frac{\partial v}{\partial y}, \frac{\partial u}{\partial y} = -\frac{\partial v}{\partial x}$$

Defferentiating with respect to x and y respectively, we get

$$\frac{\partial^2 u}{\partial x^2} = \frac{\partial^2 v}{\partial x \partial y}, \frac{\partial^2 u}{\partial y^2} = -\frac{\partial^2 v}{\partial x \partial y}$$

$$\frac{\partial^2 u}{\partial x^2} + \frac{\partial^2 u}{\partial y^2} = 0 \qquad \text{(Laplace equation)}$$

Note:

- **For a function to be regualr, the frist order partial derivations of u and v must be continuous in addition to $C - R$ equations.**
- **Mean value of any harmonic function over a circle is equal to the value of the function at the centre.**

Methods of Constructiong Analytic Functions

1. If real part of a function is given, then

$$f'(z) = \frac{\partial u}{\partial x} - i\frac{\partial u}{\partial y}$$

Integrate with points at $(z, 0)$

$$f(z) = \int\left(\frac{\partial u}{\partial x}\right)_{(z,0)} dz - i\int\left(\frac{\partial u}{\partial y}\right)_{(z,0)} dz + c$$

Similarly, in case $v(x, y)$ is known, then

$$f'(z) = \frac{\partial D}{\partial x} + i\frac{\partial v}{\partial y}$$

$$f(z) = \int\left(\frac{\partial v}{\partial x}\right)_{(z,0)} dz + i\int\left(\frac{\partial v}{\partial y}\right)_{(z,0)} dz + c$$

2. If $u(x, y)$ is known, then to find $v(x, y)$, we have

$$dv = \frac{\partial v}{\partial x}dx + \frac{\partial v}{\partial y}dy$$

$$dv = -\frac{\partial u}{\partial y}dx + \frac{\partial u}{\partial x}dy$$

Integrate this equation to find v

$$f(z) = u(x, y) + iv(x, y)$$

3. If a real part of the analytic function $f(z)$ is a given (harmonic function $u(x, y)$), then $f(z) = 2u\left(\frac{z}{2}, \frac{z}{2i}\right) - u(0, 0)$.

Complex Integration

- Line integral = $\int_C f(z)dz$, C need not be closed path

  Here, $f(z)$ = integrand, curve C = path of integration

- Contour integral = $\oint_C f(z)dz$, if C is a closed path

  If $f(z) = u(x, y) + iv(x, y)$ and $dz = dx + idy$

  $$\therefore \int_C f(z)dz = \int_C (udx - vdy) + i\int_C (vdx + udy)$$

Theorem : $f(z)$ is analytic in a simple connected domain, then

$$\int_{z_0}^{z_1} f(z)dz = f(z_1) - f(z_0).$$

- Integration is independent of the path.

Dependence on Path : In general, "Complex line inegration" depends not only on the end points but also on the path. (However, analytic function in simple connected domain is independent of path.)

Chauchy's integral Theorem

If $f(z)$ is analytic in a simple connected domain D, then for every simple closed path C in D, $\oint_C f(z)dz = 0$

Note: ....

- In other words, by Cauchy's theorem if $f(z)$ is analytic on a simple closed path C and everywhere inside C (with no exception, not even a single point) then

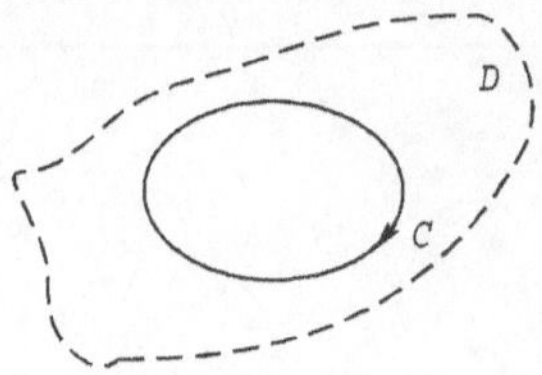

$$\oint_C f(z)dz = 0$$

....

Chauchy's Integral Formula

If $f(z)$ is analytic within and on a closed curve and if a is any point within C, then

$$f(a) = \frac{1}{2\pi i}\int_C \frac{f(z)dz}{z-a}$$

$$f'(a) = \frac{1!}{2\pi i}\int_C \frac{f(z)}{(z-a)^2}dz$$

$$f''(a) = \frac{2!}{2\pi i}\int_C \frac{f(z)}{(z-a)^3}dz$$

....

$$f^n(a) = \frac{n!}{2\pi i}\int_C \frac{f(z)}{(z-a)^{n+1}}dz$$

Morera's Theorem : If $f(z)$ is continuous in a region and $\int_C f(z)dz = 0$ around every simple closed C, then $f(z)$ is analytic in that region.

Taylor's Series : If $f(z)$ is analytic inside a circle C with centre at a then for z inside C

$$f(z) = f(a) + f'(a)(z-a) + \frac{f''(a)}{2!}(z-a)^2 + ...$$

$$f(z) = \sum_{n=0}^{\infty} a_n(z-a)^n$$

where, $$a_n = \frac{1}{2\pi i}\int \frac{f(t)}{(t-a)^{n+1}}dt$$

Other form, put $z = a + h$

$$f(a+h) = f(a) + hf'(a) + \frac{h^2}{2!}f''(a) + ...$$

Laurent's Series : If $f(z)$ is analytic in the ring-shaped region R bounded by two concentric circles C_1 and C_2 of radii r_1 and r_2 $(r_1 > r_2)$ and with centre at a then roll for all z in R.

$f(z) = a_0 + a_1(z-a) + a_2(z-a)^2 + \ldots + a_{-1}(z-a)^{-1} + a_{-2}(z-a)^{-2}$

where $$a_n = \frac{1}{2\pi i}\int \frac{f(t)}{(t-a)^{n+1}}dt$$

if $f(z)$ is analytic inside the curve, then $a_{-n} = 0$ and Laurent series reduces to Taylor's series.

Zero's of Analytic Function

The value of z for which $f(z) = 0$. If $f(z)$ is analytic in the neighbourhood of a point $z = a$, then by Taylor's theorem, we get

$$f(z) = a_0 + a_1(z-a) + a_2(z-a)^2 + \ldots + a_n(z-a)^n + \ldots.$$

$$= \sum_{n=0}^{\infty} a_n(z-a)^n$$

where $$a_n = \frac{f^n(a)}{n!}$$

if $a_0 = a_1 = a_2 = a_{n-1} = 0$, then $f(z)$ is said to have a zero of order n at $z = a$.

Singularities of an Analytic Function

A "singular point" of a function as the point at which the function ceases to be analytic.

1. **Isolated Singularity :** If $z = a$ is a singularity of $f(z)$ such that $f(z)$ is analytic at each point in its neighbourhood (i.e. there exists a circle with centre a which has not other singularity 1, then $z = a$ is called an isolated singularity).
2. **Removable Singularity :** if all the negative powers of $(z - a)$ in Laurent series are zero, then

$$f(z) = \sum_{n=0}^{\infty} a_n(z-a)^n$$

Singularity can be removed by defining $f(z)$ at $z = a$ is such a way that it becomes analytic at $z = a$. $\lim_{z\to a} f(z)$ exists finitely, then $z = a$ is a removable singularity.

Example : $f(z) = \sin z/z$, then $z = 0$ is a removable singularity.

3. **Essential Singularity :** If the number of negative power of $(z-a)$ in Laurent's series is infinite, then $z = a$ is called an essential singularity. $\lim_{n \to a} f(z)$ does not exist in this case.

4. **Poles :** If all the negative power of $(z-a)$ in Laurent's series after n^{th} are missing, then the singularity at $z = a$ is called a pole of order n, A pole of first order is called a "simple pole".

Residue Theorem

If $f(z)$ is analytic in and on a closed curve C except at a finite number of singular point within C, then $\int_C f(z)\,dz = 2\pi i$ (sum of the residue at the singualr point within C).

Calculation of Residues

1. If $f(x)$ has a simple pole at $z = a$, then

$$\text{Res } f(a) = \lim_{z \to a}[(z-a)\, f(z)]$$

2. If $f(z) = \dfrac{\phi(z)}{\phi(z)}$ where $\phi(z) = (z-a)\, f(z)$, $f(a) \neq 0$

$$\text{Res } f(z) = \frac{\phi(a)}{\phi'(a)}$$

3. If $f(z)$ has a pole of order n at $z = a$, then

$$\text{Res } f(a) = \frac{1}{(n-1)!}\left\{\frac{d^{n-1}}{dz^{n-1}}\left[(z-a)^n f(z)\right]\right\}_{z=a}$$

Here n = order of singularity.

Note:

- **If an analytic function has singularities at a finite numebr of points, then the sum of residues at these points along with infinite is zero.**

A Handbook on

Instrumentation Engineering

2

Electrical Circuits

CONTENTS

Circuit Elements and Signal Waveform

1

Electrical Network Components

Resistor

It is the property of a substance due to which it opposes the flow of current (i.e. electrons) through it.

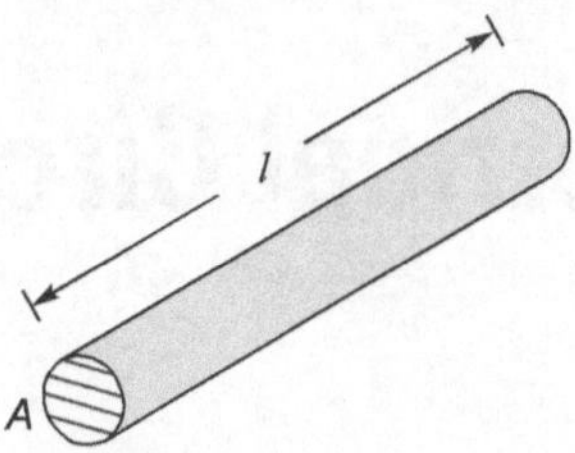

Resistance, $R = \rho \dfrac{l}{A}$, **Ohm (Ω)**

where, $l \rightarrow$ Length of conductor, metre (m)

$A \rightarrow$ Area of cross-section, m^2

$\rho \rightarrow$ Resistivity of the material, Ω-m

$$v(t) = R\,i(t)$$
$$i(t) = \frac{1}{R} v(t)$$
$\cdots$ in time domain

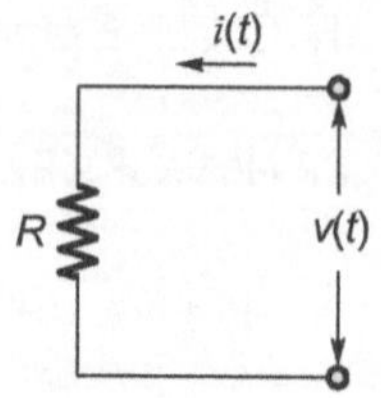

$$V(s) = R\,I(s)$$
$$I(s) = \frac{1}{R} V(s)$$
$\cdots$ in s-domain

Power loss in resistor,

$$p(t) = v(t)\,i(t) = i^2(t) = \frac{v^2(t)}{R}$$

Energy dissipated in resistor,

$$E_R = \int_{t_1}^{t_2} p(t)\,dt$$

Combination of Resistors

1. Resistors in series

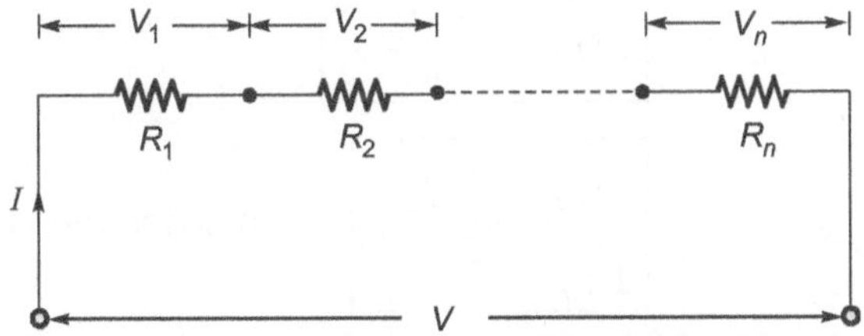

$$\boxed{V = I R_{eq}} \quad ; \quad \boxed{R_{eq} = R_1 + R_2 + \ldots.. + R_n}$$

2. Resistors in parallel

$$\boxed{I = \frac{V}{R_{eq}}}$$

$$\boxed{\frac{1}{R_{eq}} = \frac{1}{R_1} + \frac{1}{R_2} + \ldots.. \frac{1}{R_n}}$$

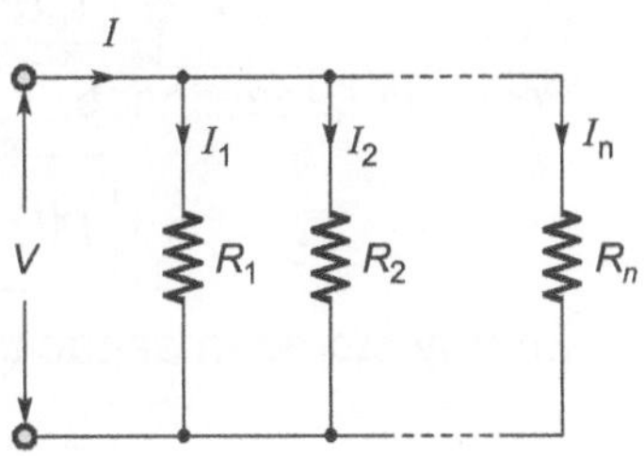

Capacitor

The circuit element that stores energy in the form of electric field is a capacitor or capacitance.

Capacitance $\boxed{C = \frac{\epsilon_0 \epsilon_r A}{d}}$ Farad (F)

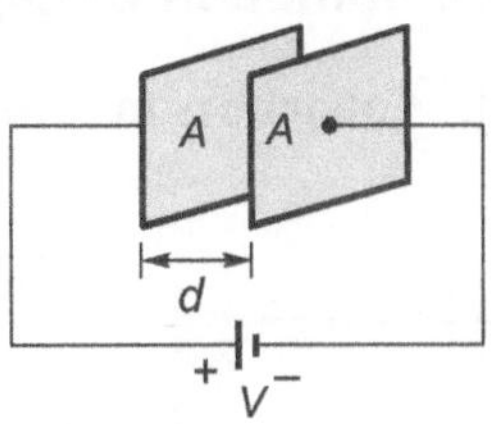

where $\epsilon_0 \rightarrow$ Permittivity of free space, F/m

$\epsilon_r \rightarrow$ Relative permittivity of the dielectric

$A \rightarrow$ Cross-sectional area of parallel plates, m^2

$d \rightarrow$ Separation of plates, m

Note:

When the capacitance is removed from the source, the capacitor retains the charge and the electric field until a discharge path is provided.

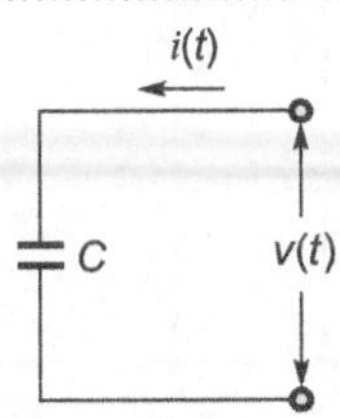

$$v(t) = \frac{1}{C}\int_0^t i(t)\,dt$$
$$i(t) = C\frac{dv(t)}{dt}$$ ... in time domain

$$V(s) = \frac{1}{sC}I(s)$$
$$I(s) = sCV(s)$$ ... in s-domain

$$Z_C(s) = \frac{1}{sC}$$
$$Y_C(s) = sC$$

Power in capacitance,

$$p(t) = v(t)\,i(t) = Cv(t)\frac{dv(t)}{dt}$$

Energy stored in capacitance,

$$E_C = \frac{1}{2}Cv^2$$

Combination of Capacitors

1. Capacitors in series

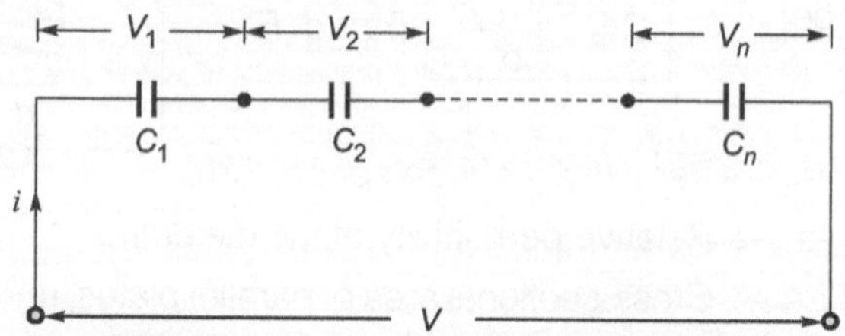

$$V = \frac{1}{C_{eq}}\int_0^t i\,dt \quad ; \quad \frac{1}{C_{eq}} = \frac{1}{C_1} + \frac{1}{C_2} + \frac{1}{C_3} + ... + \frac{1}{C_n}$$

2. Capacitors in parallel

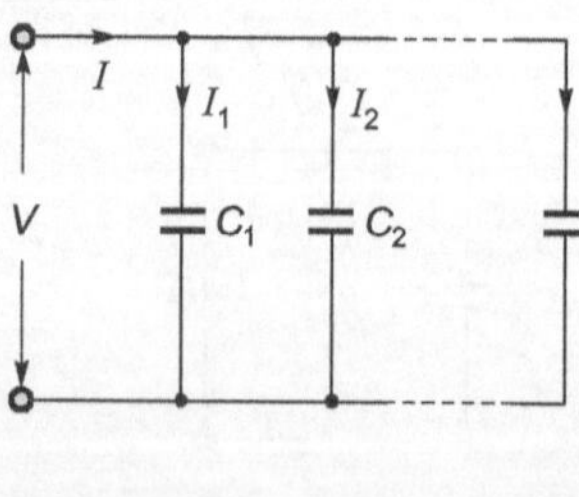

$$I = C_{eq}\frac{dv}{dt} \quad ; \quad C_{eq} = C_1 + C_2 + C_3 + \ldots\ldots + C_n$$

Note: ....

- Under steady state condition for D.C. supply capacitor acts as open circuit.
- Capacitor doesn't allow sudden change of voltage, until impulse of current is applied.
- If unit impulse of current is applied then at $t = 0^+$ it stores $\frac{1}{2C}$ Joules energy.
- Power dissipation in ideal capacitor is zero.

....

Inductor

The circuit element that stores energy in the form of a magnetic field is an inductor or inductance.

Inductance $$L = \frac{\mu_0 N^2 A}{l}$$ Henry (H)

where $\mu_0 \rightarrow$ Permeability of free space, H/m
$N \rightarrow$ Total number of turns in coil
$A \rightarrow$ Cross-sectional area of coil, m^2
$l \rightarrow$ Length of coil, m

Note: ....

When the inductance is removed from the source, the magnetic field will collapse i.e. no energy is stored without a connected sources.

....

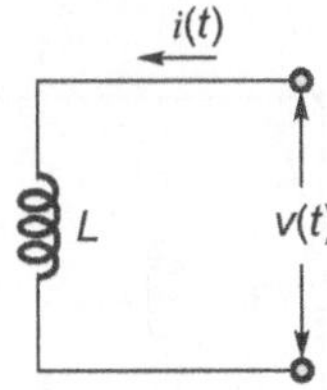

$$v(t) = L\frac{di(t)}{dt}$$

$$i(t) = \frac{1}{L}\int_0^{\infty} v(t)\,dt$$

...in time domain

$$\boxed{\begin{aligned} V(s) &= sL\,I(s) \\ I(s) &= \frac{1}{sL}V(s) \end{aligned}} \quad \cdots \text{in s-domain}$$

$$\boxed{\begin{aligned} Z_L(s) &= sL \\ Y_L(s) &= \frac{1}{sL} \end{aligned}}$$

Power in inductance,

$$\boxed{p(t) = v(t)\; i(t) = Li(t)\,\frac{di(t)}{dt(t)}}$$

Energy stored in an inductance,

$$\boxed{E_L = \frac{1}{2}Li^2(t)}$$

Combination of Inductor

1. Inductors in series

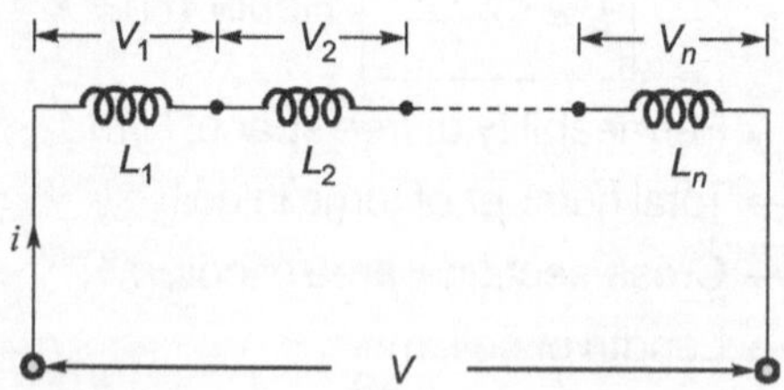

$$\boxed{V = L_{eq}\frac{di}{dt}} \quad ; \quad \boxed{L_{eq} = L_1 + L_2 + L_3 + \ldots\ldots + L_n}$$

2. Inductors in parallel

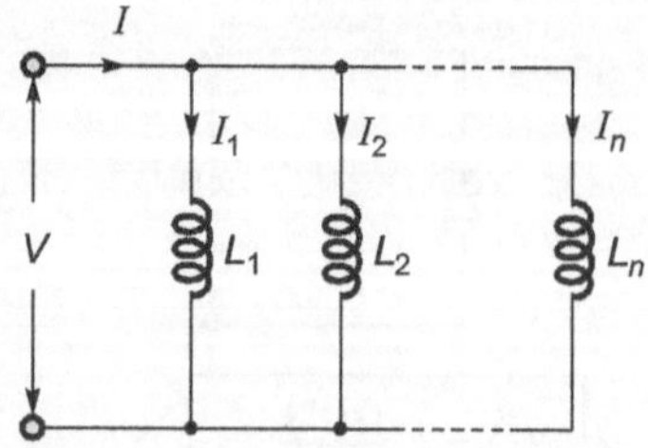

$$\boxed{I = \frac{1}{L_{eq}}\int_0^t v\,dt} \quad ; \quad \boxed{\frac{1}{L_{eq}} = \frac{1}{L_1} + \frac{1}{L_2} + \frac{1}{L_3} + \ldots\ldots \frac{1}{L_n}}$$

Note:

- Under steady state condition for D.C. supply, inductor acts as short circuit.
- Inductor doesn't allow sudden change of voltage, until impulse of voltage is applied.
- If unit impulse of voltage is applied across it then at $t = 0^+$ it stores energy

$= \frac{1}{2L}$ Joules.

- Power dissipation in ideal inductor is zero.

Ideal Transformer

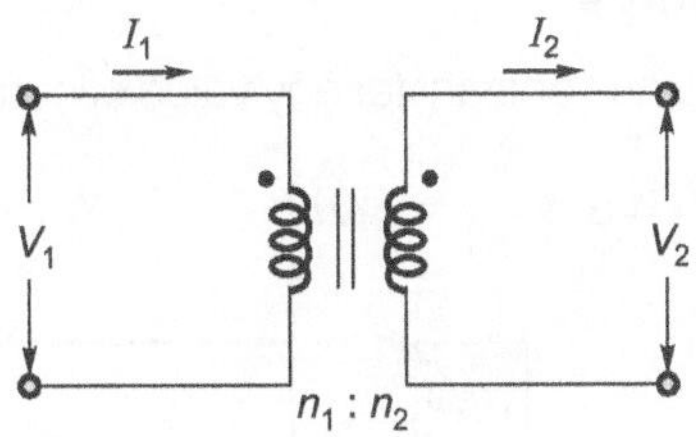

$$\frac{V_2}{V_1} = \frac{n_2}{n_1} = \frac{I_1}{I_2}$$

If $n_1 < n_2$, step up transformer

and $n_1 > n_2$, step down transformer

Gyrator

Gyrator shows an impedance inversion.

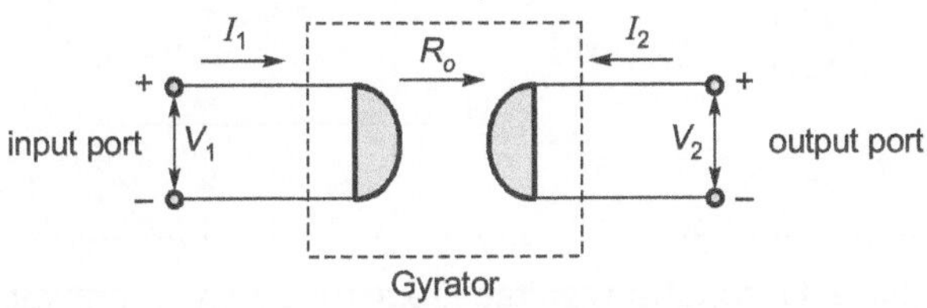

Gyrator

$$V_1 = +R_o I_2$$
$$V_2 = -R_o I_1$$

where, $R_o \rightarrow$ Coefficient of gyrator and it depends upon

(i) Op-amp parameter

(ii) Externally connected R and C values

Remember:

- Linear network is one which holds the principle of superposition and principle of homogeneity both.
- Element which conducts in both directions is called bilateral element.
- A network containing circuit elements without any energy source is called passive network. e.g. *R*, *L*, *C*.
- A network containing energy source together with other circuit elements is called active network. e.g. Op-Amp, transistors.

Ideal Sources

1. Ideal Voltage Source

Voltage always remains constant for any value of current passing through it.

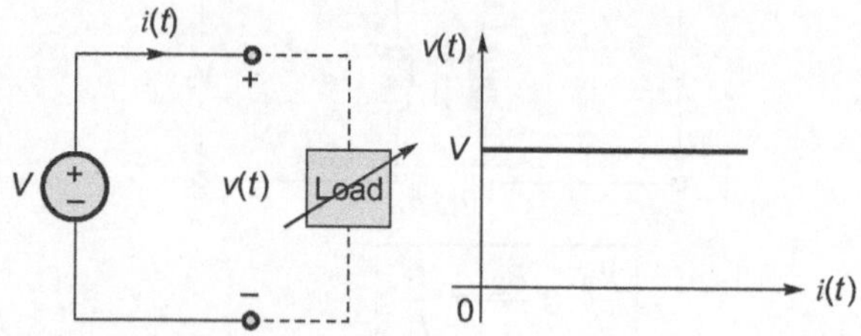

2. Practical Voltage Source

It has small internal resistance. Voltage across the element varies with respect to current.

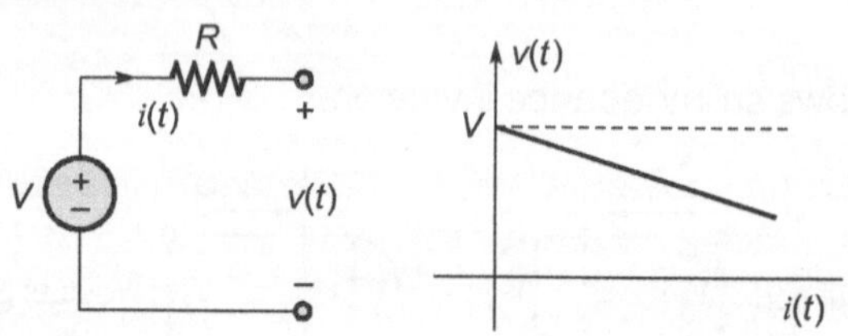

Note:

- The current through any voltage source is purely arbitrary. It will depend upon
  (a) The current source which is connected in series with it.
  (b) The load resistance which is connected in parallel with it.
- We cannot write KCL equation at a load with which any voltage source is connected as the current through this voltage source is purely arbitrary.

3. Ideal current source

Current always remains constant for any value of voltage across it

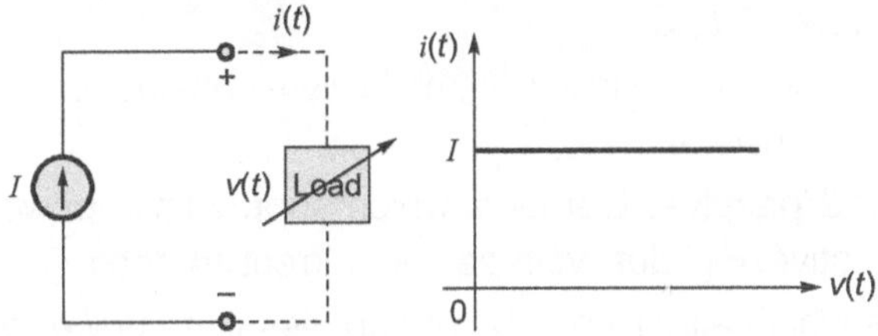

4. Practical current source

It has large value of internal resistance. Current varies with respect to the voltage across the element.

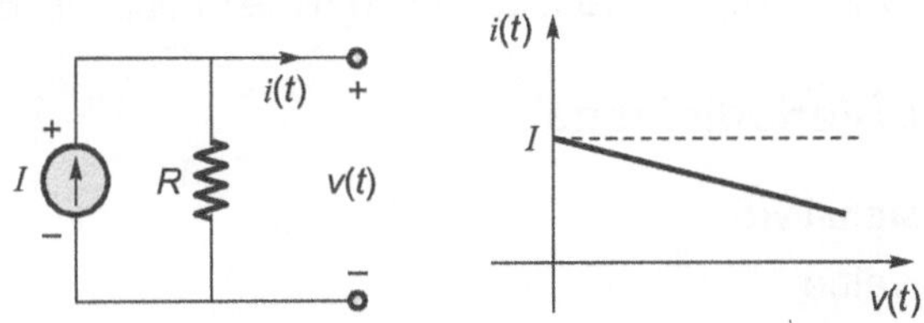

Note: ....

- The voltage across current source is purely arbitrary. It will depend upon
  (a) Voltage source which is connected in parallel with it.
  (b) The load resistance which is connected in series with it.
- We cannot write any KVL equation in a loop in which a current source is present as the voltage across the current source is purely arbitrary.

....

Dependent Sources

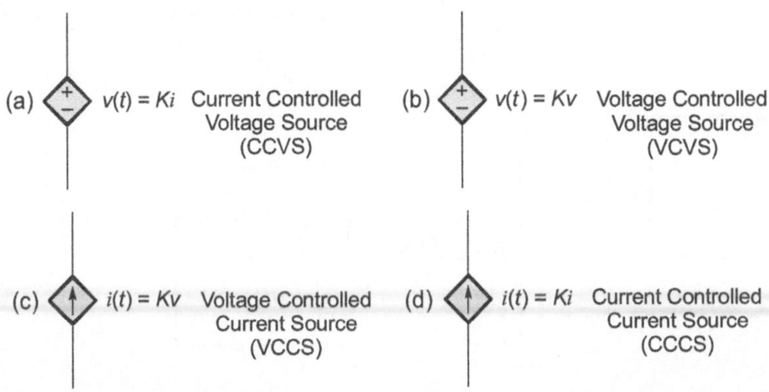

where, K is a constant.

Classification of Circuit Elements

(a) **Linear and non linear:** Elements which obey Ohm's law are linear elements. Ex: R, L, C.

Elements which do not follow Ohm's law are non-linear.

Ex. Diode, transistor.

(b) **Active and passive:** Elements which supply finite power for infinite time are active. Ex: ideal voltage and current sources.

Elements which either absorbs or supplies finite power only for finite time. Ex. R, L, C.

(c) **Unilateral and bilateral:** Elements which allow flow of current only in one direction are unilateral. Ex. diode.

Elements which allow current flow in both the directions. Ex. R, L, C.

Properties of Periodic Signal

For periodic signal *y*(*t*):

- Average value

$$Y_{av} = \frac{1}{T}\int_0^T y(t)dt$$

where T is the time period of periodic function $y(t)$

- RMS or effective value

$$Y_{rms} = \sqrt{\frac{1}{T}\int_0^T [y(t)]^2 dt}$$

If $y(t) = a_0 + (a_1 \cos\omega t + a_2 \cos 2\omega t + ...) + (b_1 \sin \omega t + b_2 \sin 2\omega t + ...)$

$$Y_{rms} = \sqrt{a_0^2 + \frac{1}{2}(a_1^2 + a_2^2 + ...) + \frac{1}{2}(b_1^2 + b_2^2 + ...)}$$

- Form factor

$$FF = \frac{Y_{rms}}{Y_{av}} = \frac{\sqrt{\frac{1}{T}\int_0^T [y(t)]^2 dt}}{\frac{1}{T}\int_0^T y(t)dt}$$

- Peak factor

$$PF = \frac{Y_{max}}{Y_{rms}}$$

Remember: ....

-

	Pure sine wave	Half-wave rectified waveform	Full-wave rectified waveform
Y_{av}	0	$\frac{Y_{max}}{\pi}$	$\frac{2Y_{max}}{\pi}$
Y_{rms}	$\frac{Y_{max}}{\sqrt{2}}$	$\frac{Y_{max}}{\sqrt{2}}$	$\frac{Y_{max}}{\sqrt{2}}$
FF	$\frac{\pi}{2\sqrt{2}}$	$\frac{\pi}{2}$	$\frac{\pi}{2\sqrt{2}}$
PF	$\sqrt{2}$	2	$\sqrt{2}$

- $\sin \omega t = \frac{e^{j\omega t} - e^{-j\omega t}}{2j}$ and $\cos \omega t = \frac{e^{j\omega t} + e^{-j\omega t}}{2}$

$\sinh \omega t = \frac{e^{\omega t} - e^{-\omega t}}{2}$ and $\cosh \omega t = \frac{e^{\omega t} + e^{-\omega t}}{2}$

....

■■■■

Network Laws and Theorems 2

Ohm's Law

The ratio of potential difference (V) between any two points on a conductor to the current (I) flowing through them is constant, provided the temperature of the conductor does not change.

$$\frac{V}{I} = \text{constant} \quad \text{or} \quad \frac{V}{I} = R$$

Where, R is the resistance of the conductor between the two points considered.

Kirchoff's Laws

1. Kirchoff's Voltage Law (KVL)

For any closed path in a network, the algebraic sum of the voltages is zero.

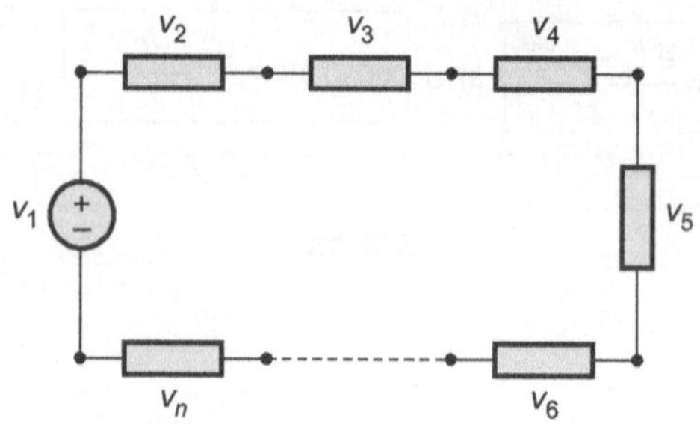

$$\sum_{k=1}^{n} v_k(t) = 0$$ ... in a closed loop

where, v_k is the voltage drop or voltage gain across k^{th} element

2. Kirchoff's Current Law (KCL)

The algebraic sum of the currents at a node is zero. Alternatively the sum of the currents entering a node is equal to the sum of the currents leaving that node.

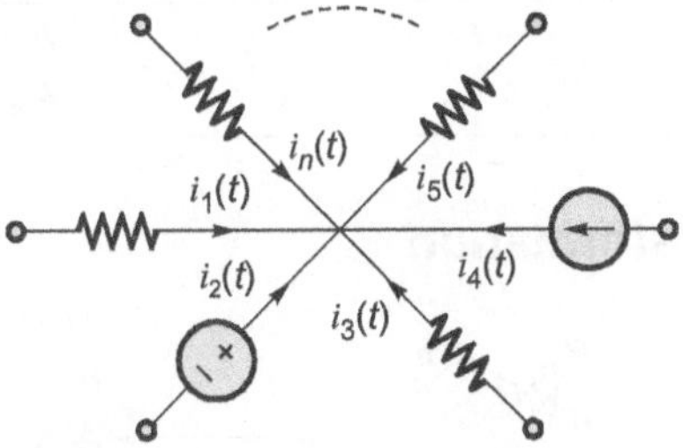

$$\boxed{\sum_{k=1}^{n} i_k(t) = 0} \quad \text{..... at any node}$$

where $i_k(t)$ is the current through k^{th} branch

Note: ....

- A network is an interconnection of elements or devices, whereas a circuit is a network providing one or more closed paths.
- Number of KVL equations = $b - (n - 1)$
- Number of KCL equations = $(n - 1)$
  where, b is number of branches and n is number of nodes.
- KVL is based on the law of conservation of energy.
- KCL is based on the law of conservation of charge.
- KVL and KCL are applicable to lumped networks.

....

Voltage Division Equations

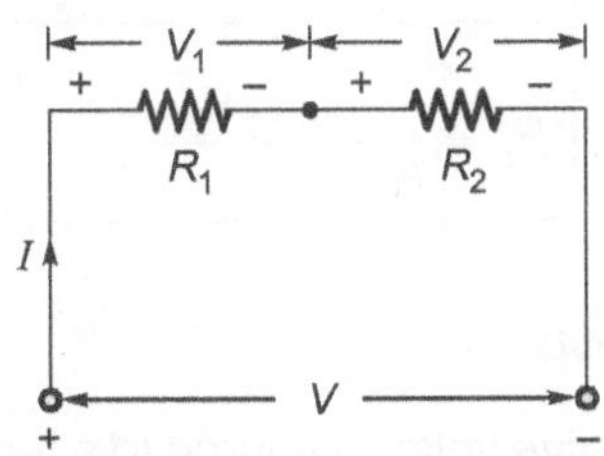

$$\boxed{V_1 = \frac{R_1}{R_1 + R_2} V} \quad ; \quad \boxed{V_2 = \frac{R_2}{R_1 + R_2} V}$$

Current Division Equations

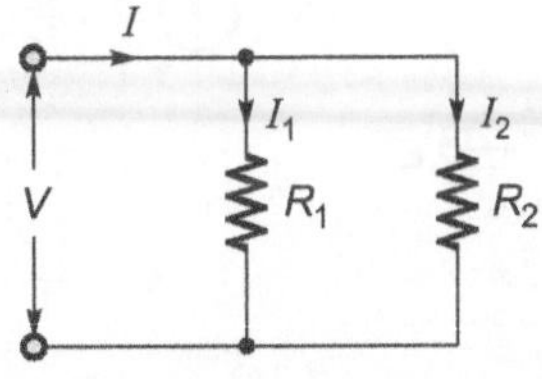

$$\boxed{I_1 = \frac{R_2\, I}{R_1 + R_2}} \quad ; \quad \boxed{I_2 = \frac{R_1\, I}{R_1 + R_2}}$$

Star to Delta Transformation

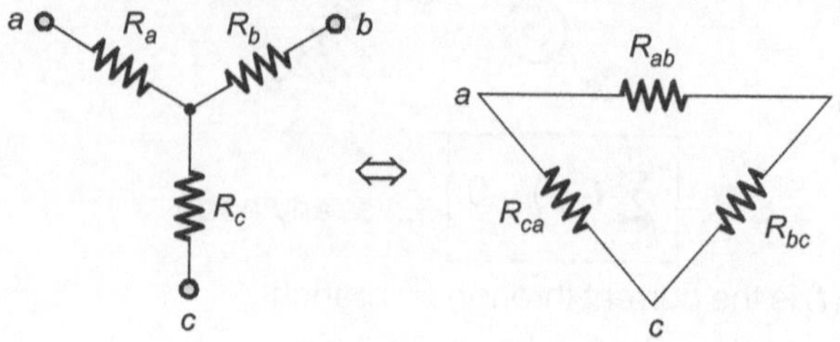

$$\boxed{R_{ab} = \frac{\Delta}{R_c}, R_{bc} = \frac{\Delta}{R_a}, R_{ca} = \frac{\Delta}{R_b}}$$

where, $\Delta \rightarrow (R_aR_b + R_bR_c + R_cR_a)$

Delta to Star Transformation

$$\boxed{R_a = \frac{R_{ca}R_{ab}}{R_{ab} + R_{bc} + R_{ca}}}$$

$$\boxed{R_b = \frac{R_{ab}R_{bc}}{R_{ab} + R_{bc} + R_{ca}}}$$

$$\boxed{R_c = \frac{R_{bc}R_{ca}}{R_{ab} + R_{bc} + R_{ca}}}$$

Source Transformation

Transformation of a resistive voltage source to a resistive current source or vice-versa.

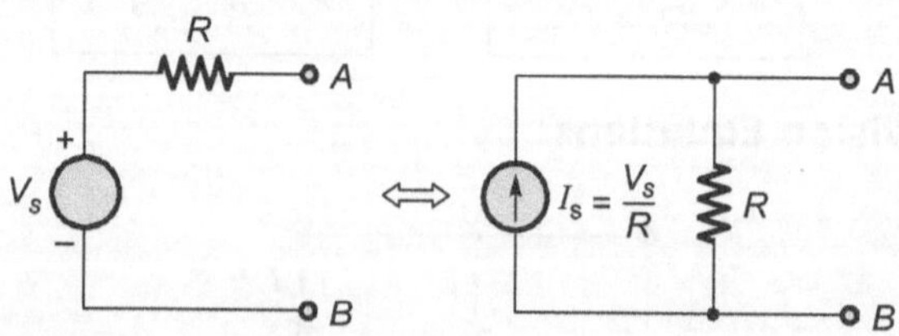

Network Theorems

1. Superposition Theorem

The response in any element of a linear, bilateral RLC network containing more than one independent voltage or current source is the algebraic sum of responses produced by the independent source when each of them is acting alone with

(a) All other independent voltage sources are short circuited (S.C.).

(b) All other independent current sources are open circuited (O.C.).

(c) All dependent voltage and current sources remain as they are and therefore, they are neither S.C. nor O.C.

Note:

- The theorem is not applicable to the network containing
  (a) Non linear elements.
  (b) Unilateral elements such as diode or BJT.
- The theorem is not applicable to power since it is a non linear parameter.
- The theorem is also applicable for circuit having initial condition.

2. Thevenin's Theorem

A linear active RLC network which contains one or more independent or dependent voltage or current sources can be replaced by a single voltage source V_{OC} in series with equivalent impedance Z_{eq}.

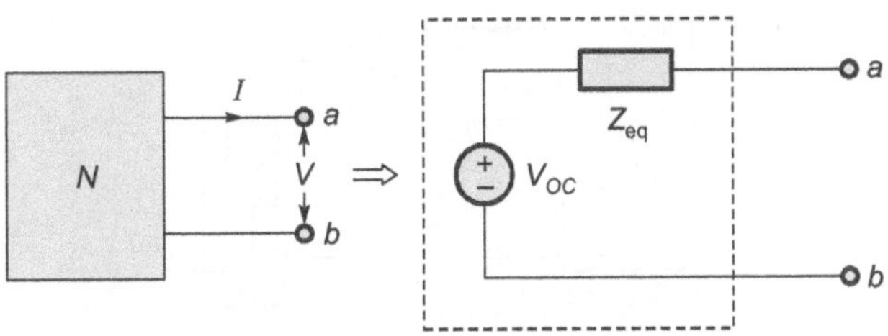

where, $V_{OC} \rightarrow$ Open circuit voltage between a and b (when $I = 0$).

$Z_{eq} \rightarrow$ Equivalent impedance between a and b, when

(a) All independent sources are replaced by their internal impedances.

(b) All dependent voltage and current sources remain as they are.

Note: ....

Theorem is not applicable to the network containing:

- Non linear element.
- Unilateral element.

....

3. Norton's Theorem

A linear, active RLC network which contains one or more independent or dependent voltage or current sources can be replaced by a single current source I_{SC} in shunt with equivalent impedance Z_{eq}.

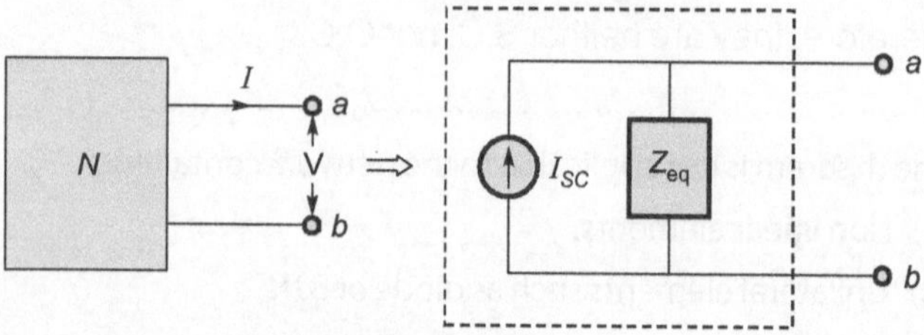

where, $I_{SC} \rightarrow$ Short circuit current between a and b (when $V = 0$)

$Z_{eq} \rightarrow$ Same as that of Thevenin's theorem

4. Maximum Power Transfer Theorem

In any linear bilateral network, the power transferred to the load will be maximum when the load resistance is equal to Thevenin's equivalent resistance.

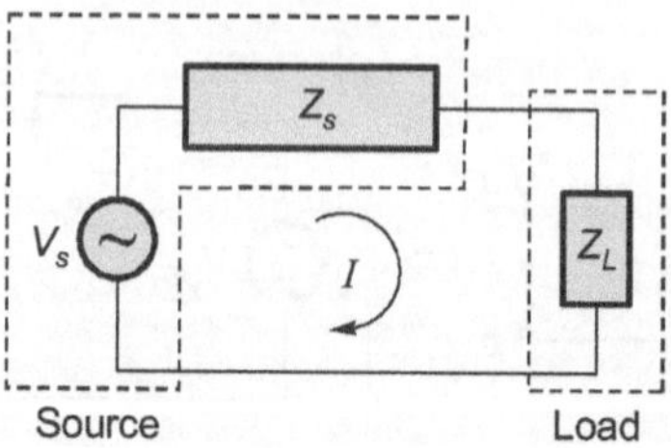

$\boxed{Z_L = Z_s^*}$ ......for maximum power transfer

Case 1: If $Z_s = R_s + jX_s$ and $Z_L = R_L + jX_L$

then $\boxed{R_L = R_s}$ and $\boxed{X_L = -X_s}$

Case 2: If $Z_s = R_s + jX_s$ and $Z_L = R_L$

then $\boxed{R_L = \sqrt{R_s^2 + X_s^2}}$

Case 3: If $Z_L = R_L$ and $Z_s = R_s$

then $\boxed{R_L = R_s}$

Note:

- Efficiency at maximum power transfer is 50%.
- If Z_s(source resistance) is varied then maximum power is transferred to the load if $Z_s = 0$.

5. Tellegen's Theorem

- In any network, the sum of instantaneous power consumed by various elements of the branches is always equal to zero.
- Total power supplied by different voltage sources is equal to total power consumed by various passive elements in various branches of the network.

$$\boxed{\sum_{k=1}^{b} v_k \cdot i_k = 0}$$

where, $b \rightarrow$ Number of branches

Note:

The theorem is valid for any type of network as long as KVL and KCL equations are valid.

6. Millman's Theorem

When number of voltage sources (E_1, $E_2 \cdots E_n$) are in parallel with internal resistances $\left(\frac{1}{Y_1}, \frac{1}{Y_2} \cdots \frac{1}{Y_n}\right)$, the arrangement can be replaced by a single equivalent voltage source E_{eq} with an equivalent series resistance $\left(\frac{1}{Y_{eq}}\right)$.

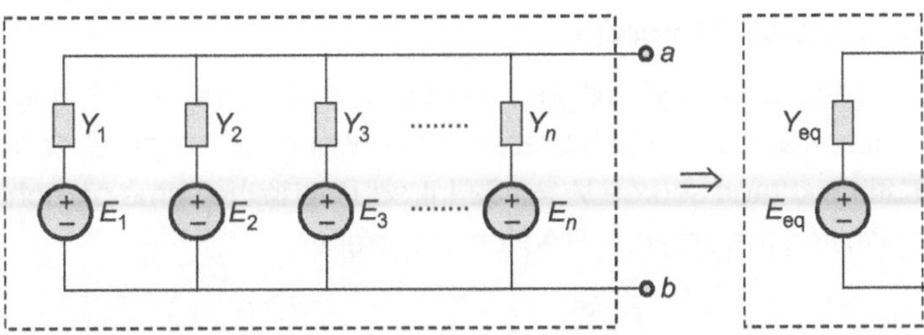

$$E_{eq} = \frac{\sum_{i=1}^{n} E_i Y_i}{\sum_{i=1}^{n} Y_i}$$

$$Y_{eq} = \sum_{i=1}^{n} Y_i$$

7. Reciprocity Theorem

In a linear bilateral single source network, the ratio of excitation to the response is constant even when the position of excitation and response are interchanged.

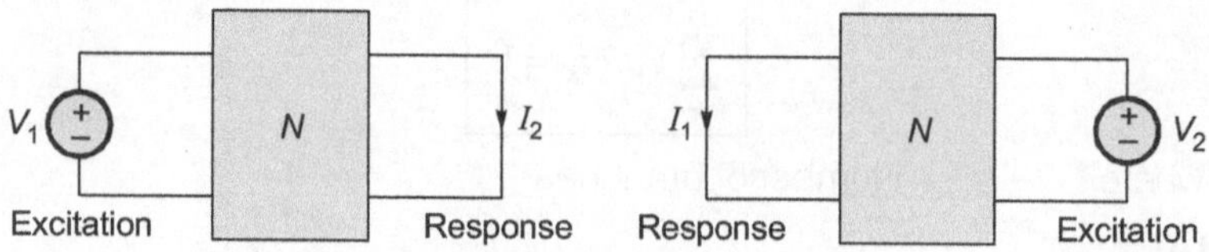

$$\frac{V_1}{I_2} = \frac{V_2}{I_1} \quad ; \quad Z_{12} = Z_{21}$$

Note:

- The basis of the theorem is the symmetry of impedance or admittance of matrix.
- The theorem is valid for network in which linear and bilateral elements are present.
- The theorem is valid only when single independent voltage or current source is present.
- The initial conditions are assumed to be zero in reciprocity theorem.

8. Compensation Theorem

If impedance 'Z' of any branch of a network is changed by 'δZ', then the incremental current 'δI' in such branch is that which will be produced by a compensating voltage source $V_c = I\,\delta Z$ introduced in the same branch with polarity opposing the original direction of current I.

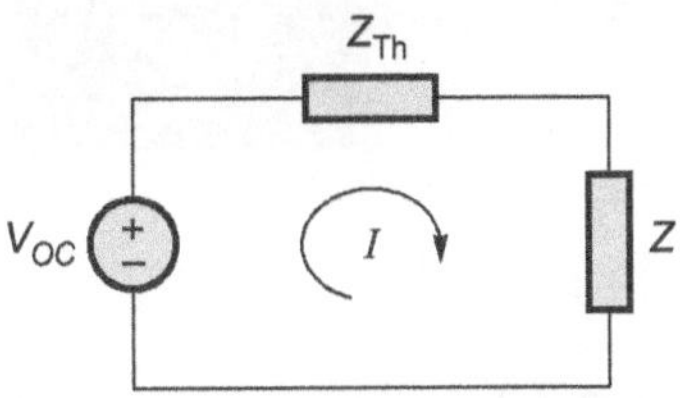

(a) Compensation network

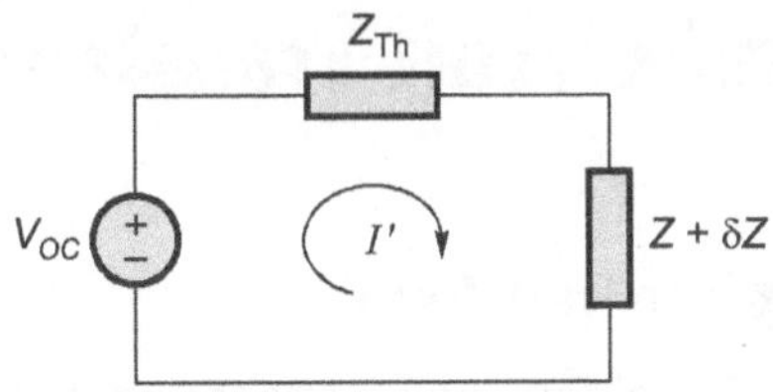

(b) Z changes to $Z + \delta Z$

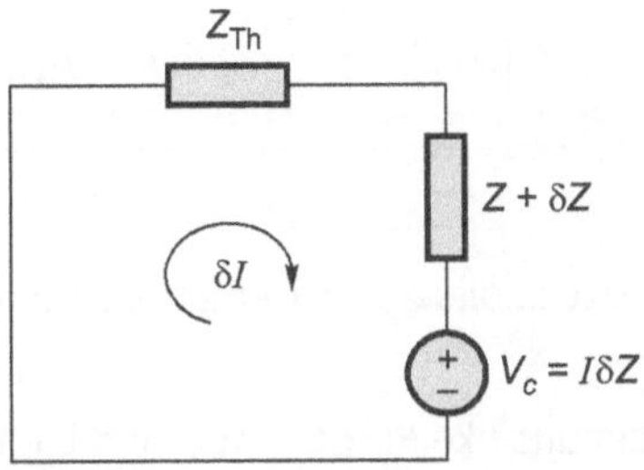

Ideal voltage source V_c connected in series

9. Substitution theorem

The voltage across the current through any branch of a bilateral network being known, this branch can be replaced by any combination of elements that will make the same voltage across and current through it.

■■■■

Laplace Transform Analysis and Circuit Transients

3

Laplace Transform

One sided Laplace transform of a function $f(t)$

$$\mathcal{L}[f(t)] = \int_0^\infty f(t)e^{-st}\,dt = F(s)$$

Note:

- For the two sided Laplace transformation, the limits on the integration are from $-\infty$ to ∞.
- In practical circuits like *RL*, *RC*, one sided Laplace transform is used for analysis purpose where initial conditions are used.

Properties of Laplace Transform

1. Linearity

$$\mathcal{L}\{af_1(t) + bf_2(t)\} = aF_1(s) + bF_2(s)$$

2. Differentiation

$$\left\{\frac{d^n f(t)}{dt^n}\right\} = s^n F(s) - s^{n-1} f(0) - s^{n-2} f'(0) - s^{n-3} f''(0) - \ \ldots$$

3. Integration

$$\mathcal{L}\left\{\int_0^t f(t)\,dt\right\} = \frac{F(s)}{s}$$

4. Scaling

$$\mathcal{L}\{f(at)\} = \frac{1}{|a|}F\left(\frac{s}{a}\right)$$

5. Time shifting

$$\mathcal{L}\{f(t-a)\} = e^{-as}F(s)$$

6. Shifting in s-domain

$$\mathcal{L}\{e^{-at}f(t)\} = F(s+a)$$

7. Convolution

$$\mathcal{L}\{f_1(t) * f_2(t)\} = F_1(s)F_2(s)$$

8. Initial value theorem

$$\lim_{t\to 0} f(t) = \lim_{s\to\infty} s\,F(s)$$

9. Final value theorem

$$\lim_{t\to\infty} f(t) = \lim_{s\to 0} s\,F(s)$$

10. Laplace transform of periodic function

$$F(s) = \mathcal{L}[f(t)] = \frac{\mathcal{L}[f_1(t)]}{1-e^{-sT}}$$

where, $f(t) \to$ Periodic function with time period T

$f_1(t) \to$ Function over one time period

11. Multiplication with '*t*'

$$\mathcal{L}\{tf(t)\} = -\frac{dF(s)}{ds}$$

Laplace Transform Pairs

$f(t)$	Laplace ⇌ Inverse Laplace	$F(s)$
$\delta(t)$	⇌	1
$u(t)$	⇌	$\frac{1}{s}$
$r(t) = tu(t)$	⇌	$\frac{1}{s^2}$
$t^n\,u(t)$	⇌	$\frac{n!}{s^{n+1}}$
$e^{-\alpha t}\,u(t)$	⇌	$\frac{1}{s+\alpha}$
$(\sin\omega t)\,u(t)$	⇌	$\frac{\omega}{s^2+\omega^2}$
$(\cos\omega t)\,u(t)$	⇌	$\frac{s}{s^2+\omega^2}$

$(\sinh \omega t)\, u(t)$	$\rightleftharpoons$	$\dfrac{\omega}{s^2 - \omega^2}$
$(\cosh \omega t)\, u(t)$	$\rightleftharpoons$	$\dfrac{s}{s^2 - \omega^2}$
$(e^{-\alpha t}\, t^n) u(t)$	$\rightleftharpoons$	$\dfrac{n!}{(s+\alpha)^{n+1}}$
$(t \sin \omega t)\, u(t)$	$\rightleftharpoons$	$\dfrac{2\omega s}{(s^2+\omega^2)^2}$
$(t \cos \omega t) u(t)$	$\rightleftharpoons$	$\dfrac{s^2-\omega^2}{(s^2+\omega^2)^2}$
$(e^{-\alpha t} \sin \omega t) u(t)$	$\rightleftharpoons$	$\dfrac{\omega}{(s+\alpha)^2+\omega^2}$
$(e^{-\alpha t} \cos \omega t) u(t)$	$\rightleftharpoons$	$\dfrac{s+\alpha}{(s+\alpha)^2+\omega^2}$

Representation of Circuit Elements in s-domain

Resistor

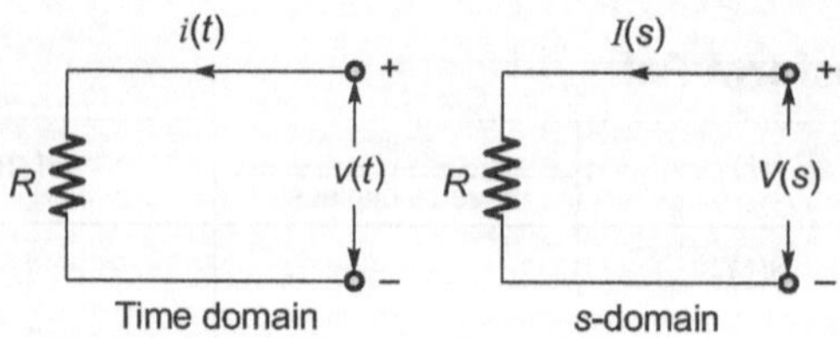

In time domain

$$\boxed{v(t) = R \cdot i(t)}$$

In s-domain

$$\boxed{V(s) = R \cdot I(s)}$$

where, $s \to \sigma + j\omega$

For sinusoidal excitation

$$\sigma = 0 \quad \text{and} \quad s = j\omega$$

Inductor

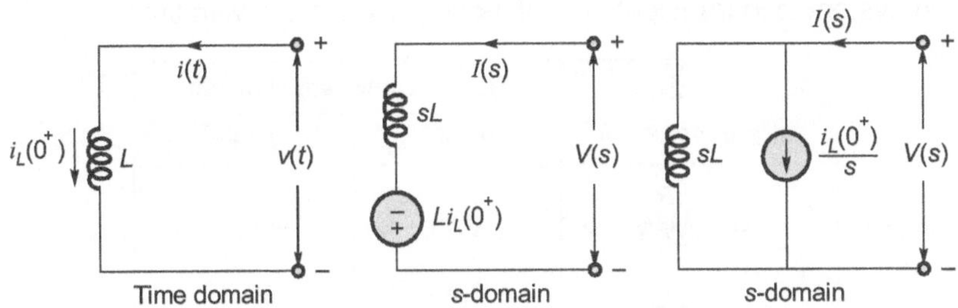

In time domain

$$v(t) = L\frac{d\boldsymbol{i}(t)}{dt}$$

$$\boldsymbol{i}(t) = \frac{1}{L}\int_{-\infty}^{t} v(t)dt$$

In s-domain

$$V(s) = sL \cdot \boldsymbol{I}(s) - L\boldsymbol{i}_L(0^+)$$

$$\boldsymbol{I}(s) = \left[\frac{V(s)}{sL} + \frac{\boldsymbol{i}_L(0^+)}{s}\right]$$

Capacitor

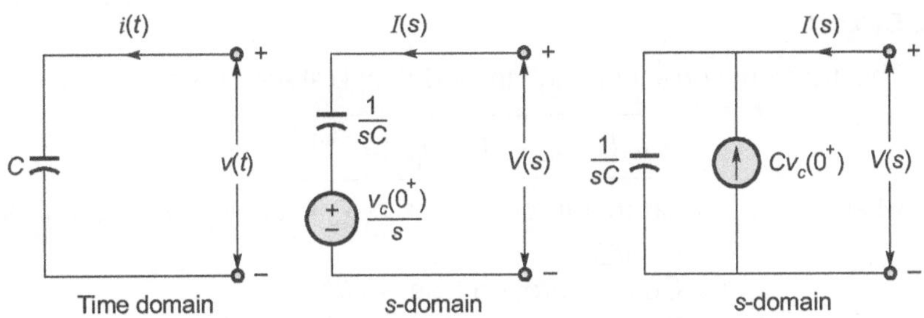

In time domain

$$\boldsymbol{i}(t) = C\frac{dv(t)}{dt}$$

$$v(t) = \frac{1}{C}\int_{-\infty}^{t} \boldsymbol{i}(t)\,dt$$

In s-domain

$$\boldsymbol{I}(s) = sCV(s) - Cv_c(0^+)$$

$$V(s) = \frac{\boldsymbol{I}(s)}{sC} + \frac{v_C(0^+)}{s}$$

Transient Response

A system is in transient state if its output changes with time.

Element ($t = 0^-$ or $t = -\infty$)	Equivalent circuit at $t = 0^+$	Equivalent circuit at $t = +\infty$
R	R	R
L, $I_o = 0$	Open	Short
L, I_o	I_o	Short
C, $+\ -$, $V_C = 0$	Short	Open
C, $+\ -$, V_C	V_C, $+\ -$	Open

Disturbance takes place at $t = 0$.

RL Circuit

The transient current through the inductor L at any time t

$$i_L(t) = i_\infty - [i_\infty - i_{0^+}]e^{-R_{eq}t/L}$$

where, $i_\infty \rightarrow$ Current through L as $t \rightarrow \infty$ i.e. steady-state current through L

$i_{0^+} \rightarrow$ Current through L as $t \rightarrow 0^+$

$R_{eq} \rightarrow$ Thevenin's equivalent resistance seen across L for $t > 0$.

RC Circuit

The transient voltage across capacitor C at any time t

$$v_c(t) = v_\infty - [v_\infty - v_{0^+}]e^{-t/R_{eq}C}$$

where, $v_\infty \rightarrow$ Voltage across capacitor as $t \rightarrow \infty$ i.e. steady state voltage across C

$v_{0^+} \rightarrow$ Voltage across C as t $\rightarrow 0^+$

$R_{eq} \rightarrow$ Thevenin's equivalent resistance seen across C for $t > 0$

■■■■

Resonance

4

Resonance

- Resonance is the condition when the voltage across a circuit becomes in phase with the current supplied to the circuit.
- At resonance, the circuit behaves like a resistive circuit.
- Power factor of the circuit at resonance is unity.

Series Resonance

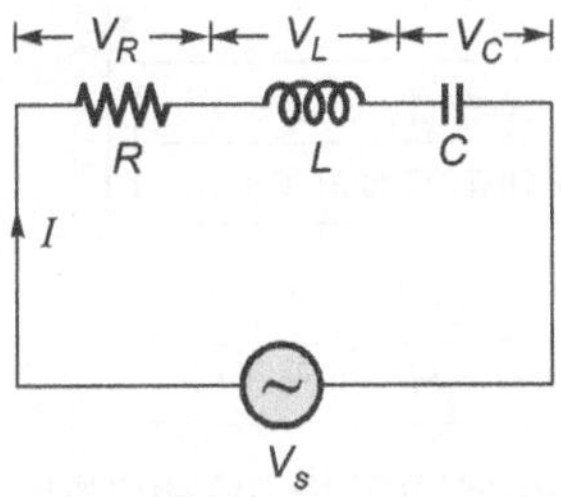

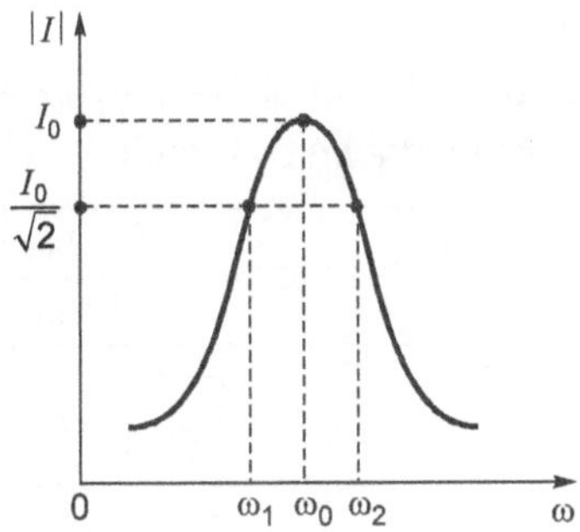

At Resonance

- $|V_L| = |V_C|$ and these are 180° out of phase
- Imaginary part of input impedance = 0

$$\boxed{Z_{in}\big|_{\omega=\omega_0} = R} \text{ ......minimum}$$

$$\boxed{I\big|_{\omega=\omega_0} = \frac{V_s}{R}} \text{ ......maximum}$$

where, $\omega_0 = \frac{1}{\sqrt{LC}} \rightarrow$ Resonant frequency in rad/sec

Remember:

- For $\omega < \omega_0$, series *RLC* circuit behaves like an *RC* circuit.
- For $\omega > \omega_0$, series *RLC* circuit behaves like an *RL* circuit.
- For $\omega = \omega_0$, series *RLC* circuit behaves like resistive circuit.

Bandwidth

The bandwidth of the network represent the range of the frequencies at which the current drawn by network becomes 0.707 of its maximum value.

$$\Delta\omega = (\omega_2 - \omega_1) = \frac{R}{L}$$

$$\omega_1 = \omega_0 - \frac{1}{2}\Delta\omega \quad \text{and} \quad \omega_2 = \omega_0 + \frac{1}{2}\Delta\omega$$

$$\omega_0 = \sqrt{\omega_1 \cdot \omega_2}$$

where, $\omega_1, \omega_2 \rightarrow$ Cut-off frequency or half power frequency

***Q*-factor**

The quality factor of the network is a measure of sharpness of the curve for the current drawn by circuit.

$$Q\text{-factor} = 2\pi\left(\frac{\text{Maximum energy stored}}{\text{Energy dissipated per cycle}}\right)$$

$$Q_0 = \frac{\omega_0}{\Delta\omega} = \frac{1}{\omega_0 RC} = \frac{\omega_0 L}{R} = \frac{1}{R}\sqrt{\frac{L}{C}}$$

- **Frequency at which voltage across capacitor is maximum**

$$f_c = \frac{1}{2\pi}\sqrt{\frac{1}{LC} - \frac{R^2}{2L^2}}$$

- **Frequency at which voltage across inductor is maximum**

$$f_L = \frac{1}{2\pi\sqrt{LC - \frac{C^2R^2}{2}}}$$

- **Selectivity**

$$\text{Selectivity} = \frac{\text{Resonance frequency}}{\text{Bandwidth}} = \frac{f_0}{f_2 - f_1}$$

Features of Series Resonance Circuit

- The resonant frequency is a geometric mean of the two half power frequency.
- The resonant frequency represent the rate at which the electrical energy stored in the capacitor is transformed to the magnetic energy stored in the inductor and vice versa.

- The quality factor represents the voltage amplification factor and must have high value for any tuned circuit.
- The quality factor of the resonant curve depends upon the numerical value of RLC components.
- For ***tuned network***, the ***quality factor*** must be ***high***, the ***bandwidth*** should be ***small*** and therefore the resistance used in the network should be small.
- Any ***series RLC*** network represents a ***band pass filter***.

Parallel Resonance

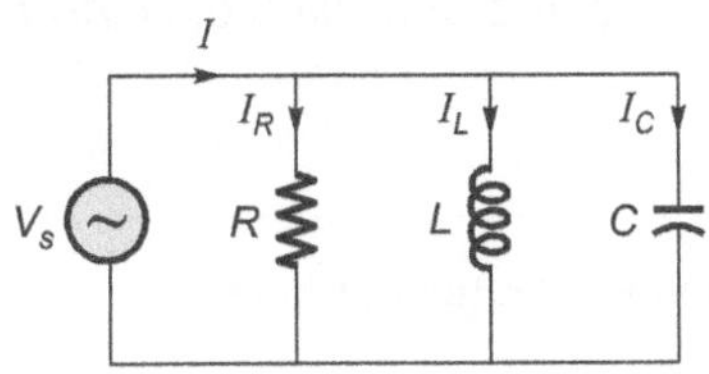

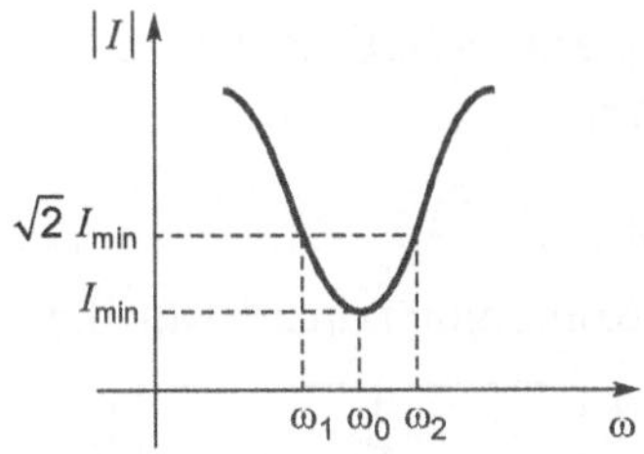

At Resonance

- $|I_L| = |I_C|$ and these are 180° out of phase
- Imaginary part of input impedance = 0

$$\boxed{Z_{in}\big|_{\omega=\omega_0} = R} \quad \text{...... maximum}$$

$$\boxed{I\big|_{\omega=\omega_0} = \frac{V_s}{R}} \quad \text{...... minimum}$$

where, $\omega_0 \rightarrow \dfrac{1}{\sqrt{LC}}$ = resonant frequency in rad/sec

Remember:

- For $\omega < \omega_0$ the circuit behaves like an *RL* circuit.
- For $\omega > \omega_0$ the circuit behaves like an *RC* circuit.
- For $\omega = \omega_0$ the circuit behaves like a resistive circuit.

Bandwidth

$$\boxed{\Delta\omega = (\omega_2 - \omega_1) = \frac{1}{RC}}$$

Q-Factor

$$Q_0 = \frac{\omega_0}{\Delta\omega} = \omega_0 RC = \frac{R}{\omega_0 L} = R\sqrt{\frac{C}{L}}$$

Features of Parallel RLC Circuit

- The ***quality factor*** of the network represents the ***current amplification*** factor and must have a high value for any tuned circuit.
- For a ***tuned network***, the ***quality factor*** should be ***high*** and therefore should have ***low bandwidth*** therefore the ***capacitance*** value is ***kept high***.
- A ***parallel RLC*** network behaves as a ***band stop filter*** or ***band reject filter***.

Note: ....

Two important circuits with their resonance frequency is given as

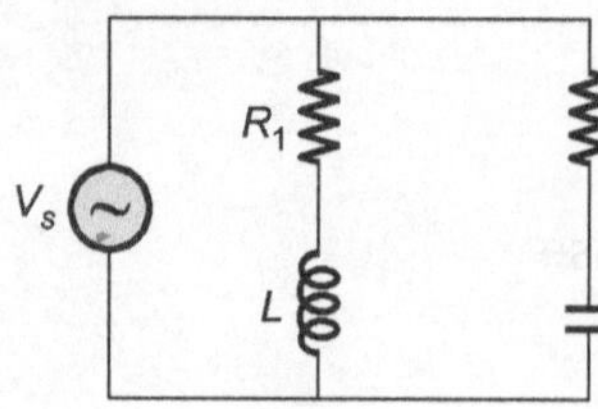

$$f_o = \frac{1}{2\pi\sqrt{LC}}\sqrt{\frac{R_1^2 - L/C}{R_2^2 - L/C}}$$

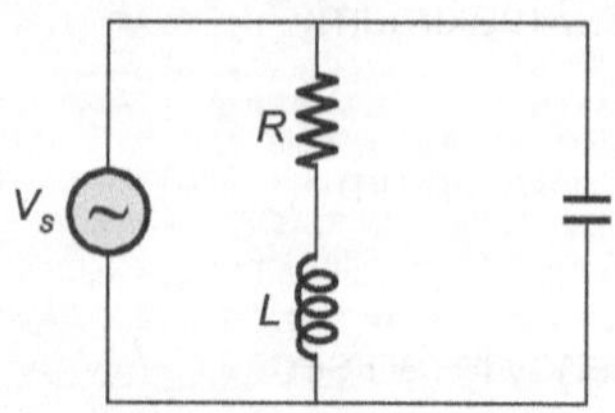

$$f_o = \frac{1}{2\pi}\sqrt{\frac{1}{LC} - \frac{R^2}{L^2}}$$

....

■■■■

Magnetically Coupled Circuit 5

Magnetic Coupling

- When the two inductor are placed physically close to each other, then due to current flow in the first inductor there will be some magnetic flux, part of which will link with the second inductor. Due to rate of change of magnetic flux and by using Faraday's law of electromagnetic induction some induced voltage is produced between the two terminals of the second inductor.
- The polarity of the induced voltage depends upon the relative sense of windings of the two inductors and therefore, the induced voltage may be additive or substractive in nature.

Cases:

(A)

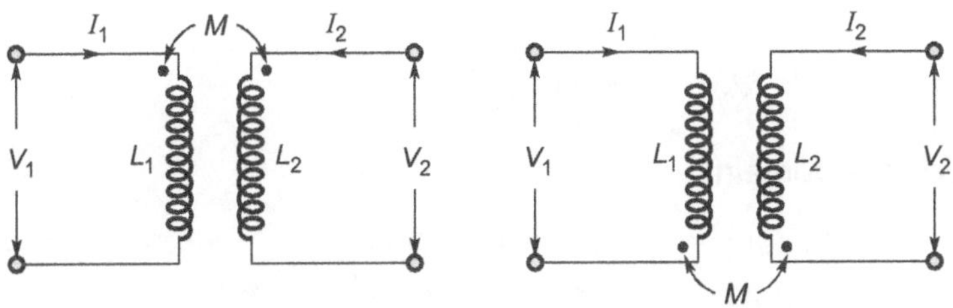

(Case-1) (Case-2)

For above two cases, the effect of mutual inductance is positive. Also induced emf (e_{ind}) is additive in nature:

KVL in time domain:

$$V_1 = L_1\frac{dI_1}{dt} + M\frac{dI_2}{dt}$$

$$V_2 = M\frac{dI_1}{dt} + L_2\frac{dI_2}{dt}$$

KVL in s-Domain:

$$V_1(s) = L_1 sI_1(s) + MsI_2(s)$$

$$V_2(s) = MsI_1(s) + L_2 sI_2(s)$$

$$\boxed{\begin{bmatrix} V_1(s) \\ V_2(s) \end{bmatrix} = \begin{bmatrix} sL_1 & sM \\ sM & sL_2 \end{bmatrix}\begin{bmatrix} I_1(s) \\ I_2(s) \end{bmatrix}}$$

For sinusoidal excitation:

$$\begin{bmatrix} V_1 \\ V_2 \end{bmatrix} = \begin{bmatrix} j\omega L_1 & j\omega M \\ j\omega M & j\omega L_2 \end{bmatrix} \begin{bmatrix} I_1 \\ I_2 \end{bmatrix}$$

(B)

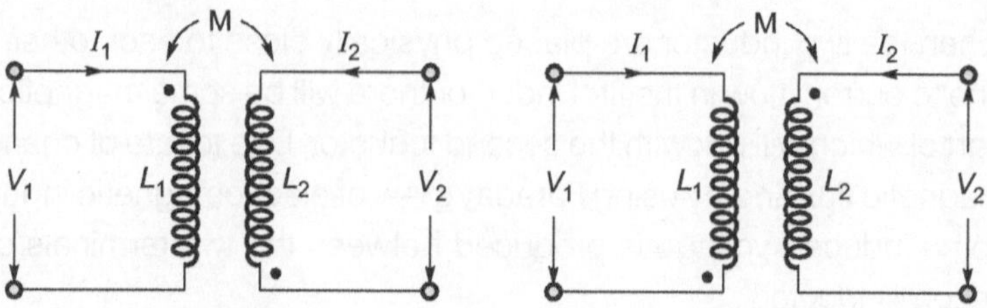

(Case-3) (Case-4)

For above two cases, effect of the mutual coupling is negative. So e_{ind} is substractive in nature.

KVL in time domain:

$$V_1 = L_1 \frac{dI_1}{dt} - M\frac{dI_2}{dt}$$

$$V_2 = -M\frac{dI_1}{dt} + L_2 \frac{dI_2}{dt}$$

KVL in s-domain:

$$\begin{bmatrix} V_1(s) \\ V_2(s) \end{bmatrix} = \begin{bmatrix} sL_1 & -sM \\ -sM & sL_2 \end{bmatrix} \begin{bmatrix} I_1(s) \\ I_2(s) \end{bmatrix}$$

For sinusoidal excitation:

$$\begin{bmatrix} V_1 \\ V_2 \end{bmatrix} = \begin{bmatrix} j\omega L_1 & -j\omega M \\ -j\omega M & j\omega L_2 \end{bmatrix} \begin{bmatrix} I_1 \\ I_2 \end{bmatrix}$$

Representation of Mutual Inductance

Case-1: In terms of M.

Case-2: In terms of 'K' i.e. coefficient of coupling.

$$K = \frac{M}{\sqrt{L_1 L_2}}$$

(a) If $K < 1$; loose coupling (under coupling circuit).

(b) If $K = 1$ i.e. $M = \sqrt{L_1 L_2}$; (critical coupling)

(c) $K > 1$; tight coupling (over coupled circuit).

Case-3: In terms of turns ratio

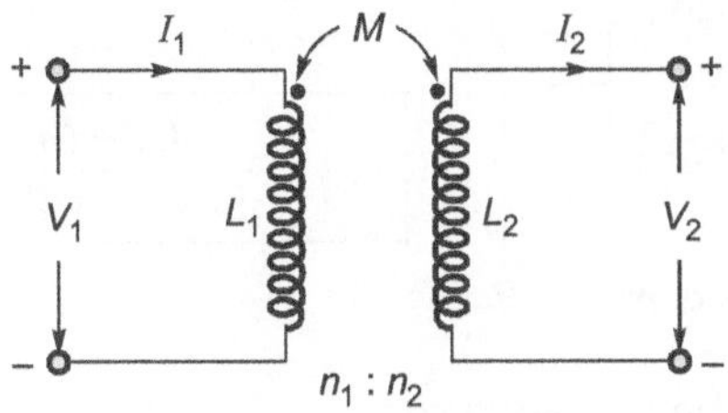

$$\boxed{\frac{V_1}{V_2} = \frac{n_1}{n_2} = \frac{I_2}{I_1}}$$

and $$\boxed{\frac{n_1}{n_2} \cong \frac{L_1}{M} \cong \frac{M}{L_2} = \sqrt{\frac{L_1}{L_2}}}$$

Note: ....

- Equivalent Circuits

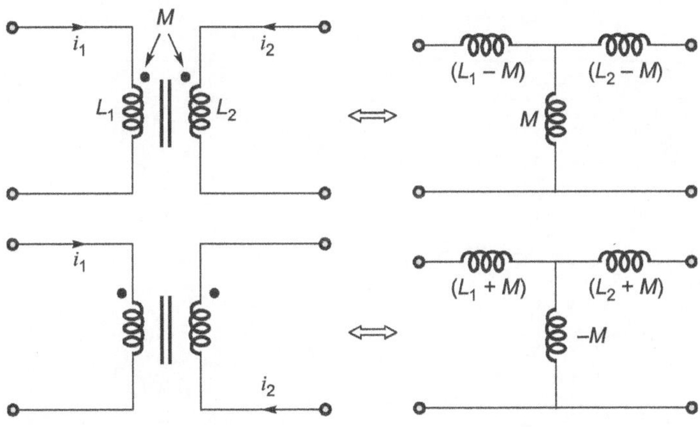

- Equivalent Inductance

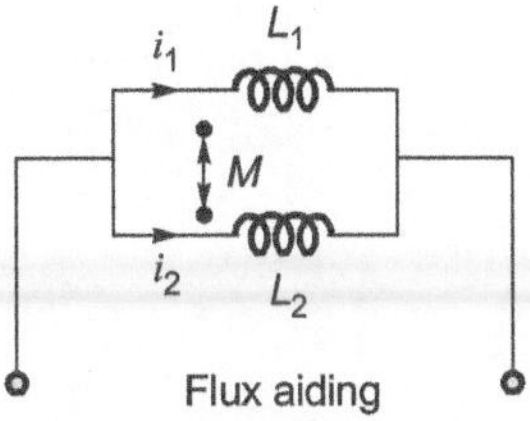

$$\boxed{L_{eq} = \frac{L_1L_2 - M^2}{L_1 + L_2 - 2M}}$$

- Equivalent Inductance

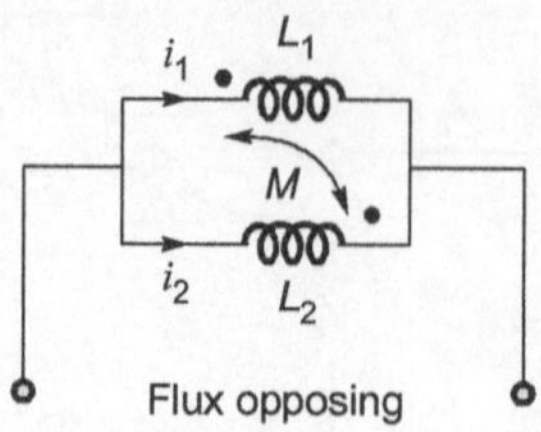

$$L_{eq} = \frac{L_1L_2 - M^2}{L_1 + L_2 + 2M}$$

- Energy stored in coupled circuit

$$\frac{1}{2}L_1 i_1^2 + \frac{1}{2}L_2 i_2^2 \pm M_1 M_2$$

(+ : Flux aiding; − : Flux opposing)

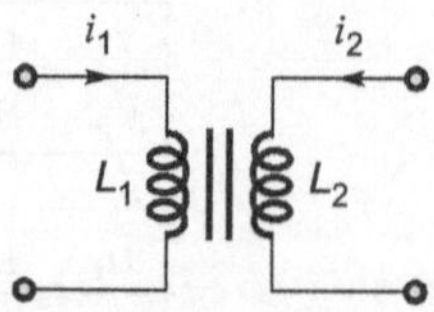

■■■■

Two Port Network

6

Two Port Network

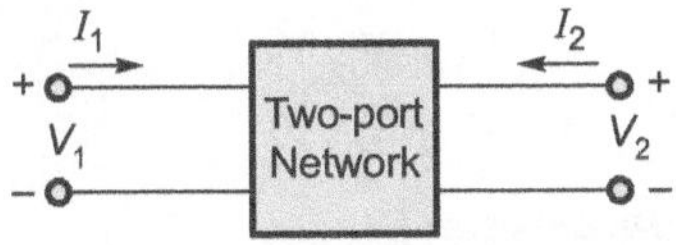

Driving Point Impedance and Admittance Functions

- Driving point impedance function

$$Z_{11}(s) = \frac{V_1(s)}{I_1(s)} \text{ and } Z_{22}(s) = \frac{V_2(s)}{I_2(s)}$$

- Driving point admittance function

$$Y_{11}(s) = \frac{I_1(s)}{V_1(s)} \text{ and } Y_{22}(s) = \frac{I_2(s)}{V_2(s)}$$

Impedance and Admittance Transfer Function

- Impedance Transfer Function

$$Z_{21}(s) = \frac{V_2(s)}{I_1(s)} \text{ and } Z_{12}(s) = \frac{V_1(s)}{I_2(s)}$$

- Admittance Transfer Function

$$Y_{21}(s) = \frac{I_2(s)}{V_1(s)} \text{ and } Y_{12}(s) = \frac{I_1(s)}{V_2(s)}$$

Voltage and Current Transfer Function

- Voltage transfer function

$$G_{12}(s) = \frac{V_1(s)}{V_2(s)} \text{ and } G_{21}(s) = \frac{V_2(s)}{V_1(s)}$$

- Current transfer function

$$\alpha_{12} = \frac{I_1(s)}{I_2(s)} \text{ and } \alpha_{21} = \frac{I_2(s)}{I_1(s)}$$

Parameters

Impedance or *Z*-parameters (Open circuit parameters)

$$V_1 = Z_{11}I_1 + Z_{12}I_2$$
$$V_2 = Z_{21}I_1 + Z_{22}I_2$$

Admittance or *Y*-parameters (Short circuit parameters)

$$I_1 = Y_{11}V_1 + Y_{12}V_2$$
$$I_2 = Y_{21}V_1 + Y_{22}V_2$$

h-parameters (Hybrid parameters)

$$V_1 = h_{11}I_1 + h_{12}V_2$$
$$I_2 = h_{21}I_1 + h_{22}V_2$$

g-parameters (Inverse hybrid parameters)

$$I_1 = g_{11}V_1 + g_{12}I_2$$
$$V_2 = g_{21}V_1 + g_{22}I_2$$

ABCD parameters (Transmission parameters)

$$V_1 = AV_2 - BI_2$$
$$I_1 = CV_2 - DI_2$$

Conditions for a network to be symmetrical and reciprocal

Reciprocal	Symmetrical
$Z_{12} = Z_{21}$	$Z_{11} = Z_{22}$
$Y_{12} = Y_{21}$	$Y_{11} = Y_{22}$
$\begin{vmatrix} A & B \\ C & D \end{vmatrix} = 1$	$A = D$
$h_{12} = -h_{21}$	$\begin{vmatrix} h_{11} & h_{12} \\ h_{21} & h_{22} \end{vmatrix} = 1$
$g_{12} = -g_{21}$	$\begin{vmatrix} g_{11} & g_{12} \\ g_{21} & g_{22} \end{vmatrix} = 1$

Interconnection Between two-port Network

Cascaded Network

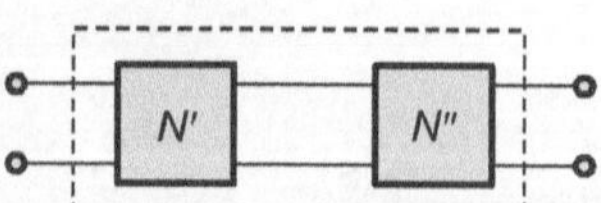

$$\begin{bmatrix} A & B \\ C & D \end{bmatrix} = \begin{bmatrix} A' & B' \\ C' & D' \end{bmatrix}\begin{bmatrix} A'' & B'' \\ C'' & D'' \end{bmatrix}$$

Series Network

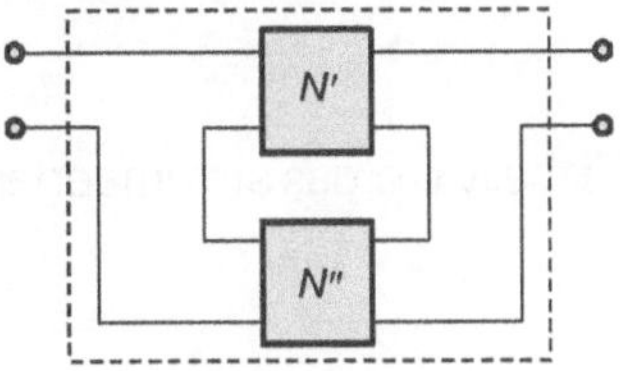

$$\begin{bmatrix} Z_{11} & Z_{12} \\ Z_{21} & Z_{22} \end{bmatrix} = \begin{bmatrix} Z'_{11} + Z''_{11} & Z'_{12} + Z''_{12} \\ Z'_{21} + Z''_{21} & Z'_{22} + Z''_{22} \end{bmatrix}$$

Parallel Network

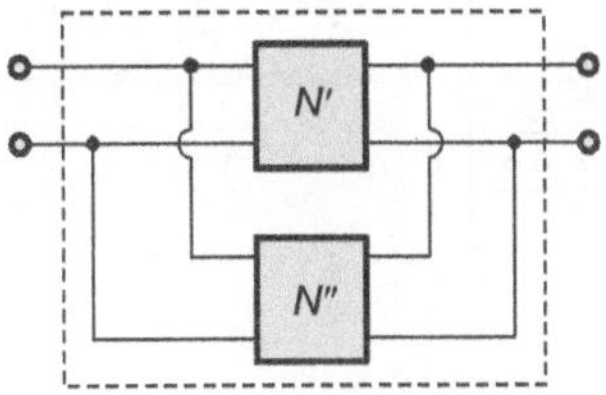

$$\begin{bmatrix} Y_{11} & Y_{12} \\ Y_{21} & Y_{22} \end{bmatrix} = \begin{bmatrix} Y'_{11} + Y''_{11} & Y'_{12} + Y''_{12} \\ Y'_{21} + Y''_{21} & Y'_{22} + Y''_{22} \end{bmatrix}$$

Series-parallel Network

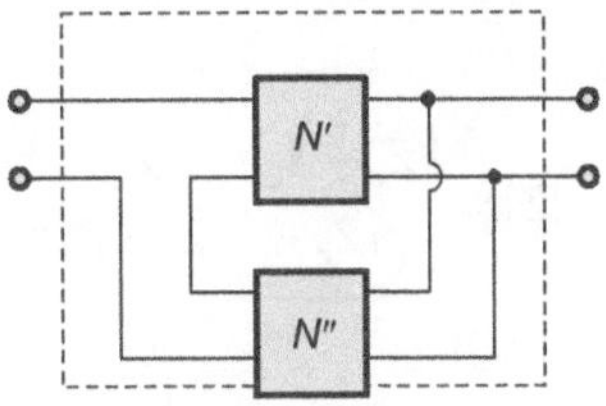

$$\begin{bmatrix} h_{11} & h_{12} \\ h_{21} & h_{22} \end{bmatrix} = \begin{bmatrix} h'_{11} + h''_{11} & h'_{12} + h''_{12} \\ h'_{21} + h''_{21} & h'_{22} + h''_{22} \end{bmatrix}$$

Parallel-series Network

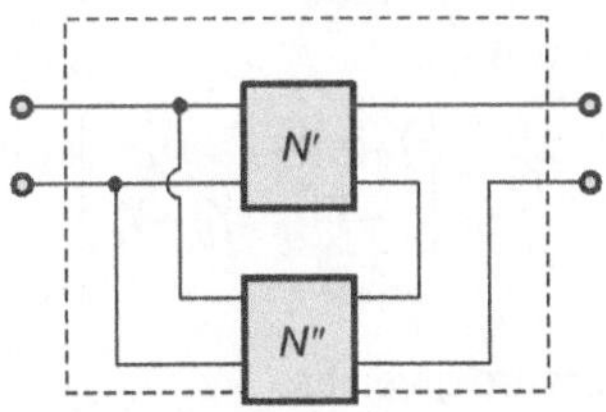

$$\begin{bmatrix} g_{11} & g_{12} \\ g_{21} & g_{22} \end{bmatrix} = \begin{bmatrix} g'_{11} + g''_{11} & g'_{12} + g''_{12} \\ g'_{21} + g''_{21} & g'_{22} + g''_{22} \end{bmatrix}$$

Lattice Network

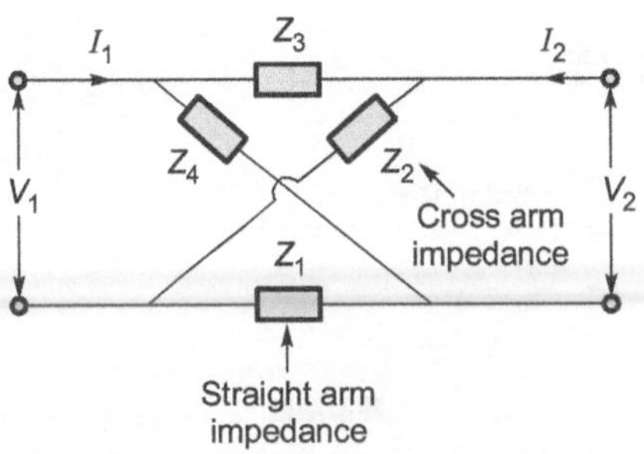

1. A lattice network is said to be symmetrical if the two straight arm impedances are equal i.e. $Z_1 = Z_3$.

   Therefore, $$Z_{11} = Z_{22}$$

2. A lattice network is said to be reciprocal if the two cross arm impedances are equal i.e. $Z_2 = Z_4$.

   Therefore, $$Z_{12} = Z_{21}$$

3. For a symmetrical and reciprocal lattice network,

$$= \begin{bmatrix} \dfrac{Z_1 + Z_2}{2} & \dfrac{Z_2 - Z_1}{2} \\ \dfrac{Z_2 - Z_1}{2} & \dfrac{Z_1 + Z_2}{2} \end{bmatrix}$$

Remember: ....

- Z-parameters of T-model are given as

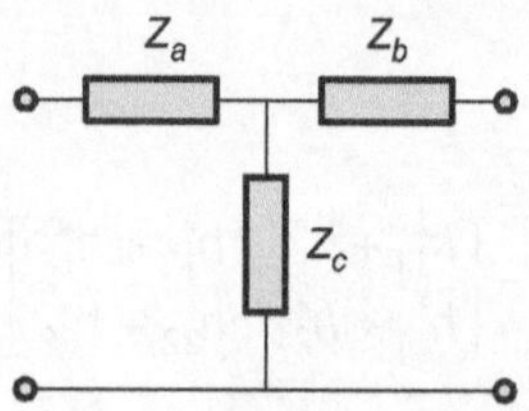

$$[Z] = \begin{bmatrix} Z_a + Z_c & Z_c \\ Z_c & Z_b + Z_c \end{bmatrix}$$

- Y-parameter of π-model are given as

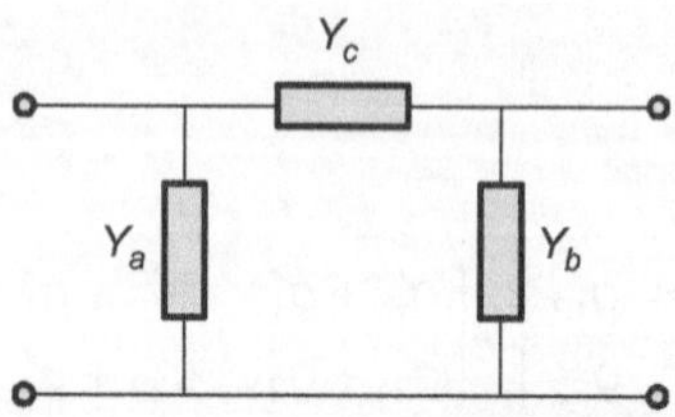

$$[Y] = \begin{bmatrix} Y_a + Y_c & -Y_c \\ -Y_c & Y_b + Y_c \end{bmatrix}$$

- For an ideal transformer, $ABCD$ parameters are given as

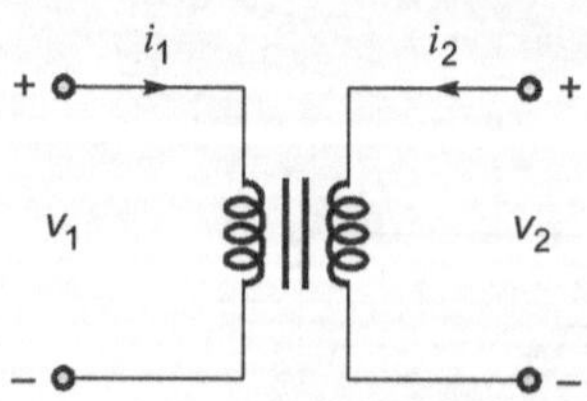

$$[T] = \begin{bmatrix} \dfrac{1}{n} & 0 \\ 0 & n \end{bmatrix}$$

....

■■■■

Miscellaneous

7

Dual Networks

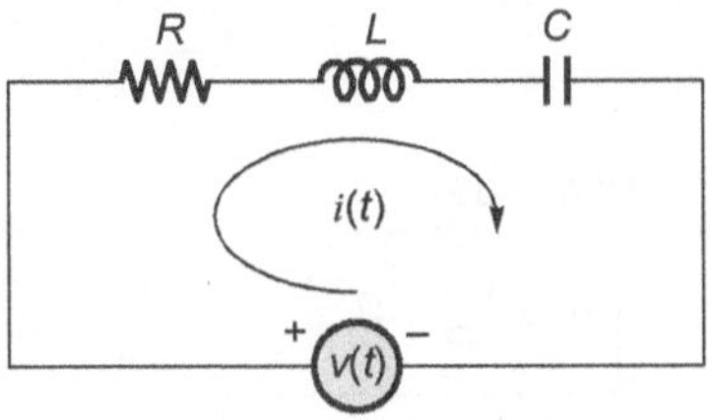

Series RLC circuit (N_1)

By applying KVL,

$$v(t) = Ri(t) + L\frac{di}{dt} + \frac{1}{C}\int_0^t i(t)\,dt$$

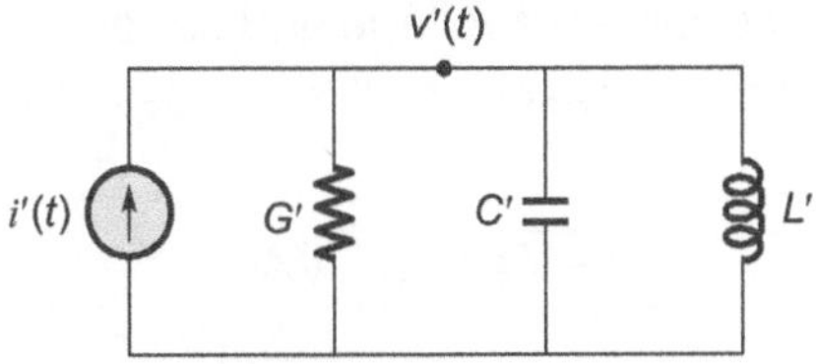

Parallel RLC circuit (N_2)

By applying KCL,

$$i'(t) = v'(t)\,G' + C'\frac{dv'(t)}{dt} + \frac{1}{L'}\int_0^t v'(t)\,dt$$

N_1	N_2
$v(t)$	$i(t)$
R	G
G	R
L	C
C	L
KVL	KCL

- The networks N_1 and N_2 are called dual networks but these networks are not equivalent networks and therefore, the network N_1 cannot be replaced by N_2 network.
- The KVL equation for network N_1 has similar format as for the KCl equation for network N_2. Therefore, if the response of the first network is known is terms of current $i(t)$ then the response of its dual network for voltage $v'(t)$ can be directly written without actually solving it by using necessary transformation by various elements as shown.

Power Relations in AC Circuits

Instantaneous Power

If v is the voltage instantaneous voltage and i is the instantaneous current, then instantaneous power P is given as

$$\boxed{P = vi}$$

Real power

Average power, P is given as

$$\boxed{P = \frac{V_m}{\sqrt{2}} \frac{I_m}{\sqrt{2}} \cos\phi} \quad \boxed{VI\cos\phi} \text{ Watts}$$

where, $V_m, I_m \rightarrow$ Maximum values of v and i.

$\phi \rightarrow$ Phase difference between voltage and current

$V, I \rightarrow$ RMS value of v and i

Reactive power

The product of rms values of voltage and current with the sine of angle between then is called reactive power. It is expressed in (VAR)-volt ampere reactive.

$$\boxed{Q = VI\sin\phi} \; VAR$$

Apparent power

The product VI^* is called apparent power. It is expressed in (VA).

$$\boxed{S = VI^*} \; VA \quad \boxed{S = P + j(Q_L - Q_C)}$$

where, $Q_L \rightarrow$ is used for inductive reactive power

$Q_C \rightarrow$ is used for capacitive reactive power

Power Factor

It is the cosine of angle between the voltage and current in any AC circuit. It is unity in resistance circuit, leading in capacitive circuit and lagging in inductive circuit.

$$\boxed{\text{Power factor} = \cos\phi}$$

Locus Diagrams

These are important to understand the variations in current for a fixed supply voltage.

(a) Series impedance locus

(i)

$Z = R + jX_L$

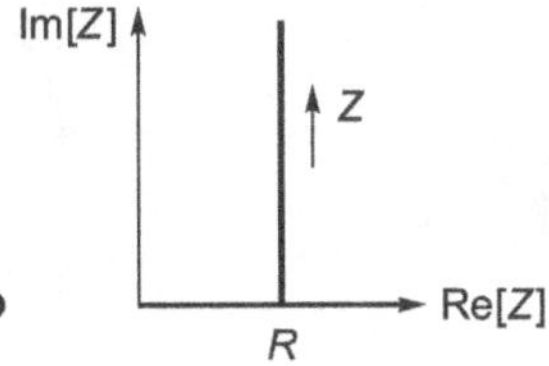

(ii) $Z = R + jX_L$

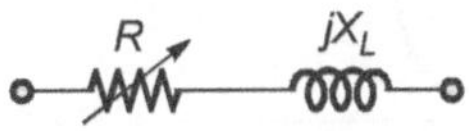

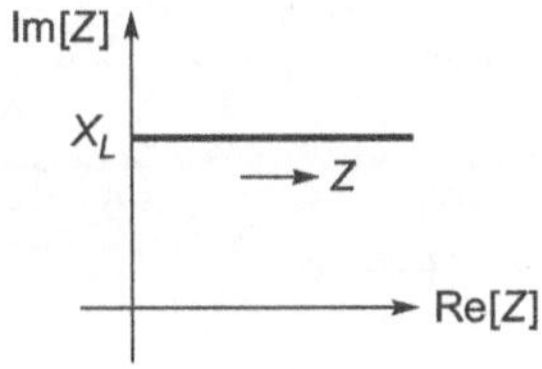

(iii) $Z = R - jX_C$

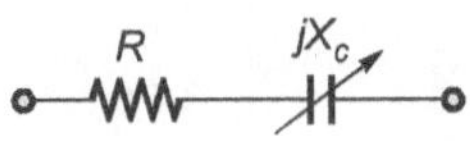

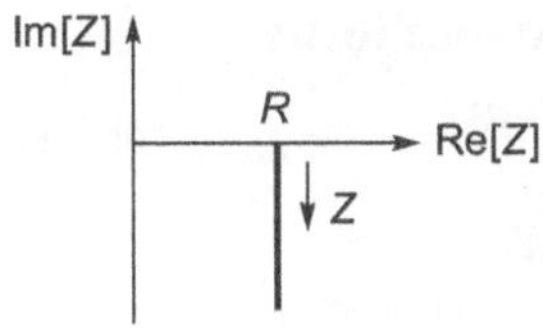

(iv) $Z = R - jX_C$

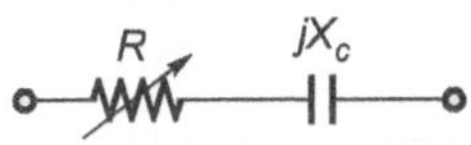

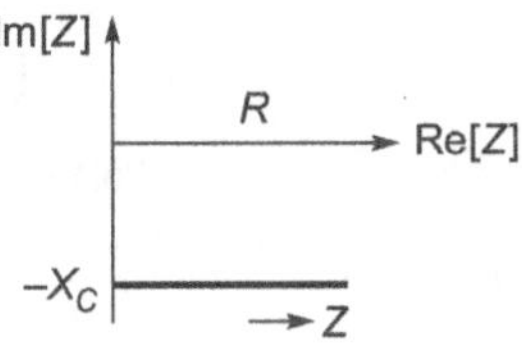

(b) Series admittance locus

(i)

$Z = R + jX_L;\ Y = \frac{1}{Z}$

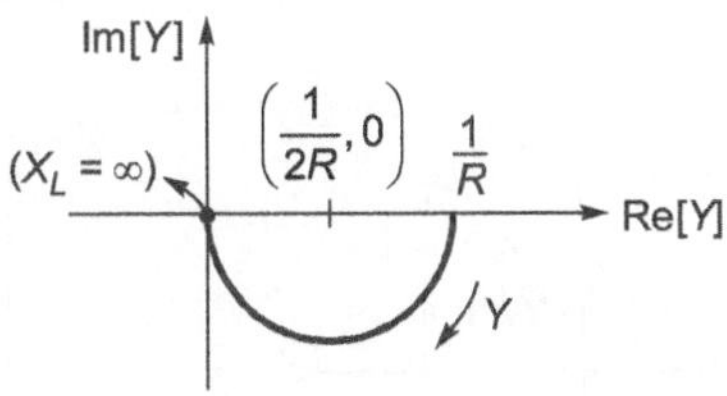

(ii)

$Z = R + jX_L;\ Y = \frac{1}{Z}$

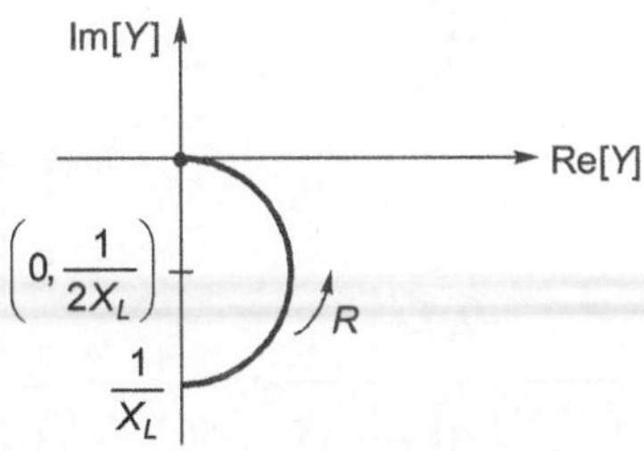

(iii)

$Z = R - jX_C$; $Y = \frac{1}{Z}$

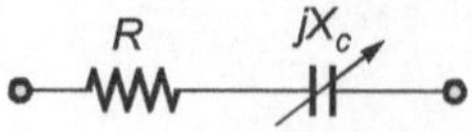

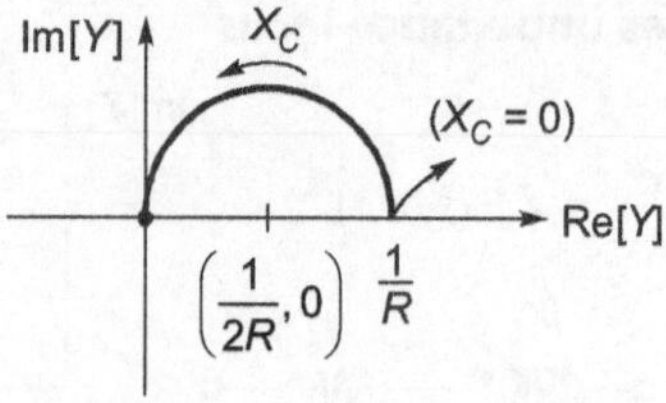

(iv)

$Z = R - jX_C$; $Y = \frac{1}{Z}$

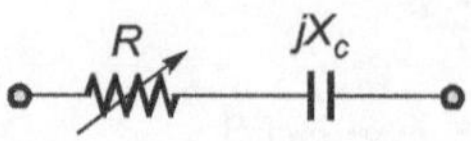

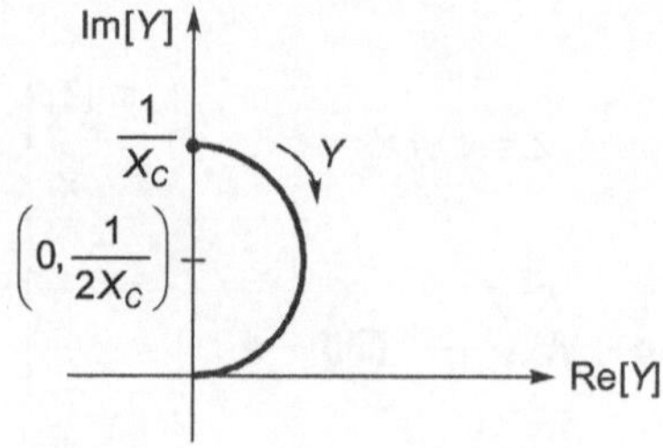

(c) Parallel admittance locus

(i) $Y = G - jB_L$

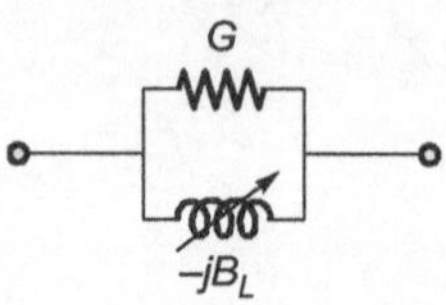

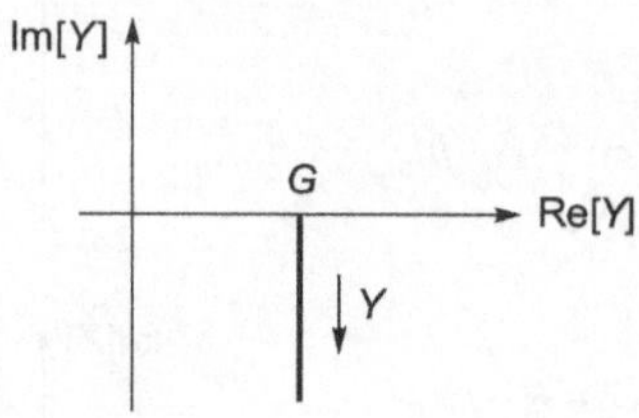

(ii) $Y = G - jB_L$

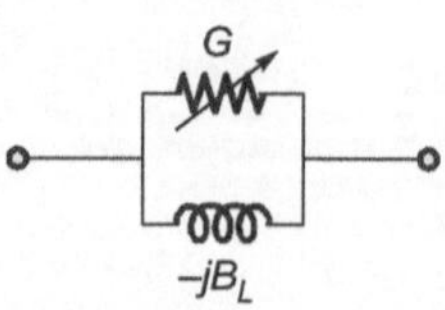

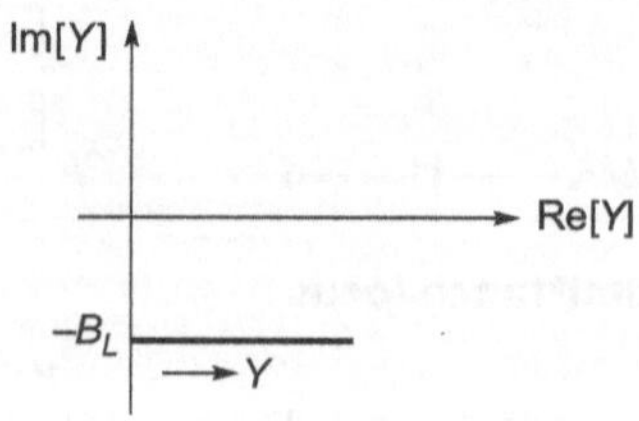

(iii)

$Y = G + jB_C$

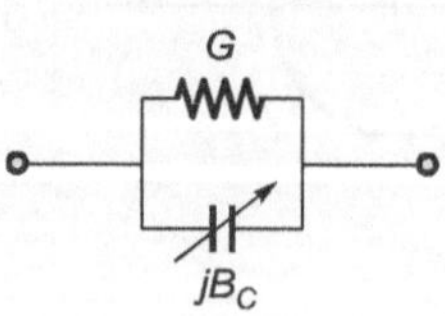

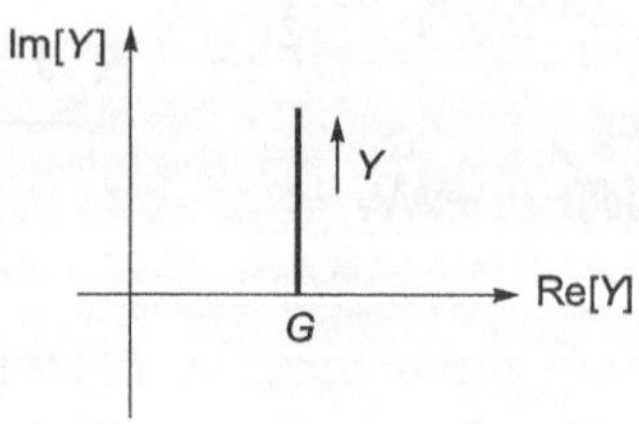

(iv)

$Y = G + jB_C$

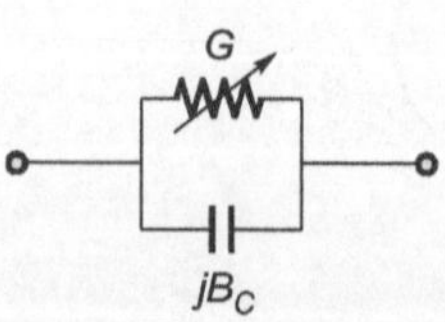

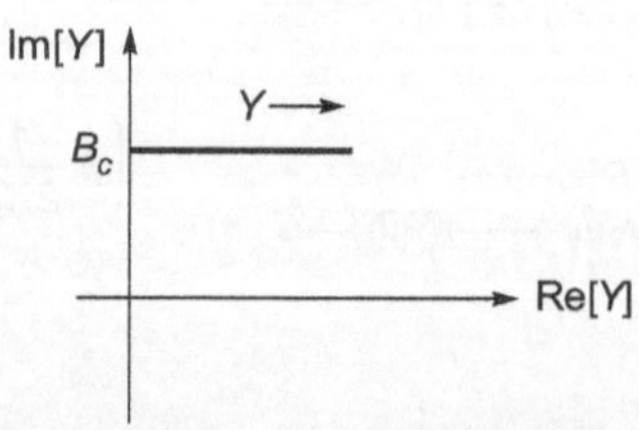

(d) Parallel impedance locus

(i) $Y = G - jB_L$

$Z = \dfrac{1}{Y}$

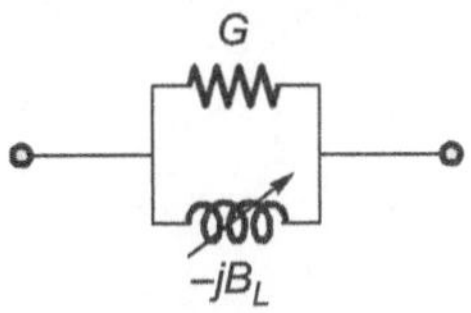

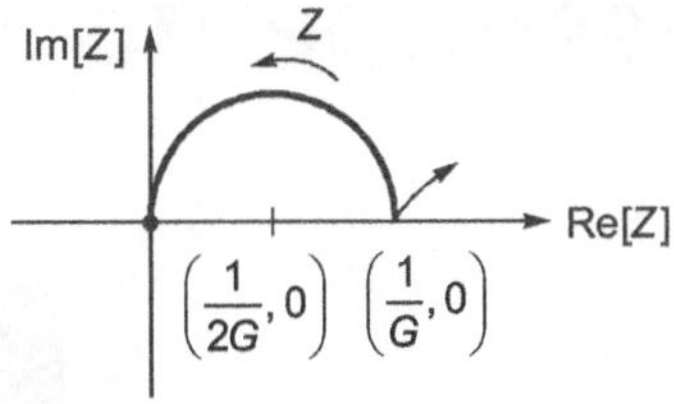

(ii) $Y = G - jB_L$

$Z = \dfrac{1}{Y}$

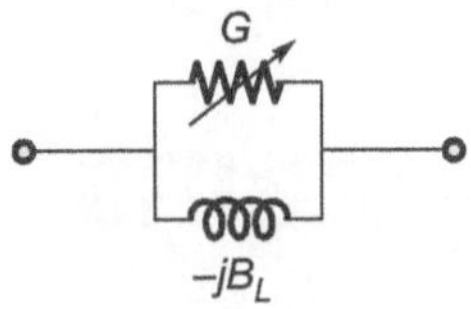

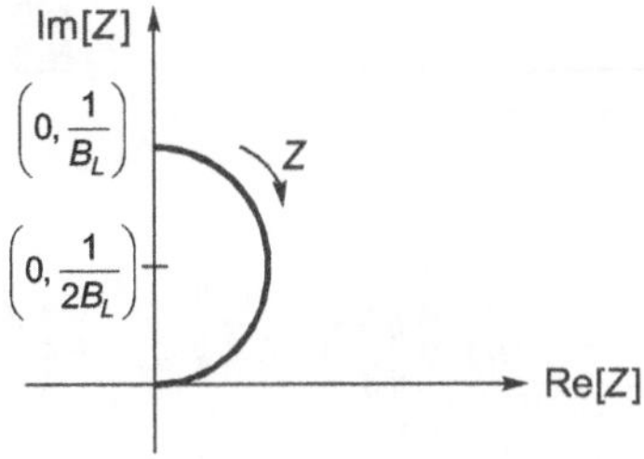

(iii) $Y = G + jB_C$

$Z = \dfrac{1}{Y}$

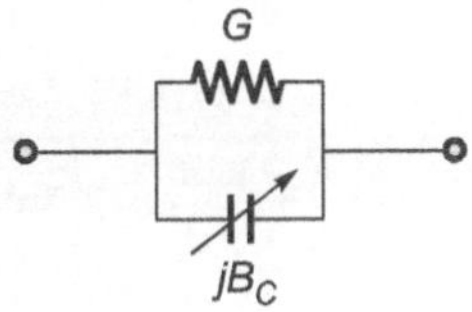

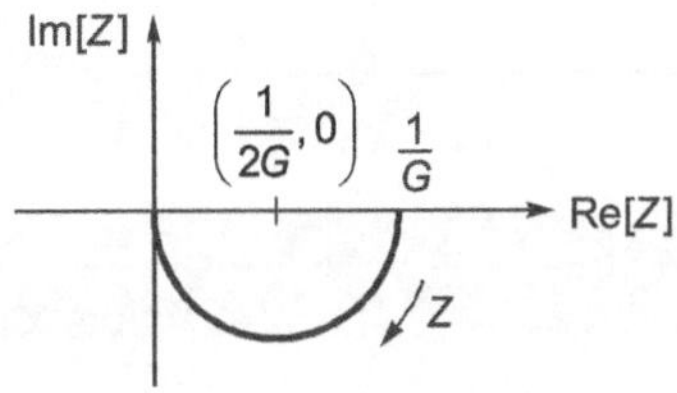

(iv) $Y = G + jB_C$

$Z = \dfrac{1}{Y}$

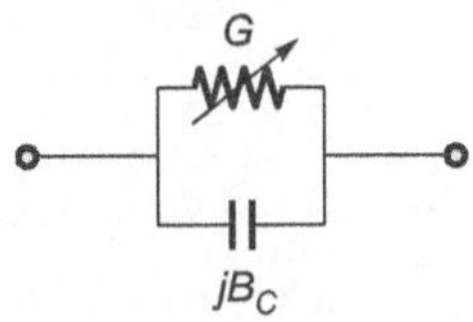

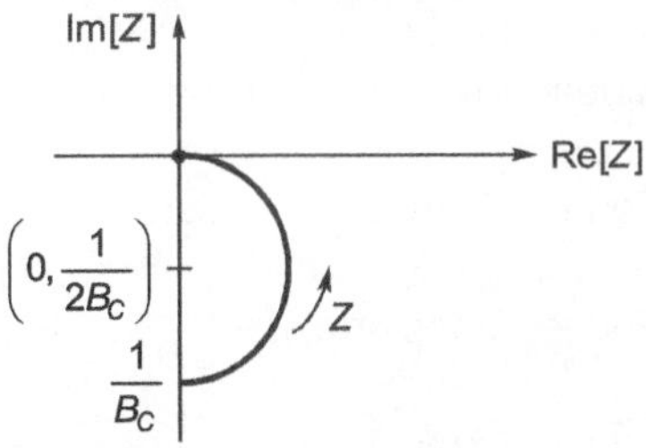

(e) *R-L-C* Series Admittance locus

$Z = R + j(X_L - X_C)$

$Y = \dfrac{1}{Z}$

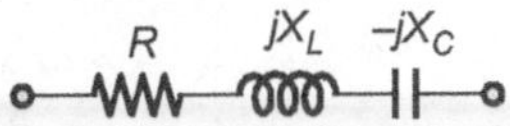

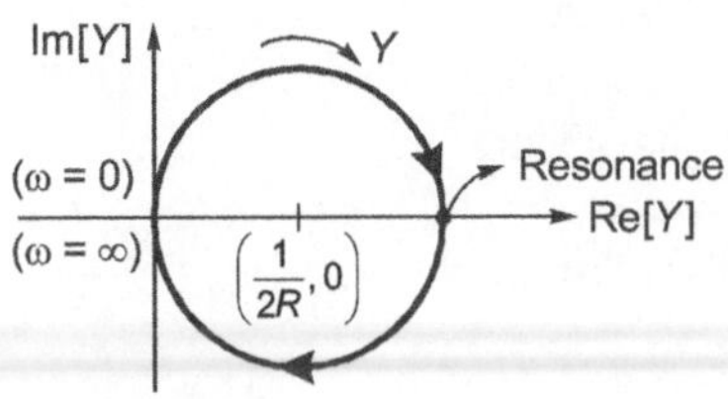

■■■■

A Handbook on
Instrumentation Engineering

3

Signals and Systems

CONTENTS

1 Introduction to Signals

Signals

A signal can be defined as a function of one or more independent variable, which conveys information about the behaviour or nature of some phenomenon.

Elementary Signals

1. Unit step function

(a) For continuous time

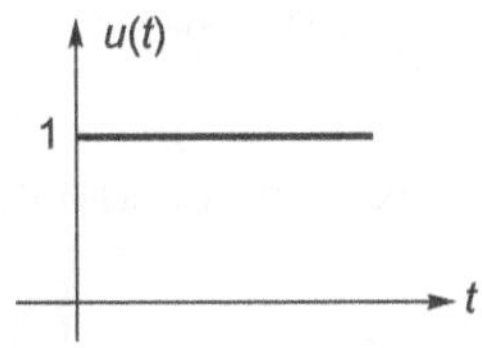

$$u(t) = \begin{cases} 1 & ; \quad t > 0 \\ 0 & ; \quad t < 0 \end{cases}$$

(b) For discrete-time

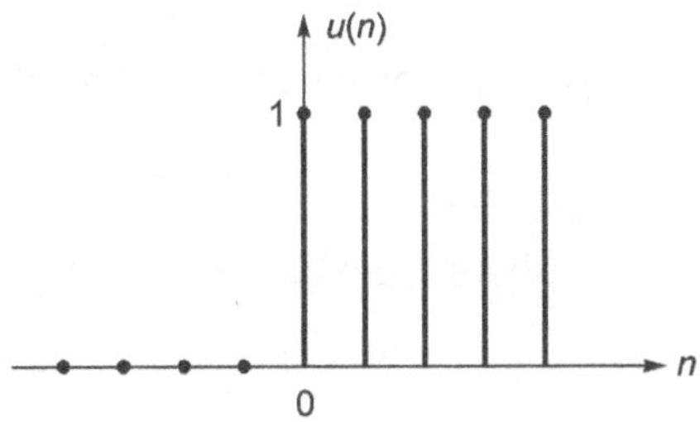

$$u[n] = \begin{cases} 1 & ; \quad n \geq 0 \\ 0 & ; \quad n < 0 \end{cases}$$

Remember:

- Unit step function is not continuous and differentiable at $t = 0$.
- Mathematically $\boxed{u(0) = \frac{1}{2}}$; average value

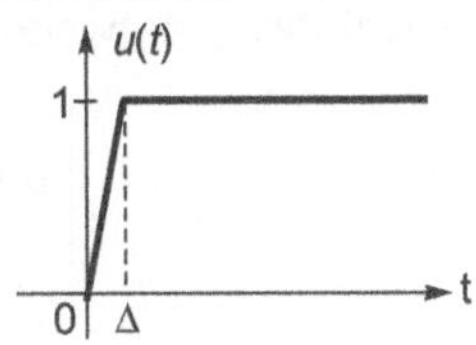

2. Unit impulse function

(a) For continuous time

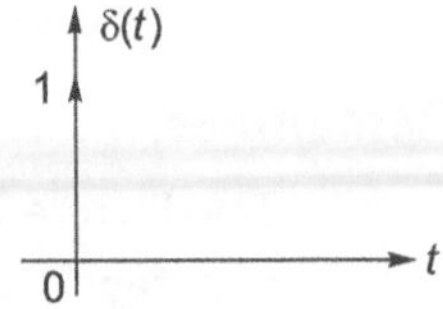

$$\delta(t) = \begin{cases} \infty & ; \quad t = 0 \\ 0 & ; \quad t \neq 0 \end{cases}$$

Properties of the impulse function

(i) Impulse function is a continuous function and the area under this function is equal to one

$$\int_{-\infty}^{\infty} \delta(t)dt = 1$$

(ii) Time reversal

$$\delta(-t) = \delta(t)$$

(iii) Scaling $\quad \delta(at) = \frac{1}{|a|}\delta(t) \;;\; \text{for } a \neq 0$

(iv) Product

$$x(t)\,\delta(t-t_0) = x(t_0)\,\delta(t-t_0);\text{where, } t_0 \text{ is time shift}$$

(v) Sampling property :

$$\int_{t_1}^{t_2} x(t)\,\delta(t-t_0) = x(t_0)\;;\quad t_1 < t_0 < t_2$$

(vi) Delta function of a function of x

$$\delta[g(x)] = \sum_i \frac{\delta(x-x_i)}{|g'(x_i)|}$$

where x_i are the roots of $g(x)$.

Ex: $\quad \delta(x^2-a^2) = \frac{1}{2|a|}[\delta(x+a)+\delta(x-a)]$

Remember: ....

Relationship between unit impulse and unit step function:

$$\boxed{u(t) = \int_{-\infty}^{t} \delta(t)dt} \text{ and } \boxed{\delta(t) = \frac{du(t)}{dt}}$$

....

(b) For discrete-time

$$\delta[n] = \begin{cases} 1 & ; \quad n = 0 \\ 0 & ; \quad n \neq 0 \end{cases}$$

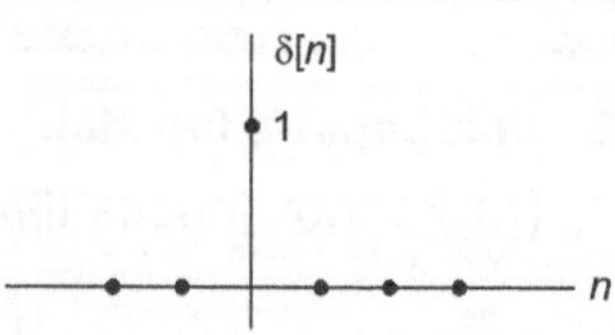

Remember: ....

- The discrete-time unit impulse is the first difference of the discrete-time unit step

$$\boxed{\delta[n] = u[n] - u[n-1]}$$

- Relationship between unit impulse and unit step

$$u[n] = \sum_{k=0}^{\infty} \delta[n-k] = \sum_{k=-\infty}^{n} \delta[k]$$

- Sampling property

$$x[n]\,\delta[n-n_0] = x[n_0]\,\delta[n-n_0]$$

3. Rectangular or Gate function

$$\text{rect}\left(\frac{t}{T}\right) = \begin{cases} 1 \; ; & -T/2 < t < T/2 \\ 0 \; ; & \text{otherwise} \end{cases}$$

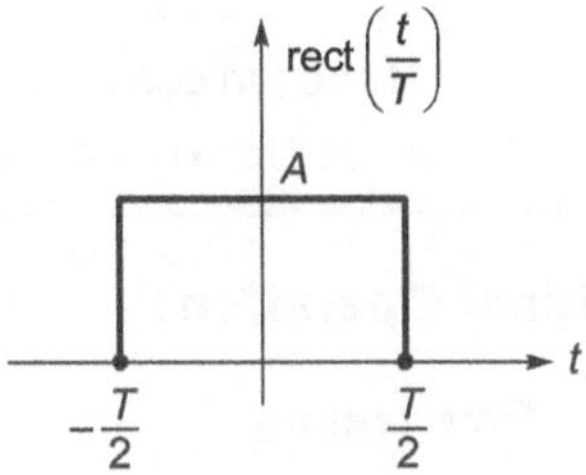

4. Unit Ramp function

$$r(t) = \begin{cases} t \; ; & t > 0 \\ 0 \; ; & t < 0 \end{cases}$$

$$\frac{d}{dt} r(t) = u(t)$$

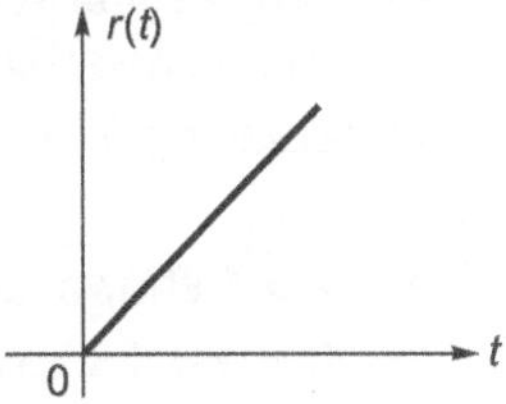

5. Signum or sgn function

$$\text{sgn}(t) = 2u(t) - 1$$

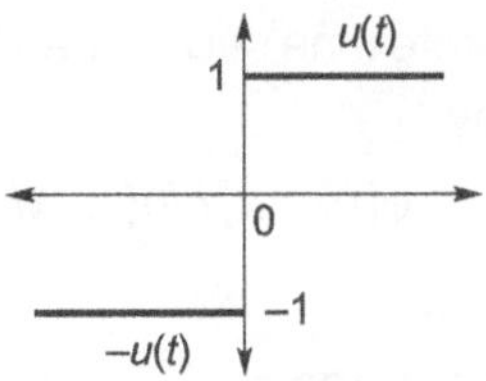

Note:

- $u(-t) = 1 - u(t)$
- Sgn function is not defined at $t = 0$ and is chosen as 0 at $t = 0$

6. Sinc and sinc² function

- The sinc and sinc² functions are defined in terms of an independent variable λ.

$$\text{sinc}(\lambda) = \frac{\sin(\pi\lambda)}{\pi\lambda}$$

- $\text{sinc}(\lambda)$ is equal to zero for $\lambda = \pm n \; (n \neq 0)$, n an integer.

7. Sine integral function

- Sine integral function is an odd function

$$Si(y) = \int_0^y \frac{\sin(\alpha)}{\alpha} d\alpha \quad ; \quad Si(y) = \frac{y}{(1)!1} - \frac{y^3}{(3)3!} + \frac{y^5}{(5)5!} - \frac{y^7}{(7)7!} + \cdots$$

Remember:

- $Si(0) = 0, Si(\pi) \simeq 1.8519, Si(\infty) = \left(\frac{\pi}{2}\right)$
- Si function converges fast and only a few terms in above equation are needed for a good approximation

Signal Operations

1. Time scaling

For analog signals

Let $x(t)$ be an arbitrary signal, a time scaled version of $x(t)$ is obtained by replacing 't' by 'at' where 'a' is scaling factor.

$$\varphi(t) = x(at)$$

- $a > 1$ shows compression of $x(t)$.
- $0 < a < 1$ shows expansion of $x(t)$.

For discrete signals

For a discrete time sequence $x[n]$, compression of a signal by factor M is given by

$$\varphi[n] = x[Mn] \;;\; M \text{ and } n \text{ both are integers}$$

2. Time Shifting

For analog signals

Shifting in time may results in time delay or time advancement.

For a continuous-time signals $x(t)$, time shifting is given as

$\varphi(t) = x(t - t_0) \cdots$ delay or shift right by 't_0'.

$\varphi(t) = x(t + t_0) \cdots$ advance or shift left by 't_0'.

For discrete signals

For a discrete time sequence x[n] time shifting is given as

$\varphi[n] = x[n - n_0] \cdots$ delay or shift right by n_0 samples.

$\varphi[n] = x[n + n_0] \cdots$ advance or shift left by n_0 samples.

3. Time Reversal

For analog signals

Time reversal $x(t)$ is achieved by rotation of signal 180° about vertical axis. This operation is also called as folding or reflection about vertical axis.

For a continuous-time signals x(t), time reversal is given as

$$\varphi(t) = x(-t)$$

For discrete signals

For discrete time sequence $x[n]$, time reversal is given as

$$\varphi[n] = x[-n]$$

Note: ....

- In priority order: Shifting > Scaling > Reversal
- There is no effect of scaling on unit step signal.
- The time scaling on ramp signal will result into magnitude scaling as

$$r(at) \longrightarrow ar(t)$$

....

Classification of Signals

1. Continuous Time and Discrete Time Signals

$x(t)$ is a continuous-time signal if 't' is a continuous variable. But if 't' is a discrete variable that is $x(t)$ is defined at discrete times, then $x(t)$ is a discrete-time signal.

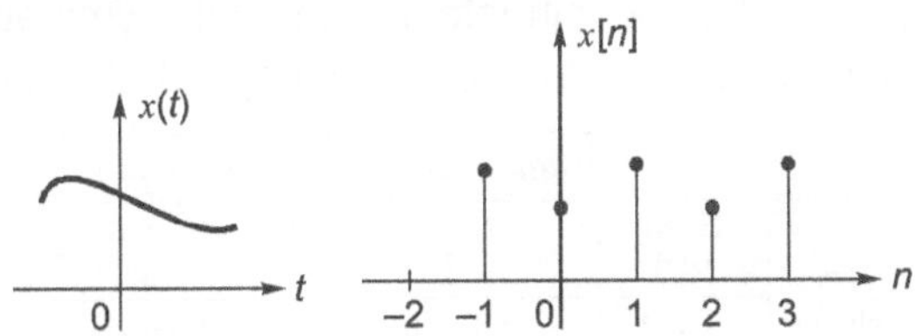

(a) Continuous-time (b) Discrete-time

Note: ....

A discrete-time signal $x[n]$ may be obtained by sampling a continuous-time signal $x(t)$.

....

2. Analog and Digital Signals

If a continuous-time signal $x(t)$ can take on any value in the continuous interval (a, b) where 'a' may be $-\infty$ and 'b' may be $+\infty$, then the continuous time signal $x(t)$ is called an analog signal. If a discrete-time signal $x[n]$ can take on only a finite number of distinct value, then this signal is called a digital signal.

3. Real and Complex Signal

A signal $x(t)$ is a real signal if its value is a real number and a signal $x(t)$ is a complex signal if its value is a complex number.

4. Deterministic and Random Signals

If $x(t)$ can be perfectly known for any time 't' then it is called deterministic signal. If $x(t)$ can not be exactly determined at any given time then it is called Random signal.

5. Even and Odd Signals

A signal $x(t)$ or $x[n]$ is an even signal

if
$$x(-t) = x(t)$$
$$x[-n] = x[n]$$

A signal x(t) or x[n] is referred to as an odd signal

if
$$x(-t) = -x(t)$$
$$x[-n] = -x[n]$$

Note:

- The product of two even signals or of two odd signals is an even signal.
- The product of an even signal and an odd signal is an odd signal.

For complex valued signal $x(t)$,

If $x(t) = x^*(-t)$; Even conjugate or Conjugate symmetric

If $x(t) = -x^*(-t)$; Odd conjugate or Conjugate Anti-symmetric

		Continuous Time	**Discrete Time**
Real valued	Even part	$\frac{x(t)+x(-t)}{2}$	$\frac{x[n]+x[-n]}{2}$
	Odd part	$\frac{x(t)-x(-t)}{2}$	$\frac{x[n]-x[-n]}{2}$
Complex Valued	Even conjugate	$\frac{x(t)+x^*(-t)}{2}$	$\frac{x[n]+x^*[-n]}{2}$
	Odd conjugate	$\frac{x(t)-x^*(-t)}{2}$	$\frac{x[n]-x^*[-n]}{2}$

6. Energy and Power signals

(a) Energy of the signal

$$E = \int_{-\infty}^{\infty} |x(t)|^2\, dt \quad ; \quad E = \sum_{n=-\infty}^{\infty} |x[n]|^2$$

(b) Average power of a signal

$$P = \lim_{T\to\infty} \frac{1}{T} \int_{-T/2}^{T/2} |x(t)|^2\, dt$$

$$P = \lim_{N\to\infty} \frac{1}{2N+1} \sum_{n=-N}^{N} |x[n]|^2$$

Note: ....

- An energy signal has zero average power always.
- A power signal has infinite energy
- Signal having infinite duration with their value tending to a non-zero constant as $t \to \infty$ and signals which are periodic are in general power signal.
- As $t \to \pm\infty$, if amplitude tends to ∞, it is neither energy nor power signal.
- All finite duration and bounded signals are energy signal.
- Energy of a signal is only affected by scaling operation as

$$E(at) \longrightarrow \frac{E(t)}{a}$$

- Power is not affected by time scaling, reversal or shifting.

....

7. Periodic and Non Periodic signals

A continuous time signal $x(t)$ is said to be periodic with period T if there is a positive non zero value of T for which

$$x(t+T) = x(t)\,; \quad \forall t$$

In discrete-time signal, a sequence x[n] is periodic with period N, if there is a positive integer N for which

$$x[n+N] = x[n]\,; \quad \forall n$$

Note: ....

- The fundamental period for analog signal is

$$T_0 = \frac{2\pi}{\omega_0}$$

- For discrete signal

$$\frac{N}{K} = \frac{2\pi}{\omega_0}; \qquad \text{where } K = 0, 1, 2, 3 \cdots$$

- The fundamental period T_0 of $x(t)$ and N of $x[n]$ is the smallest positive integer for which above equation holds good.
- Sum of two continuous-time periodic signals may not be always periodic but the sum of two discrete periodic sequences is always periodic.
- If two continuous time signals have period T_1, T_2 then for their sum to be periodic, $\frac{T_1}{T_2}$ should be rational and the period of the sum is given by $T = \text{LCM}(T_1, T_2)$.
- If two discrete time signals have period N_1, N_2 then their sum is always periodic with period $N = \text{LCM}(N_1, N_2)$.

....

For periodic signal $x(t)$

- Average value of the signal

$$x_{avg} = \frac{1}{T}\int x_T(t)dt$$

- Average signal power

$$P_x = \frac{1}{T}\int_T |x_T(t)^2| dt$$

- Effective or rms value of the signal

$$x_{rms} = \sqrt{P_x}$$

where, $x_T(t) \rightarrow$ One period of $x(t)$

Note: ....

- RMS value of $A\cos(\omega t + \phi)$ is $\frac{A}{\sqrt{2}}$.
- With DC offset, RMS value of $A_0 + A\cos(\omega t + \phi)$ is $\sqrt{A_0^2 + \frac{A^2}{2}}$.

....

Symmetry of periodic function

Odd half-wave symmetry

$$x(t) = -x\left(t \pm \frac{T}{2}\right)$$

Even half-wave symmetry

$$x(t) = x\left(t \pm \frac{T}{2}\right)$$

where, $T \rightarrow$ Period of $x(t)$

■■■■

Linear Time Invariant Systems

2

System

A system is a quantity which maps a set of input signal to a set of output signals. Linear Time Invariant (LTI) systems are used to represent signals as linear combinations of basic signals.

Continuous-Time and Discrete-Time Systems

A continuous time system (CTS) is one in which continuous time input signals are transformed into continuous time output signals.

$$y(t) = T\{x(t)\}$$

A discrete time system (DTS) is one which transform discrete time input signal into discrete time output signal.

$$y[n] = T\{x[n]\}$$

Moreover, a continuous time signal can be processed by a discrete time system. On the other hand processing of discrete time signal by a continuous time system is also possible because discrete time systems have several significant advantages over continuous time systems.

Classification of Systems

1. **Linear Systems and Non Linear Systems**

   When system satisfy principle of superposition and homogeneity, it is called linear system, else non linear system.

   (i) Superposition (Additivity)

   if $$x_1(t) \rightarrow y_1(t)$$
   $$x_2(t) \rightarrow y_2(t)$$
   then $$x_1(t) + x_2(t) \rightarrow y_1(t) + y_2(t)$$

   (ii) Homogeneity (Scaling)

   $$ax_1(t) + bx_2(t) \rightarrow ay_1(t) + by_2(t)$$

   Ex. of linear systems: $y(t) = 2x(t)$, $y(t) = \int_{-\infty}^{t} x(t)dt$

2. **Time-Invariant and Time Varying Systems**

   A system is called time-variant if a time shift in the input signal causes the same time-shift in the output signal.

If $x(t) \rightarrow y(t)$

then $x(t-t_0) \rightarrow y(t-t_0)$

Ex. of time invariant system: $y(t) = x^2(t)$

3. Causal and Non Causal systems

The output of the causal system at the present time depends on only the present and/or past value of the input, not on its future values. A system is called non causal if it is not causal.

Ex. of causal system $\rightarrow y(t) = x(\sqrt{t})$.

Note: ....

- All memory-less systems are causal, but not vice-versa.
- Causal system are referred as non-anticipative as the system output does not anticipate future values of input.

....

4. Static and Dynamic Systems

System is said to be static or memory less if the output at any instant depends on only the input at that instant otherwise, the system is a dynamic system with memory.

Ex. of static system: $y(t) = 2x(t)$.

5. Stable and Unstable Systems

A system is bounded input bounded output i.e. BIBO stable if for any bounded input x the corresponding output y is also bounded.

Ex. of stable system: $y(t) = e^{x(t)}$.

Note: ....

A consequence of the homogeneity of linear system is that a zero input yield a zero output.

....

6. Invertible and Non-Invertible System

If it is possible to define an inverse system of a given system, then the system is invertible, otherwise it is called non-invertible system.

Ex. $y(t) = \int_{-\infty}^{t} x(t)\,dt \leftarrow$ Invertible

$y(t) = \frac{d}{dt}x(t) \leftarrow$ Inverse system

Convolution

The convolution integral or the superposition integral, represent a continuous-time linear time invariant (LTI) system in terms of its response to a unit impulse.

Continuous-time LTI system

Convolution of two functions, $x_1(t)$ and $x_2(t)$

$$y(t) = x_1(t) * x_2(t) = \int_{-\infty}^{\infty} x_1(\tau)\, x_2(t-\tau)\, d\tau$$

Note:

- If two analog signals get convolved:
  (i) The resultant of convolution of two signals will have a width equal to the sum of the individual width of the two signals being convolved
  (ii) The lower extent of resultant signal will be equal to sum of individual lower extents $\boxed{l = l_1 + l_2}$, similarly upper extent is sum of individual upper extents $\boxed{u = u_1 + u_2}$.
  (iii) The area of resultant convolution is equal to the product of the area of the signals being convolved.

Properties of the convolution integral

- Commutative property

$$y(t) = x_1(t) * x_2(t) = x_2(t) * x_1(t)$$

- Distributive property

$$x_1(t) * [x_2(t) + x_3(t)] = x_1(t) * x_2(t) + x_1(t) * x_3(t)$$

- Associative property

$$x_1(t) * (x_2(t) * x_3(t)) = (x_1(t) * x_2(t)) * x_3(t)$$

- Derivative of the convolution

$$\frac{dy(t)}{dt} = \frac{dx_1(t)}{dt} * x_2(t) = x_1(t) * \frac{dx_2}{dt}(t)$$

- Convolution of two delayed functions

If $\quad y(t) = x_1(t) * x_2(t)$

$$x_1(t-t_1) * x_2(t-t_2) = y\left((t-(t_1+t_2)\right)$$

- Time scaling property

If $$y(t) = x_1(t) * x_2(t)$$

$$x_1(\alpha t) * x_2(\alpha t) = \frac{1}{|\alpha|} y(\alpha t) \quad ; \alpha \neq 0$$

Special Cases

$$\phi(t) * \delta(t) = \int_{-\infty}^{\infty} \phi(\tau)\, \delta(t-\tau)\, d\tau = \phi(t)$$

$$\delta(t) * \delta(t) = \int_{-\infty}^{\infty} \delta(\tau)\, \delta(t-\tau) d\tau = \delta(t)$$

Discrete-time LTI system

$$y[n] = \sum_{k=-\infty}^{\infty} x_1[k]\, x_2[n-k] = x_1[n] * x_2[n]$$

Note: ....

If two discrete signals get convolved:

(i) The resultant of convolution of two signals will have a length equal to the sum of the individual length of the two signals being convolved minus 1 i.e. $(L_1 + L_2 - 1)$.

(ii) Resultant of convolution has extent equal to the sum of the individual extents of the signals being convolved.

(iii) The sum of sampled values in resultant convolution is equal to the product of sum of the individual sampled values of the signals being convolved.

$$\sum_{n=-\infty}^{\infty} y[n] = \left(\sum_{n=-\infty}^{\infty} x_1[n] \right) \left(\sum_{n=-\infty}^{\infty} x_2[n] \right)$$

....

■■■■

Fourier Series

3

Introduction

It is an approximated process by which any non-standard signal is converted into a standard signal. The approximation of a given function by Fourier series gives a smooth function even when the function being approximated has discontinuities.

Eigen Function

- If the response of a system for an input signal is same as the input signal except for a scalar multiple, then this input signal is called Eigen function of the system.
- A phasor signal $e^{j\omega_0 t}$ can be defined as the Eigen function of any general LTI system.
- If the input of a system is in terms of phasors, then the response evaluation can be simplified from a more difficult convolution to simpler multiplication. This is the basic principle of Fourier series.

Convergence of Fourier Series (Dirichlet Condition)

Periodic signal $x(t)$ has a Fourier series representation if it satisfies the following Dirichlet condition.

- $x(t)$ is absolutely integrable over any period i.e.

$$\int_{T_0} |x(t)|\,dt < \infty$$

- $x(t)$ has a finite number of maxima and minima within any finite interval of t.
- $x(t)$ has a finite number of discontinuities within any finite interval of t, and each of these discontinuties is finite.

Fourier Series Representation of Continuous Time Periodic Signals

By using Fourier series, a non-sinusoidal periodic function can be expressed as an infinite sum of sinusoidal function.

1. Trigonometric Fourier series

Any practical periodic function of frequency ω_0 can be expressed as an infinite sum of sine (or) cosine functions that are integral multiples of ω_0

$$\boxed{f(t) = a_0 + \sum_{n=1}^{\infty} (a_n \cos \omega_0 t + b_n \sin \omega_0 t)}$$

where, ω_0 = Fundamental frequency

a_0, a_n, b_n = Trigonometric Fourier series coefficient

$n\omega_0$ = n^{th} harmonic of ω_0

$$a_0 = \frac{1}{T}\int_0^T f(t)dt$$

$$a_n = \frac{2}{T}\int_0^T f(t)\cos n\omega_0 t\ dt$$

$$b_n = \frac{2}{T}\int_0^T f(t)\sin n\omega_0 t\ dt$$

Note:

- Here a_0 is the value of constant component of the signal $f(t)$
- The Fourier coefficient a_n and b_n are maximum amplitude of nth harmonic component.

2. Polar form of Trigonometric Fourier Series

$$f(t) = C_0 + \sum_{n=1}^{\infty} C_n \cos(n\omega_0 t - \phi_n)$$

where, $C_0 = a_0$

$$|C_n| = \sqrt{a_n^2 + b_n^2} \quad \text{... Magnitude spectrum}$$

$$\phi_n = \tan^{-1}\left(\frac{b_n}{a_n}\right) \quad \text{... Phase spectrum}$$

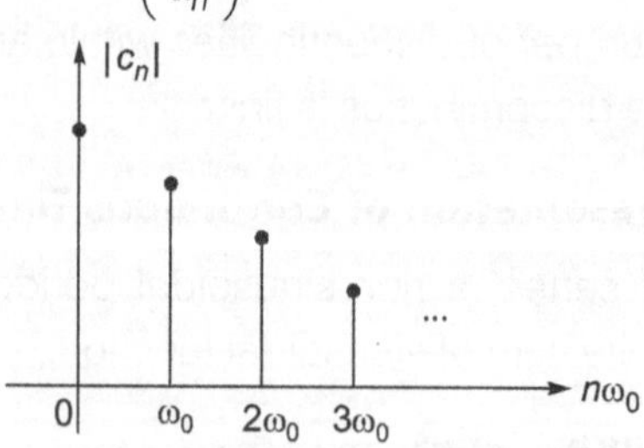

Magnitude spectrum

3. Exponential Fourier series

$$f(t) = \sum_{n=\infty}^{\infty} c_n e^{jn\omega_0 t}$$

where, $c_0 = a_0 = \frac{1}{T}\int_0^T f(t)dt$

$$c_n = \frac{1}{T}\int_0^T f(t)e^{-jn\omega_0 t}dt$$

Note: ....

Exponential Fourier series is compact form of Fourier series

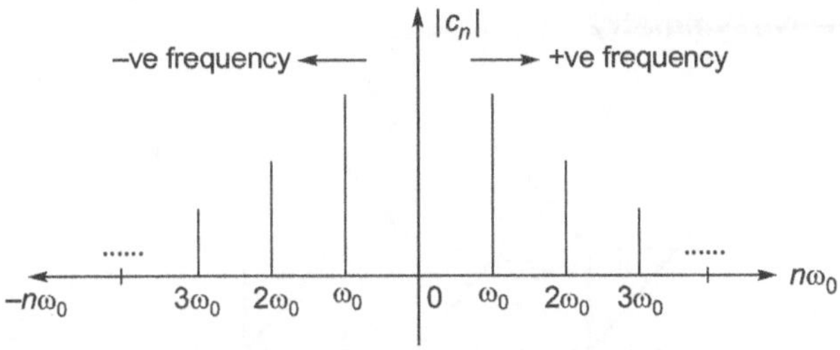

Positive and negative frequency indicate different phase of rotation, they maintain same magnitude but different phase.

....

Relation between Exponential and Trigonometric Fourier Series

$$a_n = (c_n + c_{-n}) \quad ; \quad c_n = \frac{a_n}{2} - j\frac{b_n}{2}$$

$$b_n = j(c_n - c_{-n}) \quad ; \quad c_{-n} = \frac{a_n}{2} + j\frac{b_n}{2}$$

Effect of Symmetry of Fourier Coefficients

1. Odd symmetry

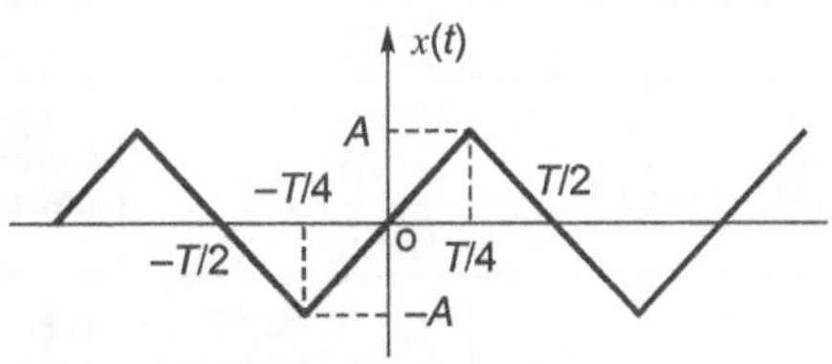

For signals with odd symmetry, the Fourier coefficients a_0 and a_n are zero.

$$x(t) = -x(-t)$$

2. Even symmetry

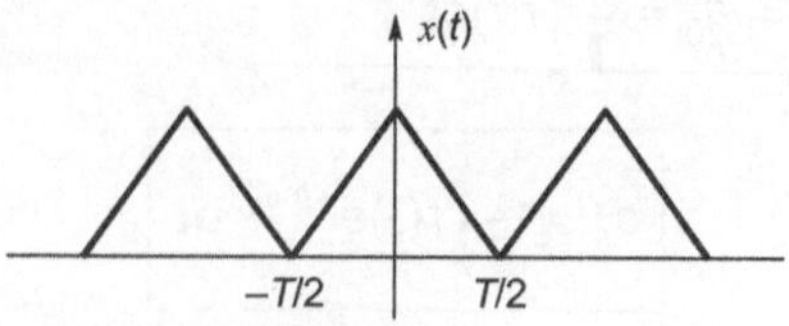

$$x(t) = x(-t)$$

For signals with even symmetry, the Fourier coefficient b_n is zero.

3. Half-wave symmetry

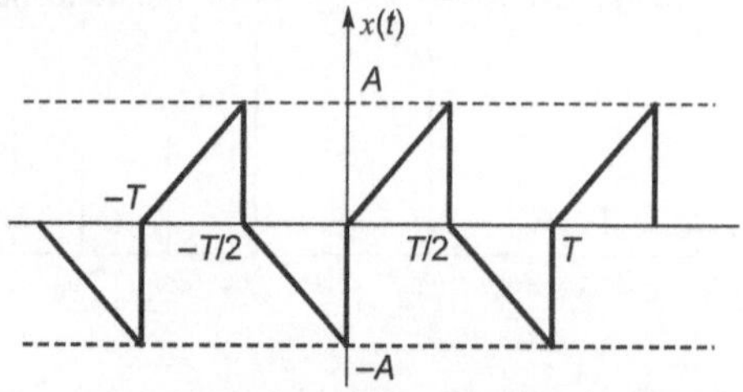

$$x(t \pm T/2) = -x(t)$$

For signals with half-wave symmetry, the Fourier series will consist of odd harmonic terms of sine and cosine signals.

Summary

Function	C_n	Fourier Coefficient	Trigonometric Fourier Series
Real (Neither even nor odd)	Generally Complex $C_n = C_{-n}^*$	$a_0 \neq 0$ $a_n = b_n \neq 0$	DC term, sine terms and cosine terms are present.
Even	Real (Even in nature)	$a_0 \neq 0$ $a_n \neq 0$ $b_n = 0$	DC term and cosine terms are present.
Odd	Imaginary (odd in nature)	$a_0 = 0$ $a_n = 0$ $b_n \neq 0$	Only sine terms are present.
Half wave symmetry	$C_n = 0$; For n = even	$\left.\begin{matrix} a_n = 0 \\ b_n = 0 \end{matrix}\right\}$ $a_0 = 0$; n = even	Odd sine and odd cosine terms are present.
Even and Half wave symmetry	C_n = Real & even $C_n = 0$; n = even	$\left.\begin{matrix} a_n = 0 \\ b_n = 0 \end{matrix}\right\}$ $a_0 = 0$; n = even	Only odd cosine terms are present.
Odd and Half wave symmetry	C_n = Imaginary & odd $C_n = 0$; n = even	$\left.\begin{matrix} a_n = 0 \\ b_n = 0 \end{matrix}\right\}$; n = even	Only odd sine terms are present.

Properties of Exponential Form of Fourier Series Coefficients

Property	Continuous Time Periodic Signal	Fourier Series Coefficients
Linearity	$Ax(t) + By(t)$	$Ac_n + Bd_n$
Time shifting	$x(t \pm t_0)$	$c_n e^{\pm jm\omega_0 t_0}$
Frequency shifting	$e^{\pm jm\omega_0 t} x(t)$	$c_{n \mp m}$
Conjugation	$x^*(t)$	c^*_{-n}
Time reversal	$x(-t)$	c_{-n}
Time scaling	$x(\alpha t)$; $\alpha > 0$ [$x(t)$ is period with period T/α]	c_n (No change in Fourier coefficient
Multiplication	$x(t)\, y(t)$	$\sum_{m=-\infty}^{+\infty} c_m d_{n-m}$
Differentiation	$\frac{d}{dt} x(t)$	$jn\omega_0 c_n$
Integration	$\int_{-\infty}^{t} x(t)\, dt$ (Finite valued and periodic only if $a_0 = 0$)	$\frac{1}{jn\omega_0} c_n$
Periodic Convolution	$\int_T x(\tau)\, y(t-\tau) d\tau$	$Tc_n d_n$
Symmetry of real signals	$x(t)$ is real	$c_n = c^*_{-n}$ $\lvert c_n \rvert = \lvert c_{-n} \rvert$; $\angle c_n = -\angle c_{-n}$ Re $\{c_n\}$ = Re $\{c_{-n}\}$ Im $\{c_n\}$ = –Im $\{c_{-n}\}$
Real and even	$x(t)$ real and even	c_n are real and even
Real and odd	$x(t)$ real and odd	c_n are imaginary and odd

where, $c_n \rightarrow$ Fourier series coefficients of $x(t)$

$d_n \rightarrow$ Fourier series coefficients of $y(t)$

Fourier Series Representation of Discrete Time Periodic Signals

Fourier series representation of $x[n]$, with period 'N' is given as

$$\boxed{x[n] = \sum_{k=0}^{N-1} c_k\, e^{jk\left(\frac{2\pi}{N}\right)n}}$$

where,
$$\boxed{c_k = \frac{1}{N} \sum_{n=0}^{N-1} x[n]\, e^{-jk\left(\frac{2\pi}{N}\right)n}}$$

c_k is periodic with period 'N', $\boxed{c_k = c_{k+N}}$

Properties of Discrete Time Fourier Series

Property	Continuous Time Periodic Signal	Fourier Series Coefficients
Linearity	$Ax[n] + By[n]$	$Ac_k + Bd_k$
Time shifting	$x[n \pm n_0]$	$c_k e^{\pm jk\left(\frac{2\pi}{N}\right)n_o}$
Frequency shifting	$e^{\pm jM\left(\frac{2\pi}{N}\right)n} x[n]$	$c_{k \mu m}$
Conjugation	$x^*[n]$	c^*_{-k}
Time reversal	$x[-n]$	c_{-k}
Time scaling	$x_{(m)}[n] = \begin{cases} x\left[\frac{n}{m}\right]; & \text{if } n \text{ is a multiple of } M \\ 0 \;; & \text{otherwise} \end{cases}$	$\frac{1}{m} c_k$ (Period mN)
Multiplication	$x[n]\, y[n]$	$\sum_{l=0}^{N-1} c_l\, b_{k-l}$
First Difference	$x[n] - x[n-1]$	$\left(1 - e^{-j\left(\frac{2\pi}{N}\right)k}\right) c_k$
Running sum	$\sum_{k=-\infty}^{n} x[k]$	$\left(\frac{1}{1 - e^{-jk\left(\frac{2\pi}{N}\right)}}\right) c_k$
Convolution	$\sum_{r=0}^{N-1} x[r]\, y[n-r]$	$Na_k d_k$

Parseval's Power Theorem

Parseval's power theorem is given by

For continuous time signals,

$$P = \frac{1}{T}\int_0^T |x(t)|^2\, dt = \sum_{n=-\infty}^{\infty} |c_n|^2 = a_0^2 + \frac{1}{2}\sum_{n=1}^{\infty} (a_n^2 + b_n^2)$$

For discrete signals,

$$P = \frac{1}{N}\sum_{n=0}^{N-1} |\, x[n]\,|^2 = \sum_{k=0}^{N-1} |\, c_k\,|^2$$

■■■■

Fourier Transform

4

Fourier Transform of Continuous Time Signals

Fourier transform provides a frequency domain description of time domain signals and is extension of Fourier series to non-periodic signals.

- Fourier transform

$$F.T\left[x(t)\right] = X(\omega) = \int_{-\infty}^{\infty} x(t)\, e^{-j\omega t} dt$$

- Inverse Fourier transform

$$I.F.T\left[X(\omega)\right] = x(t) = \frac{1}{2\pi}\int_{-\infty}^{\infty} X(\omega)\, e^{j\omega t} d\omega$$

Dirichlets Conditions of Fourier Transformation

For existence of Fourier transform

- Fourier transform is defined for all stable signals

$$\int_{-\infty}^{\infty} |x(t)|\, dt < \infty$$

- Periodic signals, which are neither absolutely integrable nor square integral over an infinite interval, can be considered to have Fourier transform if impulse functions are permitted in the transform.
- f(t) have a finite number of discontinuities and finite number of maxima and minima within any finite interval.

The Properties of Fourier Transform

Properties	$X(f)$ - form	$X(\omega)$ - form
Linearity $ax_1(t)+bx_2(t)$	$aX_1(f)+bX_2(f)$	$aX_1(\omega)+bX_2(\omega)$
Time-scaling $x(at)$	$\frac{1}{\lvert a\rvert}X\left(\frac{f}{a}\right)$	$\frac{1}{\lvert a\rvert}X\left(\frac{\omega}{a}\right)$
Time-reversal $x(-t)$	$X(-f)$	$X(-\omega)$
Time-shift $x(t\pm t_0)$	$e^{\pm j2\pi ft_0}X(f)$	$e^{\pm j\omega t_0}X(\omega)$
Frequency shift $x(t)e^{\pm j\omega_0 t}$	$X(f\mp f_0)$	$X(\omega\mp\omega_0)$
Differentiation in time	$\frac{d}{dt}x(t)\leftrightarrow j2\pi fX(f)$	$\frac{d}{dt}x(t)\leftrightarrow j\omega X(\omega)$
Frequency Differentiation	$-j2\pi tx(t)\leftrightarrow\frac{d}{df}X(f)$	$-jtx(t)\leftrightarrow\frac{d}{d\omega}X(\omega)$
Convolution in time $x(t)*h(t)$	$X(f)H(f)$	$X(\omega)H(\omega)$
Frequency convolution $x_1(t)\cdot x_2(t)$	$X_1(f)*X_2(f)$	$\frac{1}{2\pi}[X_1(\omega)*X_2(\omega)]$
Integration $\int_{-\infty}^{t}x(\tau)d\tau$	$\frac{X(f)}{j2\pi f}+0.5X(0)\delta(f)$	$\frac{X(\omega)}{j\omega}+\pi X(0)\delta(\omega)$
Parseval's theorem $\int_{-\infty}^{+\infty}x^2(t)dt$	$\int_{-\infty}^{+\infty}\lvert X(f)\rvert^2\,df$	$\frac{1}{2\pi}\int_{-\infty}^{+\infty}\lvert X(\omega)\rvert^2\,d\omega$
Conjugation $x^*(t)$	$X^*(-j2\pi f)$	$X^*(-j\omega)$
Duality	$x(t)\leftrightarrow X(f)$ $X(t)\leftrightarrow -x(f)$	$x(t)\leftrightarrow X(\omega)$ $X(t)\leftrightarrow 2\pi x(-\omega)$

where, $x_1(t)\leftrightarrow X_1(f)$ or $X_1(\omega)$

$x_1(t)\leftrightarrow X_2(f)$ or $X_2(\omega)$

Fourier Transform of Useful Signals

Signal, $x(t)$	$F(f)$ form	$F(\omega)$ form
$e^{-at}u(t)$, a>0	$\frac{1}{a+j2\pi f}$	$\frac{1}{a+j\omega}$
$e^{at}u(-t), a>0$	$\frac{1}{a-j2\pi f}$	$\frac{1}{a-j\omega}$
$\delta(t)$	1	1
A, Constant	$A\delta(f)$	$2\pi A\delta(\omega)$
$A\ \text{rect}\cdot(t/T)$	$AT\ \text{sinc}(fT)$	$AT\ \text{sinc}\left(\frac{\omega T}{2}\right)$
$e^{-a\|t\|}$	$\frac{2a}{a^2+4\pi^2 f^2}$	$\frac{2a}{a^2+\omega^2}$
e^{-at^2}	$\sqrt{\frac{\pi}{a}}\ e^{-\pi^2 f^2/a}$	$\sqrt{\frac{\pi}{a}}\ e^{-\omega^2/4a}$
$Sgn(t)$	$\frac{1}{j\pi f}$	$\frac{2}{j\omega}$
$u(t)$	$\frac{1}{j2\pi f}+0.5\delta(f)$	$\frac{1}{j\omega}+\pi\delta(\omega)$
$\cos\omega_0 t$	$\frac{\delta(f-f_0)+\delta(f+f_0)}{2}$	$\pi[\delta(\omega-\omega_0)+\delta(\omega+\omega_0)]$
$\sin\omega_0 t$	$\frac{\delta(f-f_0)-\delta(f+f_0)}{2j}$	$\frac{\pi}{j}[\delta(\omega-\omega_0)-\delta(\omega+\omega_0)]$
$t^n e^{-at}u(t)$	$\frac{n!}{(a+j2\pi f)}n+1$	$\frac{n!}{(a+j\omega)^{n+1}}$

Hilbert Transform

$$H\{x(t)\} = x(t) * \frac{1}{\pi t}$$

- Hilbert transform provides a phase shift of –90° to all signals.
- Signals and its Hilbert transform have same Energy Spectral Density or Power Spectral Density and Auto-correlation.
- Impulse response: $h(t) = \frac{1}{\pi t}$

$$H(j\omega) = -j\,\text{sgn}(\omega) = \begin{cases} -j & ; \quad \omega > 0 \\ j & ; \quad \omega < 0 \end{cases}$$

Remember: ....

- Fourier transform of a real signal is always even conjugate in nature.
- Fourier transform of real and even signal is purely real and even.
- Fourier transfer of real and odd signal is purely imaginary and odd.

- Shifting in time domain only changes the phase spectrum of signal.
- Compression in time domain leads to extension in frequency domain and vice-versa.
- Time differentiation property is only valid if $f(t) \to 0$ as $|t| \to \infty$.
- $|F(\omega)|2$ is Energy Spectral Density of signal for energy signal.

Fourier Transfer of Disceret Time Signals

DTFFT of an aperiodic signal $x[n]$ is given by

$$X(e^{j\omega}) = \sum_{n=-\infty}^{\infty} x[n]e^{-j\omega n}$$

Inverse DTFT of $X(e^{j\omega})$ is given by

$$x[n] = \frac{1}{2\pi}\int_{0}^{2\pi} X(e^{j\omega})e^{j\omega n}d\omega$$

DTFT is always periodic with period '2π'.

Properties of DTFT

Property	Aperiodic signal	Fourier Transform
Linearity	$ax[n] + by[n]$	$aX(e^{j\omega}) + bY(e^{j\omega})$
Time shifting	$x[n \mp n_0]$	$e^{\mp j\omega n_o} X(e^{j\omega})$
Frequency shifting	$e^{\pm j\omega_0 n} x[n]$	$X(e^{j(\omega \mp \omega_o)})$
Conjugation	$x^*[n]$	$X^*(e^{-j\omega})$
Time reversal	$x[-n]$	$X(e^{-j\omega})$
Time expansion	$x_{(k)}[n] = \begin{cases} x\left(\frac{n}{k}\right) & ; \text{ if } n \text{ is multiple of k} \\ 0 & ; \text{ Otherwise} \end{cases}$	$X(e^{jk\omega})$
Convolution	$x[n] * y[n]$	$X(e^{j\omega})\, Y(e^{ej\omega})$
Multiplication	$x[n]\, y[n]$	$\frac{1}{2\pi}\left[X(e^{j\omega}) * Y(e^{j\omega})\right]$
Differencing	$x[n] - x[n-1]$	$(1 - e^{-j\omega})\,(X(e^{j\omega}))$
Accumulation	$\sum_{k=-\infty}^{n} x[k]$	$\frac{1}{1-e^{-j\omega}} X(e^{(j\omega)})$
Differentiation in frequency	$nx[n]$	$j\frac{d}{d\omega} X(e^{j\omega})$
Parseval's relation	$\sum_{n=-\infty}^{\infty} \lvert x[n]\rvert^2$	$\frac{1}{2\pi}\int_{0}^{2\pi} \left\lvert X(e^{j\omega})\right\rvert^2 d\omega$

where, $x[n] \leftrightarrow X(e^{j\omega})$; $y[n] \leftrightarrow Y(e^{j\omega})$

Discrete Fourier Transform

DFT of a periodic $x[n]$ with period N is given by

$$X(k) = \sum_{n=0}^{N-1} x[n]\, e^{-jk\left(\frac{2\pi}{N}\right)n}$$

'K' varies from $k = 0$ to $N - 1$.

Inverse DFT of periodic $X(k)$ with period N is given by

$$x[n] = \frac{1}{N}\sum_{k=0}^{N-1} X(k)\, e^{jk\left(\frac{2\pi}{N}\right)n}$$

DFT Pairs

(i) $x[n] = \delta[n]\ ;\ 0 \le n \le N-1$; $X(k) = 1\ ;\ 0 \le k \le N-1$

(ii) $x[n] = 1\ ;\ 0 \le n \le N-1$; $X(k) = N\delta(k)\ ;\ 0 \le k \le N-1$

(iii) $x[n] = e^{j2\pi k_o n}\ ;\ 0 \le n \le N-1$; $X(k) = N\delta(k-k_o)\ ;\ 0 \le k \le N-1$

(iv) $x[n] = \cos(2\pi\, k_0\, n/N)\ ;\ 0 \le n \le N-1$

$$X(k) = \frac{N}{2}\left[\delta(k-k_o) + \delta\left(k-(N-k_o)\right)\right];\ 0 \le k \le N-1$$

Properties of DFT

- For real $f[n]$

$$F[k] = F^*[N-k]$$

$$F[0] = f[0] + f(1) + \ldots\ldots. + f[N-1]$$

and if N is even then $F[N/2] = f(0) - f(1) + f(2)\cdots\cdots$

- Time reversal of sequence $x[(-n)]_N \leftrightarrow X[(-k)]_N$

$$x[[-n]]_N = x[N-n]$$

- Circular time shift $x[[n-l]]_N \leftrightarrow X(k)e^{-j2\pi k\, l/N}$
- Circular frequency shift $x[n]\, e^{j2\pi l n/N} \leftrightarrow X[[k-l]]_N$
- Complex conjugate property $x^*[n] \leftrightarrow X^*[[-k]]_N$
- Circular Convolution

$$x_1[n] \leftrightarrow X_1[k]\ ;\ x_2[n] \leftrightarrow X_2[k]$$

$$x_2[n] \otimes_N x_2[n] \leftrightarrow X_1[k]\cdot X_2[k]$$

- Multiplication of two sequences

$$x_1[n]\, x_2[n] \leftrightarrow \frac{1}{N} X_1[k] \otimes_N X_2[k]$$

Circular Convolution

$$f[n] \otimes_N h[n] = \sum_{k=0}^{N-1} f[k]h[(n-k)]_N$$

$\otimes_N$ represent circular convolution of order N.
$h[(n-k)]_N$ represent circular shift of order N to $h[n]$.

Parseval's Theorem

$$\sum_{n=0}^{N-1} |x[n]|^2 = \frac{1}{N}\sum_{k=0}^{N-1} |X[k]|^2$$

Remember: ....

- **Circular convolution can be found using linear convolution. Method is divide the result of linear convolution into two parts such that first part has N samples and then add corresponding part inorder and write as circular convolution. Here N is no. of samples in each signal.**
- **Given two signals f[n] and h[n] each consisting of N-samples, if we zero pad each of the signal with N-zeros and carry out circular convolution of order 2 N this will be equal to linear convolution of two signals f[n] and $h[n]$.**

....

Fast Fourier Transform

The Fast Fourier Transform is a discrete Fourier transform algorithm which reduces the number of computations needed to find the transform. For N-point DFT

$$\text{Number of multiplications} = \frac{N}{2}\log_2 N$$

$$\text{Number of additions} = N\log_2 N$$

■■■■

Laplace Transform

5

Introduction

The main drawback of (continuous) Fourier transform is that Fourier transform (F.T) can be defined only for stable systems, where as Laplace transform (L.T) can be defined for both stable and unstable systems.

Laplace Transform

Laplace transform of a general signal f(t)

- Bilateral (or two) sided) Laplace transform

$$\boxed{F(s) = \int_{-\infty}^{\infty} f(t)e^{-st}dt}$$

- Unilateral (or one-sided) Laplace transform

$$\boxed{F(s) = \int_{0}^{\infty} f(t)e^{-st}dt}$$

Inverse Laplace Transform

$$\boxed{f(t) = \int_{-\infty}^{\infty} F(s)e^{st}ds}$$

Relation Between Laplace Transform and Fourier Transform

$$\boxed{F(\sigma + j\omega) = \int_{-\infty}^{\infty} f(t)e^{-\sigma t}e^{-j\omega t}dt = e^{-\sigma t}F(\omega)}$$

where, $s \to \sigma + j\omega$

Note:

- At $\sigma = 0, s = j\omega$, the Laplace transform become equal to Fourier transform (or) L.T. calculated on $j\omega$ axis is F.T.
- Fourier transform is used for signal analysis and Laplace transform is used for designing the system.

Region of Convergence (ROC)

The range of values of 's' for which $F(s)$ i.e. $\int_{-\infty}^{\infty} \left| f(t)e^{-\sigma t} \right| < \infty$ is define is known as ROC.

Properties of ROC

(a) ROC does not contain any poles, it is bounded by the poles.
(b) ROC of $F(s)$ contains strips parallel to imaginary axis i.e. $j\omega$ axis.
(c) If $f(t)$ is of finite duration and is absolutely integrable than the ROC is entire s-plane.
(d) If $f(t)$ is a right sided and if the line $\text{Re}\{s\} = \sigma_0$ is in the ROC then all values of s for which $\text{Re}\{s\} > \sigma_0$ will also be in ROC.

Remember:

- For stability ROC must include imaginary axis (i.e. $j\omega$ axis)
- Solution of L.T. is unique only when the ROC is given

Properties of Laplace Transform

Property	Signal	Laplace transform	ROC
Linearity	$ax_1(t)+bx_2(t)$	$aX_1(s)+bX_2(s)$	$R_1 \cap R_2$
Time shifting	$x(t-t_0)$	$e^{-st_0}X(s)$	R
Shifting in s-domain	$e^{s_0 t}x(t)$	$X(s-s_o)$	Shifted version of R
Time scaling	$x(at)$	$\frac{1}{\|a\|}X\left(\frac{s}{a}\right)$	Scaled ROC
Conjugation	$x^*(t)$	$X^*(s^*)$	R
Differentiation in time domain	$\frac{d}{dt}x(t)$	$sX(s)$	At least R
Differentiation in s-domain	$-tx(t)$	$\frac{dX(s)}{ds}$	R
Convolution	$x_1(t)*x_2(t)$	$X_1(s)X_2(s)$	$R_1 \cap R_2$
Integration	$\int_{-\infty}^{t} x(\tau)d\tau$	$\frac{1}{s}X(s)$	$R_1 \cap \{\text{Re}(s)>0\}$

where, $x(t) \rightarrow X(s)$ with ROC R
$x_1(t) \rightarrow X_1(s)$ with ROC R_1
$x_2(t) \rightarrow X_2(s)$ with ROC R_2

Initial value theorem

$$f(0) = \lim_{s \to \infty} sF(s)$$

Final value theorem

$$f(\infty) = \lim_{s \to 0} sF(s)$$

Note:

- Initial value theorem is applicable only for proper Laplace i.e. denominator power of function is more than numerator power.
- Final value theorem is not applicable if poles are conjugate or poles lie in right side of s-plane.

Characterization of LTI Systems Using Laplace Transform

Causality

- ROC associated with the system function for a causal system is a right-half plane.
- For a system with a rational transfer function, causality of the system is equivalent to the ROC being the right-half plane to the right of the rightmost pole.
- A system is anticausal if its impulse response $h(t) = 0$ for $t > 0$.

Stability

- An LTI system is stable if and only if the ROC of its system function $H(s)$ includes the $j\omega$-axis, [i.e. $\text{Re}\{s\} = 0$].
- A causal system with rational system functions $H(s)$ is stable if and only if all of the poles of $H(s)$ lie in the left-half of the s-plane-i.e. all of the poles have negative real parts.

Laplace transform of elementary function

Transform pair	Signal	Transform	ROC
1.	$\delta(t)$	1	All s
2.	$u(t)$	$\frac{1}{s}$	$\text{Re}\{s\} > 0$
3.	$-u(-t)$	$\frac{1}{s}$	$\text{Re}\{s\} < 0$
4.	$\frac{t^{n-1}}{(n-1)!}u(t)$	$\frac{1}{s^n}$	$\text{Re}\{s\} < 0$
5.	$-\frac{t^{n-1}}{(n-1)!}u(-t)$	$\frac{1}{s^n}$	$\text{Re}\{s\} < 0$
6.	$e^{-\alpha t}u(t)$	$\frac{1}{s+\alpha}$	$\text{Re}\{s\} > -\alpha$
7.	$-e^{-\alpha t}u(-t)$	$\frac{1}{s+\alpha}$	$\text{Re}\{s\} < -\alpha$
8.	$\frac{t^{n-1}}{(n-1)!}e^{-\alpha t}u(t)$	$\frac{1}{(s+\alpha)^n}$	$\text{Re}\{s\} > -\alpha$
9.	$-\frac{t^{n-1}}{(n-1)!}e^{-\alpha t}u(-t)$	$\frac{1}{(s+\alpha)^n}$	$\text{Re}\{s\} < -\alpha$
10.	$\delta(t-T)$	e^{-sT}	*All s*
11.	$[\cos\omega_0 t]u(t)$	$\frac{s}{s^2+\omega_0^2}$	$\text{Re}\{s\} > 0$
12.	$[\sin\omega_0 t]u(t)$	$\frac{\omega_0}{s^2+\omega_0^2}$	$\text{Re}(s) > 0$
13.	$\left[e^{-\alpha t}\cos\omega_0 t\right]u(t)$	$\frac{s+\alpha}{(s+\alpha^2)+\omega_0^2}$	$\text{Re}\{s\} > -\alpha$
14.	$\left[e^{-\alpha t}\sin\omega_0 t\right]u(t)$	$\frac{\omega_0}{(s+\alpha)^2+\omega_0^2}$	$\text{Re}\{s\} > -\alpha$
15.	$u_n(t) = \frac{d^n\delta(t)}{dt^n}$	s^n	*All s*

■■■■

Z -Transform

6

Introduction

- z-transform is a discrete-time counterpart of Laplace transform.
- For a discrete-time LTI system with impulse response $h[n]$, the response $y[n]$ of the system to a complex exponential input of the form z^n is

$$y[n] = z^n H(z)$$

where, $H(z) \rightarrow$ Transform function of the system

Z-transform

Z-transform of a general discrete time signal

$$X(z) = \sum_{n=-\infty}^{+\infty} x[n] z^{-n}$$

Note:

- **The range of values of 'z' for which above equation is defined give ROC of Z-transform.**
- **Z-transform are used for digital filter design, sampled data control system.**

Region of Convergence (ROC)

ROC is the region of range of values for which the summation

$\sum_{n=-\infty}^{+\infty} x[n]z^{-n}$ converges.

Properties of ROC

- The ROC of $X(z)$ consists of a circle in the z-plane centered about the origin.
- ROC does not contain any poles, it is bounded by the poles.
- If $x[n]$ is of finite duration then ROC is entire z-plane except possibly $z = 0$ and/or $z \rightarrow \infty$.
- If $x[n]$ is a right sided sequence and if circle $|z| = a$ is in the ROC then all finite values of z for which $|z| > a$ will also be in ROC.
- If $x[n]$ is a left sided sequence and if circle $|z|$ = a is the ROC then all finite values of z for which $|z| < a$ will also be in ROC.
- If $x[n]$ is two sided and if the circle $|z| = a$ is in the ROC, then the ROC will consist of a ring in the z-plane that includes the circle $|z| = a$.

Properties of Z-transform

Property	Signal	z - Transform	ROC
Linearity	$ax_1[n]+bx_2[n]$	$aX_1(z)+bX_2(z)$	$R_1 \cap R_2$
Time shifting	$x[n-n_0]$	$z^{-n_0}X(z)$	R
Scaling	$a^n x[n]$	$X\left(\dfrac{z}{a}\right)$	Scaled version of R
Time reversal	$x[-n]$	$x(z^{-1})$	Inverter R
Time expansion	$x_{(k)}[n]=\begin{cases} x[r], & n=rk \\ 0, & \text{otherwise}\end{cases}$	$X(z^k)$	$R^{1/k}$
Conjugation	$x^*[n]$	$X^*(z^*)$	R
Convolution	$x_1[n]*x_2[n]$	$X_1(z)\,X_2(z)$	$R_1 \cap R_2$
First difference	$x[n]-x[n-1]$	$(1-z^{-1})X(z)$	$R_1 \cap \{\lvert z\rvert>0\}$
Accumulation	$\sum_{k=-\infty}^{n} x[k]$	$\dfrac{1}{1-z^{-1}}X(z)$	$R \cap \{\lvert z\rvert>1\}$
Differentiation	$nx[n]$	$-\dfrac{zdX(z)}{dz}$	R

where, $x[n] \rightarrow X(z)$ with ROC R

$x_1[n] \rightarrow X_1(z)$ with ROC R_1

$x_2[n] \rightarrow X_2(z)$ with ROC R_2

Unilateral Z-transform

$$X(z) = \sum_{n=0}^{\infty} x[n]z^{-n}$$

Left shift

$$x[n+1] \leftrightarrow zX[z] - z\,x[0]$$

Right shift

$$x[n-1] \xrightarrow{Z.T.} z^{-1}X(z) + x(-1)$$

$$x[n-2] \xrightarrow{Z.T.} z^{-2}X(z) + x(-1) + x(-2)$$

First difference

$$x[n] - x[n-1] = (1-z^{-1})X(z)$$

Conjugation

$$x^*[n] = X^*(z^*)$$

Initial value theorem

If $\quad x[n] = 0 \quad$ for $n < 0$

$$x[0] = \lim_{z \to \infty} X(z)$$

Final value theorem

$$x[\infty] = \lim_{z \to 1}(1 - z^{-1})X(z) \quad \text{or} \quad x[\infty] = \lim_{z \to 1}(z - 1)X(z)$$

Characterization of LTI Systems Using Z-Transform

Causality

- A discrete-time LTI system is causal if and only if the ROC of its system function is the exterior of a circle, including infinity.
- A discrete-time LTI system with rational system function $H(z)$ is causal if and only if.
  - (a) the ROC is the exterior of a circle outside the outermost pole; and
  - (b) with $H(z)$ expressed as a ratio of polynomials in z, the order of the numerator cannot be greater than the order of the denominator.

Stability

- An LTI system is stable if and only if the ROC of its system function $H(z)$ includes the unit circle, $|z| = 1$.
- A causal LTI system with rational system function $H(z)$ is stable if and only if all of the poles of $H(z)$ lie inside the unit circle i.e., they must all have magnitude smaller than 1.

Inverse Z-transform

$$x[n] = \frac{1}{2\pi j} \oint X(z) z^{n-1} \, dz$$

where the symbol $\oint$ denotes integration around a counter clockwise circular contour centred at the origin with radius a.

Some common *z*-transform pairs

Sl.No.	Signal	Transform	ROC
1.	$\delta[n]$	1	All z
2.	$u[n]$	$\frac{1}{1-z^{-1}}$	$\lvert z\rvert > 1$
3.	$-u[-n-1]$	$\frac{1}{1-z^{-1}}$	$\lvert z\rvert < 1$
4.	$\delta[n-m]$	z^{-m}	All z, *except* 0 (if $m > 0$) *or* ∞(if $m < 0$)
5.	$\alpha^n u[n]$	$\frac{1}{1-\alpha z^{-1}}$	$\lvert z\rvert > \lvert\alpha\rvert$
6.	$-\alpha^n u[-n-1]$	$\frac{1}{1-\alpha z^{-1}}$	$\lvert z\rvert < \lvert\alpha\rvert$
7.	$n\alpha^n u[n]$	$\frac{\alpha z^{-1}}{(1-\alpha z^{-1})^2}$	$\lvert z\rvert > \lvert\alpha\rvert$
8.	$-n\alpha^n u[-n-1]$	$\frac{\alpha z^{-1}}{(1-\alpha z^{-1})^2}$	$\lvert z\rvert < \lvert\alpha\rvert$
9.	$[\cos\omega_0 n]u[n]$	$\frac{1-[\cos\omega_o]z^{-1}}{1-[2\cos\omega_o]z^{-1}+z^{-2}}$	$\lvert z\rvert > 1$
10.	$[\sin\omega_0 n]u[n]$	$\frac{[\sin\omega_0]z^{-1}}{1-[2\cos\omega_0]z^{-1}+z^{-2}}$	$\lvert z\rvert > 1$
11.	$[a^n \cos\omega_0 n]u[n]$	$\frac{1-[a\cos\omega_0]z^{-1}}{1-[2a\cos\omega_0]z^{-1}+a^2z^{-2}}$	$\lvert z\rvert > a$
12.	$[a^n \sin\omega_0 n]u[n]$	$\frac{[a\sin\omega_o]z^{-1}}{1-[2a\cos\omega_o]z^{-}+a^2z^{-2}}$	$\lvert z\rvert > a$

■■■■

7 Digital Filters

FIR Filters : Finite impulse response (FIR) filters are those filters in which the impulse response sequence is of finite duration i.e. it has a finite number of Non-zero terms.

$$h[n] = \begin{cases} 2, & |n| \leq 4 \\ 0 & \text{otherwise} \end{cases}$$

IIR Filters : Infinite impulse response (IIR) filters are those filters in which the filter which have impulse response sequence is of infinite duration. It has infinite number of non zero term.

$$h[n] = a^n u[n]$$

- IIR filter are usually implemented using structures having feedback (recursive structures poles and zeros).
- FIR filter are usually implemented using structures with no feedback (non-recursive structures-all zeros).

An IIR filter is described by the following difference equation representation with input $x[n]$ and output $y(n)$

$$\sum_{K=0}^{N} a_k \, y[n-K] = \sum_{K=0}^{M} b_k \, x[n-K]$$

An FIR filter of length M is described by following difference equation with input $x[n]$ and output $y[n]$

$$y[n] = \sum_{K=0}^{M-1} b_k \, x[n-K]$$

Where $\{b_K\}$ is the set of filter coefficient.

Design of IIR Filters

- The $j\omega$ axis in the s-plane should map on to the unit circle in the z-plane. This gives a direct relationship between the two frequency variables in the two domains.

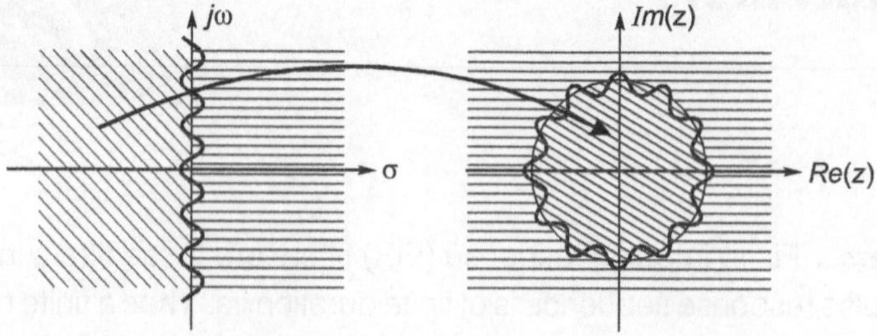

- The left-half plane of the *s*-plane should map into the inside of the unit circle in the *z*-plane to convert a stable analog filter into a stable digital filter.

  There are two methods for design of IIR filter
  (a) Impulse invariant method
  (b) Bilinear transformation

IIR Filter Design by Impulse Invariant

$$H_a(s) = \sum_{i=1}^{M} \frac{A_i}{s - P_i}$$

$$H(z) = \sum_{i=1}^{M} \frac{A_i}{1 - e^{P_i T} z^{-1}}$$

Analog poles and the digital poles are related by the relation

$$z = e^{sT}$$
$$r = e^{\sigma T}$$
$$\Omega = \omega T$$

IIR Filter Design by Bilinear Transformation

$$H(s) = \frac{b}{s+a}$$

$$H(s) = \frac{Y(z)}{X(z)} = \frac{b}{\frac{2}{T}\left[\frac{1-z^{-1}}{1+z^{-1}}\right] + a}$$

Butterworth Filters

The Butterworth low-pass filter has a magnitude response given by

$$|H(j\Omega)| = \frac{A}{\left[1 + (\Omega/\Omega_c)^{2N}\right]^{0.5}}$$

where A is the filter gain and Ω_c is the 3 dB cut-off frequency and N is the order of the filter.

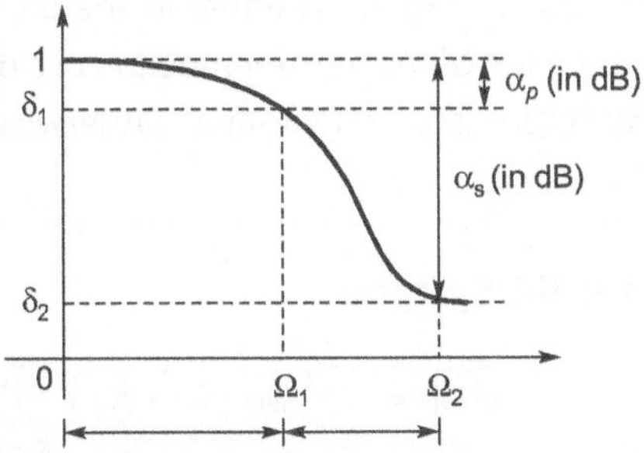

The order of the filter N is given by

$$N \geq \frac{1}{2}\frac{\log\left\{\left[(1/\delta_2^2)-1\right]/\left[(1/\delta_1)^2-1\right]\right\}}{\log(\Omega_2/\Omega_1)}$$

Poles of a Normalised Butterworth Filter

Thus the normalised poles are obtained as

$$s_n = -\sin\left[\frac{2n-1}{2N}\right]\pi + j\cos\left[\frac{2n-1}{2N}\right]\pi$$

Chebyshev Filters

The Chebyshev low-pass filter has a magnitude response given by

$$|H(j\Omega)| = \frac{A}{\left[1+\epsilon^2 C_N^2(\Omega/\Omega_c)\right]^{0.5}}$$

where A is the filter gain, ϵ is a constant and Ω_c is the 3 dB cut-off frequency. The Chebyshev polynomial of the I kind of Nth order, $C_N(x)$ is given by

$$C_N(x) = \begin{cases} \cos(N\cos^{-1}x) & \text{for } |x| \leq 1 \\ \cosh(N\cosh^{-1}x) & \text{for } |x| \geq 1 \end{cases}$$

The magnitude response of the Chebyshev filter is shown in the given figure. The order of the filter N is given by

$$N \geq \frac{\cosh^{-1}\left\{\frac{1}{\epsilon}\left[\frac{1}{\delta_2^2}-1\right]^{0.5}\right\}}{\cosh^{-1}(\Omega_2/\Omega_1)}$$

Note:

- **Canonic and Non-Canonic Structures :** If the number of delays in the realisation block diagram is equal to the order of the difference equation or the order of the transfer function of a digital filter, then the realisation structure is called canonic. Otherwise, it is non-canonic structures.

Basic Structures for IIR Systems

$$y[n] = \sum_{K=1}^{N} a_K\, y[n-K] + \sum_{K=0}^{M} b_K\, x[n-K]$$

where a_K and b_K are constants with

$$a_o \neq 0 \quad \text{and} \quad M \leq N$$

$$H(z) = \frac{Y(z)}{X(z)} = \frac{\sum_{K=0}^{M} b_k z^{-k}}{1 - \sum_{K=1}^{N} a_K\, z^{-K}}$$

Realisation of IIR filters

1. Direct Form-I Realisation

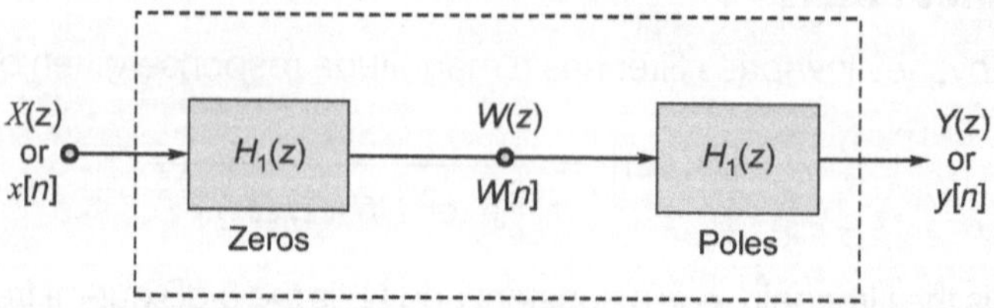

$$w[n] = \sum_{K=0}^{M} b_K\, x[n-K]$$

and

$$y[n] = \sum_{K=1}^{N} a_K\, y[n-K] + w[n]$$

$$w(z) = X(z) \sum_{K=0}^{M} b_K\, z^{-K}$$

$$Y(z) = \frac{w(z)}{1 - \sum_{K=1}^{N} a_K\, z^{-K}}$$

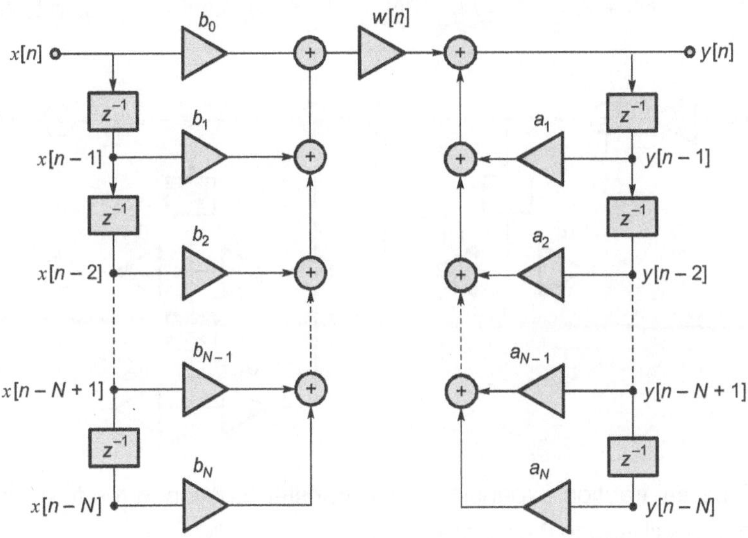

2. Direct Form-II

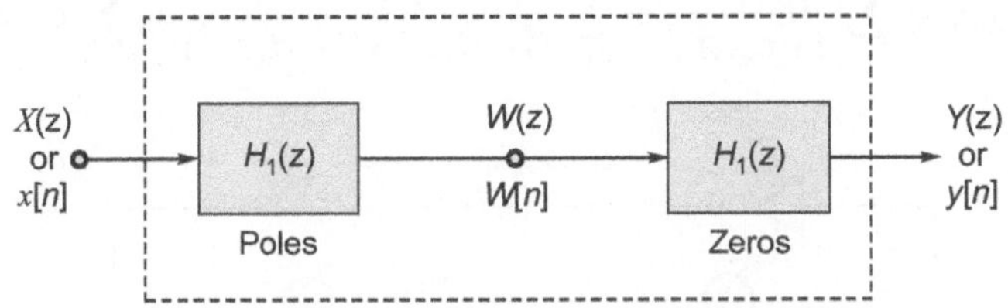

$$w[n] = \sum_{K=1}^{N} a_K\, w[n-K] + x[n]$$

$$y[n] = \sum_{K=0}^{M} b_K\, w[n-K]$$

$$w(z) = \frac{X(z)}{1 - \sum_{K=1}^{M} a_K\, z^{-K}} \quad \text{and} \quad Y(z) = W(z) \sum_{K=0}^{M} b_K\, z^{-k}$$

3. Cascade Realisation of IIR Systems

- In cascade realisation, the transfer function $H(z)$ is broken into a product of transfer functions $H_1(z)$, $H_2(z)$, ... $H_K(z)$.

$$H(z) = G\,\frac{\prod_{K=1}^{M_1}(1-g_K z^{-1}) \prod_{K=1}^{M_2}(1-h_K z^{-1})(1-h_K^{*} z^{-1})}{\prod_{K=1}^{N_1}(1-c_K z^{-1}) \prod_{K=1}^{N_2}(1-d_K z^{-1})(1-d_K^{*} z^{-1})}$$

where, $M = M_1 + 2M_2$, $N = N_1 + 2N_2$ and the constant G is a gain term.

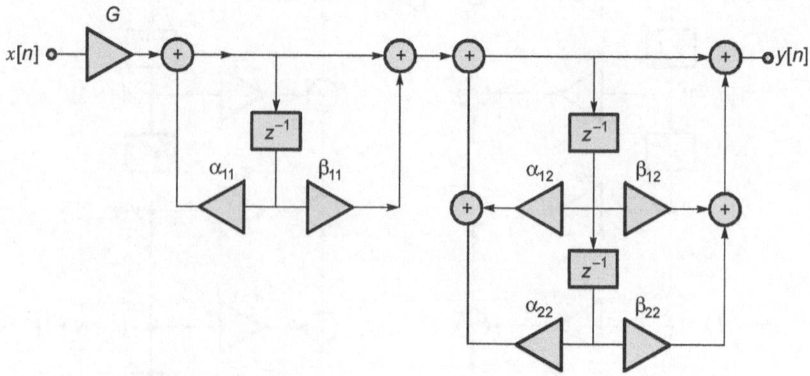

- A partial fraction expansion of the transfer function in the form given below will lead to the parallel form.

$$H(z) = \sum_{K=1}^{N_1} \frac{A_K}{1 - c_K z^{-1}} + \sum_{K=1}^{N_2} \frac{B_K(1 - e_K z^{-1})}{(1 - d_K z^{-1})(1 - d_K^* z^{-1})} + \sum_{K=0}^{M-N} c_K z^{-K}$$

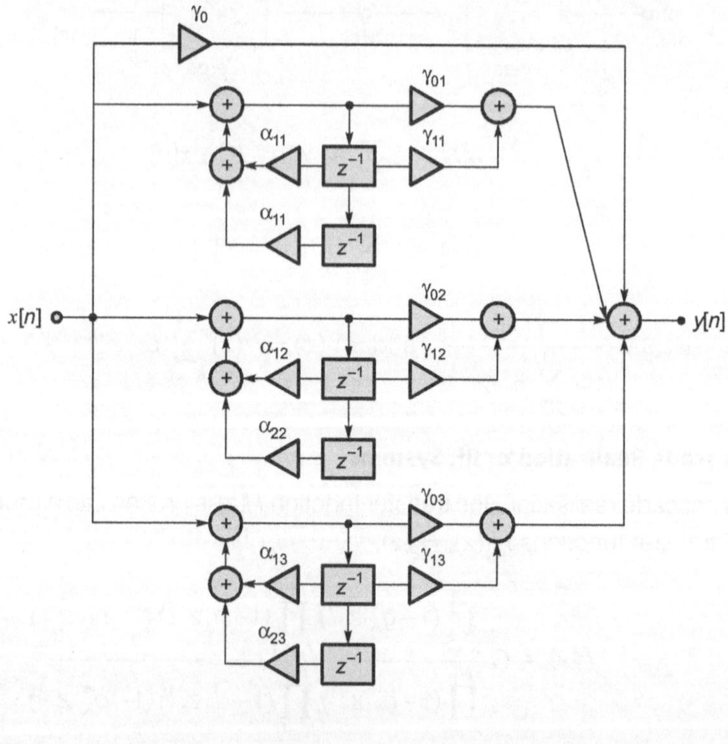

Basic Structures for FIR System

$$H(z) = \sum_{n=0}^{M-1} h[n]\, z^{-n}$$

Realisation of FIR System

1. Direct Form

$$y[n] = \sum_{K=0}^{M-1} h[k]\, x[n-k]$$

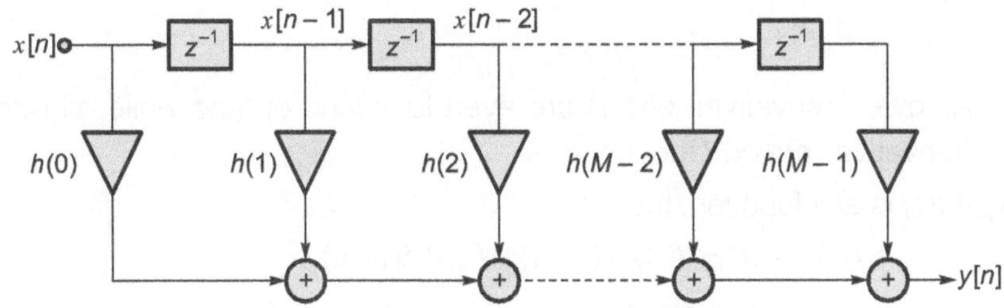

2. Cascade Form

$$H(z) = \prod_{K=1}^{M/2} (B_{0K} + B_{1K}\, z^{-1} + B_{2K} z^{-2})$$

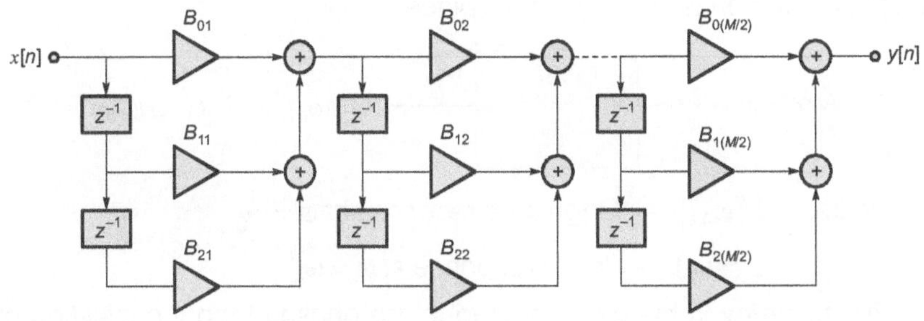

8 Miscellaneous

Unit Doublet Function

The derivative of a unit impulse function is called a unit doublet $(\delta'(t))$.

Properties

(i) $\delta'(t) = \dfrac{d\delta(t)}{dt}$

(ii) $\delta'(at) = \dfrac{1}{a|a|}\delta'(t)$

All even derivatives of $\delta(t)$ are even functions of time while all odd derivatives are odd functions.

(iii) If $f(t)$ is any function, then

$$f(t)\,\delta'(t-a) = f(a)\,\delta'(t-a) - f'(a)\,\delta(t-a)$$

(iv) $\displaystyle\int_{-\infty}^{\infty} f(t)\delta'(t-a) = -\frac{d}{dt}f(t)\Big|_{t=a}$

Response of LTI System to a Sinusoidal Input

Any LTI system does not effect the frequency of input signal, only amplitude and phase change takes place.

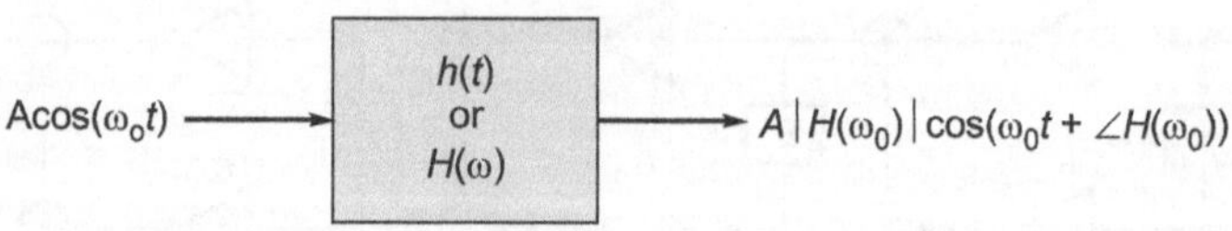

where, $|H(\omega_o)| \rightarrow$ Magnitude response at $\omega = \omega_0$

$\angle H(\omega_0) \rightarrow$ Phase response at $\omega = \omega_0$

Phase delay is the delay offered to the phase of input consisting of single frequency.

$$\boxed{\text{Phase delay} = -\frac{\phi(\omega)}{\omega}}$$

Group delay is the delay offered to phase of input consisting of multiple frequencies.

$$\text{Group delay} = -\frac{d\phi(\omega)}{d\omega}$$

For a distortionless LTI system, both group delay and phase delay are constant and are equal.

Sampling

Any bandlimited signal $x(t)$ is uniquely determined by its samples if the sampling frequency (ω_s) is greater than twice the maximum frequency (ω_m) present in the signal. This is sampling theorem.

$$\omega_s > 2\omega_m$$

There are two types of sampling:

(a) Ideal/Instantaneous sampling (zero pulse width).

(b) Practical sampling (non-zero pulse width).

  (i) Natural sampling

  (ii) Flat top sampling

Remember:

- **The reconstruction of a continuous time signal from the samples can be done using a low pass filter.**
- **For a bandpass signal with the spectral components from ω_L to ω_H, the sampling frequency (ω_s) is given by**

$$\frac{2\omega_H}{n} \le \omega_s \le \frac{2\omega_L}{n-1}$$

For any integer n satisfying

$$1 \le n \le \left\lfloor \frac{\omega_H}{\omega_H - \omega_L} \right\rfloor \quad (\lfloor\ \rfloor \rightarrow \text{floor function})$$

Correlation

Correlation is a measure of relation between two signals. For two signals $f(t)$ and $g(t)$, correlation is defined as

$$R_{fg}(\tau) = \int_{-\infty}^{\infty} f(t)\, g(t-\tau)dt = [f(t) \otimes g(-t)]_{t=\tau}$$

Autocorrelation function is a special case where both the signals are same

For energy signals

$$R_{ff}(\tau) = \int_{-\infty}^{\infty} f(t)\, f(t-\tau)dt = [f(t) \otimes f(-t)]_{t=\tau}$$

For power signals

$$R_{ff}(\tau) = \lim_{X\to\infty} \frac{1}{T}\int_{-\infty}^{\infty} f(t)\, f(t-\tau)d\tau$$

Properties of auto-correlation function:

(i) $R_{ff}(\tau)$ is even ; $R_{ff}(\tau) = R_{ff}(-\tau)$.

(ii) $R_{ff}(0) > R_{ff}(\tau)$

(iii) For energy signals, energy = $R_{ff}(0)$

For power signals, power = $R_{ff}(0)$

(iv) For energy signals, Fourier transform of $R_{ff}(\tau)$ gives Energy Spectral Density, i.e. $R_{ff}(\tau) \leftrightarrow F(\omega)\, F^*(\omega)$ [ESD$_f$]

For power signals, Fourier transform of $R_{ff}(\tau)$ gives power spectral density,

i.e. $R_{ff}(\tau) \leftrightarrow \lim_{T\to\infty} \frac{|F_T(\omega)|^2}{T}$ [PSD_f]

Remember: ....

The autocorrelation function of a binary rectangular random pulse train $x(t) = \sum_{-\infty}^{\infty} a_n p(t - nT + t_o)$ where, an $a_n's$ are independent variables with values ±1 (equal probability), t_o (time offset, uniformly distributed between $\pm T/2$) and $p(t)$ is pulse with amplitude 'A' and width 'T' is given by

$$R_{xx}(\tau) = A^2\left[1 - \frac{|\tau|}{T}\right]; |\tau| \le T.$$

....

Pre-envelope and Complex Envelope

Let $x(t)$ be a real bandpass signal with Fourier transform as $X(\omega)$, then $x_p(t)$ is its pre-envelope and $x_c(t)$ is its complex envelope.

(a) Pre-envelope

$$\boxed{x_p(t) = x(t) + jx_h(t)}$$

$$\boxed{X_p(\omega) = 2X(\omega)\,u(\omega)}$$

where, $x_h(t) \rightarrow$ Hilbert transform of $x(t)$

$X_p(\omega) \rightarrow$ Fourier transform of $x_p(t)$

$u(\omega) \rightarrow$ Unit step signal

(b) Complex Envelope

$$\boxed{x_c(t) = x_p(t)e^{-j\omega_c t} = x_i(t) + jx_Q(t)}$$

$$\boxed{x(t) = \text{Re}\left\{x_p(t)\right\} = x_i(t)\cos\omega_c t - x_Q(t)\sin\omega_c t}$$

where,

$x_i(t), x_Q(t) \rightarrow$ In phase and quadrature components respectively

$\omega_c \rightarrow$ Centre frequency of carrier

Remember:

- Complex envelope is a low pass signal.
- Pre-envelope is a band pass signal.
- Complex envelope and pre-envelope are frequency shifted version of each other.

Digital Filters

Digital filter is a system that performs mathematical operations on a sampled, discrete time signal to reduce or enhance certain parameters of that signal.

There are two types:

(a) Finite Impulse Response ilter

The impulse response is of finite duration. The difference equation is given by

$$\boxed{y[n] = \sum_{i=0}^{N} b_i\, x[n-i]}$$

where, $x[n], y[n] \rightarrow$ Input and output respectively
$N \rightarrow$ Filter order
$b_i \rightarrow$ Value of impulse response at i^{th} instant

Properties:

(i) No feedback is required.

(ii) They are inherently stable.

(iii) They can be easily designed to be linear phase by making the coefficient sequence symmetric.

(iv) Computationally complex.

Ex: Moving average filter, $y[n] = b_0 x[n] + b_1 x[n-1] + b_2 x[n-2]$.

(b) Infinite Impulse Response filter

The impulse response of IIR filters continues indefinitely. The difference equation is given as

$$y[n] = \frac{1}{a_0}\left(\sum_{i=0}^{P} b_i x[n-i] - \sum_{j=1}^{Q} a_j y[n-j]\right)$$

where, $P \rightarrow$ Feedforward filter order
$Q \rightarrow$ Feedback filter order
$b_i \rightarrow$ Feedforward filter coefficients
$a_i \rightarrow$ Feedback filter coefficients

Properties:

(i) Feedback is required.

(ii) Stability depends on the filter coefficients.

(iii) They are preferred over FIR filters for their efficiency in implementation.

Ex: $y[n] = y[n-1] + x[n]$.

■■■■

A Handbook on
Instrumentation Engineering

4

Control Systems & Process Control

CONTENTS

1 Introduction

Control System

A control system is a combination of elements arranged in a planned manner wherein each element causes an effect to produce a desired output.

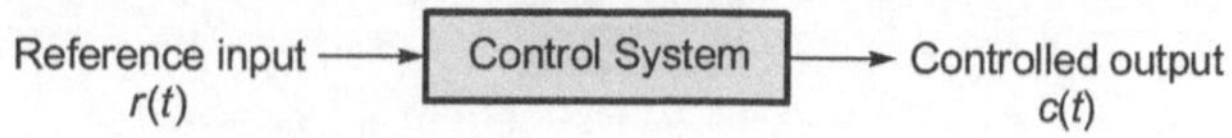

Classification of Control System

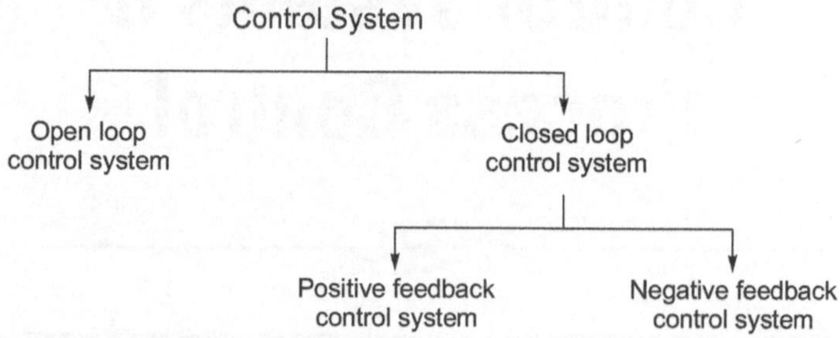

Open Loop Control System

It is a conditional control system, formulated under the basic condition that, the system is not subjected to any type of the disturbances. Control characteristics of such systems are independent of output of the system. The output is neither measured nor fedback for comparison with the input.

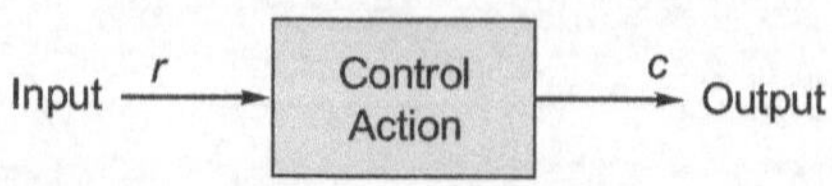

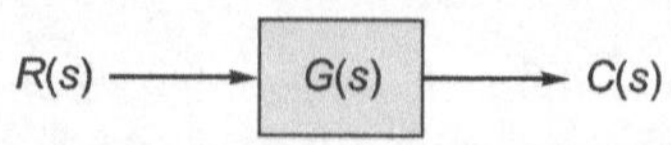

Note:

- Performance analysis is not applicable to open loop system because they are highly stable system.
- Faithfulness of an open loop control system depends upon the accuracy of input calibration.

Closed Loop Control System

Control characteristics of the system depends upon the output of the system. It is also termed as feedback control system. The control action is actuated by an error signal 'e', which is the difference between input signal and output signal. The purpose of feedback is to reduce the error between the reference input and the system output.

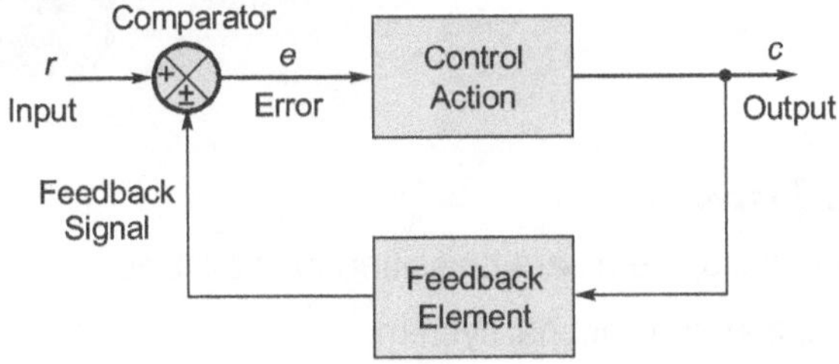

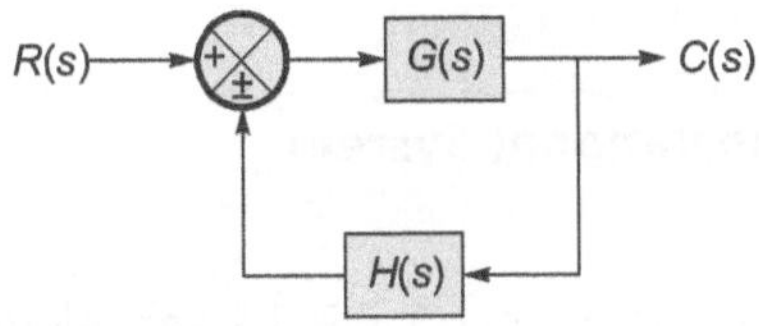

Effect of Negative Feedback

1. Effect of parameter variation reduces.
2. The gain of system reduces by a factor $(1 + GH)$.
3. The bandwidth of the system increases.
4. Effect of internal disturbance reduces.

Comparison of Open Loop and Closed Loop Control System

Open Loop System	Close Loop System
1. As long as the calibration is good, open-loop system will be accurate	1. Due to feedback, the closed-loop system is more accurate
2. Organization is simple and easy to construct	2. Complicated and difficult
3. Generally stable in operation	3. Stability depends on system components
4. If non-linearity is present, system operation degrades	4. Comparatively, the performance is better than open-loop system if non-linearity is present

■■■■

Mathematical Modelling

2

Mechanical Systems

All mechanical systems are classified in two types.

(i) Mechanical translational system.

(ii) Mechanical rotational system.

1. Mechanical Translational System

Input : Force

Output : Linear displacement (x) OR Linear velocity (v)

(a) Intertia force

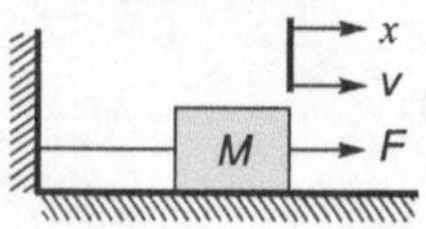

$$F = M\frac{d^2x}{dt^2} = M\frac{dv}{dt}$$

where, $F \rightarrow$ Force on block M

$x \rightarrow$ Displacement of block M

$v \rightarrow$ Velocity of block M

$M \rightarrow$ Mass of block M

(b) Damping force

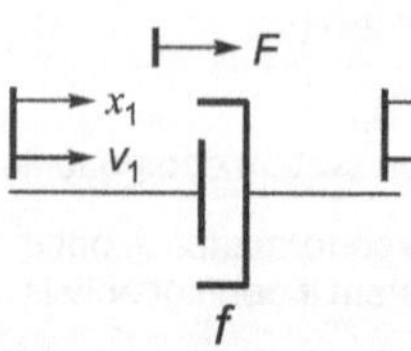

$$F = f\frac{d(x_1 - x_2)}{dt} = f(v_1 - v_2)$$

where, $F \rightarrow$ Damper force

$x_1, x_2 \rightarrow$ Displacement at side 1 and 2 of damper

$v_1, v_2 \rightarrow$ Velocity at side 1 and 2

$f \rightarrow$ Damper constant

(c) Spring force

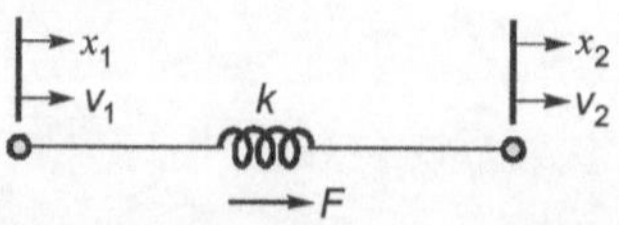

$$F = k(x_1 - x_2) = k\int(v_1 - v_2)\,dt$$

where $F \rightarrow$ Spring force

$x_1, x_2 \rightarrow$ Displacement at side 1 and 2 of spring

$v_1, v_2 \rightarrow$ Velocity at side 1 and 2

$k \rightarrow$ Spring constant

2. Mechanical Rotational System

Input : Torque (T)

Output : Angular displacement (θ) OR Angular velocity (ω)

(a) Inertia torque

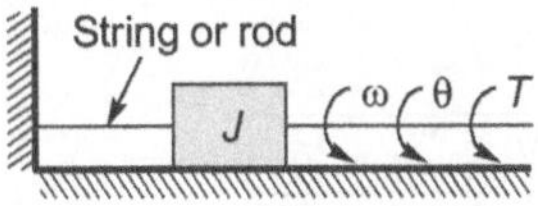

$$T = J\frac{d^2\theta}{dt^2} = J\frac{d\omega}{dt}$$

where, $T \rightarrow$ Torque

$\theta \rightarrow$ Angular displacement

$\omega \rightarrow$ Angular velocity

$J \rightarrow$ Inertia

(b) Damper torque

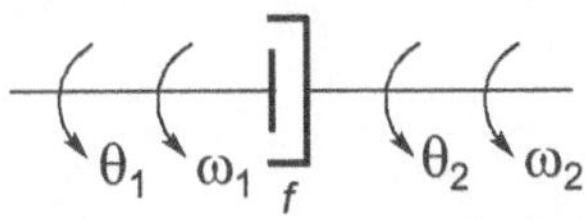

$$T = f\frac{d\theta}{dt} = f(\omega_1 - \omega_2)$$

where, $f \rightarrow$ Damping coefficient

(c) Spring torque

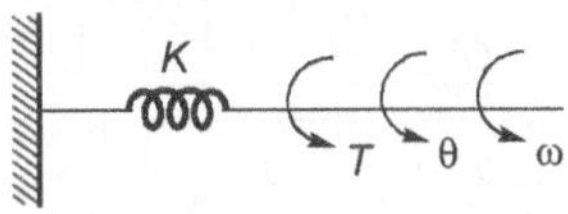

$$T = K\theta = K\int \omega dt$$

where, $K \rightarrow$ Torsional stiffness

Analogous System

Electrical equivalent of mechanical system is called analogous system

1. Mechanical Translation of Systems

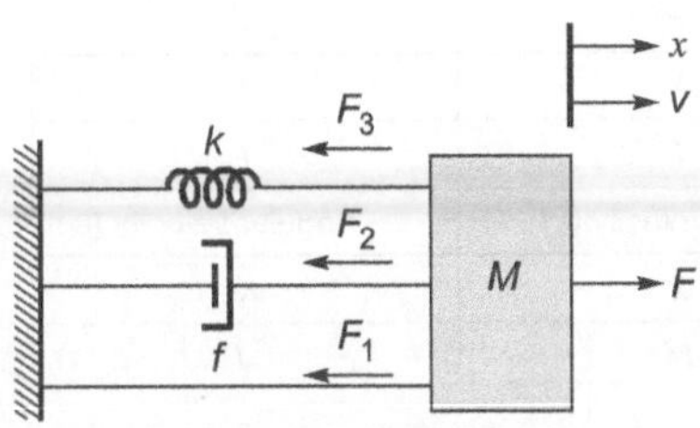

$$F = F_1 + F_2 + F_3$$

$$F = M\frac{dv}{dt} + fv + k\int vdt$$

$$F = M\frac{d^2x}{dt^2} + f\frac{dx}{dt} + kx$$

2. Mechanical Rotational Systems

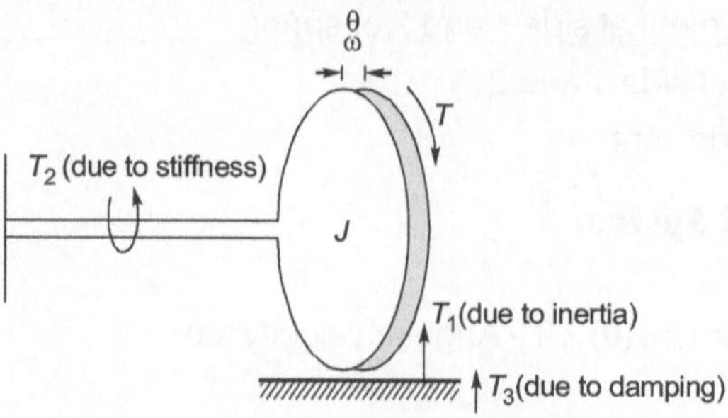

$$T = T_1 + T_2 + T_3$$

$$T = J\frac{d\omega}{dt} + f\omega + K\int \omega dt$$

$$T = J\frac{d^2\theta}{dt^2} + f\frac{d\theta}{dt} + K\theta$$

3. Electrical Series RLC Systems

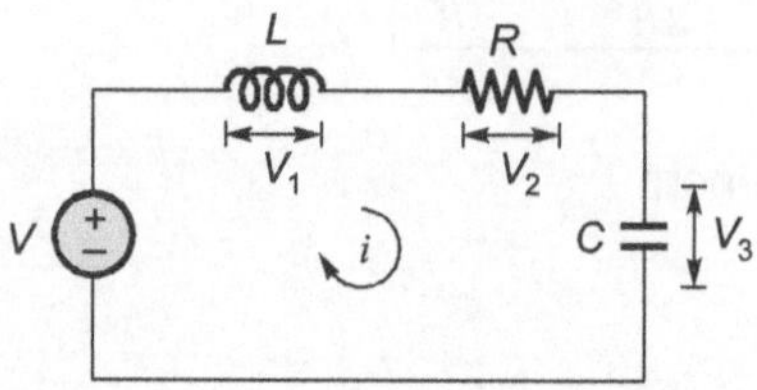

$$V = V_1 + V_2 + V_3$$

$$V = L\frac{di}{dt} + iR + \frac{1}{C}\int idt$$

$$V = L\frac{d^2q}{dt^2} + R\frac{dq}{dt} + \frac{q}{C}$$

where, $q \rightarrow$ Charge

4. Electrical Parallel RLC Systems

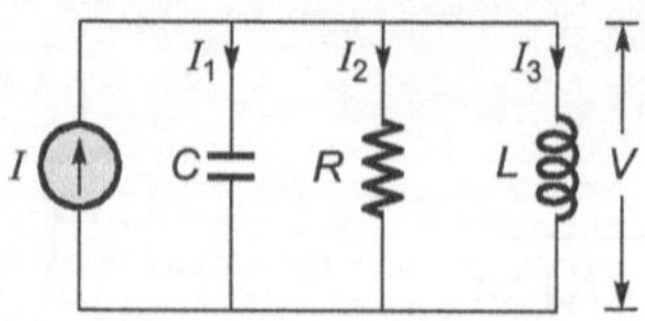

$$I = I_1 + I_2 + I_3$$

$$I = C\frac{dV}{dt} + \frac{V}{R} + \frac{1}{L}\int Vdt$$

$$I = C\frac{d^2\phi}{dt} + \frac{1}{R}\frac{d\phi}{dt} + \frac{\phi}{L}$$

where, $\phi \rightarrow$ Flux linkage

Force Voltage Analogy & Force Current Analogy

Series RLC	Parallel RLC	Mechanical Translation System	Mechanical Rotational System
V	I	F	T
q	ϕ	x	θ
R	$1/R$	f	f
I	V	Linear velocity (v)	Angular velocity (ω)
$1/C$	$1/L$	k	K
L	C	M	J

Nodal Method

1. Number of nodes = Number of displacement.
2. Take an additional node which is reference node.
3. Connect the mass or inertia element between principal node and reference node only.
4. Connect the spring and damping elements either between the principal nodes or between the principal node and reference node, depending on their position.
5. Obtain the nodal diagram and write the differential equation at each node.

Servomotors

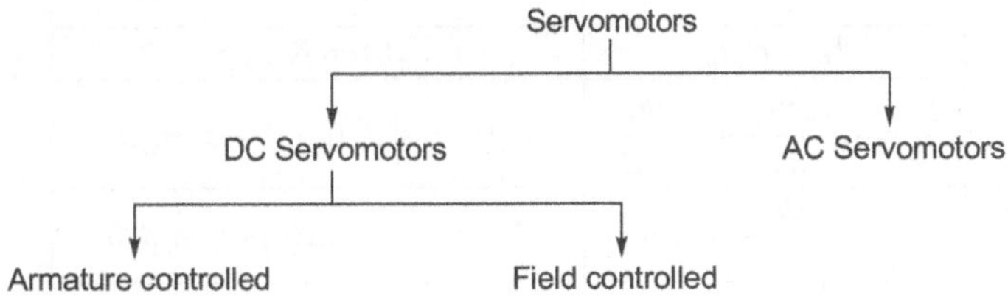

1. Armature Controlled DC Servo Motors

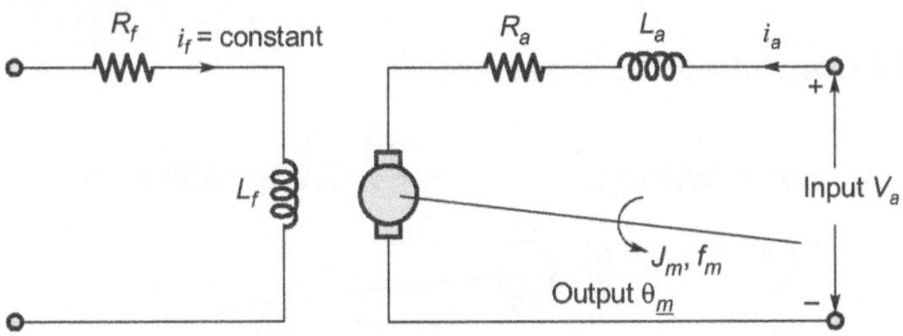

Block diagrams

(a) Relating $\theta_m(s)$ and $V_a(s)$

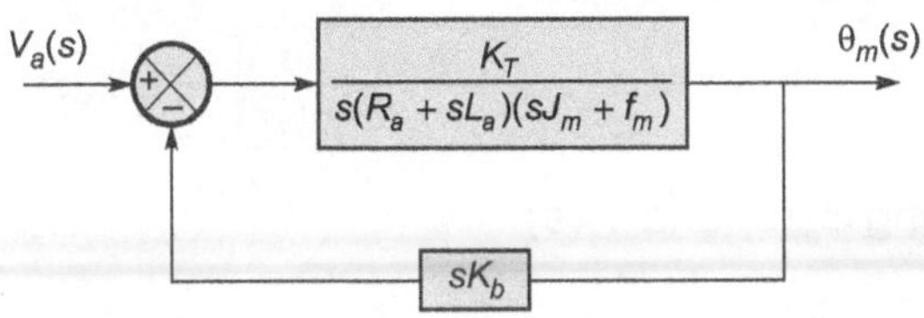

(b) Relating $\omega_m(s)$ and $V_a(s)$

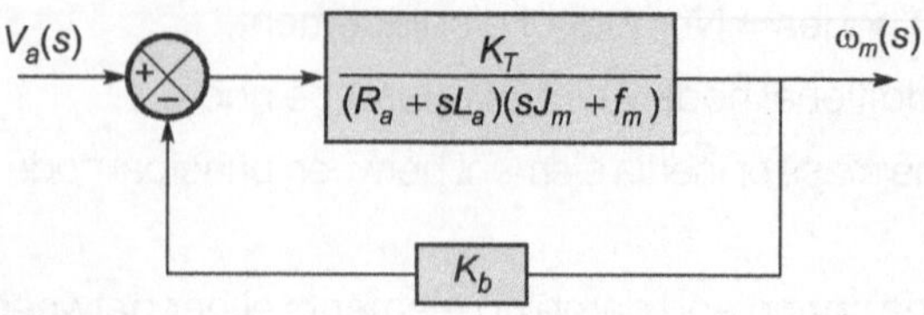

Dynamic equations

Dynamic equations in time domain	Corresponding equation in s-domain
$V_a - e_b = R_a i_a + L_a \frac{di_a}{dt}$	$V_a(s) - E_b(s) = R_a I_a(s) + sL_a I_a(s)$
$e_b = K_b \frac{d\theta_m}{dt} = K_b\omega_m$	$E_b(s) = sK_b\theta_m(s) = K_b\omega_m(s)$
$T_M = K_T i_a$	$T_M(s) = K_T I_a(s)$
$T_M = J_m \frac{d^2\theta_m}{dt^2} + f_m \frac{d\theta_m}{dt}$	$T_M(s) = s^2 J_m\theta_m(s) + sf_m\theta_m(s)$
$T_M = J_m \frac{d\omega_m}{dt} + f_m\omega_m$	$T_M(s) = sJ_m\omega_m(s) + f_m\omega_m(s)$

where, $e_b \rightarrow$ Back emf
$T_m \rightarrow$ Torque

2. Field Controlled DC Servomotor

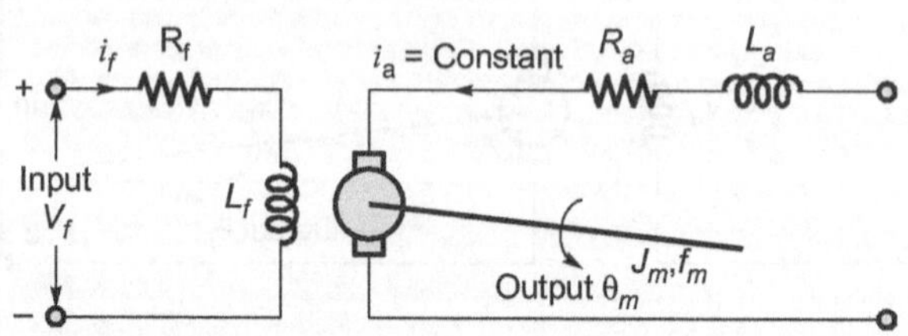

Block diagrams

(a) Relating $\theta_m(s)$ and $V_f(s)$

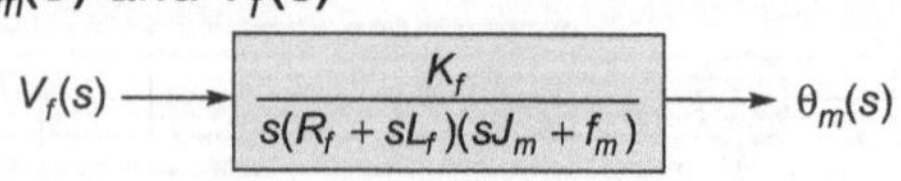

(b) Relating $\omega_m(s)$ and $V_f(s)$

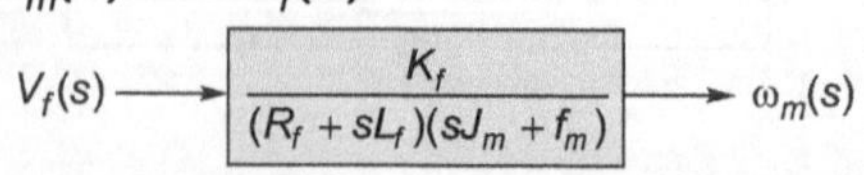

Dynamic equations

Dynamic equations in time domain	Corresponding equation in s-domain
$V_f = R_f i_f + L_f \frac{di_f}{dt}$	$V_f(s) = R_f I_f(s) + sL_f I_f(s)$
$T_M = K_f i_f$	$T_M(s) = K_f I_f(s)$
$T_M = J_m \frac{d^2\theta_m}{dt^2} + f_m \frac{d\theta_m}{dt}$	$T_M(s) = s^2 J_m \theta_m(s) + sf_m \theta_m(s)$
$T_M = J_m \frac{d\omega_m}{dt} + f_m \omega_m$	$T_M(s) = sJ_m \omega_m(s) + f_m \omega_m(s)$

3. Two Phase AC Servomotor

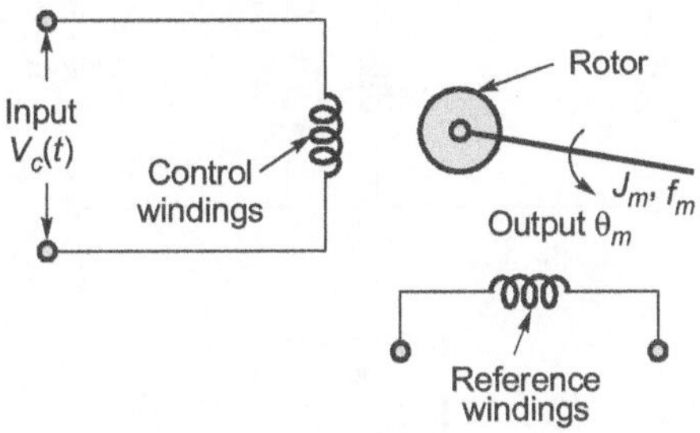

Block diagrams

(a) Relating $\theta_m(s)$ and $V_c(s)$.

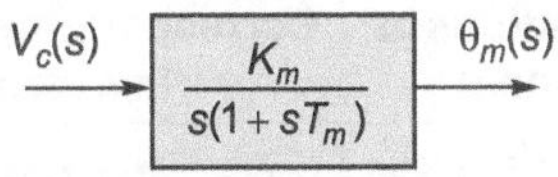

(b) Relating $\omega_m(s)$ and $V_c(s)$.

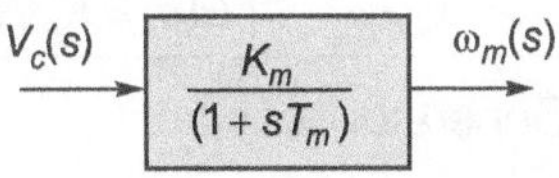

Dynamic equations

Dynamic equations in time domain	Corresponding equations in s-domain
$T_M = m\frac{d\theta_m}{dt} + KV_c$	$T_M(s) = sm\theta_m(s) + KV_c(s)$
$T_M = J_m \frac{d^2\theta_m}{dt^2} + f_m \frac{d\theta_m}{dt}$	$T_M(s) = s^2 J_m \theta_m(s) + sf_m \theta_m(s)$

where, $m = \frac{T_o}{\omega_o}$ (slope of torque speed characteristic)

$K = \frac{T_o}{V_c}$

$T_o \rightarrow$ Stalling torque

$\omega_0 \rightarrow$ No load speed

Generators

1. Separately Excited DC Generator

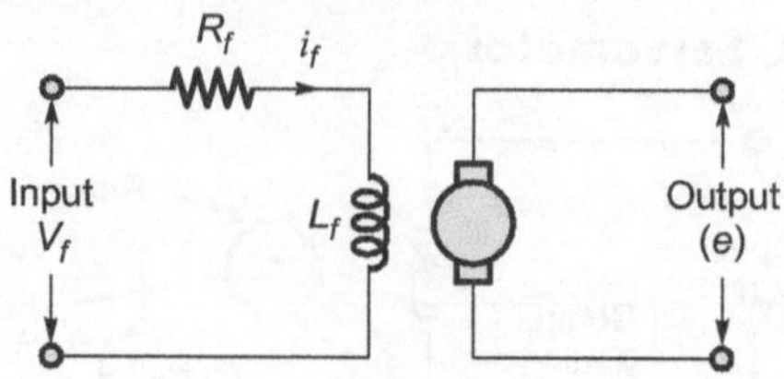

Block diagram

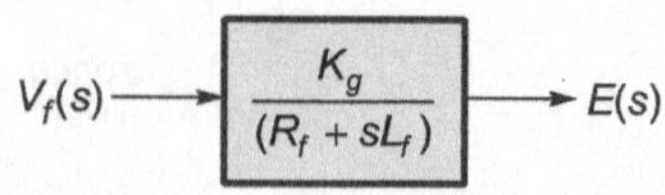

Dynamic equations

Related equations in time domain	Corresponding equations in s-domain
$V_f = R_f i_f + L_f \frac{di_f}{dt}$	$V_f(s) = R_f I_f(s) + sL_f I_f(s)$
$e = K_g i_f$	$E(s) = K_g I_f(s)$

where, $K_g \rightarrow$ Generator constant

2. Separately Excited DC Generator Connected to Load

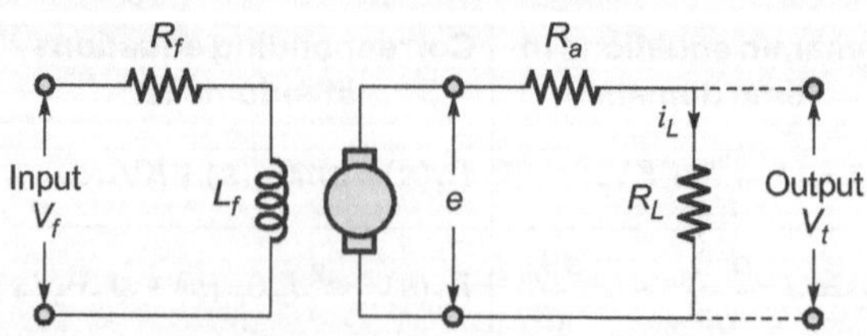

Block diagram

Relating $V_t(s)$ and $V_f(s)$

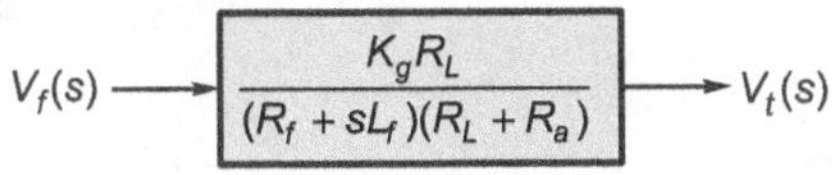

Dynamic equations

Related equations in time domain	Corresponding equations in s-domain
$V_t = e - R_a i_L$	$V_t(s) = E(s) - R_a I_L(s)$
$I_L = \frac{V_t}{R_L}$	$I_L(s) = \frac{V_t(s)}{R_L}$

Note:

- Servomechanism is electromechanical system whose input is electrical ***voltage*** and output is mechanical ***position*** or its time derivative.
- Field control DC servo motor is used for low power applications while armature control DC servomotor is used for high power applications.
- AC servomotor is generally used for low power application.
- Torque speed characteristic is linear for AC servomotor whereas, it is nonlinear for general purpose induction motor.
- For the fast response of the servomotor system, the inertia of the rotor should be kept to minimum. This is achieved by reducing the diameter and increasing the length of the rotor.

Tachometer

Tachometer is a speed transducer, used as feedback element in control system.

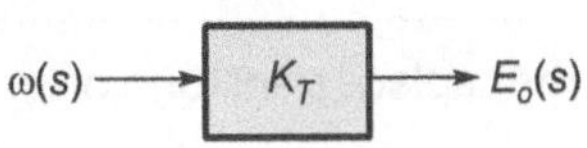

(or)

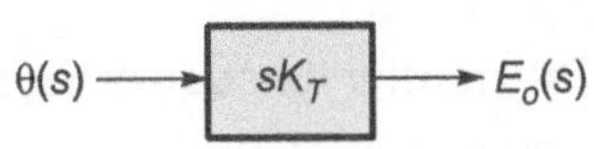

where, $sK_T \rightarrow$ Tachometer constant

3 Transfer Function

Transfer function

The transfer function of a linear, time-invariant, differential equation system is defined as the ratio of the Laplace transform of the output to the Laplace transform of the input, under the assumption that all initial conditions are zero.

Note:

- A system is said to be linear if principle of superposition and principle of homogeneity applies.
- A differential equation is linear if the coefficients are constants or function only of independent variable.

Open Loop Transfer Function

- Transfer function (T.F.)

$$\text{T.F.} = \frac{C(s)}{R(s)} = G(s)$$

where, $C(s) \rightarrow$ Laplace transform of output
$R(s) \rightarrow$ Laplace transform of input

- Impulse response of a system

$$g(t) = c(t)$$

where, $g(t) \rightarrow$ Impulse response
$c(t) \rightarrow$ Output of the system

Note:

- The transfer function of a system is the Laplace transform of its impulse response.
- Transfer function is always calculated with zero initial conditions.

Closed Loop Transfer Function

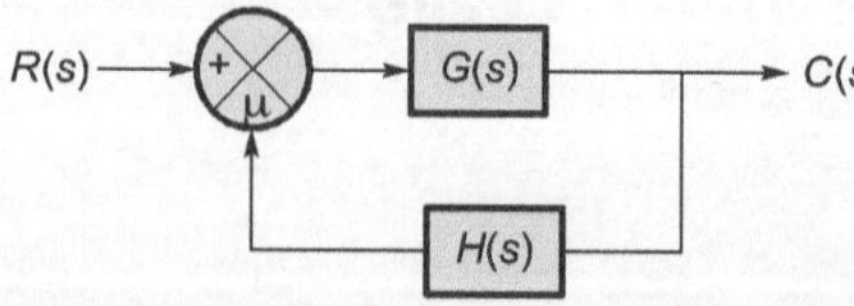

- Transfer function

$$\text{T.F.} = \frac{G(s)}{1 \pm G(s)H(s)}$$

Block Diagram Reduction

Original Block Diagram	Equivalent Block Diagram
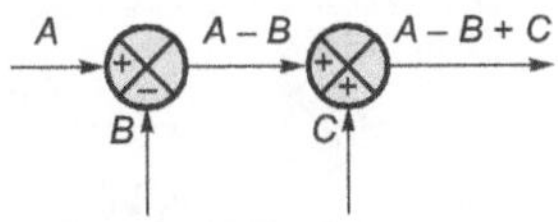	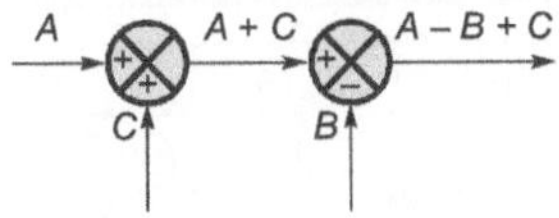
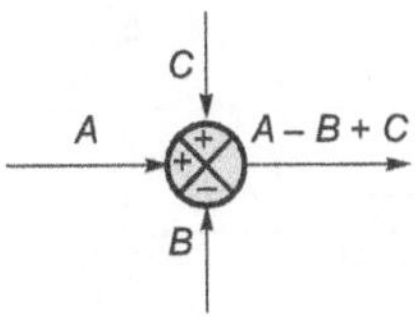	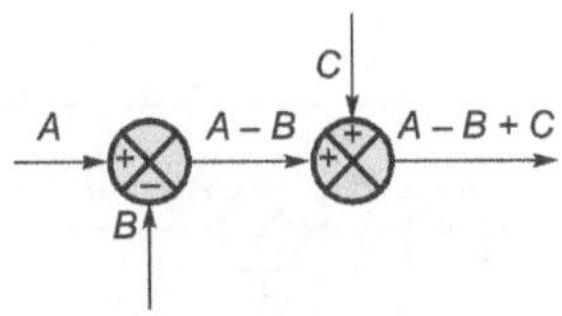
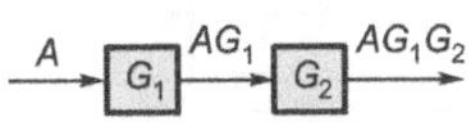	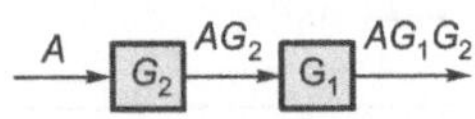
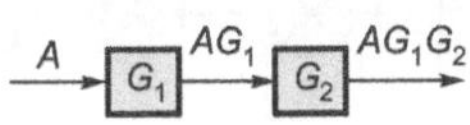	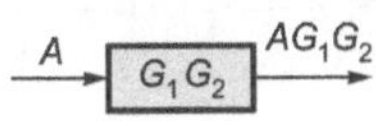
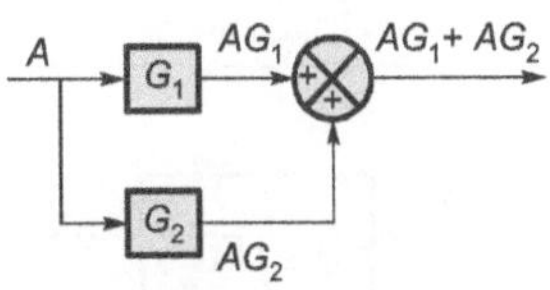	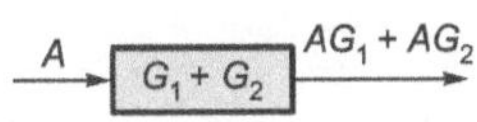
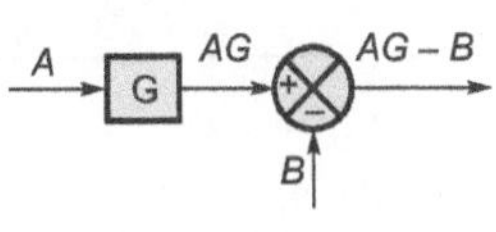	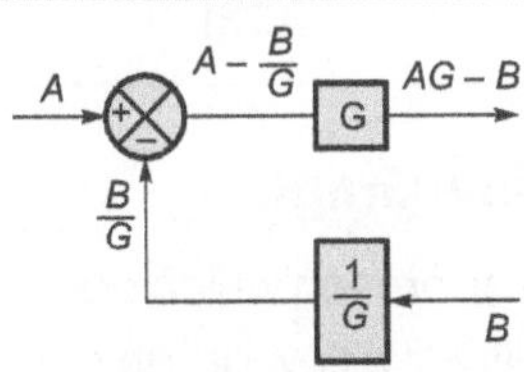
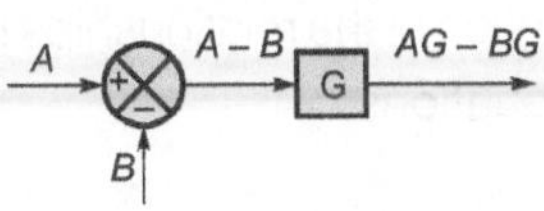	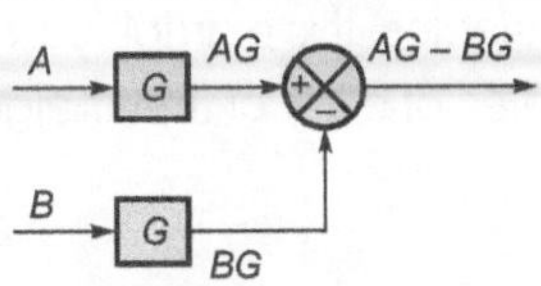

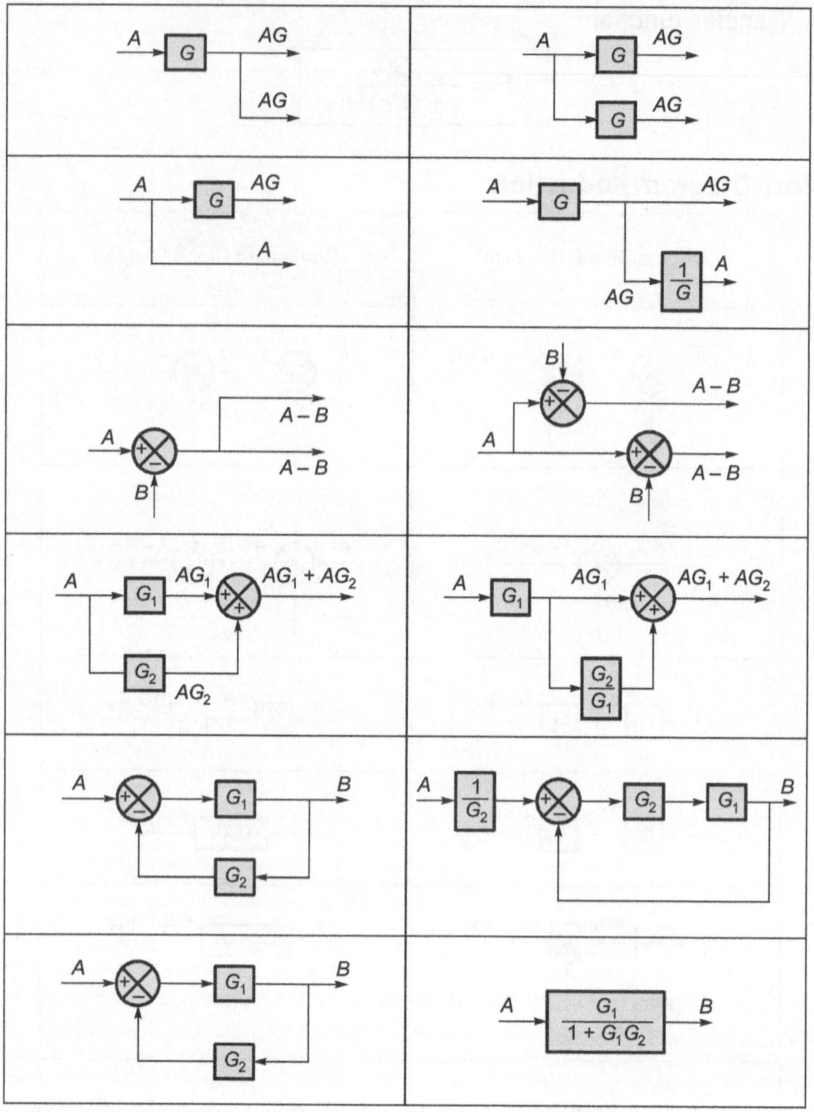

Signal Flow Graph

It is an approach which does not require any reduction process because of availability of a flow graph gain formula which relates the input and output system variables. It is a graphical representation of the relationships between the variables of a set of linear algebraic equations.

Rules for Drawing Signal Flow Graph From a given Block Diagram

1. While drawing SFG from a given block diagram, the adjacent summing points and take-off points (but not a take-off point preceding a summing point in the direction of SFG) are represented by a node and the block transfer function is represented by a line joining the respective nodes. The direction of signal flow is indicated by an arrow on the line.
2. If in the direction of signal flow, a take-off point precedes a summing point then such points are represented by two separate nodes with a transmittance of unity between them.

Original Block Diagram	Equivalent SFG
	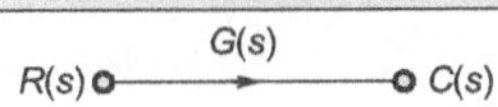
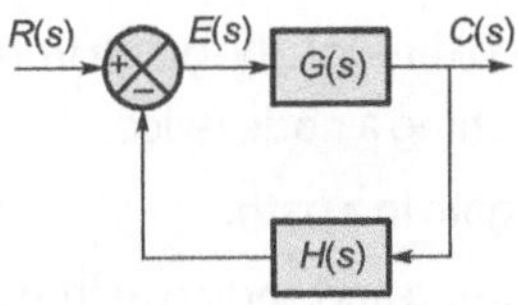	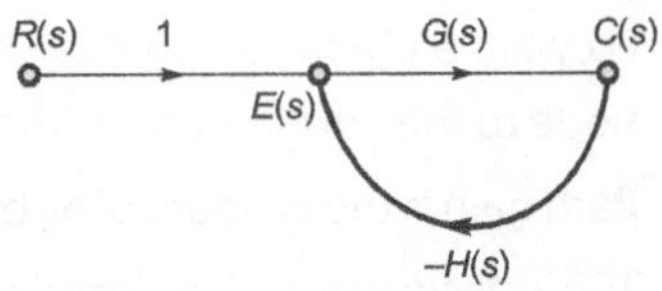
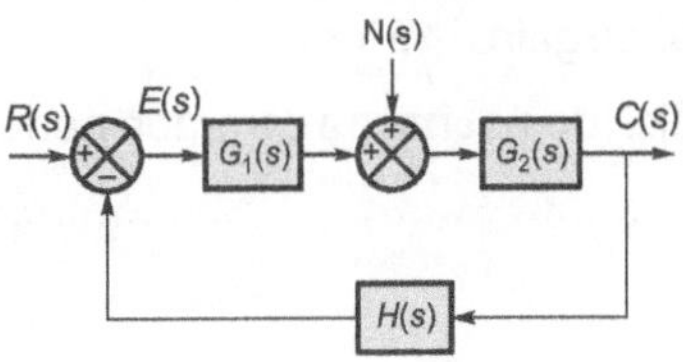	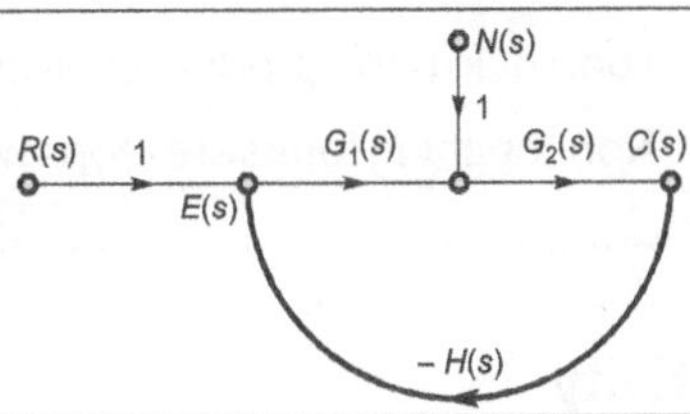
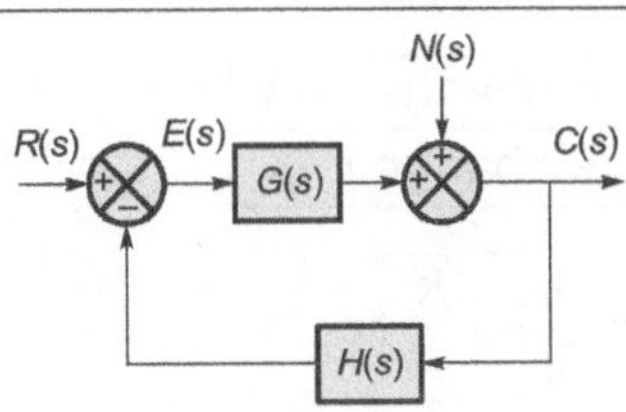	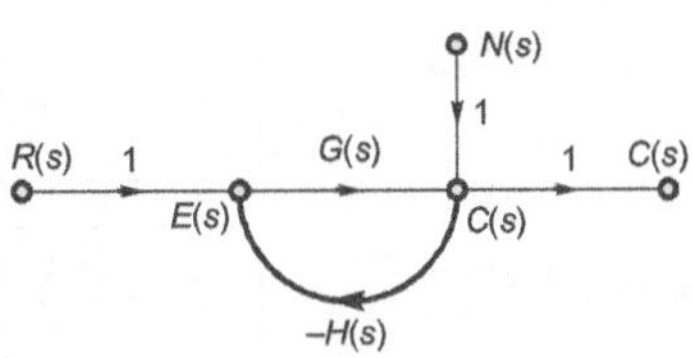
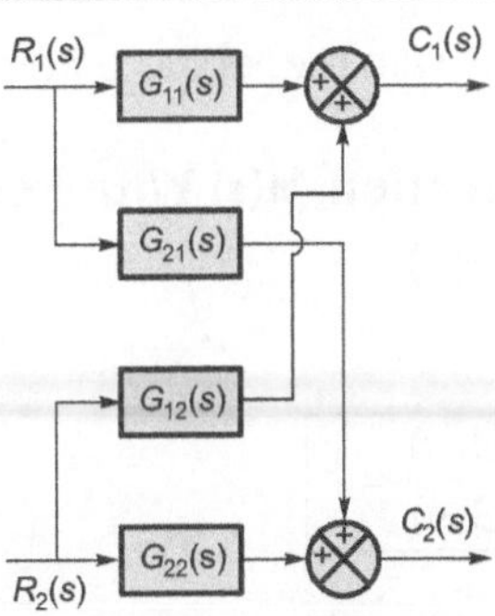	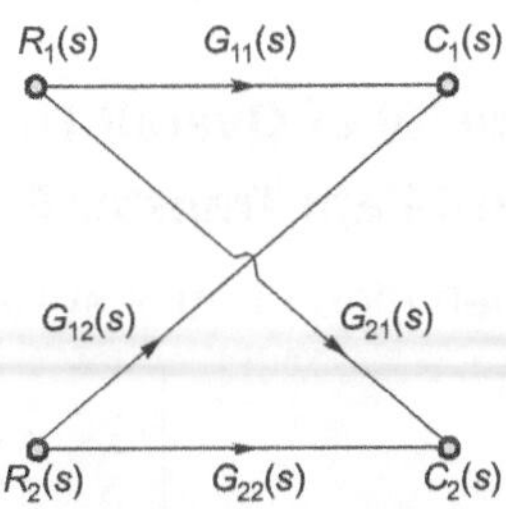

Mason's gain formula

$$P = \frac{1}{\Delta}\sum_k P_k \Delta_k$$

where, $P \rightarrow$ Overall gain or transfer function

$P_k \rightarrow$ Path gain of kth forward path

$\Delta \rightarrow$ 1 – (sum of loop gains of all individual loops) + (sum of gain products of all possible combinations of two non-touching loops) – (sum of gain products of all possible combinations of three non-touching loops) + ...

$\Delta_k \rightarrow$ Value of Δ obtained by removing all the loops touching kth forward path

Note: ....

- Forward path is a path that can be traced through the graph from the input node to the output node without touching a node twice.
- Path gain is the product of all branch gain in a path.
- The individual loop is a closed path starting and ending at the same node.
- Loop gain is the product of all branch gains in a loop.
- Non touching loops are loops which do not have a common node.

....

Sensitivity

$$\text{Sensitivity} = \frac{\%\text{ change in } A}{\%\text{ change in } K}$$

$$S_K^A = \frac{\partial A / A}{\partial K / K} = \frac{\partial A}{\partial K} \times \frac{K}{A}$$

where, $S_K^A \rightarrow$ Sensitivity of variable A with respect to parameter K

1. Sensitivity of Overall Transfer Function M(s) With Respect to Forward Path Transfer Function G(s)

- For open loop control system

$$S_G^M = \frac{G(s)}{M(s)} \cdot \frac{\partial M(s)}{\partial G(s)} = 1$$

- For closed loop control system

$$S_G^M = \frac{1}{1+G(s)\,H(s)}$$

2. Sensitivity of Overall Transfer Function $M(s)$ With Respect to Feedback Path Transfer Function $H(s)$

$$S_H^M = -\frac{G(s)\,H(s)}{1+G(s)\,H(s)}$$

Note: Closed loop system is less sensitive to parameter variation. Hence, closed loop system is better.

■■■■

4

Time Response Analysis of Control System

Introduction

The time response of a control system means as to how a system behaves in accordance with time when a specified input test signal is applied.

The time response of a control system is divided in two parts:

(i) Transient response

(ii) Steady state response

Transient part of time response reveals the nature of response and its speed, where as steady state part of time response reveals the accuracy of a control system.

Standard Input Test Signals

(a) Step function

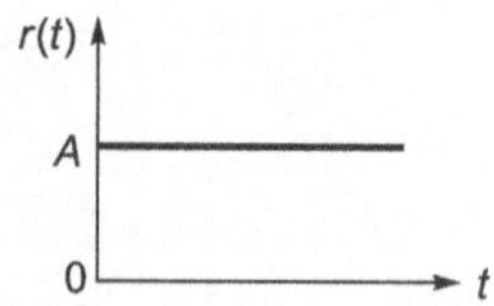

$$r(t) = Au(t)$$

where, $$u(t) = \begin{cases} 1 \; ; & t \geq 0 \\ 0 \; ; & t < 0 \end{cases}$$

(b) Ramp function

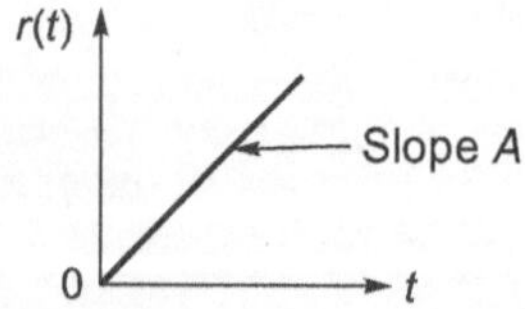

$$r(t) = Atu(t)$$

(c) Parabolic function

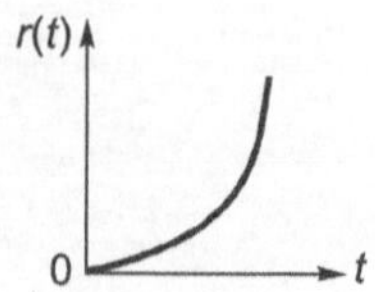

$$r(t) = \frac{At^2}{2} u(t)$$

(d) Impulse Function

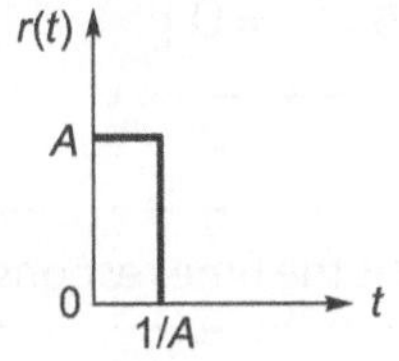

$$\delta(t) = \dot{u}(t)$$

$$\delta(t) = 0;\ t \neq 0$$

$$\lim_{\alpha \to 0} \int_{-\alpha}^{+\alpha} \delta(t)dt = 1$$

Time response of a first order control system

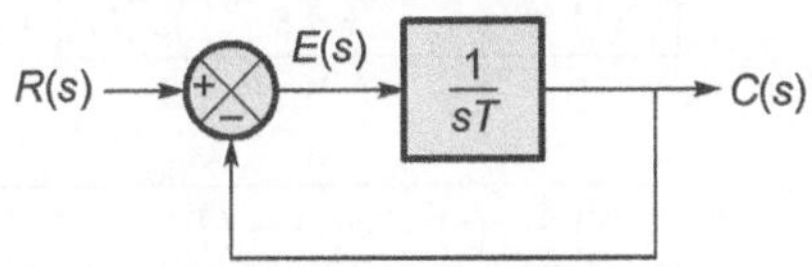

Transfer Function

$$\frac{C(s)}{R(s)} = \frac{1}{sT + 1}$$

(i) Response to Unit Impulse Input Function

Time response expression

$$c(t) = \frac{1}{T} e^{-t/T}\ u(t)$$

(ii) Response to Unit Step Input Function

$$C(s) = \frac{1}{s(sT + 1)}$$

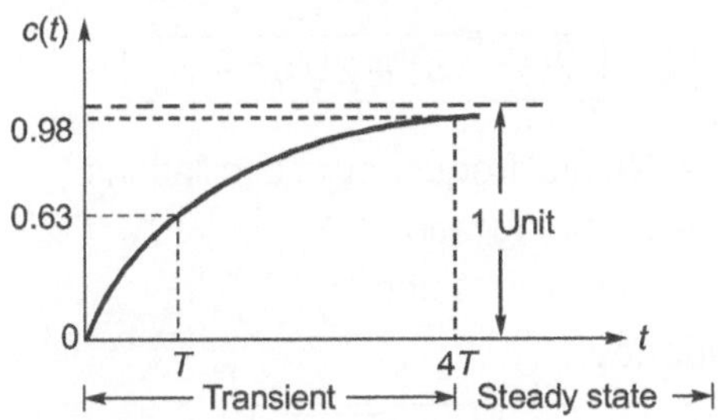

- Time response expression

$$c(t) = \mathcal{L}^{-1}\{C(s)\} = (1 - e^{-t/T})\ u(t)$$

- Error

$$e(t) = r(t) - c(t) = e^{-t/T} u(t)$$

- Steady state error

$$e_{ss} = \lim_{t\to\infty} e^{-t/T} = 0$$

Note: Lower the time constant, faster is the time response of a control system.

(iii) Response to Unit Ramp Input Function

- Time response expression

$$c(t) = (t - T + Te^{-t/T})u(t)$$

- Error

$$e(t) = r(t) - c(t) = (T - Te^{-t/T})u(t)$$

- Steady State Error

$$e_{ss} = \lim_{t\to\infty}(T - Te^{-t/T}) = T$$

Time Response of Second Order Control System

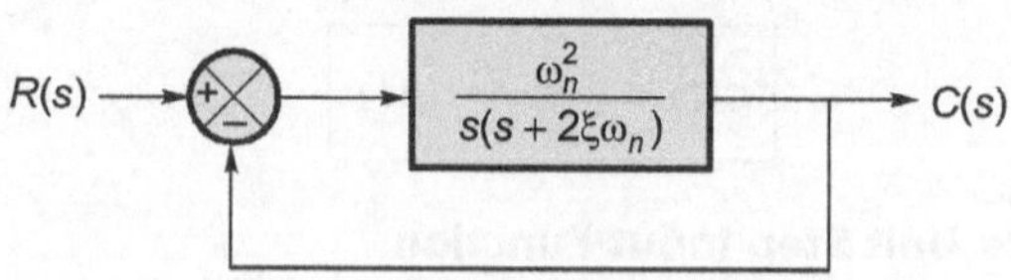

- Transfer function

$$\frac{C(s)}{R(s)} = \frac{\omega_n^2}{s^2 + 2\xi\omega_n s + \omega_n^2}$$

where, $\omega_n \to$ Natural frequency of oscillations

$\xi \to$ Damping ratio

$\xi\omega_n \to$ Damping factor

- Damped frequency

$$\omega_d = \omega_n\sqrt{1-\xi^2}$$

where, $\omega_d \to$ Damped frequency of oscillations

Response to Unit Step Input Function

- Time response expression

Case-1: $\xi < 1$ i.e. Underdamped oscillations

$$c(t) = \left[1 - \frac{e^{-\xi\omega_n t}}{\sqrt{1-\xi^2}}\sin(\omega_d t + \phi)\right]u(t)$$

$$\phi = \tan^{-1}\left(\frac{\sqrt{1-\xi^2}}{\xi}\right)$$

Note:

Response settles within 2% of the desired value (1 unit) after damping out the oscillations in a time 4T (or $4/\xi\omega_n$).

Case-2: $\xi = 0$ i.e. Sustained (undamped) Oscillations

$$c(t) = (1 - \cos\omega_n t)u(t)$$

Case-3: $\xi = 1$ i.e. Critically Damped Oscillations

$$c(t) = [1 - e^{-\omega_n t}(1 + \omega_n t)]$$

Case-4: $\xi > 1$ i.e. Overdamped Oscillations

$$c(t) = 1 - \frac{e^{-\left(\xi - \sqrt{\xi^2 - 1}\right)\omega_n t}}{\left(2\sqrt{\xi^2 - 1}\right)\left(\xi - \sqrt{\xi^2 - 1}\right)}$$

Effect of ξ on Time Response

S.No.	Value of x	Type of closed poles	Nature of response	System classification
1.	$\xi = 0$	Purely imaginary	Oscillations with constant frequency and amplitude	Undamped
2.	$0 < \xi < 1$	Complex conjugate	Damped oscillation	Underdamped
3.	$\xi = 1$	Real, equal and negative	Critical and pure exponential	Critically damped
4.	$1 < \xi < \infty$	Real, unequal and negative	Purely exponential slow and sluggish	Overdamped

Pole Locations for Different Cases, for a Second Order System

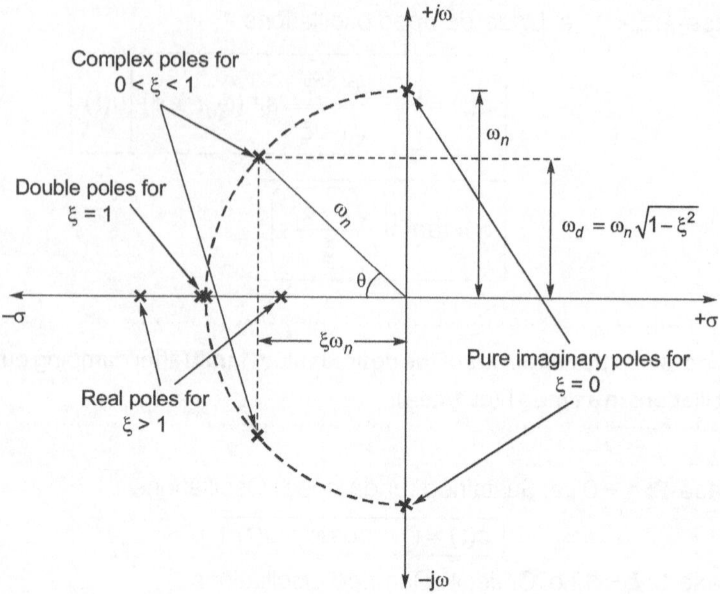

7

Note: ....

- As ξ increases from 0 to 1 then the frequency of oscillation reduces.
- For all positive value of ξ, system are stable.
- Pole away from origin are called insignificant pole and does not affect the stability of system.
- $\theta = \cos^{-1} \xi$.

....

Some Practical Example of Second Order System

Series RLC Circuit	Parallel RLC Circuit	Translatory System	Rotational System
Characteristic equation	Characteristic equation	Characteristic equation	Characteristic equation
$s^2 + \frac{R}{L}s + \frac{1}{LC} = 0$	$s^2 + \frac{1}{RC}s + \frac{1}{LC} = 0$	$s^2 + \frac{f}{M}s + \frac{K}{M} = 0$	$s^2 + \frac{f}{J}s + \frac{K}{J} = 0$
$\omega_n = \frac{1}{\sqrt{LC}}$	$\omega_n = \frac{1}{\sqrt{LC}}$	$\omega_n = \sqrt{\frac{K}{M}}$	$\omega_n = \sqrt{\frac{K}{J}}$
$\xi = \frac{R}{2}\sqrt{\frac{C}{L}}$	$\xi = \frac{1}{2R}\sqrt{\frac{L}{C}}$	$\xi = \frac{f}{2\sqrt{KM}}$	$\xi = \frac{f}{2\sqrt{KJ}}$

Transient Response Specifications of Second Order Control System

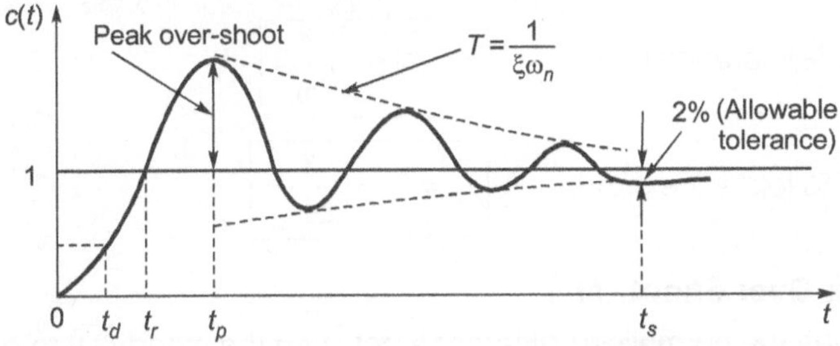

1. Delay Time (t_d)

It is the time required for the response to rise from 0 to 50% of the final value.

$$t_d = \frac{1+0.7\xi}{\omega_n}$$

2. Rise Time (t_r)

It is the time required for the response to rise from

0 to 100% (underdamped)

5 to 95% (critical damped)

10 to 90% (over damped)

of the final value

$$t_r = \frac{\pi - \tan^{-1}\left(\frac{\sqrt{1-\xi^2}}{\xi}\right)}{\omega_d} = \frac{\pi - \cos^{-1}\xi}{\omega_d}$$

3. Peak Time (t_p)

It is the time required for the response to rise from zero to peak of the time response.

$$t_p = \frac{n\pi}{\omega_d}$$

where, $n \rightarrow 1, 2, 3, \ldots$

for $n \rightarrow 1$, first overshoot

$n \rightarrow 2$, first undershoot

4. Settling Time (t_s)

It is the time required for response to rise and reach to the tolerance band

For 2% tolerance band,
$$t_s = 4\tau = \frac{4}{\xi\omega_n}$$

For 5% tolerance band,
$$t_s = 3\tau = \frac{3}{\xi\omega_n}$$

5. Peak Over Shoot (M_p)

It gives the normalised difference between the steady state output to first peak of the time response.

$$\%M_p = e^{-\left(\frac{\pi\xi}{\sqrt{1-\xi^2}}\right)} \times 100\%$$

Note: ....

- The time period of the oscillation before reaching the steady state.

$$T_{\text{oscillation}} = \frac{2\pi}{\omega_d}$$

- Number of oscillation before reaching steady state is

$$N = \frac{T_{\text{settling}}}{T_{\text{oscillation}}}$$

....

Type and Order of System

- Every Transfer function representing the control system has certain type and order.
- The steady state analysis depends on type of the system.
- The type of the system is obtained from open loop transfer function i.e. $G(s)\,H(s)$.
- The number of open loop poles occuring at origin determines the type of the system.

Let
$$G(s)\,H(s) = \frac{K(1+sT_a)}{s^p(1+sT_1)}$$

For
$p = 0$ for Type-0 system

$p = 1$ for Type-1 system

$p = n$ for Type-n system

- Transient state analysis depends on order of the system.
- The order of system is obtained from closed loop transfer function i.e. $\frac{G(s)}{1+G(s)H(s)}$.
- The highest power of characteristic equation, i.e. $1 + G(s)H(s) = 0$ determines the order of control system.

Steady State Error

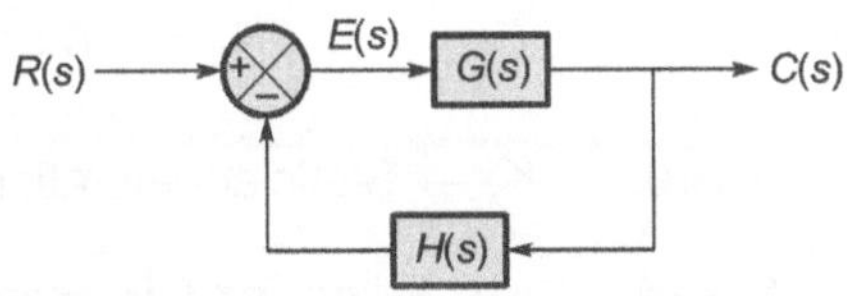

$$E(s) = \frac{R(s)}{1+G(s)H(s)}$$

Steady state error, $e_{ss} = \lim_{t\to\infty} e(t)$ **(in time-domain)**

$$e_{ss} = \lim_{s\to 0} sE(s) \quad \textbf{(in } s\textbf{-domain)}$$

$$e_{ss} = \lim_{s\to 0} s\frac{R(s)}{1+G(s)H(s)}$$

Note: ....

Steady state error depends on two factor:

(i) Type of input applied i.e. $R(s)$.

(ii) Type of system i.e. $G(s)H(s)$.

....

1. Steady State Error for Different Types of Input

(i) Step input

$$R(s) = \frac{A}{s}$$

$$e_{ss} = \frac{A}{1+K_p} \quad ; \quad K_p = \lim_{s\to 0} G(s)H(s)$$

where, $K_p \to$ Static position error constant

(ii) Ramp input

$$R(s) = \frac{A}{s^2}$$

$$e_{ss} = \frac{A}{K_v} \quad ; \quad K_v = \lim_{s\to 0} sG(s)H(s)$$

where, $K_v \to$ Static velocity error constant

(iii) Parabolic input

$$R(s) = \frac{A}{s^3}$$

$$\boxed{e_{ss} = \frac{A}{K_a}}$$

$$\boxed{K_a = \lim_{s \to 0} s^2 G(s) H(s)}$$

where, $K_a \rightarrow$ Static acceleration error constant

2. Steady State Error for Different Types of System

Type	Step input	Ramp input	Parabolic input
Type-0	$\frac{A}{1+K}$	∞	∞
Type-1	0	$\frac{A}{K}$	∞
Type-2	0	0	$\frac{A}{K}$

Where, K is the system gain.

Observation

(i) $e_{ss} \propto \frac{1}{K}$, so, as system gain increases, steady state error decreases.

(ii) For Linear Time Invariant (LTI) system, the maximum type number is two. Beyond type-2, the system exhibits non-linear characteristics.

Note:

- Steady state error is valid only for closed loop stable system.
- Steady state error are calculated for the closed loop system by using open loop transfer function.

■■■■

Stability in Time-Domain

5

Stability

A Linear Time Invariant (LTI) system is said to be stable if the following conditions are satisfied:

(i) If the system is excited by a bounded input, the output must be bounded.

(ii) If input to the system is zero, the output must be zero, irrespective of all the initial condition.

Marginal Stable System

A LTI system is said to be marginal stable if for the bounded input, the output maintains the constant amplitude and frequency.

Absolute Stable System

System is stable for all the value of system gain (K) from 0 to ∞.

Conditional Stable System

System is stable for a certain range of K.

Routh Hurwitz (R-H) Criteriaon

Purpose

To find:

1. Closed loop system stability.
2. Number of closed loop pole in right side or left side of the s-plane.
3. Range of K for conditionally stable system.
4. Value of K required for marginal stability and thereby, determine the frequency of oscillations.

Note:

To find a closed loop system stability by using *R-H* criteria we need characteristic equation.

Routh Array

- General n^{th} order characteristics equation is
  $a_0s^n + a_1s^{n-1} + a_2s^{n-2} + \cdots + a_{n-1}s^1 + a_n = 0$
- Routh array

$$\begin{array}{c|ccc} s^n & a_0 & a_2 & a_4 \ \cdots \\ s^{n-1} & a_1 & a_3 & a_5 \ \cdots \\ s^{n-2} & \dfrac{a_1a_2 - a_0a_3}{a_1} & \dfrac{a_1a_4 - a_0a_5}{a_1} & \cdots \\ \vdots & \vdots & & \\ s^0 & a_n & & \end{array}$$

1. For closed loop system to be stable, all the coefficient in the first column must be positive and there should not be any missing coefficient in the first column.
2. If there is any sign change in the first column of the Routh array then system is unstable.
3. Number of sign changes is equal to number of roots lying in the right side of the s-plane.

Note: ....

- **Magnitude and sign of every term in each row and column depend on first column element, therefore we always see first column term.**
- **The presence of a zero in the first column of the Routh's tabulation leads to following conclusions**
  - **(i) Equal real roots with opposite signs.**
  - **(ii) Pair of conjugate roots on imaginary axis.**

....

Difficulties and Limitations of Routh Array

1. The routh array is applicable to LTI system only.
2. It determines poles in LHS or RHS of s-plane but not their exact location.
3. Whenever any one coefficient is zero in first column then replace zero by smallest positive constant ($\in$) and continue the Routh array, finally substitute $\in = 0$ and check for the sign change.
4. When routh array ends abruptly, construct an auxiliary equation and differentiate it to get new coefficient to complete the routh array.

5. When the system is marginally stable, find the frequency of sustained oscillations from auxiliary equation. The auxiliary equation should be an even polynomial of order two only.

Root Locus

Root locus is defined as the locus of closed loop poles obtained when system gain K is varied from 0 to ∞.

Angle Condition

Angle condition is used for checking whether certain points lie on root locus or not and hence, the validity of root locus for closed loop poles.

For a point to lie on root locus, the angle evaluated at that point must be an odd multiple of $\pm 180°$ i.e. $\pm(2q + 1)180°$.

Magnitude Condition

The magnitude condition is used for finding the system gain K at any point on root locus.

$$|G(s)H(s)| = 1$$

Construction Rules of Root Locus

1. Root locus is symmetrical about real axis.

2. Let P = Number of open loop poles

Z = Number of open loop zeros

If $P > Z$

then (a) The number of branches of root locus = P

(b) The number of branches terminating at zero = Z

(c) The number of branches terminating at infinity = $P - Z$

3. A point on real axis is said to be on root locus if, to the right side of this point, the sum of number of open loop poles and zeros is an odd number.

4. Angle of Asymptotes

- Number of asymptotes = $P - Z$
- Angle of asymptotes = $\dfrac{(2q + 1)180°}{P - Z}$ (where, $q = 0, 1, 2, \cdots$)

5. Centroid

Centroid is the point of intersection of asymptote on the real axis.

$$\text{Centroid} = \frac{\Sigma\text{Real part of open loop poles} - \Sigma\text{Real part of open loop zeros}}{P - Z}$$

6. Break away point

They are the points where multiple roots of characteristic equation occurs.

Procedure to find out break away point:

(a) Construct characteristic equation i.e. $1 + G(s)\,H(s) = 0$.

(b) Write K in terms of s.

(c) Find $\frac{dK}{ds}$.

(d) The root of $\frac{dK}{ds} = 0$ give break away points.

7. Intersection of root locus with imaginary axis

Roots of auxiliary equation at $K = K_{(\text{marginal})}$ from routh array gives intersection of root locus with imaginary axis.

8. Angle of departure and arrival

The angle of departure is tangent to the root locus at the complex poles.

$$\boxed{\phi_D = 180° + \phi}$$

The angle of arrival is tangent to the root locus at the complex zero.

$$\boxed{\phi_A = 180° - \phi}$$

$$\boxed{\phi = \phi_z - \phi_p}$$

where, $\phi_z \rightarrow$ Sum of all the angles subtended by remaining zeros

$\phi_p \rightarrow$ Sum of all the angle subtended by poles

Note: ....

- **Whenever the system transfer function consist of the poles at origin then the root locus is nothing but angle of asymptotes.**
- **Value of system gain K on root locus can be found out by**

$$K = \frac{\Pi(\text{Phasor length from open loop poles})}{\Pi(\text{Phasor length from open loop zeros})}$$

....

Effect of Addition of Pole

1. Operating range of K decreases for system to be stable.
2. Relative stability reduces.
3. System becomes more oscillatory.
4. The break away point shift towards the imaginary axis.
5. Damping factor decreases and system becomes unstable.
6. Settling time (T_s) and bandwidth increases.
7. Rise time (T_r) decreases.

Effect of Addition of Zero

1. System become stable.
2. Relative stability improves.
3. Range of K for stability increases.
4. System becomes less oscillatory.
5. Break away point shift towards the left side of the s-plane.
6. Damping factor increases.
7. Bandwidth decreases.

■■■■

Industrial Controller

6

Controller

It is a device that is used usually in series with a plant to drive the plant towards decreased output.

Types of Industrial Controller

1. Proportional Controller

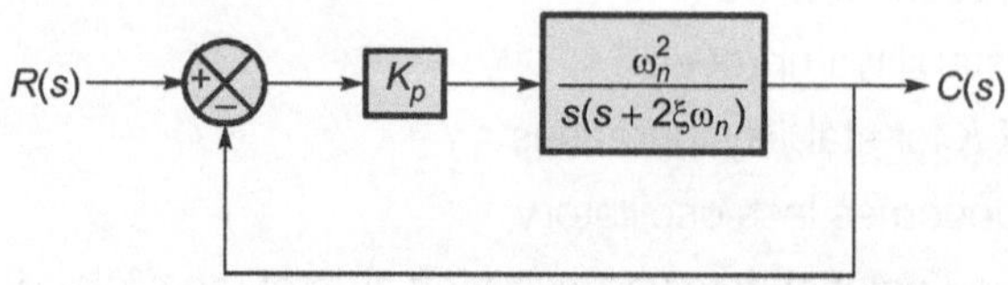

$$\boxed{\omega_n' = \omega_n\sqrt{K_p}} \quad \text{and} \quad \boxed{\xi' = \frac{\xi}{\sqrt{K_p}}}$$

where, $K_p \rightarrow$ Proportional gain

Effect

(i) Natural frequency of oscillation (ω_n) increases by $\sqrt{K_p}$.

(ii) Damping ratio (ξ) decreases by $\sqrt{K_p}$.

(iii) Peak overshoot (M_p) increases.

(iv) Steady state error reduces.

2. Derivative Controller

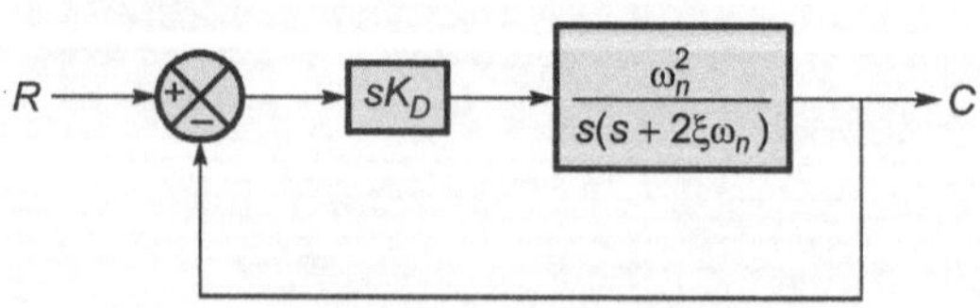

where, $K_D \rightarrow$ Rate constant

Effect

(i) Type and order of the system reduces by one.

(ii) Oscillations dies out, hence transient response improves.

(iii) Not used in isolation.

3. Integral Controller

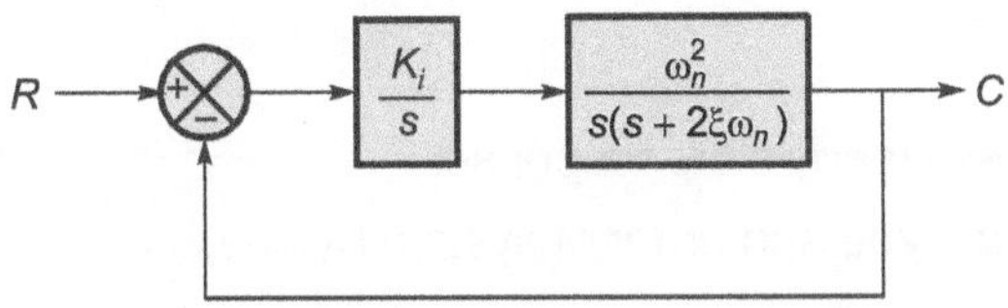

where, $K_i \rightarrow$ Integral scaling

Integral controller is a memory based controller.

Effect:

(i) It increases type and order by one.

(ii) Makes the system less stable.

(iii) Steady state error reduces.

(iv) It improves the steady state response.

4. P-D Controller

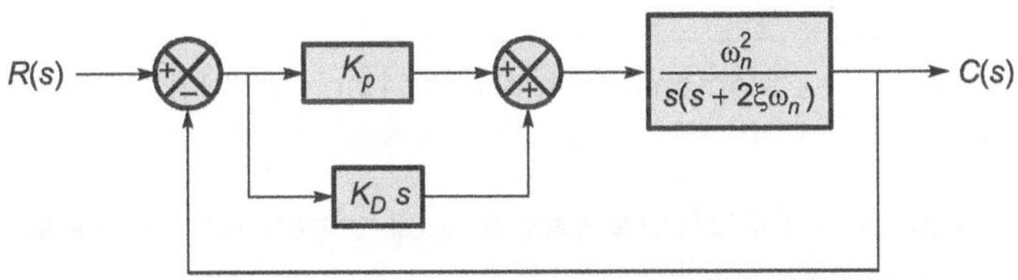

Effect:

(i) Transient response improves.

(ii) Damping ratio increases.

(iii) Peak overshoot decreases.

(iv) Bandwidth increases.

(v) Noise level increases.

(vi) Improves gain margin, phase margin and resonant peak.

5. P-I Controller

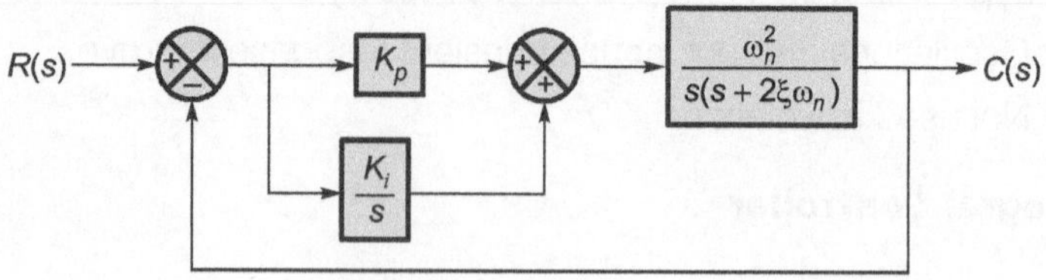

Effect

(i) Improves steady state response.

(ii) Improves type and order of system by one.

(iii) Noise level reduces.

(iv) Increases error constant.

(v) Bandwidth reduces.

6. PID Controller

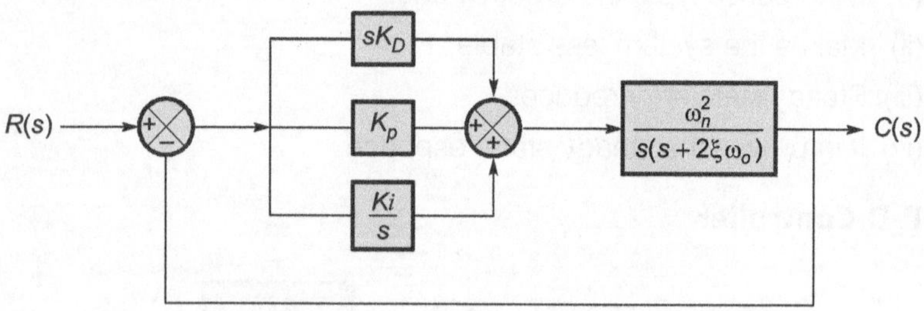

It is similar to lead-lag compensator and band reject filter.

Effect

(i) It improves both steady state as well as transient response.

(ii) It reduces rise-time.

(iii) It increases bandwidth.

(iv) It increases stability.

(v) If eliminate steady state error between input and output.

(vi) It increases type and order of system by one.

■■■■

Compensator

7

Introduction

Compensator in control system are used for improving the performance specifications i.e. the transient and steady state response characteristics.

Lead Compensator

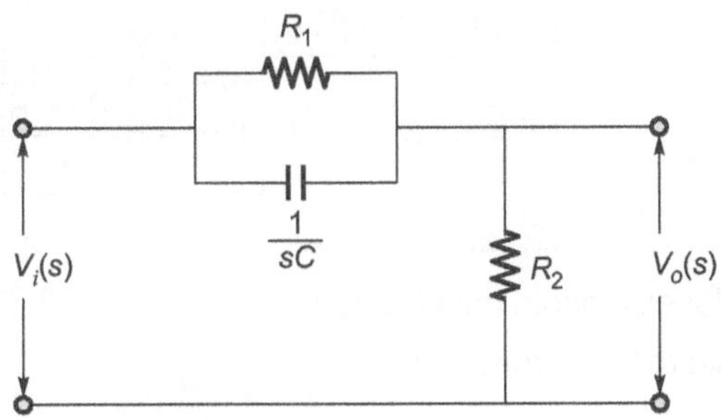

$$\frac{V_o(s)}{V_i(s)} = \frac{\alpha(1+Ts)}{(1+\alpha Ts)}$$

where, $T = R_1 C$ and $\alpha = \frac{R_2}{R_1 + R_2}$ $(\alpha < 1)$

Note:

- Zero is closer to origin than the pole.
- It is similar to the PD controller.

Effect of Lead Compensator

1. It improves the transient response.
2. It increases the margin of stability.
3. It increases the bandwidth.
4. $\left(\frac{\text{Signal}}{\text{Noise}}\right)_{\text{output}} < \left(\frac{\text{Signal}}{\text{Noise}}\right)_{\text{input}}$
5. It helps to increase error constant upto some extent.
6. It allows to pass high frequencies and attenuates low frequencies.
7. Increase the phase shift.

Lag Compensator

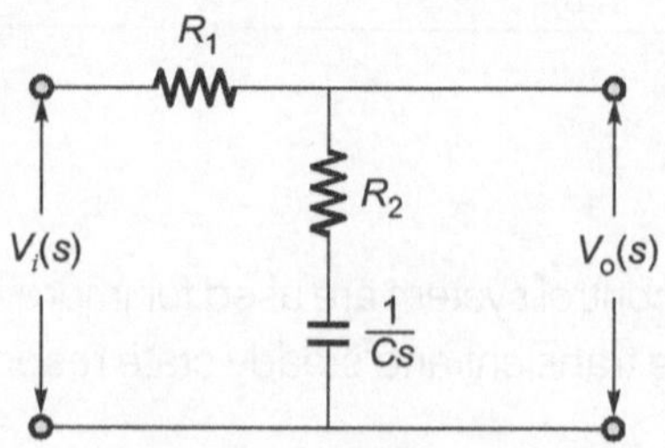

$$\frac{V_o(s)}{V_i(s)} = \frac{1+Ts}{1+\beta Ts}$$

where, $T = R_2 C$ and $\beta = \frac{1}{\alpha} = \frac{R_1 + R_2}{R_2}$ $(\beta > 1)$

Note:

- Pole is closer to origin than the zero.
- It is similar to the PI controller.

Effect of Lag Compensator

1. It improves the steady state response.
2. It increases the error constant.
3. It decreases the bandwidth.
4. It reduces the effect of noise.
5. It reduces the stability margin.
6. It does not affect the transient response.
7. System become lesser stable.
8. It allows to pass low frequencies and attenuates the high frequencies.
9. Decreases the phase shift.

Lead-Lag/Lag-Lead Compensator

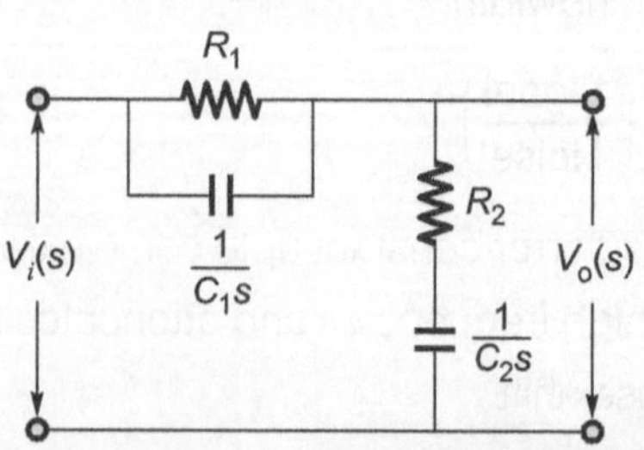

$$\boxed{\frac{V_o(s)}{V_i(s)} = \frac{\alpha(1+T_1 s)(1+T_2 s)}{(1+\alpha T_1 s)(1+\beta T_2 s)}}$$

where, $\boxed{T_1 = R_1 C_1}$ and $\boxed{T_2 = R_2 C_2}$

$\boxed{\alpha = \dfrac{R_2}{R_1 + R_2}}$ and $\boxed{\beta = \dfrac{R_1 + R_2}{R_2}}$

Effect of lead-lag/lag-lead compensator

Lead-Lag network improves both steady state and transient response of the system.

Note: ....

- Compensator in the form $G(s) = \dfrac{s+a}{s+b}$ will work as lead compensator if $a < b$, and as lag compensator if $a > b$.
- Frequency at which maximum phase occurs is $\boxed{\omega = \sqrt{ab}}$ rad/s.
- Maximum phase (ϕ_m) is given as

$$\boxed{\sin\phi_m = \frac{1-\alpha}{1+\alpha}}$$

....

■■■■

Frequency Response Analysis

8

Frequency Response Analysis for Second Order System

Transfer Function

$$\frac{C(s)}{R(s)} = \frac{\omega_n^2}{s^2 + 2\xi\omega_n s + \omega_n^2}$$

Resonant Frequency

It is defined as, the frequency at which the magnitude has maximum value.

$$\omega_r = \omega_n\sqrt{1 - 2\xi^2}$$

Resonant Peak

It is maximum value of magnitude occuring at resonant frequency ω_r.

$$|M_r| = \frac{1}{2\xi\sqrt{1 - \xi^2}}$$

Phase Angle at Resonant Frequency

$$\phi_r = -\tan^{-1}\left[\frac{\sqrt{(1 - 2\xi^2)}}{\xi}\right]$$

Remember:

- As ξ approaches zero, ω_r approaches ω_n and M_r approaches to infinity.
- For $0 < \xi \leq 1/\sqrt{2}$, the resonant frequency always has a value less than ω_n and the resonant peak has a value greater than 1.

Cut-off Frequency and Bandwidth

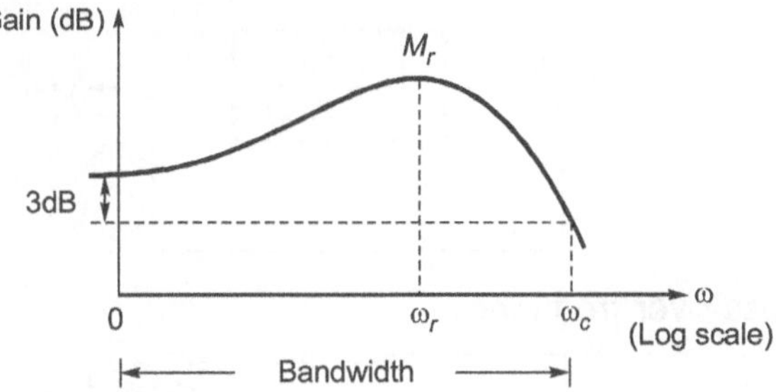

- Cut-off frequency

$$\omega_c = \omega_n\left[1 - 2\xi^2 + \sqrt{4\xi^4 - 4\xi^2 + 2}\right]^{1/2}$$

- Bandwidth

$$\text{B.W.} = \omega_c$$

Stability from Frequency Response Plot

- Phase Margin (P.M.)

$$\text{P.M.} = 180° + \angle G(j\omega)\, H(j\omega)$$

....at gain cross over frequency

- Gain Margin (G.M.)

$$G.M. = \frac{1}{|G(j\omega)\, H(j\omega)|_{\omega=\omega_{pc}}} = \frac{1}{X} \quad \cdots \text{ at phase crossover frequency}$$

$$\text{G.M. } (dB) = 20\log\left(\frac{1}{X}\right)$$

Remember: ....

- **For stable systems**
  (i) Gain cross over frequency < phase cross-over frequency
  (ii) G.M. and P.M. both are positive
- **For unstable systems**
  (i) Gain cross over frequency > phase cross-over frequency
  (ii) G.M. and P.M. both are negative
- **For marginally stable systems**
  (i) Gain cross over frequency = Phase cross-over frequency
  (ii) G.M. and P.M. both are zero

....

G.M. and P.M. for Second Order System

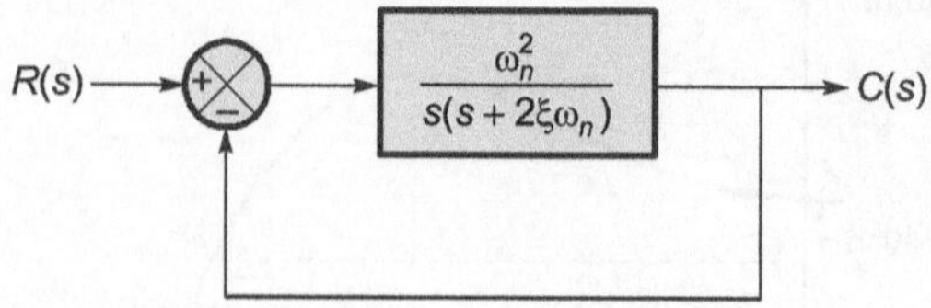

- Gain cross-over frequency

$$\omega_{gc} = \omega_n\sqrt{-2\xi^2 + \sqrt{4\xi^4 + 1}}$$

- Phase margin

$$\text{P.M.} = \tan^{-1}\left[\frac{2\xi}{\sqrt{-2\xi^2 + \sqrt{4\xi^4 + 1}}}\right] \approx 100\xi$$

Gain margin of first and second order system is ∞.

Polar Plot

It is a plot of absolute values of magnitude and phase angle in degree of open loop transfer function $G(j\omega)\ H(j\omega)$ versus ω drawn on polar coordinates.

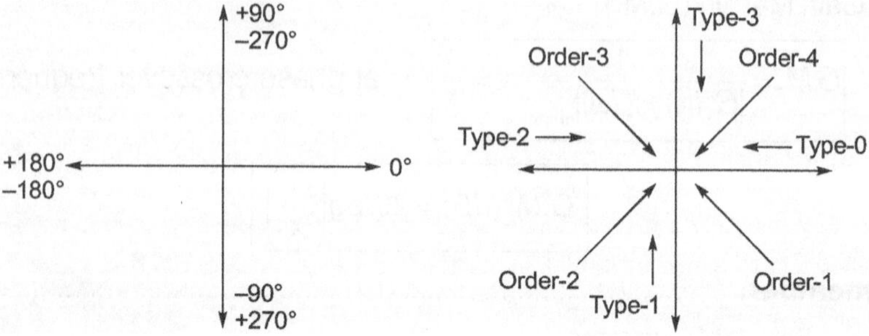

General Shapes of Polar Plot

1. Type-0/Order-1

$$G(s) = \frac{1}{1+Ts}$$

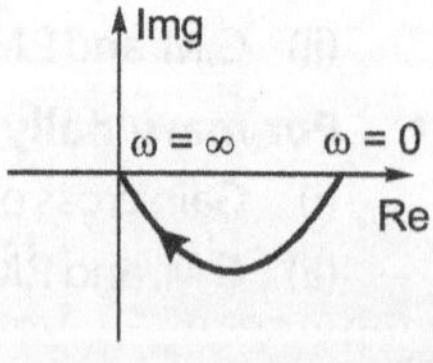

2. Type-0/Order-2

$$G(s) = \frac{1}{(1+T_1 s)(1+T_2 s)}$$

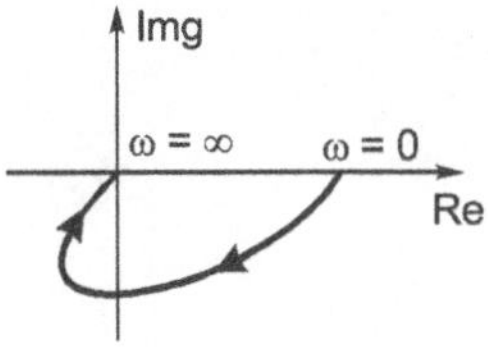

3. Type-0/Order-3

$$G(s) = \frac{1}{(1+T_1 s)(1+T_2 s)(1+T_3 s)}$$

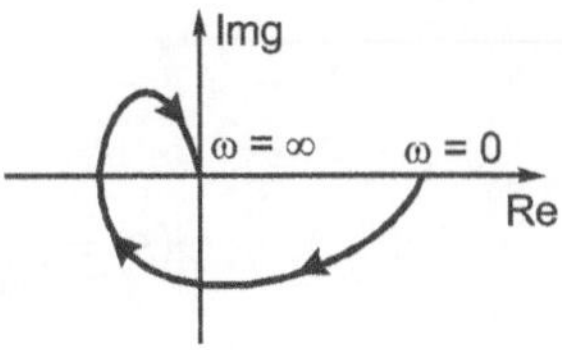

4. Type-0/Order-4

$$G(s) = \frac{1}{(1+T_1 s)(1+T_2 s)(1+T_3 s)(1+T_4 s)}$$

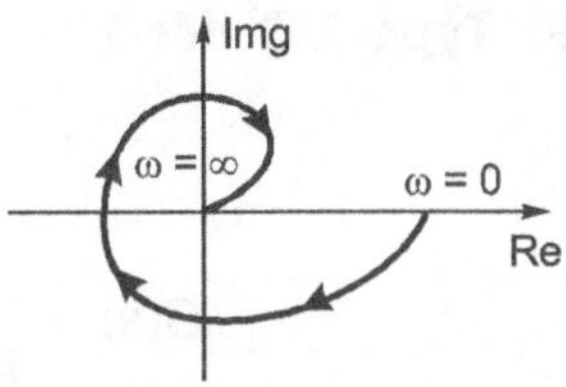

5. Type-1/Order-1

$$G(s) = \frac{1}{s}$$

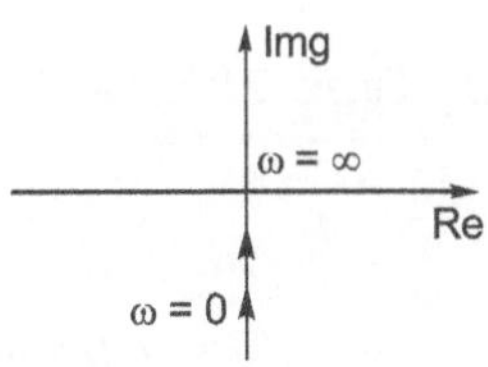

6. Type-1/Order-2

$$G(s) = \frac{1}{s(1+Ts)}$$

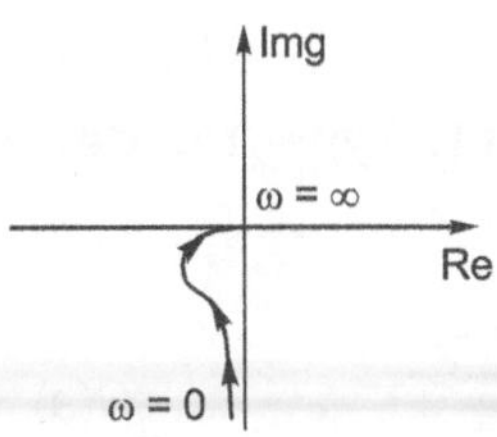

7. Type-1/Order-3

$$G(s) = \frac{1}{s(1+T_1 s)(1+T_2 s)}$$

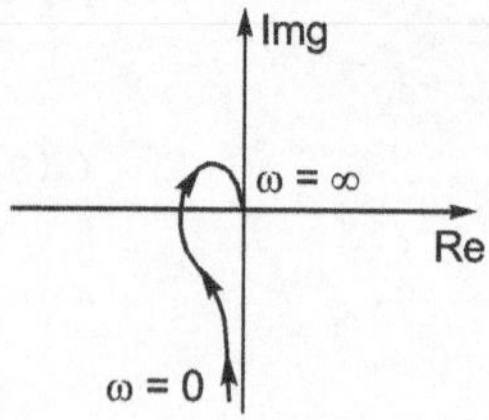

8. Type-1/Order-4

$$G(s) = \frac{1}{s(1+T_1 s)(1+T_2 s)(1+T_3 s)}$$

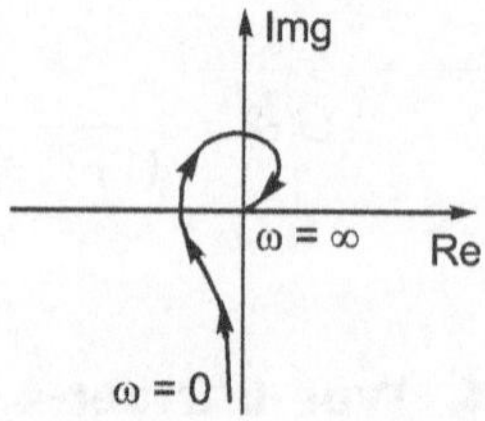

9. Type-2/Order-2

$$G(s) = \frac{1}{s^2}$$

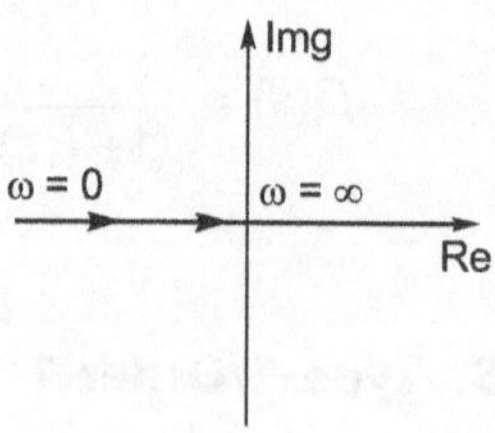

10. Type-2/Order-3

$$G(s) = \frac{1}{s^2(1+Ts)}$$

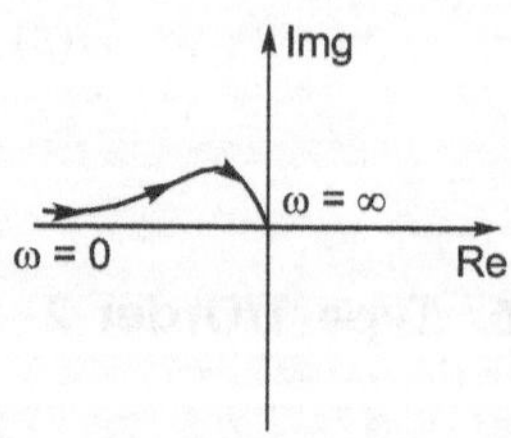

11. Type-2/Order-4

$$G(s) = \frac{1}{s^2(1+T_1 s)(1+T_2 s)}$$

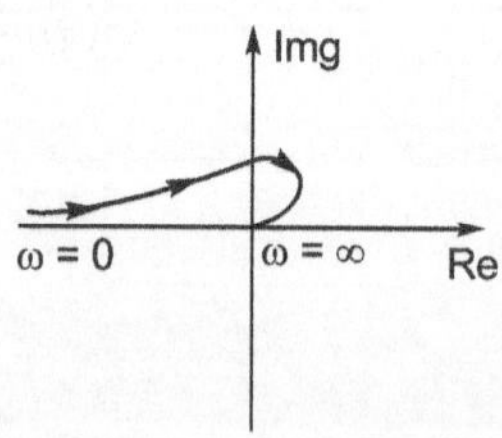

12. Type-3/Order-3

$$G(s) = \frac{1}{s^3}$$

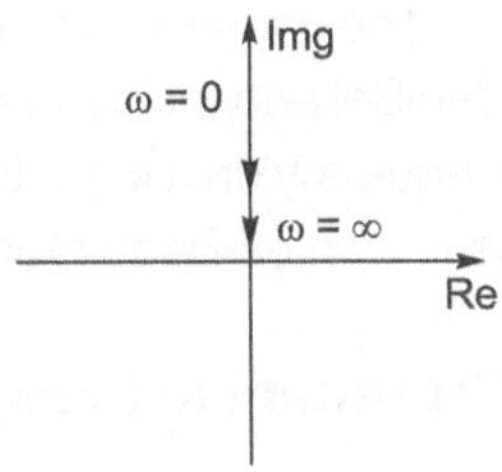

13. Type-3/Order-4

$$G(s) = \frac{1}{s^3(1+Ts)}$$

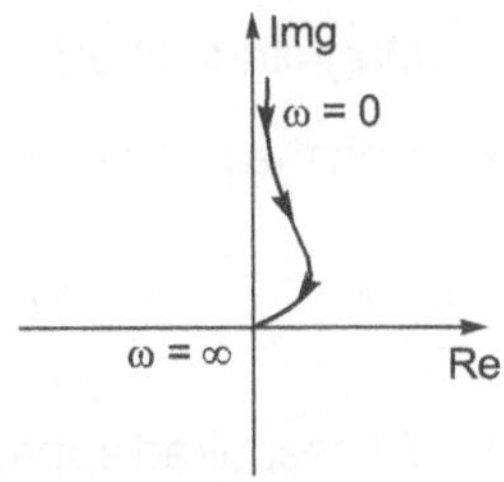

Note: For standard transfer function of type-2 and type-3 systems, the polar plot intersects negative real axis as many times as there are zeros in open loop transfer function.

Nyquist Stability Criteria

Nyquist stability criteria states that the number of encirclements about the critical point (–1 + j0) should be equal to poles of open loop transfer function $G(s)\ H(s)$, which are in the right half of s-plane.

i.e. $\boxed{N = P - Z}$

For stability (i) $Z = 0$

and (ii) $N = P$

where, P = Number of open loop poles in RHS of s-plane

Z = Number of closed loop poles in RHS of s-plane

N = Number of encirclements about of the point (–1 + j0) by $G(s)H(s)$ plot, the positive direction of encirclements being anti-clockwise.

Bode Plot

The variation of magnitude of sinusoidal transfer function expressed in deciBel and the corresponding phase angle in degrees being plotted w.r.t. frequency on a logarithmic scale in rectangular axis, the plot thus obained is known as Bode plot.

Procedure to Draw Bode Plot

- s is replaced by jω to convert into frequency domain.
- Find magnitude in terms of ω and write in terms of dB.

  Magnitude in dB $\boxed{M_{dB} = 20\log|G(j\omega)\,H(j\omega)|}$
- Find the phase angle ∠ϕ

$$\boxed{\angle\phi = \tan^{-1}\left(\frac{\text{imaginary part}}{\text{real part}}\right)}$$

- With required approximation and by varying the frequency from minimum to maximum, draw the magnitude and phase plot.
- If $G(s)\,H(s) = K$

  then $\boxed{|M_{dB}| = 20\log K}$

  System gain K shift the magnitude plot either in upward or downward by "20 logK".
- **Slope**

$$\boxed{\text{Slope} = \frac{d\,|M|_{dB}}{d\log\omega}}$$

Note:

The magnitude plot should start at the frequency of 0.1 with opposite sign of the slope and it should pass through 0 dB line and intersect at ω = 1, when $K = 1$.

- ***n* poles at origin gives**

$$\text{Slope} = -20n \text{ dB/dec}$$
$$\angle\phi = -90n^\circ$$

- ***n* zeros at origin gives**

$$\text{Slope} = +20n \text{ dB/dec}$$
$$\angle\phi = +90n^\circ$$

Remember:

The initial slope of the magnitude plot is given by the poles or zero located at origin.

Corner frequency

The frequency at which slope changes from one level to another level.

- If $G(s)H(s) = \dfrac{1}{(1 + sT)^n}$

Frequency	Slope	Phase (ϕ)
Below corner frequency	0	0
Above corner frequency	$-20n$ dB/dec	$-90n°$

- If $G(s)H(s) = (1 + sT)^n$

Frequency	Slope	Phase (ϕ)
Below corner frequency	0	0
Above corner frequency	$20n$ dB/dec	$90n°$

Remember:

- The error is nothing but difference between actual value and asymptotic value.
- Error is maximum at corner frequency.
- Error is even function w.r.t. corner frequency.
- 6 dB/oct = 20 dB/dec.

Classification of system

1. Minimum Phase System (MPS)

A system in which all finite poles and finite zeros are located in left half of s-plane then it is called as minimum phase system.

2. Non Minimum Phase System (NMPS)

A system in which one or more zeros lies in the right half of s-plane, remaining all the poles and zeros lies on the left half of s-plane.

3. All Pass System (APS)

A system in which zeros lies on the right half of s-plane, poles lies on the left half of *s*-plane and which are symmetrical about imaginary axis. For all pass system magnitude should be 1 and phase angle is –180°.

Note: ....

- $\text{Gain}_{NMPS} = \text{Gain}_{MPS} \times \text{Gain}_{APS}$
- $\angle\phi_{NMPS} = \angle\phi_{MPS} + \angle\phi_{APS}$

....

Constant Magnitude Loci (M-circles)

M-circles represents magnitude of closed loop frequency response plotted against frequency.

- Radius of *M*-circles

$$\text{Radius} = \frac{M}{1-M^2}$$

- Centre of M-circles

$$\text{Centre} = \left(\frac{M^2}{1-M^2}, 0\right)$$

Constant Phase Angles Loci (N-circles)

N-circles represent phase of closed loop frequency response plotted against frequency.

- Radius of N-circles

$$\text{Radius} = \sqrt{\frac{1}{4} + \frac{1}{4N^2}}$$

- Centre of N-circles

$$\text{Centre} = \left(-\frac{1}{2}, \frac{1}{2N}\right)$$

$$N = \tan\phi$$

■■■■

9 State Space Analysis

State

It describes the future behaviour of a system based on present and past data. The knowledge of present is given by input and knowledge of part is given by state variables.

State Variables

These are the minimum set of variables such that knowledge of these variables alongwith the input can comletely describe behaviour of the system.

State Space

The set of possible values that a state vector can assume is called state space.

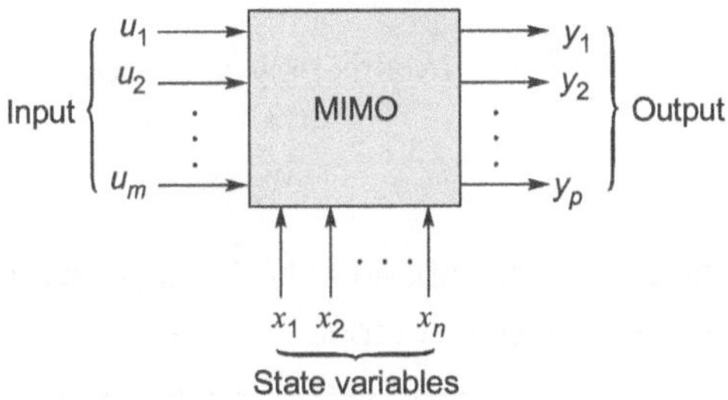

General Representation of State Model

- State equation

$$\boxed{\dot{x} = Ax + Bu}$$

- Output equation

$$\boxed{y = Cx + Du}$$

where, $\dot{x} \rightarrow$ Velocity vector $(n \times 1)$

$x \rightarrow$ State vector $(n \times 1)$

$u \rightarrow$ Input vector $(m \times 1)$

$y \rightarrow$ Output vector $(p \times 1)$

$A \rightarrow$ State matrix $(n \times n)$

$B \rightarrow$ Input matrix $(n \times m)$
$C \rightarrow$ Output matrix $(p \times n)$
$D \rightarrow$ Transmission matrix $(p \times m)$
$n \rightarrow$ Number of state variables
$p \rightarrow$ Number of outputs
$m \rightarrow$ Number of inputs

Controllability

A system is said to be controllable if it is possible to transform the initial state of the system to any desired state with application of input within a finite interval of time.

$$\boxed{Q_c = \left[B : AB : A^2B : ...... : A^{n-1}\,B\right]}$$

where, $Q_c \rightarrow$ Controllability test matrix

Condition for State Controllability

$\boxed{|Q_c| \neq 0}$ .... (Matrix be non singular)

Observability

A system is said to be the observable if it is possible to measure all state variables by measuring the output.

$$\boxed{Q_o = \left[C^T : A^TC^T : (A^T)^2C^T : ..... : (A^T)^{n-1}C^T\right]}$$

where, $Q_o \rightarrow$ Observability test matrix

Condition for State Observability

$\boxed{|Q_o| \neq 0}$ .... (Matrix be non singular)

Note: ....

- If AB is controllable, A^TB^T is observable.
- If AC is observable, A^TC^T is controllable.
- If input-output transfer function of a linear time invariant system does not have pole-zero cancellation, the system can always be represented by completely controllable and observable state model.

....

Transfer function (T.F.)

$$\text{T.F.} = C\,[s\boldsymbol{I} - A]^{-1}\,B + D$$

State transition matrix (STM)

$$e^{At} = \phi(t) = L^{-1}\left[(s\boldsymbol{I} - A)^{-1}\right]$$

Properties of STM

(i) $\phi(0) = I$ (identity matrix)

(ii) $\phi^{-1}(t) = \phi(-t)$

(iii) $[\phi(t)]^K = \phi(Kt)$

(iv) $\phi(t_1 + t_2) = \phi(t_1)\,\phi(t_2)$

(v) $\phi(t_2 - t_1)\,\phi(t_1 - t_0) = \phi(t_2 - t_0)$

■■■■

Process Dynamics & Control

10

Terminology of Process Control

Process

A process denotes an operation or series of operation on fluid or solid materials by which the materials are placed in a more useful state.

In the process mainly three variables are involved :

1. Controlled variable (CV)
2. Manipulated variable (MV)
3. Load Variable (LV)

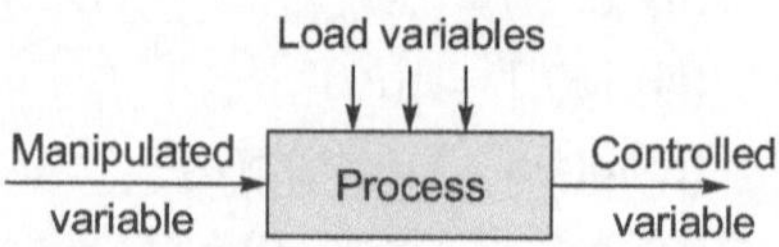

Controlled Variable (CV)

- The variable which we want to measure, monitor and control exactly at the set point.
  Ex.: Temperature, pressure, flow, level, speed, etc.

Manipulated Variable (MV)

- The variable which is manipulated to make the controlled variable remain at the set point value.

Load Variable (LV)

- The set of all other variables that affects the controlled variable to deviate from set point value.

Example of some process

1. Liquid Level System
   q_i = inlet flow rate (MV)
   h = level of liquid in tank (CV)
   q_0 = outlet flow rate (LV)

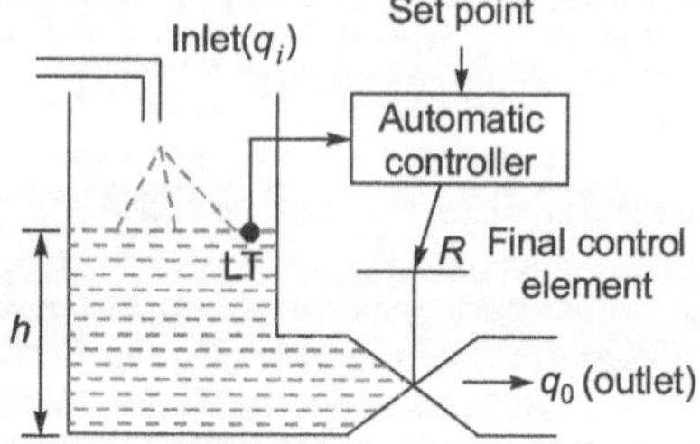

2. Temperature Control System

   $CV = T_L$

   $MV = q_s, T_s$

   $LV = q_i, T_i, q_o, T_A$

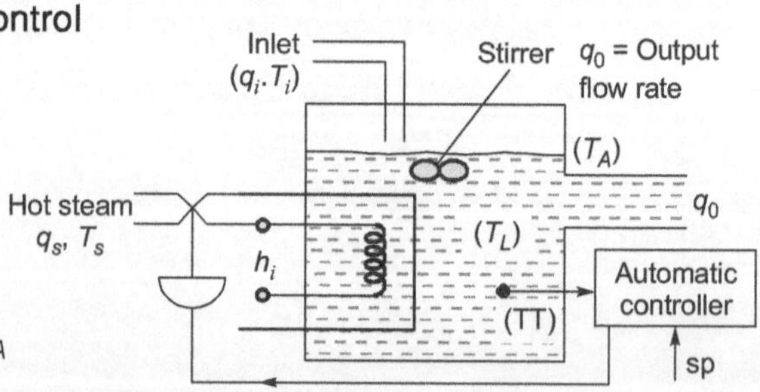

Process Control

Process control refers to "Regulation and Manipulation of variables influencing the conduct of a process in such a way as to obtain a product of desired quality and quantity in an efficient manner.

1. Manual control
2. Automatic control

Block diagram of automatic process control :

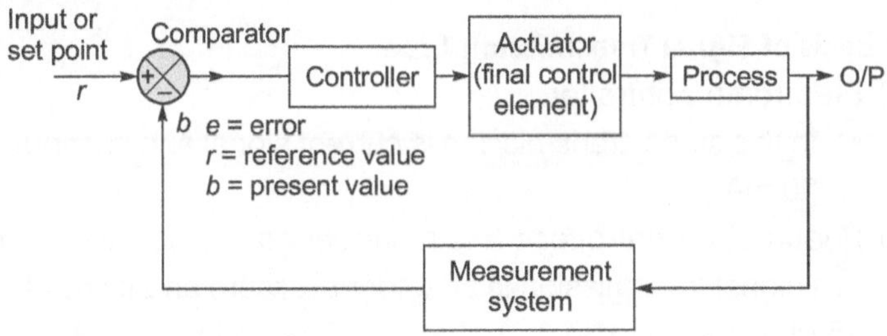

1. **Measurement System :** it is used to determine the magnitude of the existing physical variable i.e. controlled variable. A transducer is normally employed for the measurement.
2. **Controller :** It is an automatic device having an output that varies to regulate a controlled variable in a specific manner. A controller performs operations such as comparison, evaluation and operation upon the next stage to meet the demands of control objective.
3. **Final Control System :** The device that directly controls the value of manipulated variable in a process control loop. Often the final control element is a control valve.
4. **Error Detector / Comparator**

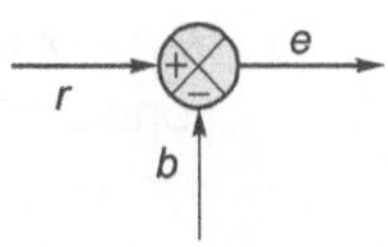

$$e = r - b$$

Error $$\%e = \frac{r-b}{b_{max} - b_{min}} \times 100$$

b = actual measured value; b_{max} = maximum of measured value

b_{min} = minimum of measured value

Controller Output

$$P\% = \frac{C - C_{min}}{C_{max} - C_{min}} \times 100\%$$

C = actual controller output ;
C_{max}, C_{min} = standard signal transmission range

Elements of Process Dynamics

- **Process resistance** : It is the opposition to energy flow.
- **Capacitance** : It is the energy storage element.

- **Dead time :** It is also called as a distance / velocity lag.
  - It depends on the placement of sensors, actuator and final control elements.
- **Process lag time :** It is a function of resistance, capacitance and dead time. If process is slow it takes more lag time.

Classification of Controllers

On the basis of Signal Transmission Type

(a) Electronic controller :
- In this signal transmission is current signal and its range is 4 to 20 mA.

(b) **Pneumatic controller** : It is a flapper-nozzle-bellows system. In this signal transmission is air under pressure and its range is 3 to 15 psi.

(c) **Hydraulic controller** : It is designed by using a power system cylinder. Signal transmission is in the form of shaft movement by fluid under pressure and range is 0 - 2 inches.

On the basis of Modes of Control Action

Discontinuous Controller Mode

Two position controller : This mode is mathematically expressed as

$$P = 100\%,\ e > 0$$
$$P = 0\%,\ e < 0$$

- Whenever there is error ($e > 0$) the controller output is 100%. In terms of both ON (or) OFF states. When there is no error $e < 0$. The controller output can not be defined.

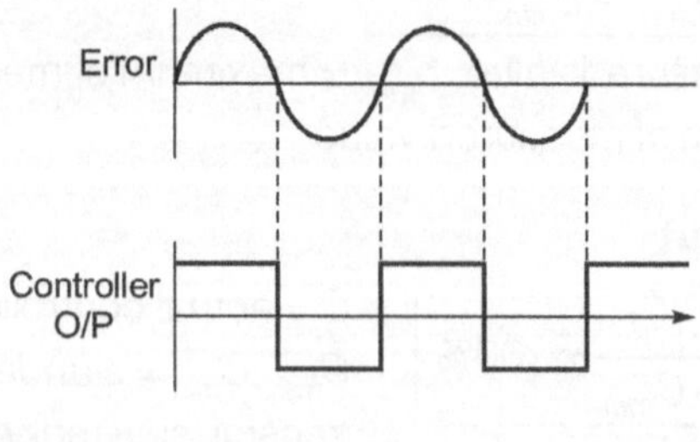

- It leads to damage the hardware part because of frequent ON-OFF. So in practical case, always it is designed with differential gap/ dead zone.

- **With differential gap :** Whenever the controller output has to change 0% → 100% or viceversa,There is a differential gap known as neutral zone around zero error point where controller output remains constant.

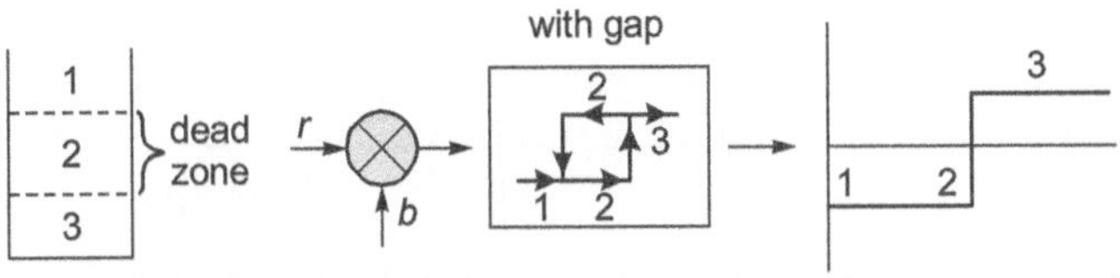

- In this way ON-OFF frequency drastically reduced, so protect hardwares.

Characteristics of Two position controller

1. Non-linear controller
2. Proportional band / B.W. = 0
3. In actual cases, 2 position controllers are designed by switches, relays or alarms.
4. Simple, cheapest and easy to design.

Disadvantage

- It leads to oscillations, overshoot and undershoot problems.

Applications

It is suitable for the large systems

- Room heating or air conditioning systems.
- Liquid bath temperature control.
- Level control in large volume tanks.

Multiposition Mode

The most common example is a three position controller

$$P = \begin{cases} 100\% & e > e_2 \\ 50\% & -e_1 < e < e_2 \\ 0\% & e < -e_1 \end{cases}$$

As long as the error is between e_2 and e_1 of the set point, the controller stays at a nominal setting of 50% if the error exceeds the set point by e_1 or more, the output is increased by 100%. If it is less than the set point by $-e_1$ or more the controller output is reduced to zero.

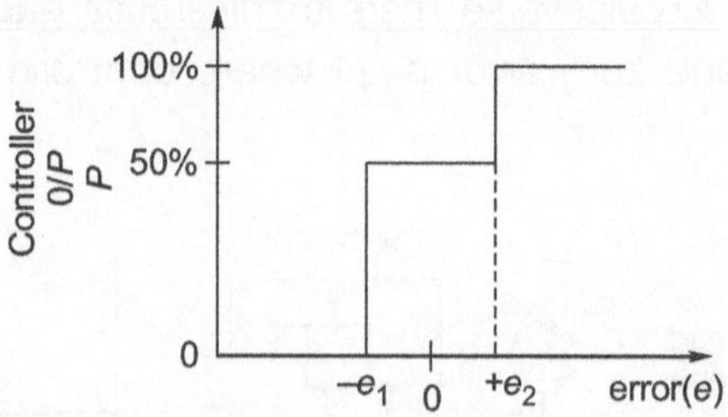

Continuous Controller Modes

1. Proportional Controller Mode

- In this mode there is a one-to-one correspondence existing between controller output and error.
  $P = k_p\, e_p + P_0$
  P = controller output, K_p = proportional gain
  e_p = present error or error position
  P_0 = initial value of controller output

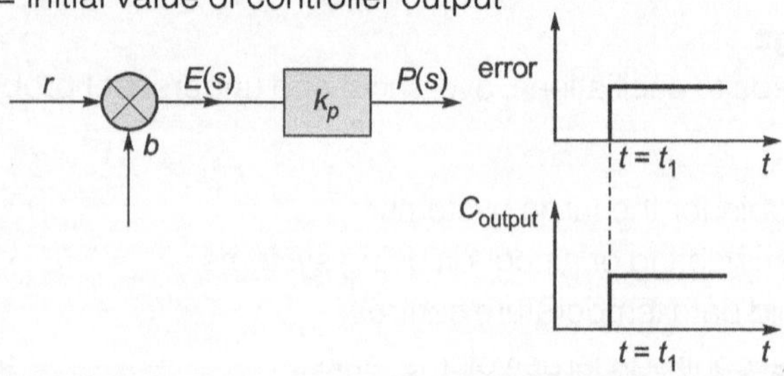

- Range of errors to cover 0 to 100% of the controller output is known as **proportional band.**
- Based on the gain of controller, proportional band (PB) is decided.

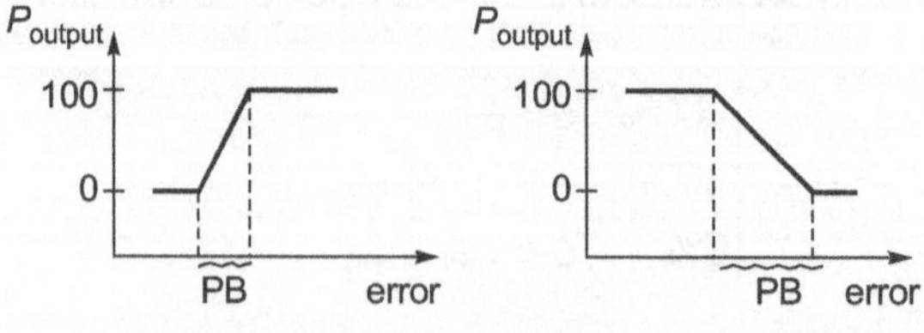

$$\boxed{\text{P.B.} = \frac{100}{k_p}}$$

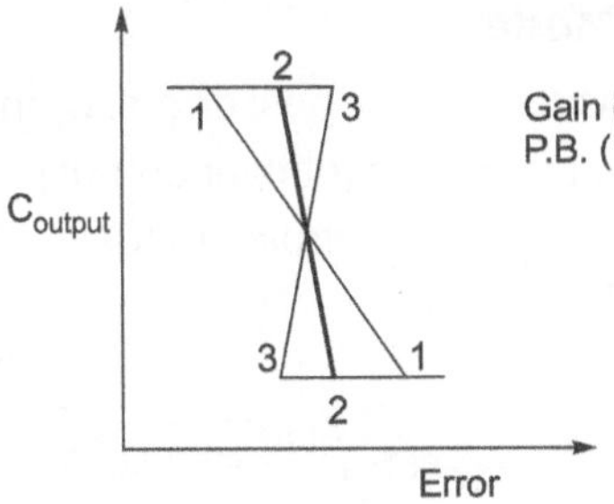

Offset

One of the main disadvantage of proportional control action is that is produces a permanent residual error in the operating point of the controlled variable when a process load change results in error. This residual error is known as OFFSET.

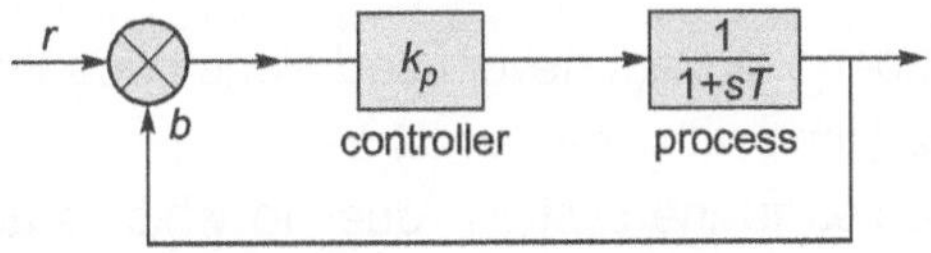

- Steady state error (offset) for 1st order process

$$e_{ss} = \frac{A}{1+k_p}$$

k_p = controller gain; T = time constant

$$k_p \uparrow \Rightarrow \text{PB} \downarrow \text{ error } \downarrow$$

- The P controllers offset never becomes zero, so this is used in the applications wherever manual reset is permitted.

Electronic Design

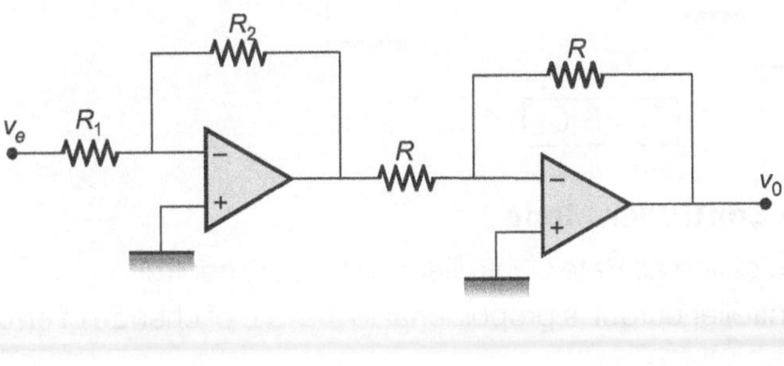

$$k_p = \frac{R_2}{R_1}$$

$$V_0 = \frac{R_2}{R_1} V_e$$

2. Integral Controller Mode

- In this, mode it considers all the past errors, by integrating the total past errors and after that the controller output is generated.
- It is also called as a RESET controller mode.
- It eliminates the offset error.

$$\boxed{P = K_I \int e\,dt + P(0)}$$

$$\boxed{P(s) = \frac{1}{T_I}\frac{e}{s} + P(0)}$$

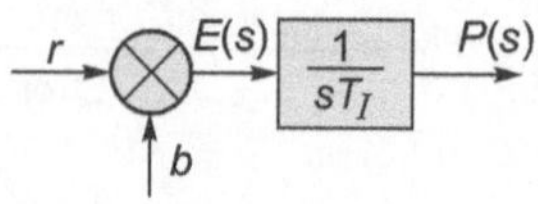

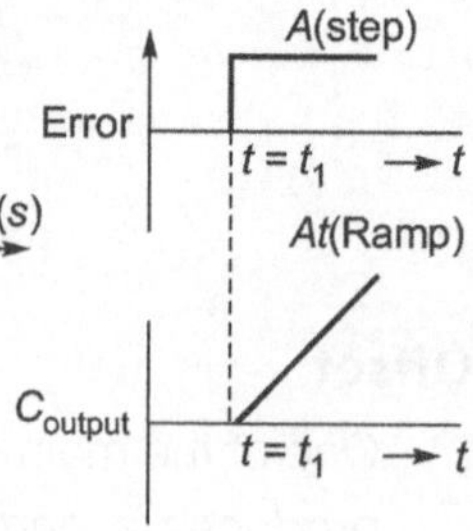

$P(0)$ = Initial value of controller output when error is zero.

K_I = Integral gain T_I = reset time

Characteristics

- Integral controllers are never used alone because they have very large dead time with slow response.
- It adds poles to the system due to which stability decreases because root locus shift to the right side.
- It does not respond for transient state.
- It improves the steady state response by eliminating steady state error.
- It act as a lag controller or **low pass filter**.

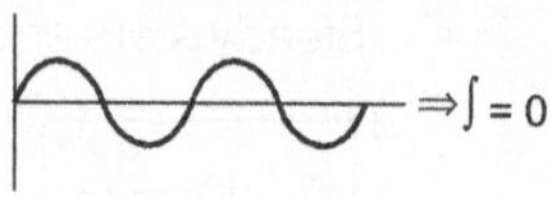

Electronic Design

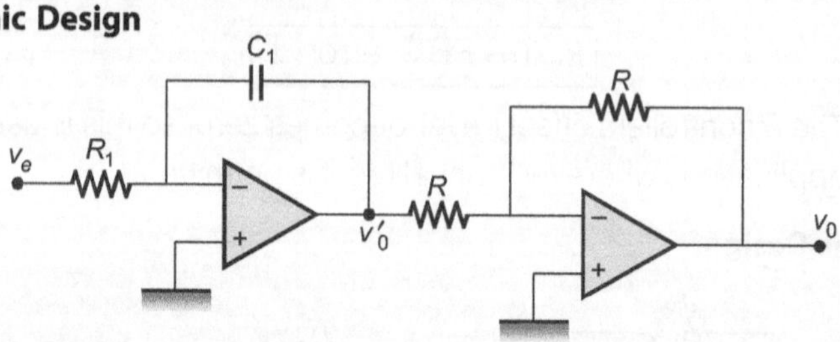

$$\boxed{V_0 = \frac{V_e}{sR_1C_1}} \qquad \boxed{T_i = R_1C_1}$$

3. Derivative Controller Mode

- It is also called as Rate Controller/anticipatory control.
- The controller output is proportional to the rate of change of errors

$$P \propto \frac{de}{dt} \qquad P = t_d \frac{de}{dt} \qquad \boxed{P(s) = T_d\, s\, E(s)}$$

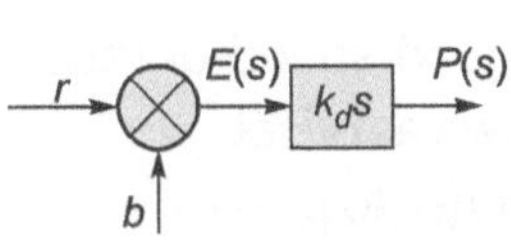

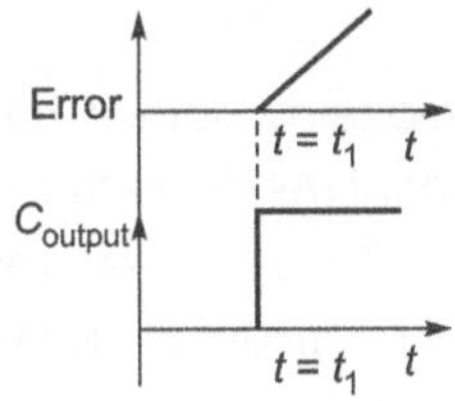

T_d = rate time

Characteristics

- If the error is step change i.e. sudden, the controller output will be zero.
- When D controller is used, it adds zeros to the system, So stability increases oscillations decreases because root locus shifts to left.
- The derivative action's response is much faster.
- It act as a lead controller and high pass filter.
- This mode is suitable if process reaction rate itself is sluggish, as it considerably improves the speed of response.

Electronic Design

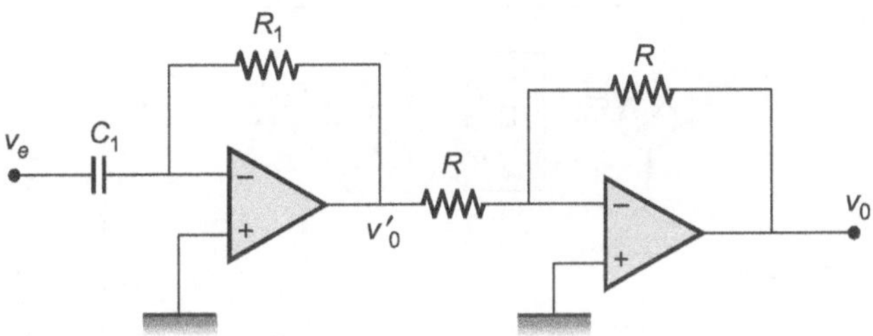

$\boxed{v_0 = v_e sR_1 C_1}$ $\boxed{T_d = R_1 C_1}$

Steady state errors for integral and derivative controllers for 1st order process

$\boxed{e_{ss} = 0}$

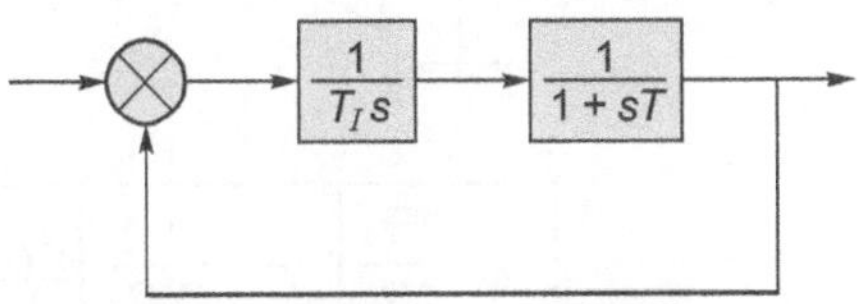

$\boxed{e_{ss} = A}$

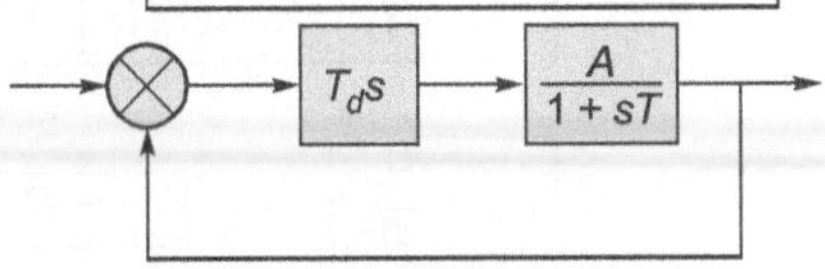

A = process gain

Note: ....

- Proportional controller has good speed of response but has a permanent residual error known as offset.
  P : $P \propto e$ → response to present errors
- Integral controller is capable of eliminating error, but is response characteristics are sluggish.
  I : $P \propto \int e$ → response to past errors
- Derivative controller's output is zero for sudden error (constant) and it may even drive the process to instability.
  D : $P \propto \frac{de}{dt}$ → future

....

4. P+I Controller Mode

$$\boxed{P = k_p e \frac{k_p}{T_i}\int edt + P(0)}$$

$$\boxed{P(s) = k_p\left(1 + \frac{1}{T_I s}\right)E(s)}$$

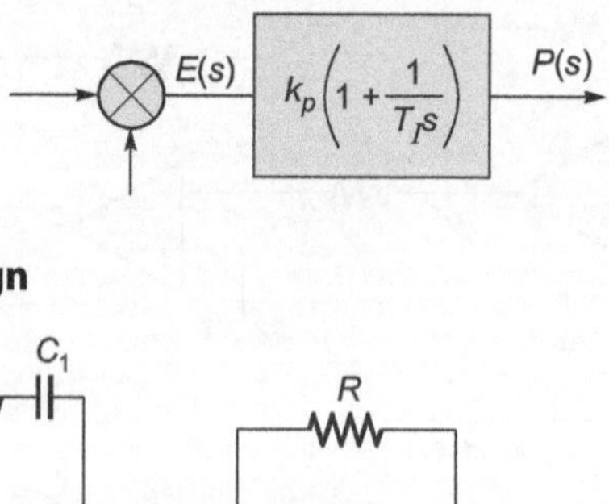

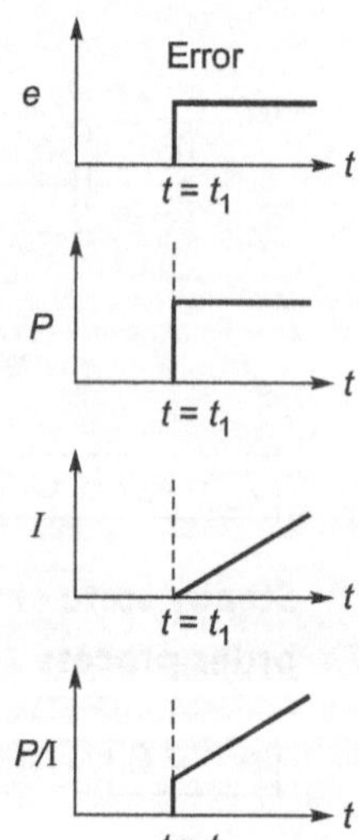

Electronic Design

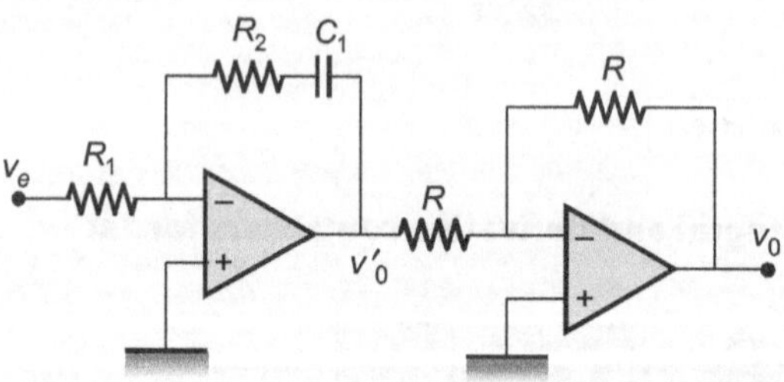

$$\boxed{v_0 = \frac{R_2}{R_1}\left[1 + \frac{1}{sC_1R_1}\right]E(s)}$$

So, $\boxed{k_p = \frac{R_2}{R_1}}$ $\boxed{T_I = R_1C_1}$

Characteristics

- $P + I$ controller characteristics are similar to lag compensator or low pass filter. It is capable of improving only the steady state response characteristics i.e. the presence of integral action eliminates the offset of proportional action and also any type of error.
- This control action also has good speed of response.

5. *P+D* Controller Mode

$$P = k_p e + k_p T_d \frac{de}{dt} + P_0$$

$$P(s) = k_p(1 + T_d s)\, E(s)$$

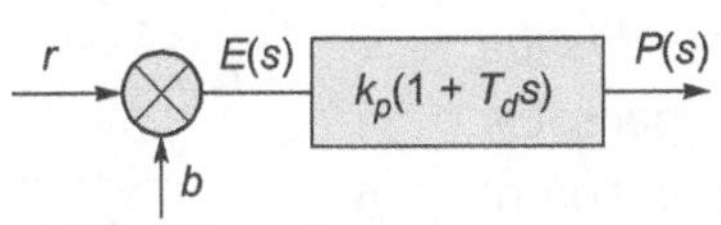

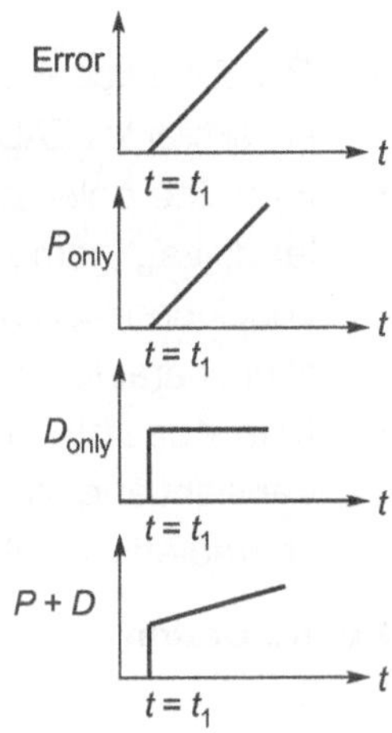

Electronic Design

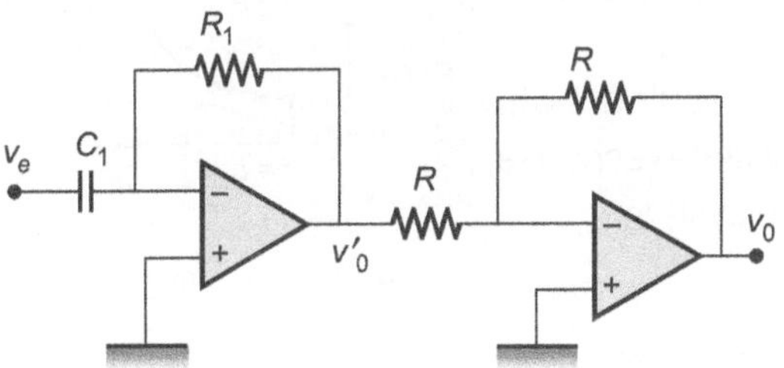

$$k_p = \frac{R_2}{R_1}$$
$$T_d = R_1 C_1$$

$$v_0 = \frac{R_2}{R_1}(1 + sR_1C_1)\, v_e$$

Characteristics

- *P+D* controller characteristics are similar to lead compensators or high pass filter.
- It is capable of improving only transient sate characterstics i.e. speed of response of the system.
- It is best suitable for processes having very slow response.
- It improves the damping factor and reduces the maximum overshoot and also decreases peak time, rise time and settling time.

6. *P+I+D* Controller Mode

$$P = k_p e + \frac{k_p}{T_i}\int e\,dt + k_p T_d + P(0)$$

$$P(s) = k_p\left(1 + \frac{1}{T_i s} + T_d s\right)E(s)$$

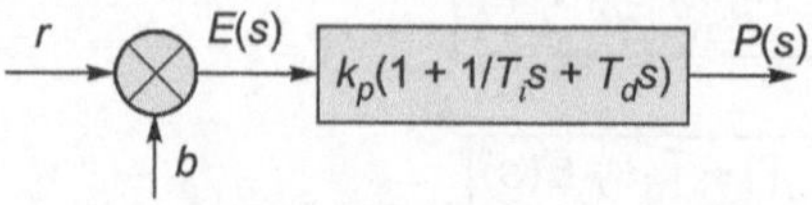

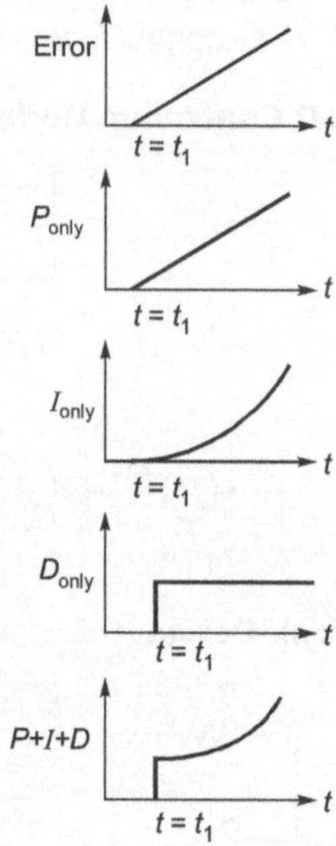

- $P + I + D$ controller characteristics are similar to LEAD-LAG compensator.
- From the pole-zero configuration of lead. Lag compensator it may be concluded that is exhibits band reject filter characteristics.
- It is capable of improving both transient and steady state response characteristics of the system.

Electronics Design

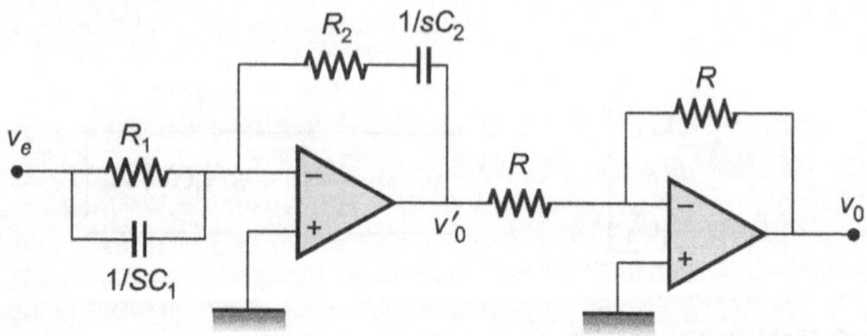

$$V_0 = \frac{R_1C_1 + R_2C_2}{R_1C_2}\left\{1 + \frac{1}{sR_1C_2} + \frac{R_1C_2}{R_1C_1 + R_2C_2} + \frac{sC_1R_1(R_1C_2)}{R_1C_1 + R_2C_2}\right\}V_e \quad \text{...(i)}$$

Standard transfer function of PID controller $= k_p\left(1 + \frac{1}{T_I s} + T_d s\right)$ ...(ii)

So, $\boxed{k_p = \dfrac{R_1C_1 + R_2C_2}{R_1C_2}}$ $\boxed{T_I = R_1C_1 + R_2C_2}$ $\boxed{T_d = \dfrac{R_1R_2C_1C_2}{R_1C_1 + R_2C_2}}$

- In this, expression of k_p, T_I and T_d are complicated and inter dependent, so tuning of PID controller is complicated.

Comparison between different Controllers

ON/OFF	*P*	*I*	*D*
Simple cheap, oscillation increases, used where precise control is not required, slow rate, large process.	Tuning is easy, but offset (fast) error is present, used where load change is small and manual reset is accepted.	Eliminates offset error, (slow) but oscillation increases, stability decreases, so used in fast process, small lag time.	Oscillations decreases, stability increases, fast process used in slow processes (sudden/fast change of controller output)

PI	*PD*
Long period of oscillations, Settling time increases	Oscillations decreases, Settling time decreases, fast and quick settling

Tuning of Controllers

1. Process Reaction Curve Method (Open Loop Method)

- To draw the process reaction curve, controller is disconnected within the process loop.
- Some disturbance signal is applied to the process.
- Process is reacted and control variable in the process is recorded.
- A typical process reaction curve "*s*" shaped (sigmoidal shape) is obtained.

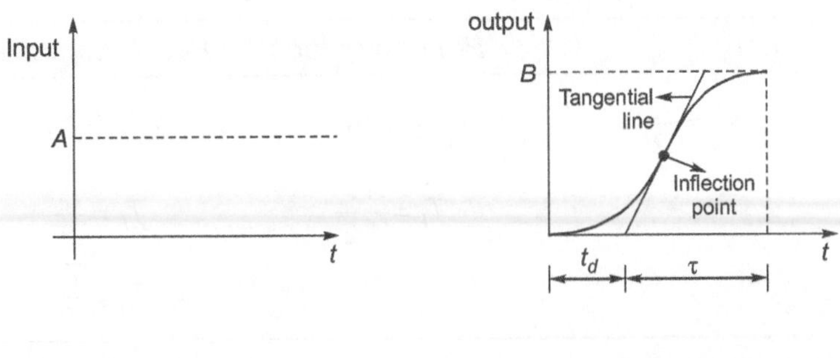

where S = slope of tangnt line at inflection point (or) It is also called reaction rate (B/τ).
t_d = dead time or lag
B = Ultimate value of output

- Using semi-empirical rules the controller parameter setting are:

P-Mode	*P*+*I* Mode	*P*+*I*+*D* Mode
$k_p = \frac{A}{St_d}$	$k_p = \frac{0.9A}{St_d}$	$k_p = 1.2\frac{A}{St_d}$
	$T_i = 3.33t_d$	$T_i = 2\ t_d$
		$T_D = 0.5\ t_d$

where, A = step change at final control elements

2. Ultimate Gain Method (Closed Loop Method)

- The integral and derivative actions are reduced to the minimum effect after the process reaches steady state at the normal level of operation.
- The value of proportional gain k_p is gradually increased while providing periodic small disturbances to the system. If the response decays a higher value of k_p is selected. This process is repeated till the response exhibits sustained oscillations about the steady state value of input.
- Note the value of proportional gain at which the response exhibits conterminous and sustained oscillations. This is called as ultimate (or) critical gain k_c. The frequency and hence the period of oscillations is ultimate (or) critical period T_c.
- The optimal values of k_p, T_d and T_i can be obtained from critical gain and period using Ziegler-Nicol settings.

P	*P*+*I* controller	*P*+*I*+*D* controller
$k_p = \frac{k_c}{2}$	$k_p = 0.45\ k_c$	$k_p = 0.6\ k_c$
	$T_I = \frac{T_c}{1.2}$	$T_I = \frac{T_c}{2}$
		$T_d = \frac{T_c}{8}$

Process Control System Configurations

1. Feedback Control Configuration

- In a feedback control scheme the output of the process is continuously compared with the input of the process and if there is any deviation the error signal is generated which is sent to the controller.
- The controller then tries to bring back the process to the desired level with the help of the final control element. A generalized process showing a feedback control loop is shown below.

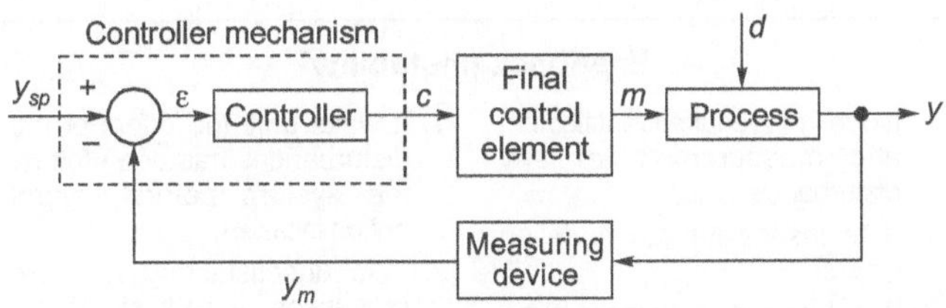

2. Feed Forward Control Configuration

- A feedforward control configuration measures the disturbance (load) directly and takes the control action to eliminate its impact on the process output.
- The figure below shows the feedforward control action.

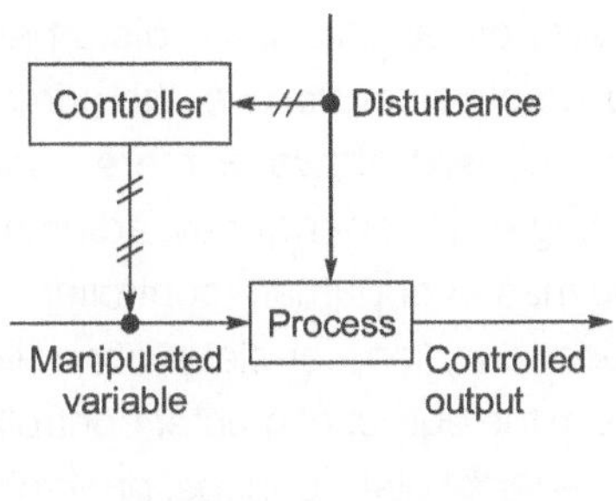

(Feedforward Control)

- In case of a feedforward control, the effect of the disturbance is completely eliminated as it is already sensed and corrected before reaching the process.
- The feedforward acts beforehand in an anticipatory manner.

Comparision between Feedforward and Feedback Control

Advantages	Disadvantages
Feedforward (stability)	
1. Acts before the effect of a disturbance has been felt by the system. 2. Is good for slow systems. (multicapacity) or with significant dead time. 3. It does not introduce instability in the closed-loop	1. Requires identification of all possible disturbances and their direct measurement. 2. Cannot cope with unmeasured disturbances 3. Sensitive to process parameter variations. 4. Requires good knowledge of the process model.
Feedback (instability)	
1. It does not require identification and measurement of any disturbance. 2. It is insensitive to modeling errors. 3. It is insensitive to parameter changes.	1. It waits until the effect of the disturbances has been felt by the system before control action is taken. 2. It is unsatisfactory for slow processes or with significant dead time. 3. It may create instability in the closed-loop response.

3. Cascade Control Configuration

- In cascade control loop configuration we have one manipulated variable and more than one measurement.
- An intermediate process variable that responds both to the manipulated variable and to some disturbances can be used to achieve more effective control over the primary process variable.
- Cascade loops consist of two or more controllers in series, but have only a single, independently adjustable set point by the operator to the master or primary controller.
- The use of secondary loop or slave controller whose set point is determined from the output of master controller is to act as the first line of defense against disturbances, preventing these upsets from entering the primary process.
- The secondary loop must have faster dynamics than the primary loop for the cascading effect to yield better results.

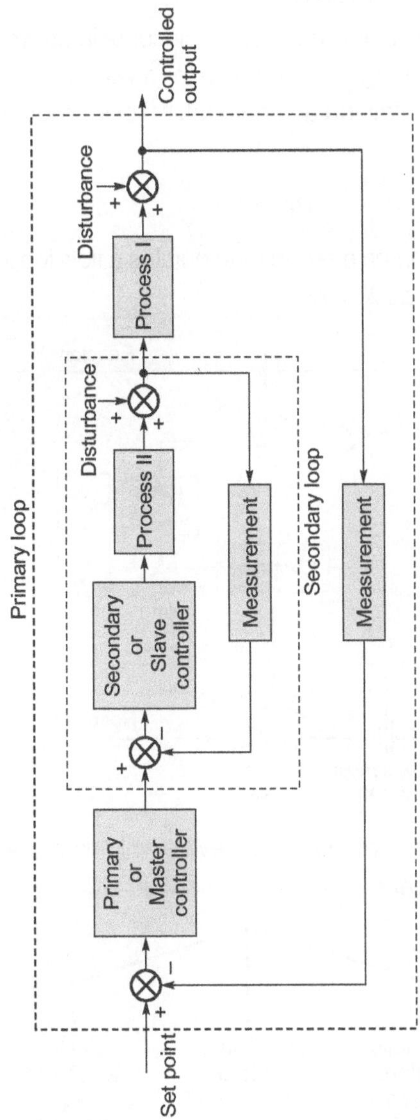

Cascade Control System

Note: ....

- In cascade control configuration the cross over frequency for the secondary loop is higher than that for the primary loop.
- This allows us to use higher gains in the secondary controller in order to regulate more effectively the effect of a disturbance occurring in the secondary loop without endangering the stability of the system.

....

4. Ratio Control Configuration

- Ratio control systems maintain a relationship between two variables to monitor and regulate/control a third variable.
- Consider maintaining a certain ratio R of ingredient X to ingredient Y.

$$\text{Ratio } R = \frac{X}{Y}$$

- The most common method manipulates a flow loop whose set point is calculated as $X = YR$.

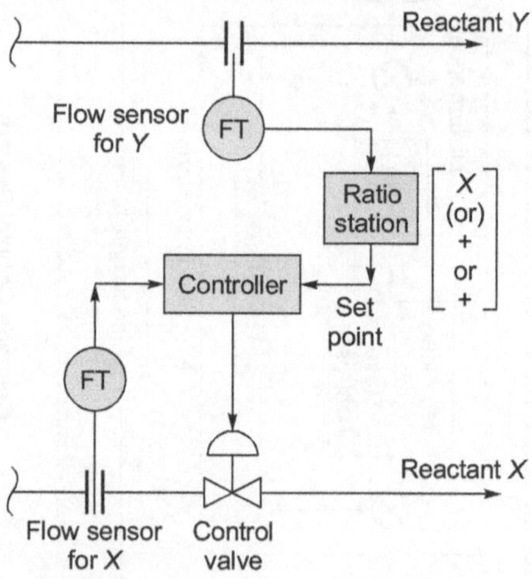

Actuators: The actuated are the device which are used to operate the final control element.

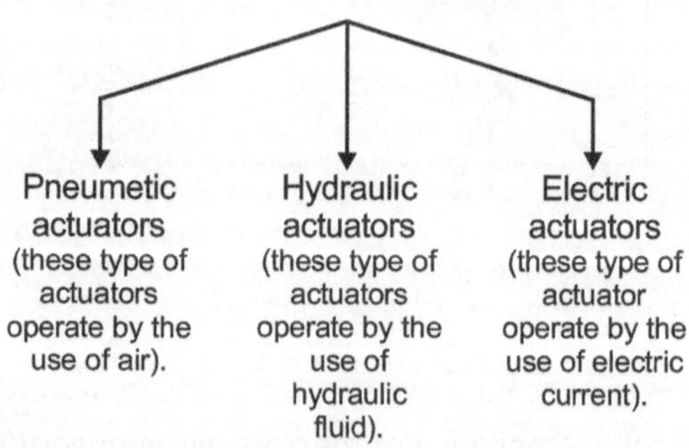

- These all type of actuating signals are used to achieve a control mechanism by operating a stem connected valve either directly or through an electric motor.

Fluid Flow Through Control Valves

- A control valve placed in a pipeline acts like a variable restriction as can be seen from the figure below. As the mass flow rate of the air (m_1) changes it leads to the vertical motion of the plug and the stem of the control value which changes the area of the opening of the port.

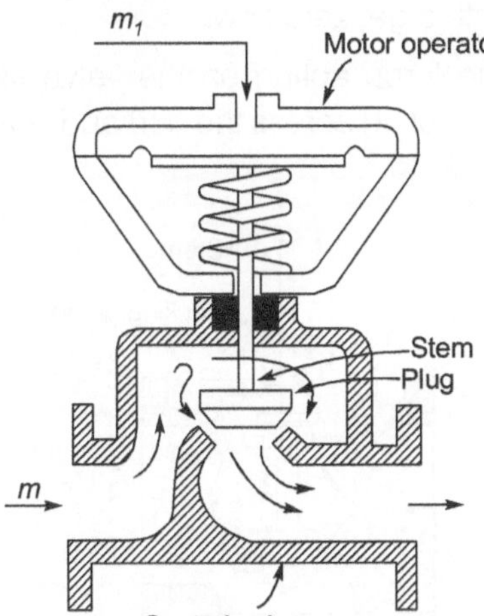

Control valve

- Thus, the flow rate of the fluid passing through the port is throttled by the positioning of the valve stem.
- The fluid flow rate (m) through the valve can not be denoted as:

$$m = K_1 a\sqrt{2g(h_1 - h_2)}$$

K_1 = Flow rate coefficient
a = Area of the control value (ft^2)
g = Acceleration due to gravity (ft/sec^2)
h_1 = Upstream static head of flowing fluid (ft)
h_2 = Downstream static head of flowing fluid (ft)

Types of Stem-Control Valve

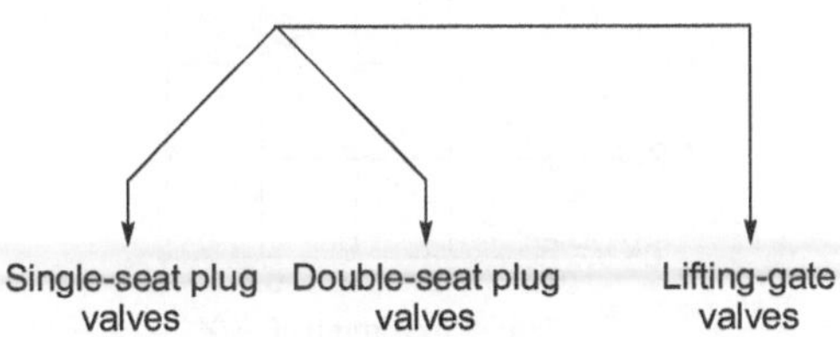

Single-Seat Plug Valves

- Single-seat type control valve has only one port opening between seat and plug and the entire flow passes through.
- Features of single seat type control valve:
  - It is simple in construction.
  - It can be shut off to get zero flow.
  - There is a large force acting on the valve stem because of the differential head acting across the port and seat area.

Double Seat Valve

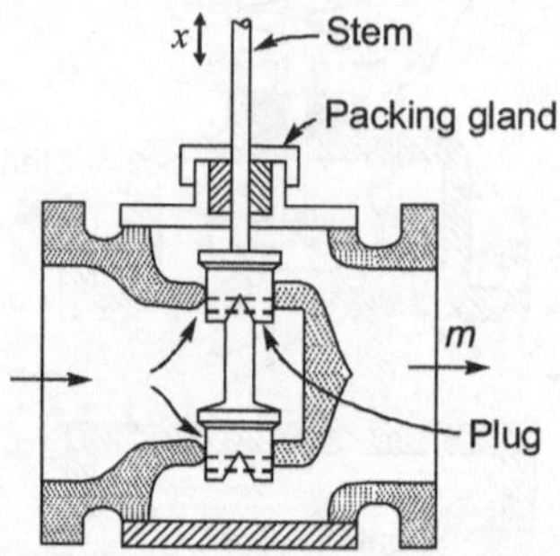

- The double seat valve has two port openings and two seats and two plugs. The port opening are not usually identical in size. The double seat type control valve has generally these features:
  - Net force acting on the valve stem is generally small.
  - They cannot be shut off tightly because of differential temperature expantion of value plug and valve body.

Characteristics of a Control Value

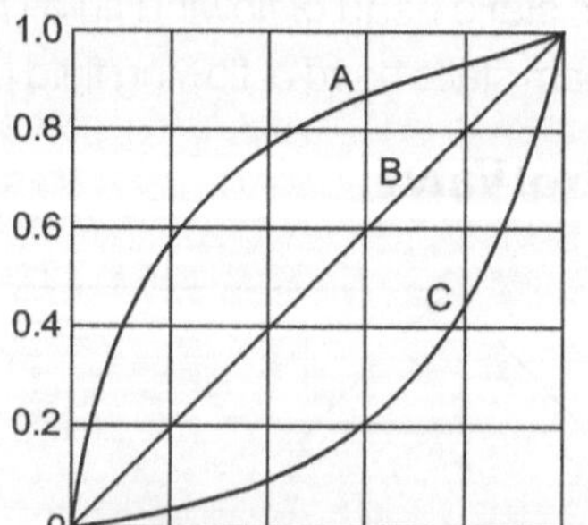

Fraction of maximum lift, x/X
A Decreasing sensitivity type
B Linear type
C Equal-percentage type

Curve-A: It is the characteristics of the ***decreasing sensitivity type*** *of* control valve, as the x increases the flow rate ($\Delta m/\Delta x$) once increases but finally rate (i.e. $\Delta m/\Delta x$) decreases.

Curve-B: The valve sensitivity ($\Delta m/\Delta x$) here in constant. So it the characteristics of the ***linear type*** of control valve.

Curve-C: It is the ***increasing sensitivity valve.*** The valve sensitivity ($\Delta m/\Delta x$) increases with increasing flow rate.

This type of control valve is termed the equal-percentage, logarithmic, parabolic or the characterized type of valve because the flow-lift curve plotted on semilogarithmic coordination is approximately a straight line.

Rotating-Shaft Control Valves

- In rotating-shaft type control valve the restriction is accomplished by the rotation of plug or vane.

Rotating-plug valves | Butterfly valves | Louvers

(i) Rotating-Plug Valve:

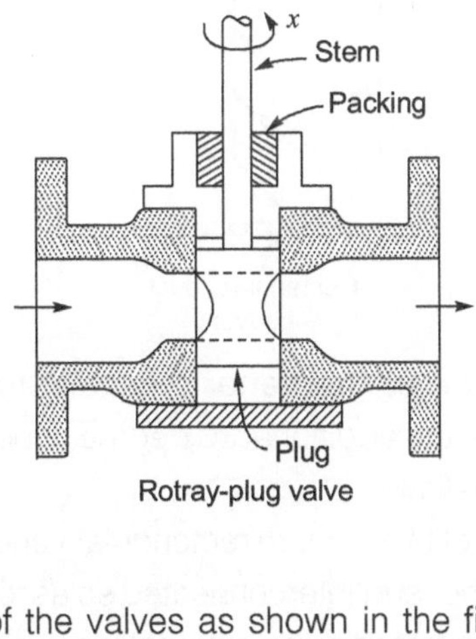

Rotray-plug valve

- In those type of the valves as shown in the figure, the plug is the cylindrical or the conial element with transverse opening. It is rotated in the valve body by an external power such that the opening on one side of the plug is gradually covered or incovered.

(ii) Butterfly Valve

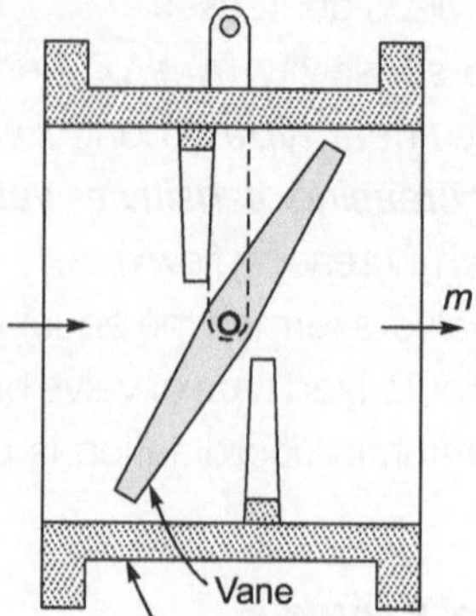

Butterfly valve

- A butterfly valve consists of a single vane rotating inside a circular or rectangular pipe or casing.

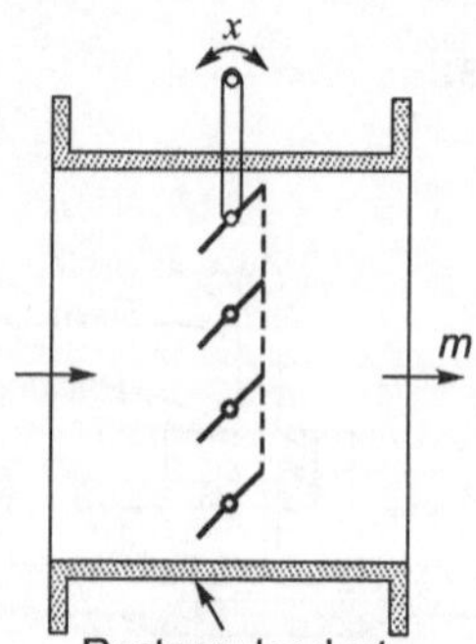

Louver

- The total rotation of the vane is usually restrinted to about 60 degrees because the additional 30 degrees does not produce much further increase in flow.
- Louvers consists of two or more rectangular vanes mounted on shafts one above the other and interconnected so as to rotate together.
- The louvers are operated by an external force.
- In the unidirectional louver the vanes remain parallel at all positions, whereas in case of counter rotational louver alternate vanes rotate in an opposite directional.

Control Valve Sizing (C_v)

- C_v is the coefficient used to define the size of a control valve. The proper size of the control valve is important because of the effect on the operation of automatic controller.
- If the control valve is ***oversized***, example the control valve must operate at low lift and minimum controllable flow is too large.
- If the control valve is ***undersized***, the maximum flow desired for operation of the process may not be provided.

The flow of liquid a fully open control valve is given by

$$\boxed{m = K_a\sqrt{\frac{2g\,\Delta P}{\gamma}}} \quad \ldots(i)$$

where,

m = Flow rate (ft^3/sec)

K = Flow coefficient

a = Area of port opening (ft^2)

g = Acceleration due to gravity (ft/sec^2)

ΔP = Pressure differential (lb/ft^2)

γ = Fluid density (lb/ft^3)

Stepper Motor (Open Loop)

- It is an electromechanical device designed to convert a series of input electrical pulses into a discrete angular steps.
- It complete rotation by step by step movements only.
- For one pulse, how much rotation it takes place is known as step angle.
- Step (SA) angle is inversly proportional to the resolution of stepper motor.

$$\boxed{\text{Resolution} = \frac{360}{\text{step angle}}} \qquad R = \frac{360}{SA(\theta)}$$

- It is used for precise positioning and precise speed control without any closed loop feedback.
- Generally range of stepper motor SA ($\theta = 0.72° - 90°$)

- Step angle θ is decided based on the rotor poles and stator phases (pole pairs)

$$SA = \frac{360}{\text{Rotor poles} \times \text{stator phases}}$$

For Example

Rotor poles 4, statorphase 3

$$SA = \frac{360}{12} = 30°$$

- In this stator 3 phases (A, B, C) are selected such that each phase is at 120° to each other and 4 rotor poles are selected such that each pole is at 90° to each other.

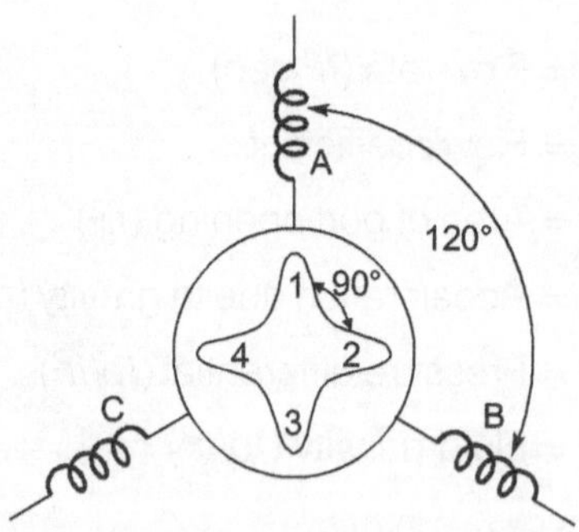

- When phase A is excited. Rotor pole 1 is aligned with Phase A

Pulse	Excited phase	Rotor pole aligned
1	A	1
2	Phase B	$2 \rightarrow 30°$
3	Phase C	$3 \rightarrow 30°$
4	Phase A	$4 \rightarrow 30°$
5	Phase B	$1 \rightarrow 30°$
6	Phase C	$2 \rightarrow 30°$
7	Phase A	$3 \rightarrow 30°$
8	Phase B	$4 \rightarrow 30°$

9	Phase C	$1 \rightarrow 30°$
10	Phase A	$2 \rightarrow 30°$
11	Phase B	$3 \rightarrow 30°$
12	Phase C	$4 \rightarrow 30°$
13	Phase A	$1 \rightarrow 30°$

- In the 13th pulse, again initial condition reached i.e. rotor pole 1 is again aligned with phase A.

■■■■

A Handbook on
Instrumentation Engineering

5

Analog Electronics

CONTENTS

Junction Diode Characteristics

1

Junction Diode

- A diode is the second generation semiconductor family.
- It is a single layer, two region, single junction, two terminal unilateral active semiconductor device.
- A *p-n* junction is formed from a piece of semiconductor by diffusing *p*-type material to one half side and *n*-type material to other half side.
- The plane dividing the two zones is known as a junction.
- A region near the junction is without any free charge particles called depletion region or charged free region or space charge region because there is no charge available for conduction.

Note:

- **Depletion layer consists of immobile charged particles i.e. ions only.**
- **Depletion layer consists of negative ion (Acceptor ions) on the *p*-side and positive ion (Donor ions) on the *n*-side.**
- **Depletion layer opposes majority carriers in crossing the junction.**

Current Equation of Diode

Boltzman's equation of diode current

$$\boxed{I = I_o \left[\exp\left(\frac{V}{\eta V_T} \right) - 1 \right]}$$

where, I = Diode current

I_o = Diode reverse current

V = Diode voltage (Positive for forward bias and negative for reverse bias)

η = 1 for germanium and 2 for silicon

V_T = Thermal voltage

T = Temperature in Kelvin

In forward bias

$$\boxed{I_F = I_o \exp\left(\frac{V_F}{\eta V_T} \right)}$$

In reverse bias

$$I_R \cong -I_o$$

Note:

Reverse current I_o is temperature dependent. It gets double for every 10°C rise.

Resistance of Diode

Resistance of diode can be calculated only when the diode is in forward bias.

(a) **Static resistance (DC resistance):** It is ratio of voltage and current at any point.

$$R_{DC} = \frac{V}{I}$$

(b) **Dynamic resistance (AC resistance):** It is define as reciprocal of slope, which is the smallest linear region in the entire non-linear curve.

$$R_{AC} = \frac{1}{\text{Slope}} \quad ; \quad R_{AC} = \frac{dV}{dI}$$

$$R_{AC} = \frac{\eta V_T}{I_F} = \frac{\eta kT}{I_F q}$$

Note:

Static resistance is always greater than AC or dynamic resistance.

Zener Diode

- It is a heavily-doped *p-n* junction diode which is operated under reverse bias condition in the breakdown region.
- Zener current is independent of the supply voltage and only depends on external circuit resistance. Therefore Zener diode is known as constant voltage or voltage reference device.
- At reverse voltages less than 6 V, Zener breakdown predominates while at about 8 V, Avalanche breakdown predominates.
- Zener breakdown diodes are negative temperature coefficient of voltage and Avalanche breakdown diodes are positive temperature coefficient of voltage.
- These are used for voltage regulators.

Rectifiers

Form factor (*F*)

$$F = \frac{\text{rms value}}{\text{DC value}} = \frac{V_{rms}}{V_{DC}}$$

Half Wave Rectifier

Average value of current and voltage

(a) Ideal case

$$I_{DC} = \frac{I_m}{\pi} \quad \text{and} \quad I_m = \frac{V_m}{R_L}$$

$$V_{DC} = \frac{V_m}{\pi}$$

where, I_{DC} = Average value of current
V_{DC} = Average value of voltage
I_m = Maximum value of current
V_m = Maximum value of voltage
R_L = Load resistance

(b) Practical case

$$I_{DC} = \frac{I'_m}{\pi} \quad ; \quad V_{DC} = \frac{V'_m}{\pi}$$

$$I'_m = \frac{V_m}{R_s + R_f + R_L} \quad ; \quad V'_m = I'_m R_L$$

where, R_s = Coil resistance
R_f = Diode forward resistance
R_L = Load resistance

RMS value of current and voltage

$$I_{rms} = \frac{I_m}{2} \qquad V_{rms} = \frac{V_m}{2}$$

Full Wave Centre Tap Rectifier

Average value of current and voltage

$$I_{DC} = \frac{2 I_m}{\pi}$$, ideal case $$V_{dc} = \frac{2V_m}{\pi}$$, ideal case

RMS value of current and voltage

$$I_{rms} = \frac{I_m}{\sqrt{2}}$$, ideal case $$V_{rms} = \frac{V_m}{\sqrt{2}}$$, ideal case

For practical case I_m is replaced by I'_m.

where, $$I'_m = \frac{V_m}{\frac{R_s}{2} + R_f + R_L}$$ and $$V'_m = I'_m R_L$$

Full Wave Bridge Rectifier

For full wave rectifier (bridge type) in all the formulas $\frac{R_S}{2}$ is replaced by R_S and R_f is replaced by $2R_f$.

Clamper

- A circuit which can add DC voltage to an AC waveform.
- A clamper does not change shape of the waveform, it only shift the waveform up or down.

  Clamper is also called DC reinserter or DC restorer.

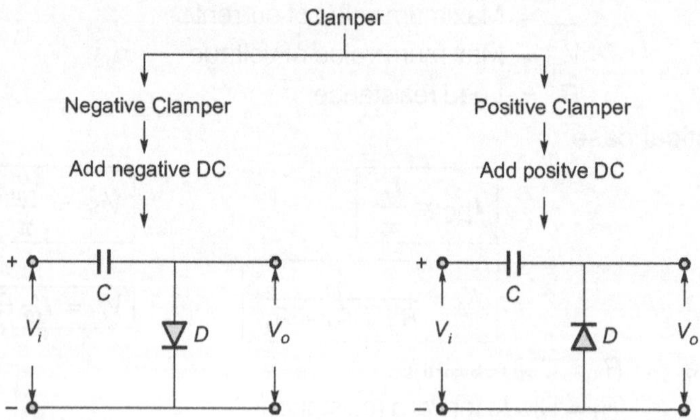

Application

A clamper is used to reinsert or restore DC component into a waveform which has lost DC component after passing through a processing circuit such as amplifier.

Voltage Multiplier

A circuit whose input is an AC waveform of peak V_m and output is DC voltage which is an integer multiple of peak AC input. $V_o = 2\,V_m, 3\,V_m, 4\,V_m$ etc.

if V_o = DC voltage of 2 V_m; Voltage doubler

V_o = DC voltage of 3 V_m; Voltage tripler

V_o = DC voltage of 4 V_m; Voltage quadrupler

Voltage doubler

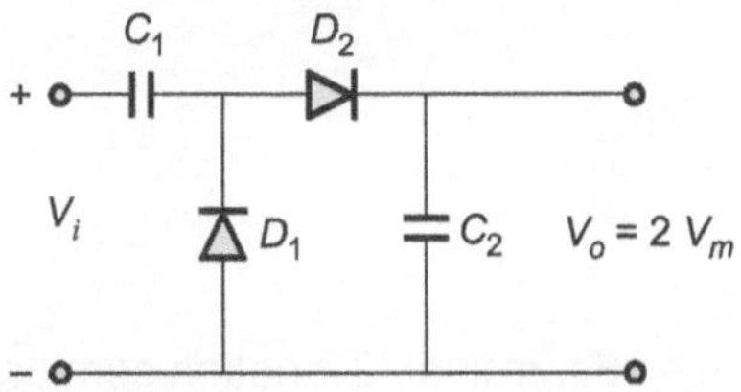

Voltage Tripler/Quadrupler

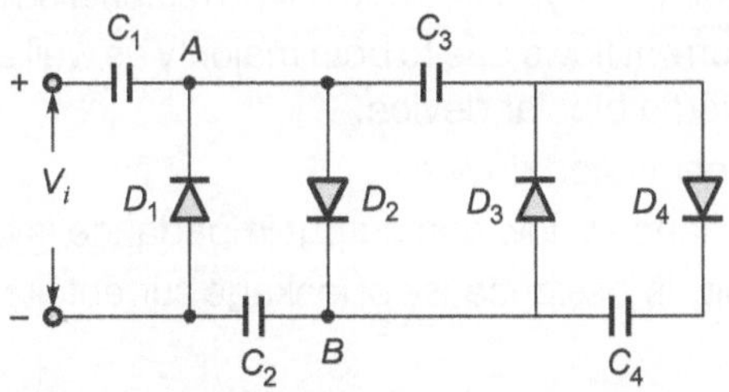

- DC voltage of 4 V_m can be obtained if V_o is measured across C_4 and C_2.
- DC voltage of 3 V_m can be obtained if V_o is measured across C_3 and C_1 combination.

■■■■

Bipolar Junction Transistor

2

Introduction

- A transistor is 3 layer, 3 terminal, 2 junction semiconductor device.
- It consist of either two *n*-type and one *p*-type layers of material called *npn* transistor or two *p*-type and one *n*-type material called *pnp* transistor.
- In transistor current flows due to both majority as well as minority carrier that's why called a bipolar device.
- It is a current controlled device.
- Its input impedance is low and output impedance is high.
- Thermal stability is less because of leakage current or reverse saturation current.
- In transistor all the major currents are diffusion currents.
- Base current is recombination current.
- A transistor represented by two diode connected back to back cannot work as transistor as there is no bonding between base and collector.
- Current conduction in *pnp* transistor is carried out by hole whereas in *npn* transistor it is due to electrons.

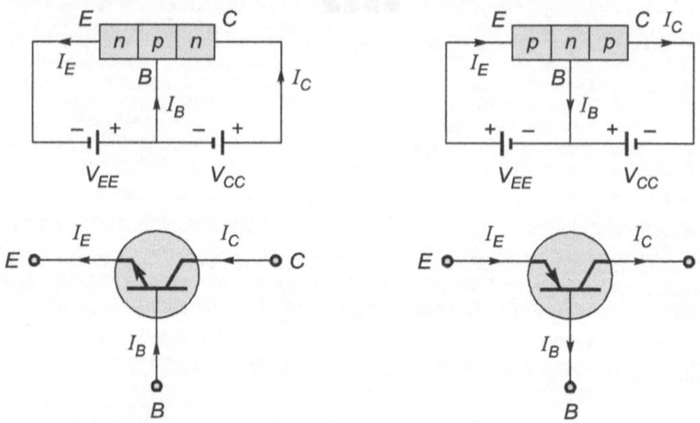

(a) *npn* transistor **(b) *pnp* transistor**

Transistor Sections

1. **Emitter:** It is heavily doped as compared to other regions. The main function of this region is to supply majority charge carrier i.e. either electrons or holes.

2. **Base:** This section is very lightly doped and very thin as compared to other sections. Its main function is to transfer the majority carriers from emitter to collector.
3. **Collector:** This section is moderately doped and has largest area compared to other two regions because it collects the charge carriers from emitter and base. There is a large amount of heat liberation and so it is provided with large area to dissipate the heat.

BJT Configuration

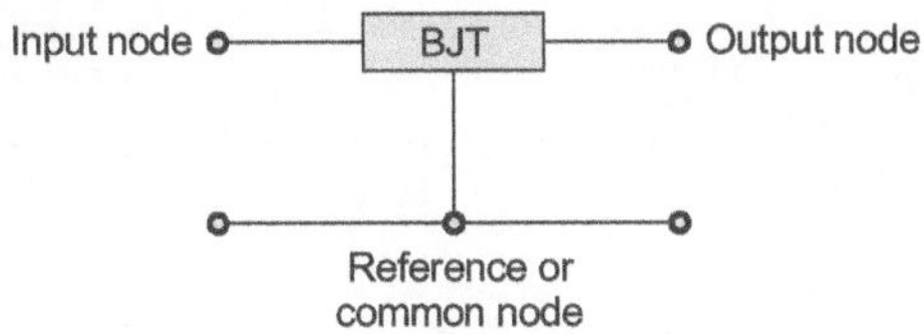

Based on the reference node a BJT can be used in 3 configuration as given in table below:

Configuration	Input node	Output node
Common Base (CB)	*E*	*C*
Common Emitter (CE)	*B*	*C*
Common Collector (CC)	*B*	*E*

1. Common Base Configuration

$$\boxed{I_E = I_B + I_C + I_{CBO}}$$

or $\boxed{I_E \approx I_B + I_C}$

where, $I_E \rightarrow$ Emitter current

$I_B \rightarrow$ Base current

$I_C \rightarrow$ Collector current

$I_{CBO} \rightarrow$ Collector to base leakage current

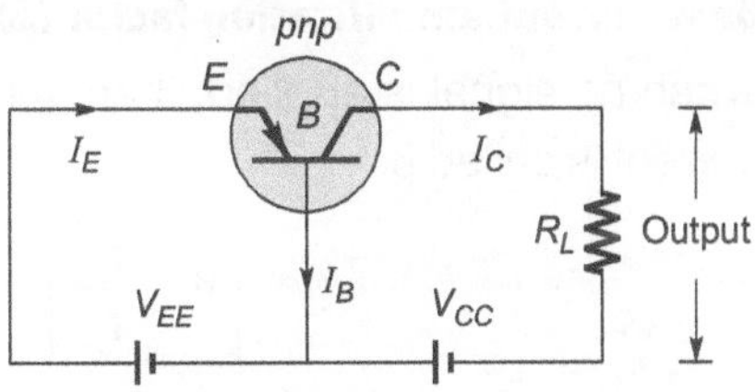

Common base *pnp* transistor

Note:

$I_B \approx 2\%$ of I_E ; $I_C \approx 98\%$ of I_E.

Current amplification factor (α)

When no signal is applied then the ratio of the collector current to emitter current is called current amplification factor or current gain.

$$\boxed{\alpha = \frac{I_C}{I_E}}$$

Note:

α of a transistor is a measure of the quality of transistor. Higher is the value of α better is the transistor as I_C approaches I_E.

Collector current can also be expressed as

$$\boxed{I_C = \alpha I_E + I_{CBO}}$$

2. Common Emitter Configuration

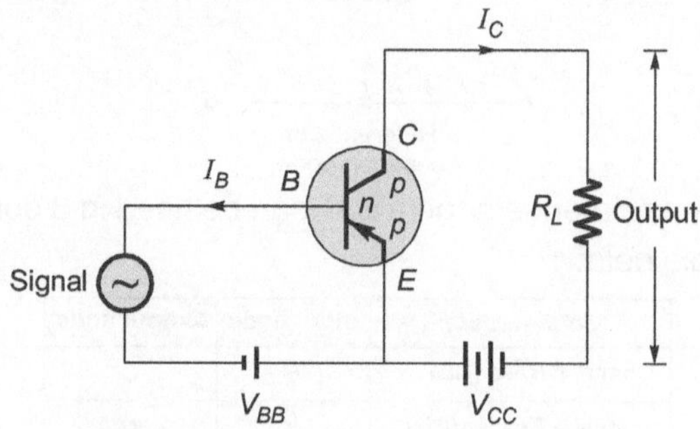

Common emitter *pnp* transistor

Base current amplification factor (β)

When no signal is applied, then the ratio of collector current to the base current is called β.

$$\boxed{\beta = \frac{I_C}{I_B}} \quad ; \quad \text{Also} \quad \boxed{\alpha = \frac{\beta}{1+\beta}}$$

Note:

- β ranges from 20 to 500.
- This configuration is used when appreciable current gain as well as voltage gain is required.

Total collector current

$$\boxed{I_C = \beta I_B + I_{CEO}} \quad \text{(or)} \quad \boxed{I_C = \beta I_B + (1+\beta) I_{CBO}}$$

where, I_{CEO} → Collector to emitter leakage current

$$\boxed{I_{CEO} = (1+\beta) I_{CBO}}$$

3. Common Collector Configuration

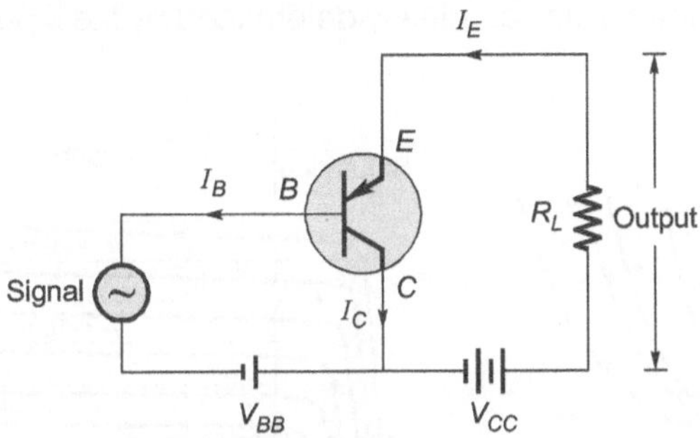

Common collector *pnp* transistor

Current amplification factor (γ)

When no signal is applied then the ratio of emitter current to the base current is called γ of the transistor.

$$\gamma = \frac{I_E}{I_B} \quad ; \quad \gamma = 1 + \beta = \frac{1}{1-\alpha}$$

Total emitter current

$$I_E = (1+\beta)I_B + (1+\beta)I_{CBO}$$

Note:

- This configuration has very high input resistance ≈ 750 kΩ and very low output resistance.
- Voltage gain is always less than one so never used for amplification purpose.
- This configuration is used for impedance matching i.e. driving a low impedance load from a high impedance source.

Characteristics	CB Configuration	CE Configuration	CC Configuration
Input resistance	Very low (40 Ω)	Low (50 kΩ)	Very high (750 kΩ)
Output resistance	Very high (1 MΩ)	High (10 kΩ)	Low (50 Ω)
Current gain	Less than unity	High (100)	High (100)
Voltage gain	Small (150)	High (500)	Less than unity
Leakage current	Very small	Very large	Very large
Applications	For high frequency applications	For audio frequency applications	For impedance matching
Phase shift between input and output	0°	180°	0°

Input and Output Characteristics of Different Configuration

The collector current I_C is completely determined by the input current I_E and V_{CB} voltage.

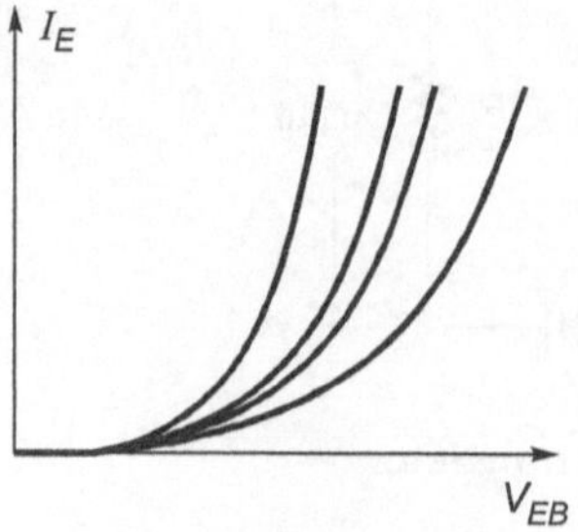

Input characteristics of CB configuration

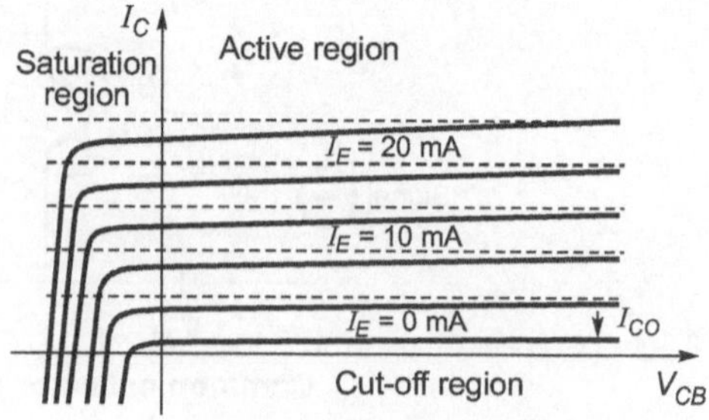

Output characteristics of CB configuration

The curve between I_B and V_{BE} for different values of V_{CE} are shown in figure.

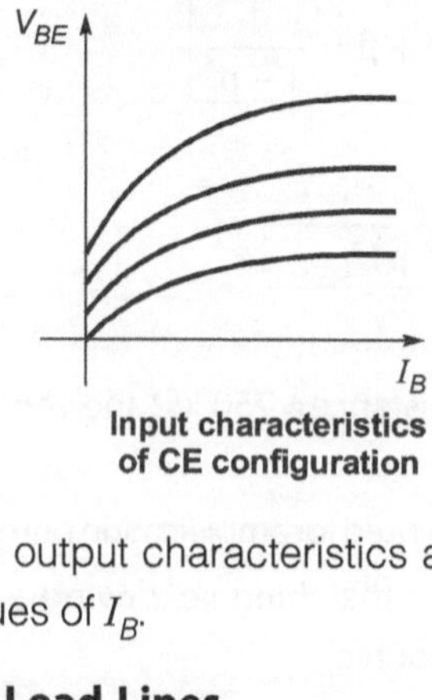

Input characteristics of CE configuration

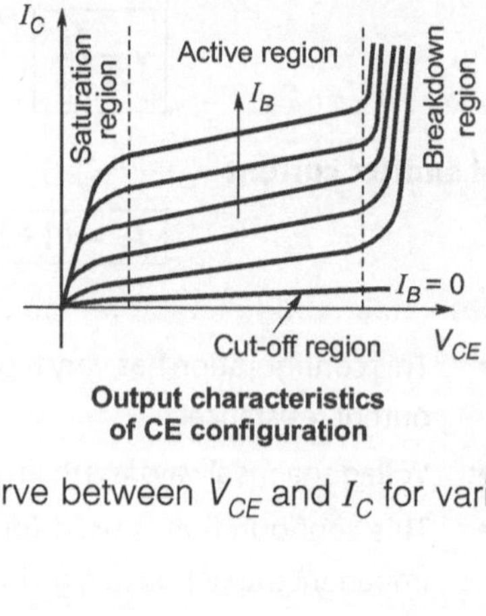

Output characteristics of CE configuration

The output characteristics are the curve between V_{CE} and I_C for various values of I_B.

DC Load Lines

- DC load line is the locus of all possible operating point at which BJT remains in active region.

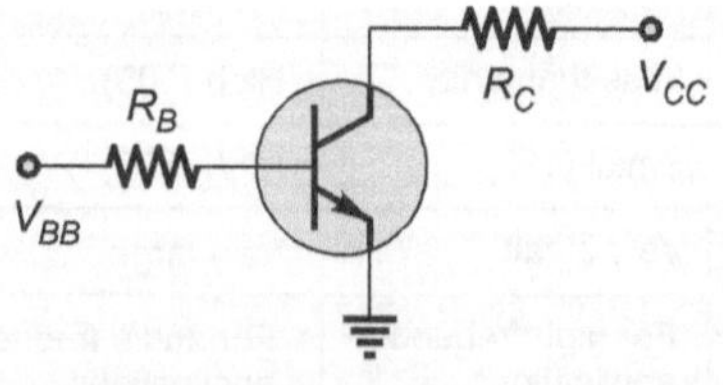

- If base current $I_B < \frac{I_{C\,sat}}{\beta}$ then transistor operate in active region.

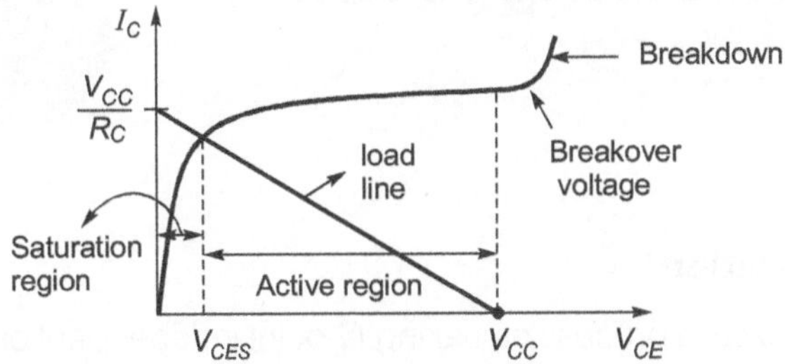

- If $I_B > \frac{I_{C\,sat}}{\beta}$ then it operates in saturation region.
- Maximum value of current

$$I_{C\max} = \frac{V_{CC}}{R_C} \qquad \text{[Taking } V_{CE} = 0 \text{ at saturation]}$$

  This is the ideal case.
- ***Power dissipated*** in a transistor is

$$P_D = I_C V_{CE}$$

Remember:

Power dissipation is maximum in active region and minimum in cutoff region and saturation region.

■■■■

Transistor Biasing Circuits

3

Bias Stabilization

- Stabilization is a process of making *Q*-point independent of changes in temperature and changes in transistor parameter.
- If I_{CO}, V_{BE} and β changes simultaneously then net change in I_C.

$$\Delta I_C = \frac{\partial I_C}{\partial I_{CO}}\Delta I_{CO} + \frac{\partial I_C}{\partial V_{BE}} \times \Delta V_{BE} + \frac{\partial I_C}{\partial \beta}\Delta\beta$$

where, $\frac{\partial I_C}{\partial I_{CO}} = S \rightarrow$ Current stability factor

$\frac{\partial I_C}{\partial V_{BE}} = S' \rightarrow$ Voltage stability factor

$\frac{\partial I_C}{\partial \beta} = S_\beta = S'' \rightarrow$ Amplification stability factor

Note:

Out of three stability factor *S* is most significant reason being, if *S* is within tolerable limit then other *S′* and *S″* are guaranteed to remain within tolerable limit.

$$S_{ideal} = 1$$

Practically *S* should be less than 20.

$$\text{Alternate evaluation of } S = \frac{1+\beta}{1-\beta\frac{\partial I_B}{\partial I_C}}$$

BJT Biasing

Biasing refers to providing appropriate DC voltage and DC current to an electronics device to operate it in a desired way.

A BJT is biased:

(i) To operate the BJT in active region so that it can be used as amplifier.

(ii) To maintain I_C stable so that the operating point does not drift and thermal run away does not happen.

Commonly Used BJT Biasing Circuit are:

(i) Fixed biased circuit

(ii) Collector to base bias circuit.

(iii) Self bias circuit

(i) Fixed biased circuit

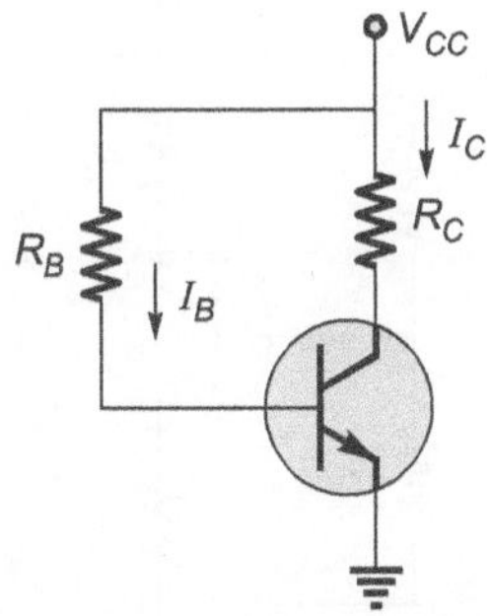

$$I_B = \frac{V_{CC} - V_{BE}}{R_B} \quad ; \quad I_C = \frac{V_{CC} - V_{CE}}{R_C}$$

I_B is constant or fixed. Hence circuit is called fixed biased circuit.

$S = 1 + \beta$; since S is very large I_C will be unstable.

(ii) Collector to Base bias circuit

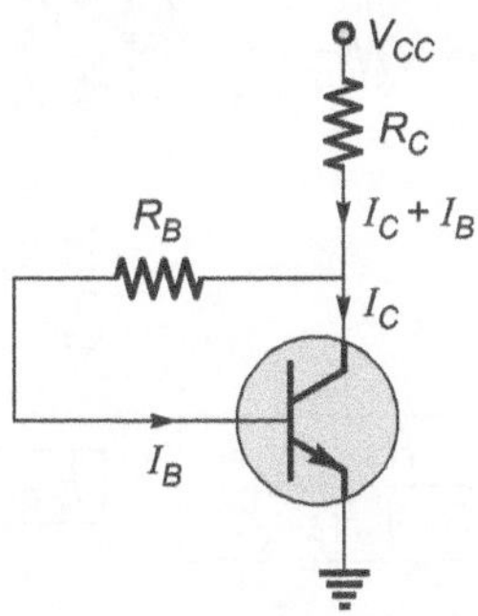

$$I_C = \frac{V_{CC} - V_{BE}}{R_C}$$ (Assuming β to be large or $I_B \approx 0$)

$$I_C = \left[\frac{V_{CC} - V_{BE}}{R_C}\right] \times \left[\frac{\beta}{\beta + 1}\right]$$

$$S = \frac{1 + \beta}{1 + \beta\left(\frac{R_C}{R_B + R_C}\right)}$$

Advantage: It gives more stable I_C than fixed biased circuit.

Drawback: Stability factor depend upon R_C, if R_C becomes smaller or zero, stability factor becomes very large and I_C does not remain stable.

(iii) Self-Bias Circuit (Voltage Divider Circuit)

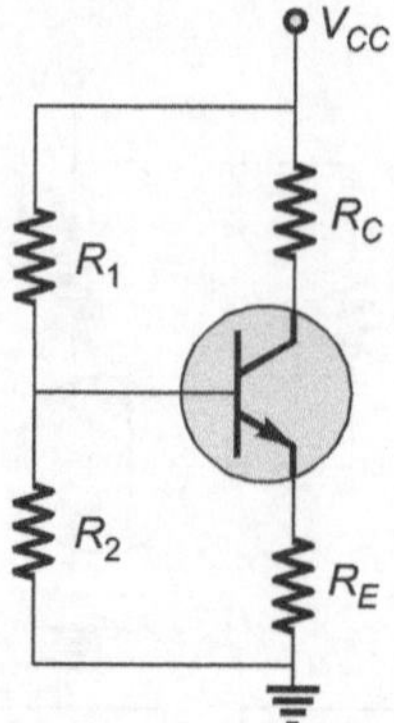

Thevenin Equivalent

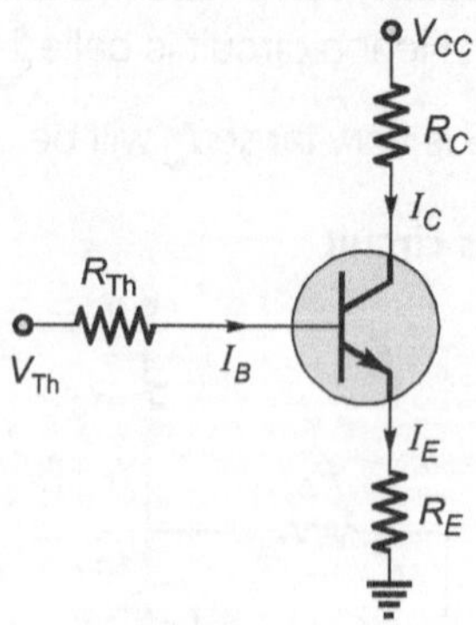

$$\boxed{V_{Th} = \frac{R_2}{R_1 + R_2} V_{CC}} \quad ; \quad \boxed{R_{Th} = R_1 \| R_2}$$

$$\boxed{I_C \cong \frac{V_{CC} - V_{CE}}{R_C + R_E}} \quad ; \quad \boxed{I_E = \frac{V_{Th} - V_{BE}}{\left(R_E + \frac{R_{Th}}{\beta + 1}\right)}} \quad \text{(Assuming } \beta \text{ to be large)}$$

$$\boxed{S = \frac{1+\beta}{1+\beta\left(\frac{R_E}{R_{Th} + R_E}\right)}} \quad \text{if } (1 + \beta)\, R_E >> R_{Th}.$$

Advantages:

(i) Self bias circuit has least stability factor among all biasing circuits.

(ii) Stability factor is independent of R_C.

(iii) Self bias circuit can be used to bias BJT for any configuration.

Drawback:

(i) Resistance R_E creates negative feedback due to which voltage gain

$\frac{V_o}{V_i}$ or $\frac{V_o}{V_s}$ becomes smaller.

Basic Current Mirror

$$I_{out} = \frac{I_{in}}{1+\frac{2}{\beta}}$$

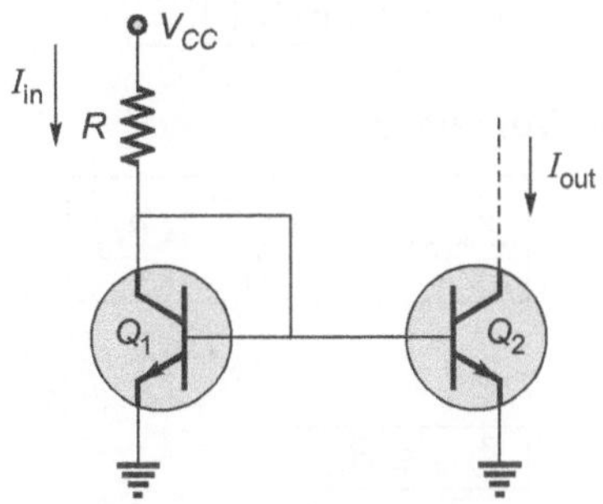

Current mirror is symbolically shown as constant current source.

Applications of current mirror

(i) A current mirror is used to bias BJT in IC amplifiers.

(ii) Current mirror can be used as active load in IC amplifier.

Advantage of current mirror biasing

(i) It require smaller chip area than self bias circuit.

(ii) A current mirror is equivalent to constant current source hence the DC collector current of bias transistor remains highly stable.

■■■■

4 BJT as an Amplifier

Common Emitter Amplifier

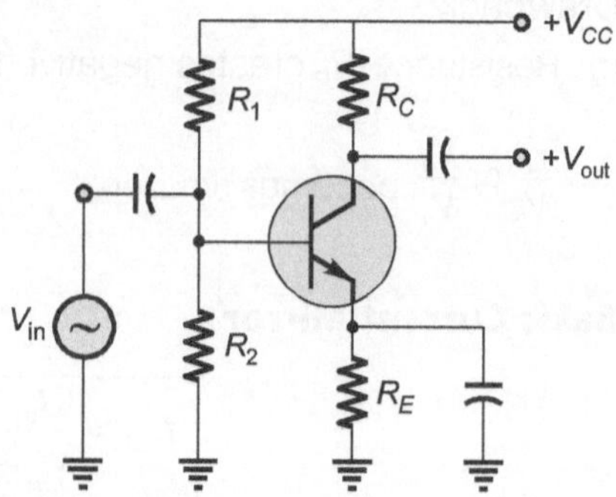

Output voltage

$$V_{out} \cong -I_E R_C$$

Voltage gain

$$A = \frac{V_{out}}{V_{in}} \cong -\frac{R_C}{r'_e}$$

Input impedance

$$Z_{in} \cong R_1 \parallel R_2 \parallel \beta r'_e$$

Input impedance seen from base

$$Z_{in(base)} \cong \frac{\beta I_b r'_e}{I_b} = \beta r'_e$$

Output impedance

$$Z_{out} \cong R_C$$

where, $$r'_e = \frac{1}{g_m} = \frac{V_T}{I_C}$$

Common Base Amplifier

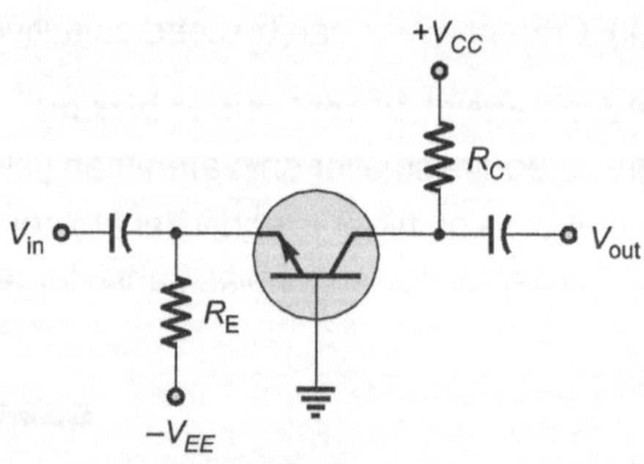

Input impedance

$$Z_{in} \cong r'_e$$

Voltage gain

$$A = \frac{V_{out}}{V_{in}} \cong \frac{R_C}{r'_e}$$

Output impedance

$$Z_{out} \cong R_C$$

Common Collector Amplifier (or Emitter Follower)

DC output voltage

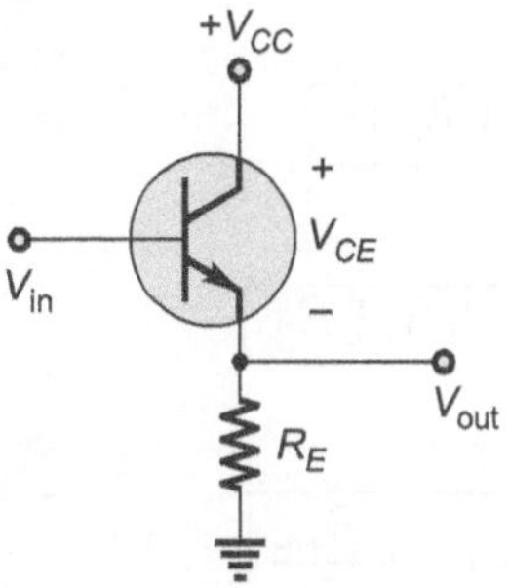

$$V_{out} = V_{in} - V_{BE}$$

Input impedance

$$Z_{in} = R_1 \parallel R_2 \parallel \beta R_E$$

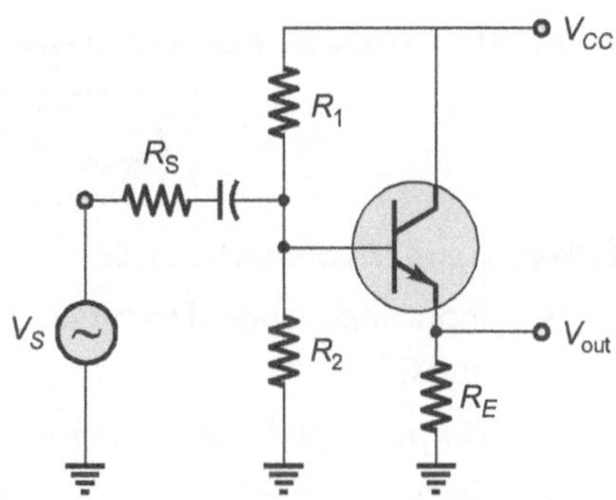

Input impedance seen from base

$$Z_{in(base)} = \beta(R_E + r_e') \cong \beta R_E$$

Output impedance

$$Z_{out} = r_e' + \frac{R_s \parallel R_1 \parallel R_2}{\beta}$$

Voltage gain

$$A = \frac{R_E}{R_E + r_e' + (R_s \parallel R_1 \parallel R_2)/\beta} \cong 1$$

Darlington Amplifier

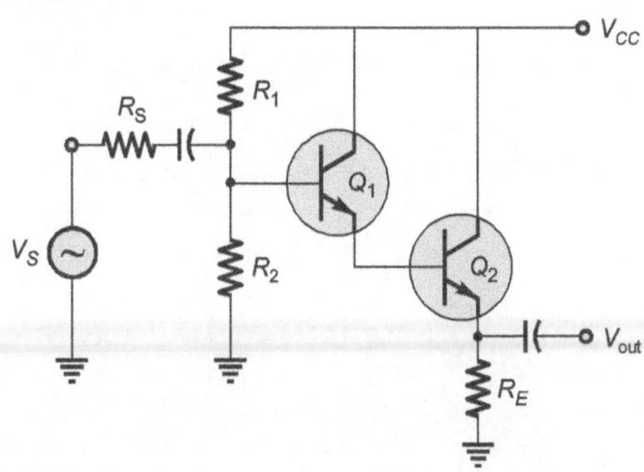

Input impedance of second stage

$$Z_{in(2)} \cong \beta_2 R_E$$

Input impedance of first stage

$$Z_{in(1)} = \beta_1 \beta_2 R_E$$

AC Thevenin impedance at input

$$R_{Th} = R_s \parallel R_1 \parallel R_2$$

Output impedance of first stage

$$Z_{out(2)} = r'_{e1} + \frac{R_{Th}}{\beta_1}$$

Output impedance of second stage

$$Z_{out(2)} = r'_{e2} + \frac{r'_{e1} + R_{Th}/\beta_1}{\beta_2}$$

Note:

- Input impedance of common collector amplifier is highest among CE, CB and CC.
- Output impedance of common collector is lowest among CE, CB and CC.
- Darlington pair is known as super β circuit.
- Darlington pair is used where very high input impedance and high current gain is required.
- Voltage gain of CB is highest and current gain of CC is highest.
- CE has highest power gain.

■■■■

Analysis of Small Signal Amplifier

5

Introduction

In the analysis of small signal amplifier, transistor is replaced with an equivalent circuit known as small signal model.

Small Signal Model

- Low frequency
  - (i) Hybrid parameter model
  - (ii) π-model
  - (iii) *T*-mode (*r*-model)

  } used in analysis of BJT amplifier at medium or low frequency
- High frequency
  - (i) π-model
  - (ii) *T*-model

  } H.F. analysis of BJT amplifier

Hybrid parameter model

The *h*-parameter model of BJT is defined by two-port network as

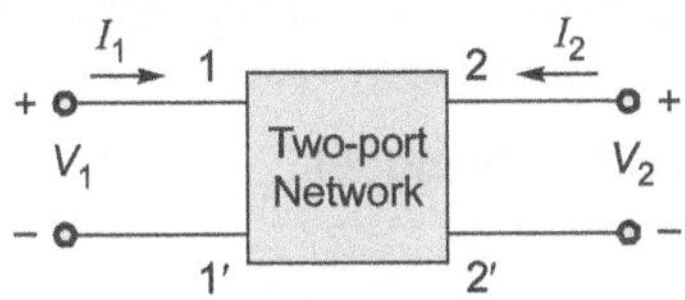

$$V_1 = h_{11} I_1 + h_{12} V_2$$

$$I_2 = h_{21} I_1 + h_{22} V_2$$

Input impedance h_{11}/h$_i$

$$h_{11} = \left.\frac{V_1}{I_1}\right|_{V_2=0}$$

Current gain h_{21}/h_f

$$h_{21} = \left.\frac{I_2}{I_1}\right|_{V_2=0}$$

Reverse voltage gain h_{12}/h_r

$$h_{12} = \left.\frac{V_1}{V_2}\right|_{I_1=0}$$

Output admittance h_{22}/h_o

$$h_{22} = \left.\frac{I_2}{V_2}\right|_{I_1=0}$$

Typical h-parameter values for transistor

Parameters	CE	CC	CB
$h_{11} = h_i$	1.1 kΩ	1.1 kΩ	21.6 Ω
$h_{12} = h_r$	2.5×10^{-4}	≈ 1	2.9×10^{-4}
$h_{21} = h_f$	50	–51	–0.98
$h_{22} = h_o$	24 μA/V	24 μA/V	0.49 μA/V
$1/h_o$	40 kΩ	40 kΩ	2.04 MΩ

Approximated conversion formulae for h-parameters

CE to CC	CE to CB
$h_{ic} = h_{ie}$	$h_{ib} = \dfrac{h_{ie}}{1 + h_{fe}}$
$h_{rc} = 1 - h_{re}$	$h_{rb} = \dfrac{h_{ie} h_{oe}}{1 + h_{fe}} - h_{re}$
$h_{fc} = -(1 + h_{fe})$	$h_{fb} = \dfrac{-h_{fe}}{1 + h_{fe}}$
$h_{oc} = h_{oe}$	$h_{ob} = \dfrac{h_{oe}}{1 + h_{fe}}$

Hybrid model or low frequency model

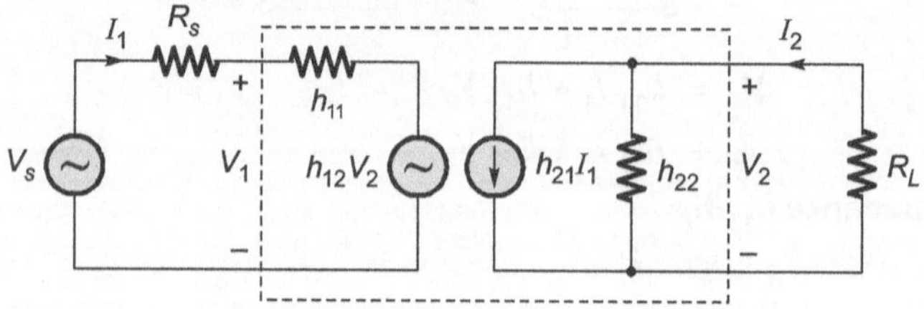

Current gain

$$\boxed{A_I = -\frac{h_f}{1 + h_o R_L}}$$

$$\boxed{A_I' = \frac{A_I R_s}{R_i + R_s}}$$

Input impedance

$$Z_{in} = R_s + h_i + A_I h_r R_L$$

$$\boxed{R_i = h_i + A_I h_r R_L}$$

Voltage gain

$$\boxed{A_V = A_I \frac{R_L}{R_i}}$$

$$\boxed{A_{VS} = \frac{V_o}{V_{in}} = A_V\left(\frac{R_i}{R_i + R_s}\right)}$$

Output admittance

$$\boxed{Y_o = h_o - \frac{h_r h_f}{R_s + h_i}}$$

Under assumption $R_L \to \infty$ and $V_{in} = 0$

CE Amplifier with unbypassed R_E

$$R_i = h_{ie} + (1 + h_{fe})R_E$$

If by-pass capacitor is disconnected then A_V will decrease, input and output impedance will increase.

Small Signal Amplifier Analysis

	Medium Frequency	Low Frequency	High Frequency
Value of large (μF) capacitor	Short circuit	Active	Short circuit
Value of small (pF) capacitor	Open circuit	Open circuit	Active

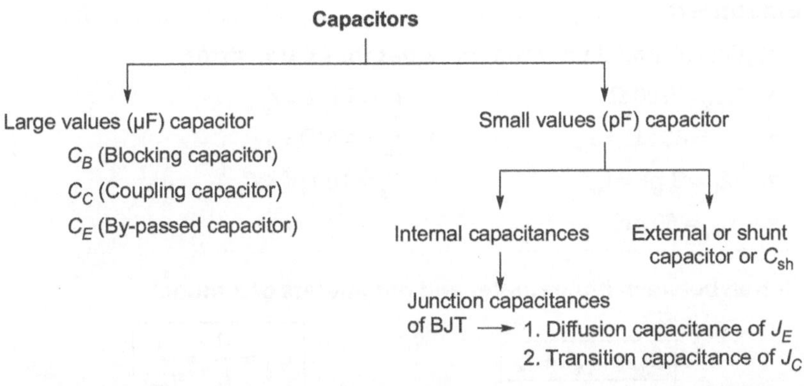

(i) Blocking capacitor (C_B) and coupling capacitor (C_C) provide DC isolation to biasing circuit. So that no DC current flows out of the biasing circuit and no DC current enters the biasing circuit from the outside.

(ii) By-pass capacitor (C_E) prevents decrease in voltage gain.

Capacitor	Lower cut-off frequency (f_L)
Coupling capacitor (C_C)	$\frac{1}{2\pi C_c(R_C + R_L)}$
Bypass capacitor (C_E)	$\frac{g_m}{2\pi C_E}$
Blocking capacitor (C_B)	$\frac{1}{2\pi C_B(R_s + R_i)}$

π-model or High Frequency Model

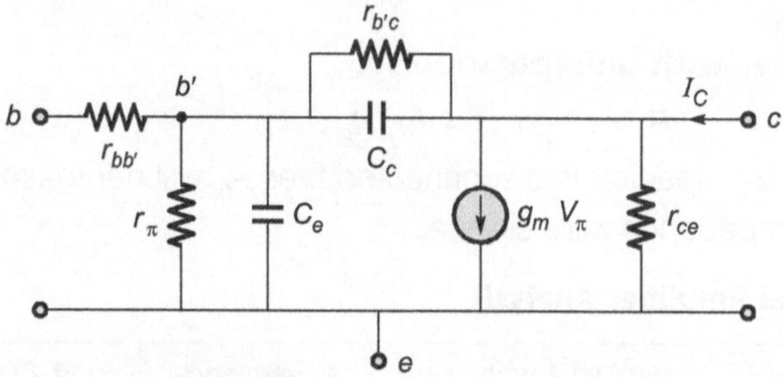

where, $V_{b'e} = V_\pi$

$r_{bb'} \rightarrow$ Ohmic base spreading resistance

$C_c \rightarrow$ Collector junction barrier capacitance

$C_e \rightarrow$ Diffusion capacitance between b' and e

Remember: ....

Typical π-model parameters values for CE transistor

- $r_{bb'} = 100\ \Omega$
- $r_\pi = 1\ k\Omega = r_{b'e}$
- $r_o = 80\ k\Omega = r_{ce}$
- $r_\mu = 4\ M\Omega = r_{b'c}$
- $C_\mu = 3\ pF = C_c$
- $C_\pi = 100\ pf = C_e$
- $g_m = 50\ mS$

....

Relation between *h*-parameter and parameters of π-model

$$\boxed{h_{ie} = r_{bb'} + r_\pi}$$

$$\boxed{g_0 = \frac{1}{r_o} = \frac{1}{r_{ce}}}$$

$$\boxed{h_{re} = \frac{r_\pi}{r_\mu}}$$

$$\boxed{g_\mu = \frac{1}{r_\mu} = \frac{1}{r_{b'c}}}$$

$$\boxed{h_{oe} = g_o + (1 + h_{fe})g_u}$$

$$\boxed{r_\pi = \frac{h_{fe}}{g_m}}$$

$$\boxed{g_m = \frac{|I_c|}{V_T}}$$

$$\boxed{r_{b'c} = \frac{r_{b'e}}{h_{re}} = \frac{r_\pi}{h_{re}}}$$

$$\boxed{r_{bb'} = r_b = h_{ie} - r_{b'e} = h_{ie} - r_\pi}$$

Common emitter short circuit current gain

- $$A_I = \frac{h_{fe}}{\sqrt{1 + \left(\frac{f}{f_\beta}\right)^2}}$$

  where, $f_\beta \rightarrow$ 3 db upper cut-off frequency

  $$f_\beta = \frac{1}{2\pi r_{b'e}(C_e + C_c)}$$

- Frequency at which short circuit current gain becomes unity (f_T)

  $$f_T = f_\beta\sqrt{h_{fe} - 1} \quad \text{or} \quad f_T = h_{fe} f_\beta \ (\because \ h_{fe} >> 1)$$

  $$f_T = \frac{h_{fe}}{2\pi r_{b'e}(C_e + C_c)} \quad \text{or} \quad f_T = \frac{g_m}{2\pi(C_e + C_c)}$$

- Common emitter gain with resistive load at output

  $$A_I = \frac{h_{fe}}{\sqrt{1 + \left(\frac{f}{f_H}\right)^2}} \qquad \text{where, } f_H = \frac{1}{2\pi r_{b'e}[C_e + C_c(1 + g_m R_L)]}$$

Note: ....

$$\boxed{f_T > f_\beta > f_H}$$

....

Multistage Amplifier

- Upper cut-off frequency of overall configuration is

  $$f_H^* = f_H\sqrt{2^{1/n} - 1} \quad \text{(For identical amplifiers in cascade)}$$

  where, $n \rightarrow$ Number of amplifiers in cascade

  $f_H \rightarrow$ Upper cut-off frequency of one amplifier

- Lower cut-off frequency of overall configuration

$$f_L^* = \frac{f_L}{\sqrt{2^{1/n} - 1}}$$

Approximate bandwidth of amplifier in cascade is f_H^*.

- When amplifier are non-identical then

$$f_L^* = 1.1\sqrt{f_{L_1}^2 + f_{L_2}^2 + \cdots + f_{L_n}^2} \text{ and } \frac{1}{f_H^*} = 1.1\sqrt{\frac{1}{f_{H_1}^2} + \frac{1}{f_{H_2}^2} + \cdots \frac{1}{f_{H_n}^2}}$$

- Rise time, $t_r^* = 1.1\sqrt{t_{r_1}^2 + t_{r_2}^2 + \cdots t_{r_n}^2}$

Special Multistage Amplifier

1. Darlington amplifier 2. Cascode amplifier

1. Darlington amplifier

- It is cascade of two common collector amplifier or cascade of two common drain amplifier.
- It has unit voltage gain, very high input resistances, low output resistance and high current gain.
- It is used as buffer.

2. Cascode amplifier

- It is cascade of *CE* and *CB* amplifiers (or) cascade of common source and common gate amplifier.
- It can provide both voltage and current amplification therefore it is preferred over common base amplifier in radio frequency amplification and wideband amplification.

■■■■

6 Amplifier Circuit Using FET and MOSFET

All the formulae are same whether it is designed using FET or MOSFET.

Common Source Configuration

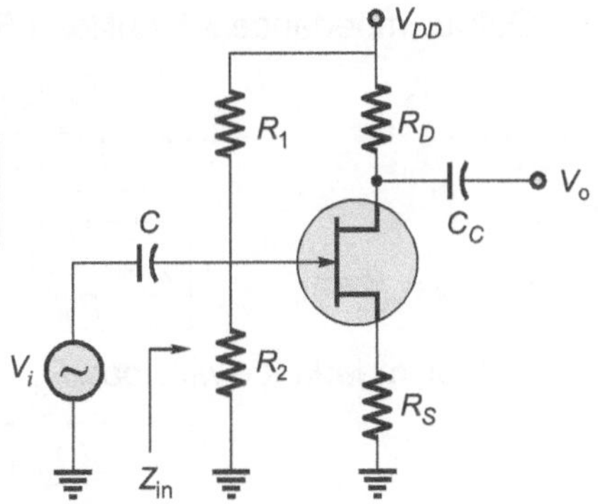

- Voltage gain

$$A_v = \frac{V_o}{V_i} = \frac{-\mu R_D}{r_d + R_D + (1+\mu) R_s}$$

If $r_d \to \infty$ then

$$A_v = \frac{-g_m R_D}{1 + g_m R_s}$$

if $R_s = 0$ then

$$A_v = -g_m R_D$$

where, $A_v \to$ Voltage

- Input impedance $Z_{in} = R_1 \| R_2$

- Output impedance Z_o(with load R_D) = $R_D \| [r_d + (\mu + 1)R_s]$

without load $R_D \Rightarrow$ $Z_o \Rightarrow r_d + (\mu + 1)R_s$

If $r_d \to \infty$ then $Z_o = R_D$ where, $r_d \to$ Drain resistance

Common Drain Amplifier

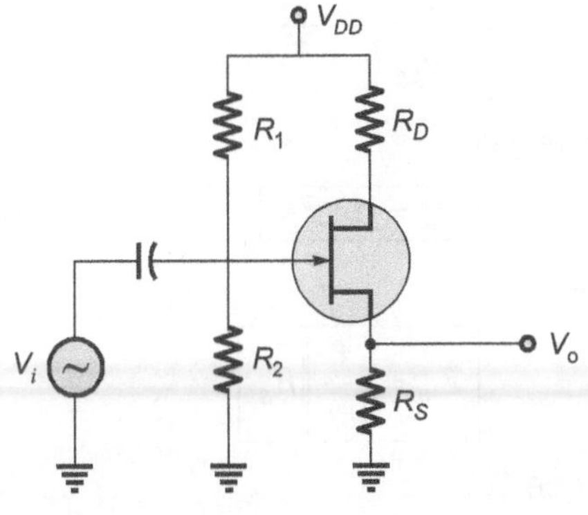

- Voltage gain

$$A_v = \frac{\mu R_s}{r_d + R_D + (\mu + 1)R_s} = \frac{V_o}{V_i}$$

- If $r_d \to \infty$ then $$A_v = \frac{g_m R_s}{1 + g_m R_s}$$

- Output impedance without load R_s

$$Z = \left(\frac{r_d + R_D}{\mu + 1}\right)$$

If $r_d \to \infty$ then $$Z = \frac{1}{g_m}$$

- Output impedance with load R_s

$$Z_o = \left(\frac{r_d + R_D}{\mu + 1}\right) \Bigg\| R_s$$

If $r_d \to \infty$ then $$Z_o = \frac{1}{g_m} \Bigg\| R_s$$

Common Gate Amplifier

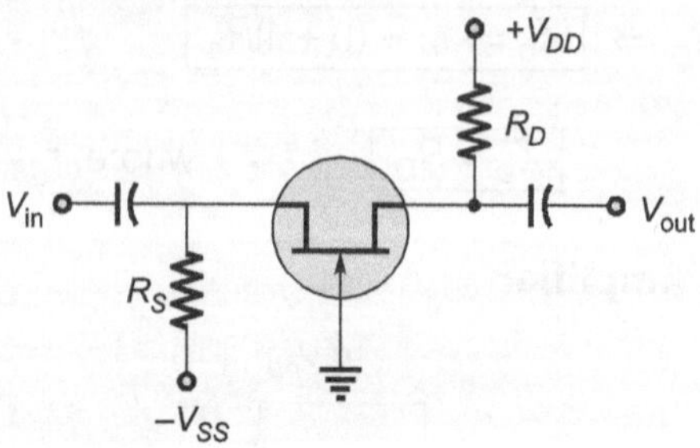

- Unloaded voltage gain

$$A_v = g_m R_D$$

- Input impedance

$$Z_{in} = \frac{1}{g_m} \Bigg\| R_s$$

- Output impedance

$$Z_0 = R_D$$

- DC ON state resistance

$$r_{DS(ON)} = \frac{V_{DS}}{I_D}$$

where, $r_{ds\,(ON)} \rightarrow$ DC resistance in saturation region
$V_{DS} \rightarrow$ DC drain source voltage
$I_D \rightarrow$ DC drain current

Note: ....

- Common source amplifier is equivalent to *CE* amplifier of BJT.
- Common drain amplifier is equivalent to common collector of BJT.
- Common gate amplifier is equivalent to common base amplifier BJT.
- Common drain amplifier is also known as source follower.
- Common gate amplifier has unity current gain, low input resistances, high output resistance and high voltage gain.

Low Frequency Analysis of FET Amplifier

Small signal low frequency equivalent of JFET

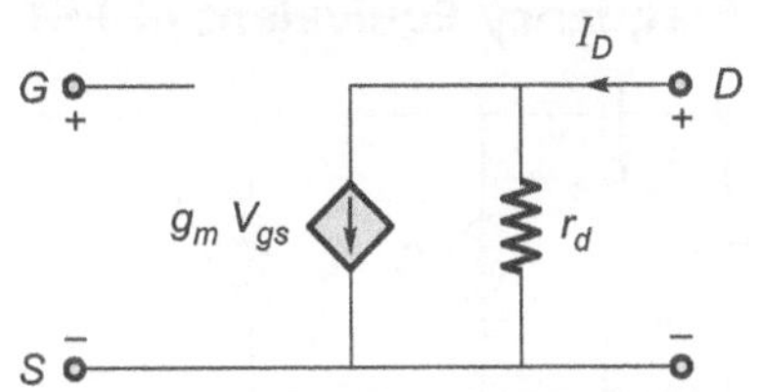

- Amplifier are analyzed at low frequency by considering the effect of large capacitances like C_B, C_C.
- Low frequency analysis of *CS* amplifier is same as low frequency analysis of *CE* amplifier.

1. $$|A_{VL}| = \frac{|A_{vm}|}{\sqrt{1+\left(\frac{f_L}{f}\right)^2}}$$

2. $$\angle A_{VL} = 180° + \tan^{-1}\left(\frac{f_L}{f}\right)$$

3. If C_C is consider $$f_L = \frac{1}{2\pi C_C\,(R'_D + R_L)}$$

4. If bypass capacitor (C_S) is considered,

$$f_L = \frac{g_m}{2\pi C_s}$$

5. If C_B is considered

$$f_L = \frac{1}{2\pi C_B (R_s + R_i')}$$

High Frequency Analysis of FET Amplifier

It is performed by considering the effect of small capacitor.

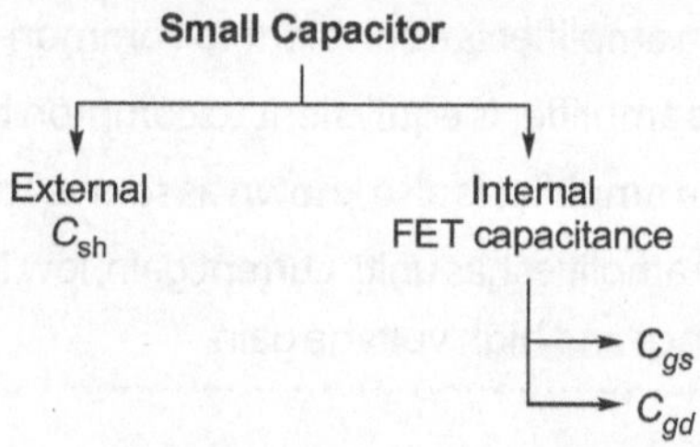

C_{gs}, C_{gd}: Transition capacitances of gate channel junction.
C_{ds}: Internal capacitance of channel.

Small Signal High Frequency Equivalent of FET Amplifier

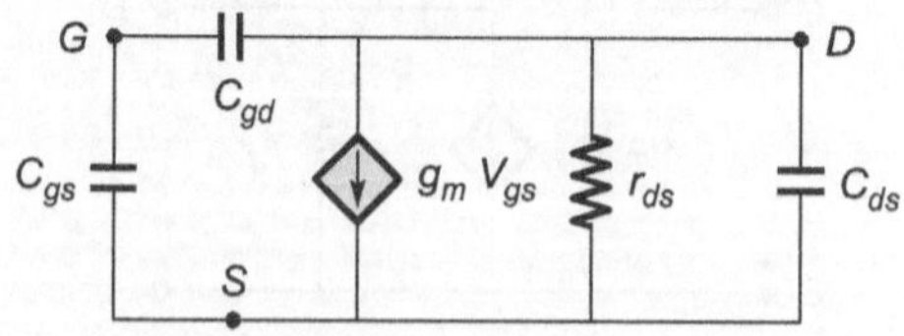

- In common source configuration C_{gd} will be present between input node gate and output node drain. Therefore it can be replaced with C_m and C_n using Miller theorem.

$$C_m = C_{gd}(1 - A_v) \quad ; \quad C_n = C_{gd}\left(1 - \frac{1}{A_v}\right)$$

- C_m appears in parallel with C_{gs} therefore net capacitance between gate and source or net input capacitance will be

$$C_i = C_{gs} + C_m \quad ; \quad C_i = C_{gs} + C_{gd}(1 - A_v)$$

- C_n appears parallel to C_{ds} therefore net output capacitance will be

$$C_o = C_{ds} + C_n \quad ; \quad C_o = C_{ds} + C_{gd}$$

- Miller effect occurs in *CE* and *CS* configuration, it causes increase in input capacitance.
- HF analysis of *CS* amplifier is same as that of *CE* amplifier.

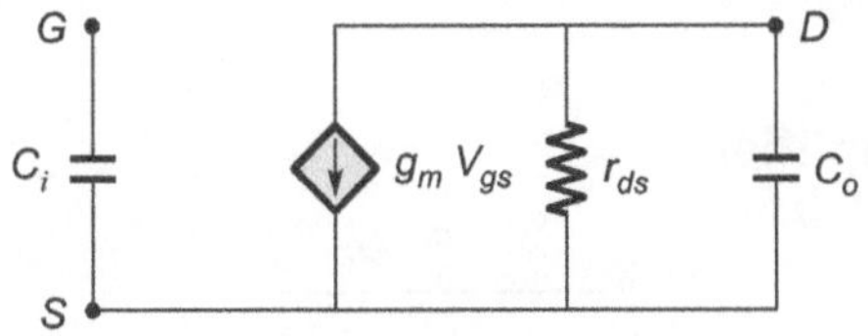

- $$|A_{VH}| = \frac{|A_{vm}|}{\sqrt{1+\left(\frac{f}{f_H}\right)^2}}$$
- $$\angle A_{VH} = 180° - \tan^{-1}\frac{f}{f_H}$$
- If C_{sh} is consider $$f_H = \frac{1}{2\pi R_L'' C_{sh}}$$
- If C_i is consider $$f_H = \frac{1}{2\pi C_i (r_{ds} \| R_s)}$$

where, $R_L' = R_D \| R_L$

$R_L'' = R_L' \| r_{ds}$

■■■■

7

Operational Amplifier

Differential Amplifier

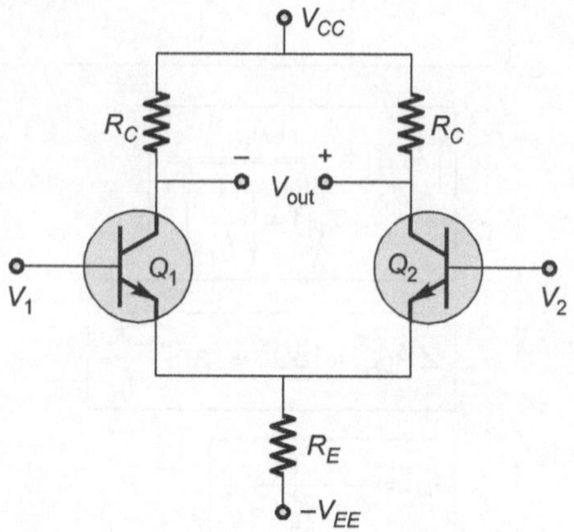

- Here,

$$V_{out} = A(V_1 - V_2)$$

where, $V_{out} \rightarrow$ Voltage between collectors

$$\boxed{A = \frac{R_c}{r_e} = g_m R_c}$$

where, $r_e \rightarrow$ Emitter resistance

$V_1 \rightarrow$ Non-inverting input voltage

$V_2 \rightarrow$ Inverting input voltage

DC analysis of a differential amplifier

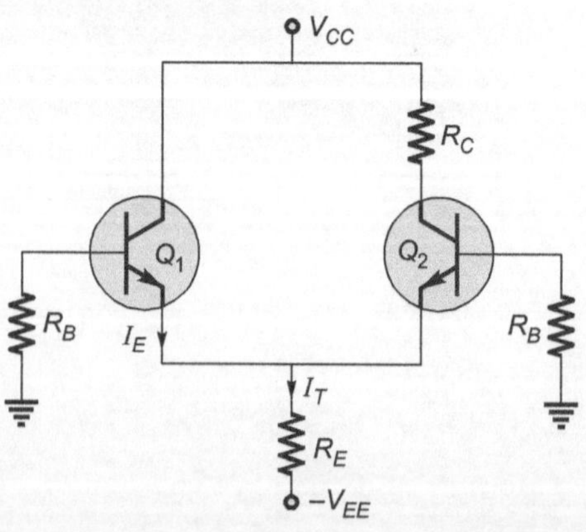

- Emitter bias current

$$I_E = \frac{V_{EE} - V_{BE}}{2R_E}$$

- Tail current

$$I_T = \frac{V_{EE} - V_{BE}}{R_E}$$

AC analysis of a differential amplifier

1. Non inverting input:
   - AC emitter current

$$i_e = \frac{V_1}{2r_e}$$

where, $r_e \rightarrow$ Emitter resistance

   - AC output voltage

$$V_{out} = i_C R_C = \frac{V_1}{2r_e} R_C$$

   - Voltage gain for non-inverting input

$$\frac{V_{out}}{V_1} = \frac{R_c}{2r_e}$$

2. Inverting input:
   - AC emitter current

$$i_e = \frac{V_2}{2r_e}$$

   - AC output voltage

$$V_{out} = -i_C R_C = \frac{-V_2}{2r_e} R_C$$

   - Voltage gain for non-inverting input is

$$\frac{V_{out}}{V_2} = \frac{R_c}{2r_e}$$

   - Differential voltage gain

$$A = \frac{R_C}{2r_e}$$

   - Input impedance

$$r_{in} = 2\beta r_e$$

Common mode voltage gain

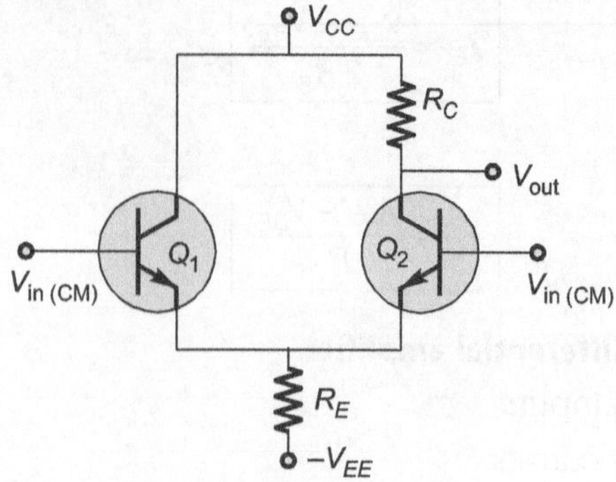

$$A_{CM} = \frac{V_{out}}{V_{in(CM)}} = \frac{-R_C}{r_e + 2R_E} = \frac{-R_C}{2R_E}$$

$$R_E >> r_e$$

$$\boxed{\text{CMRR} = \left|\frac{A}{A_{CM}}\right| = 2g_m R_E}$$

where, $A_{CM} \rightarrow$ Common mode gain
$A \rightarrow$ Differential mode gain

Operational Amplifier

An op-omp is basically a very high gain, direct-coupled amplifier with high input impedance and low output impedance.

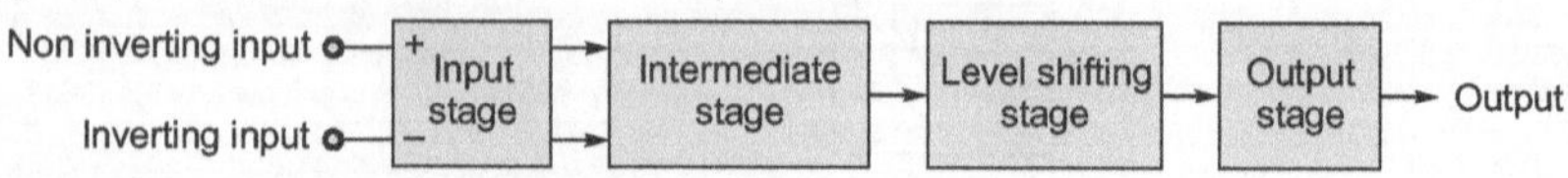

Equivalent circuit of an op-amp

Op-amp amplifies the between the two input signals applied at non-inverting and inverting input terminals.

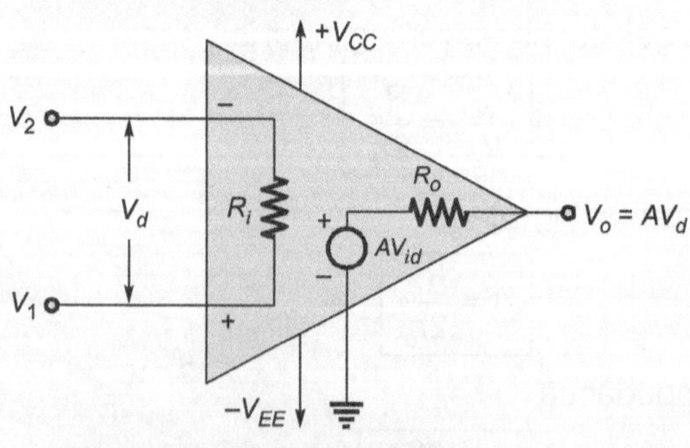

Equivalent circuit of op-amp

where, Difference input voltage → $V_d = V_1 - V_2$

A → Open loop gain of op-amp

Output voltage → $V_o = AV_d$

R_i → Differential input resistance

R_o → Output resistance

Properties of Op-amp

Parameters	Ideal value	Practical value
Voltage gain	∞	10^6
Input resistance	∞	$10^6\ \Omega$ or 1 MΩ
Output resistance	0	10 Ω to 100 Ω
B.W.	∞	10^6 Hz or 1 MHz
CMRR	∞	10^6 or 120 dB
Slew rate	∞	80V/μ sec.

Slew rate

For input $V_m \sin \omega t$

$$f_{max} = \frac{SR}{2\pi A_{CL} V_m}$$

where, SR → Slew rate

f_{max} → Maximum frequency of operation

A_{CL} → Closed loop gain of OP-AMP.

Bias Currents and Voltage

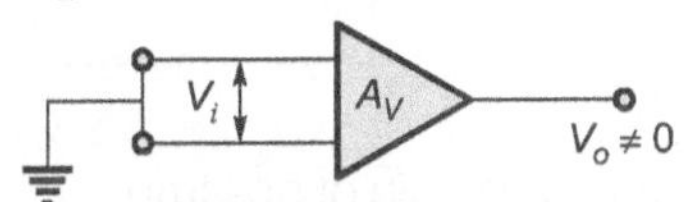

$V_0 \neq 0$ → Called O/P offset voltage.

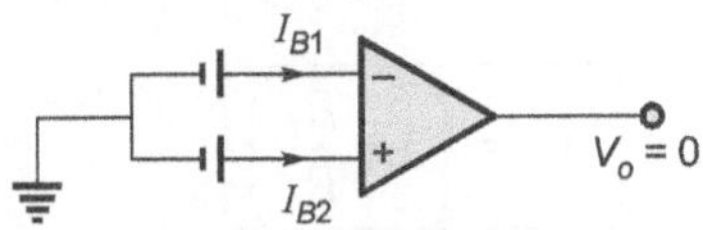

Balanced condition.

Input bias current (I_B)

$$I_B = \frac{I_{B_1} + I_{B_2}}{2}$$

Input offset current

$$I_{io} = (I_{B1} - I_{B2})$$

Note:

Input offset voltage is the voltage which must be applied between input terminals to balance amplifier ($V_0 = 0$).

Linear Op-Amp Circuits

Inverting Amplifier

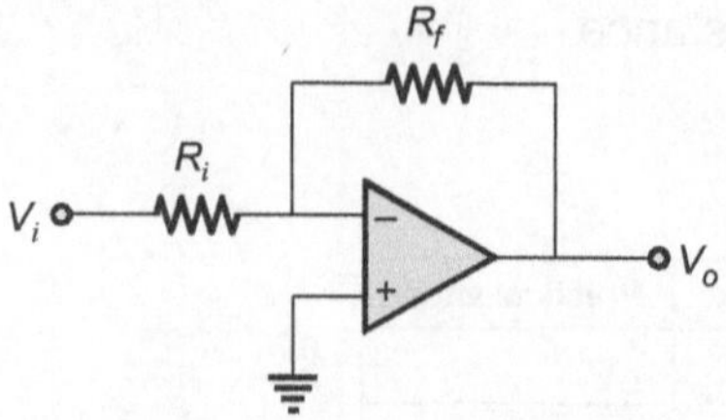

$$\text{Gain} = A_V = \frac{V_o}{V_i} = \frac{-R_f/R_1}{1+\left(\dfrac{1+R_f/R_1}{A_{OL}}\right)}$$

where, $A_{OL} \rightarrow$ Open loop gain of op-amp
ideally; $A_{OL} = \infty$

$$A_V = -\frac{R_f}{R_1}$$

Non-inverting Amplifier

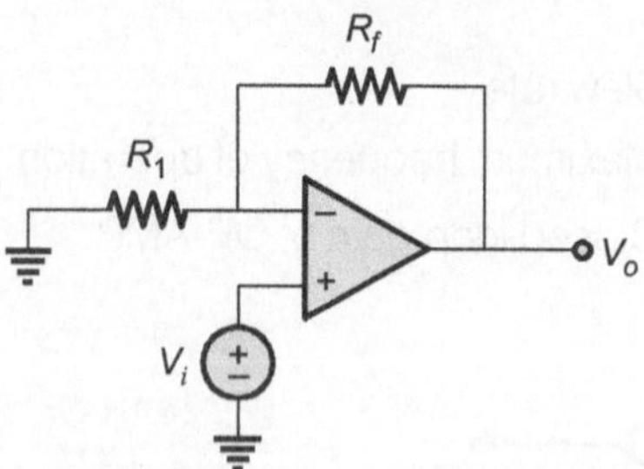

$$\text{Gain} = A_V = \frac{V_o}{V_i} = \frac{\left(1+\dfrac{R_f}{R_1}\right)}{1+\left(\dfrac{1+R_f/R_1}{A_{OL}}\right)}$$

where, $A_{OL} \rightarrow$ Open loop gain of op-amp
ideally; $A_{OL} \rightarrow \infty$

$$A_V = \left(1+\frac{R_f}{R_1}\right)$$

Difference Amplifier

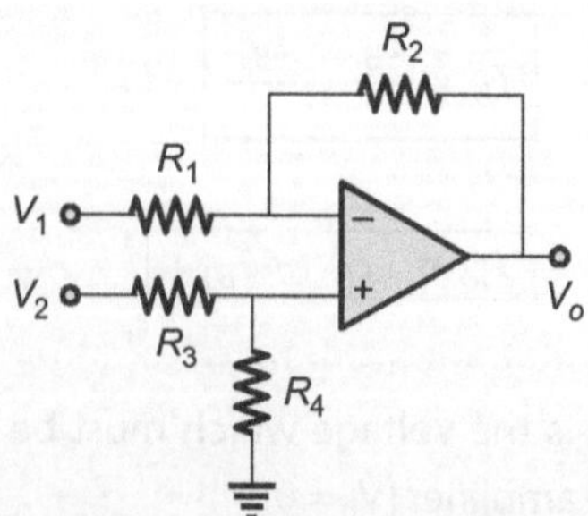

$$V_o = -\frac{R_2}{R_1}V_1 + \left(\frac{R_4}{R_3+R_4}\right)\left(1+\frac{R_2}{R_1}\right)V_2$$

if $\boxed{\frac{R_2}{R_1} = \frac{R_4}{R_3}}$ then, $V_0 = \frac{R_4}{R_3}[V_2 - V_1] = \frac{R_2}{R_1}(V_2 - V_1)$

Note: ....

In this case when $\frac{R_2}{R_1} = \frac{R_4}{R_3}$,

$A_c = 0 \rightarrow$ Common mode gain and CMRR $= \infty$

....

Nonlinear Op-Amp circuits

Integrator

$$\boxed{V_0 = -\frac{1}{RC}\int V_i\,dt + V_c(0^+)}$$

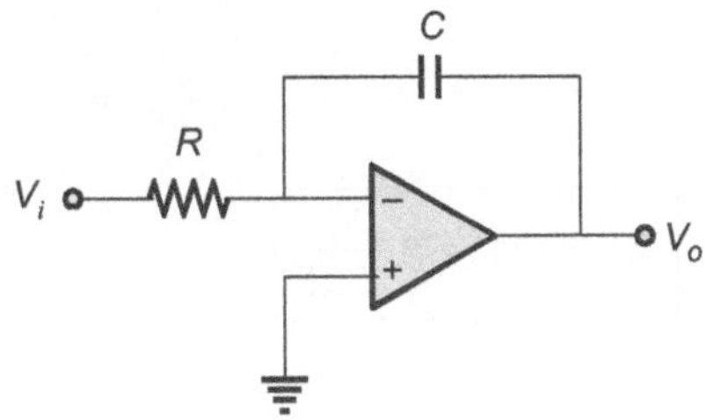

Note: ....

- This integrator becomes unstable at low frequency due to finite gain.
- The drawback of instability can be eliminated by connecting a resistor R_f in feedback path and such integrator circuit is called practical integrator.

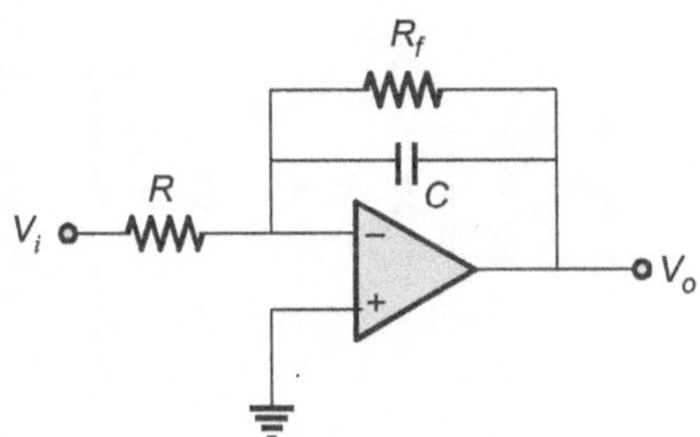

....

Differentiator

$$\boxed{V_0 = -RC\frac{dV_i}{dt}}$$

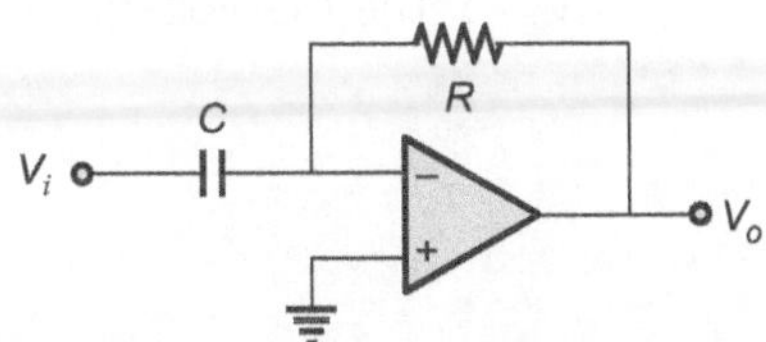

Note:

- This differentiator becomes unstable at high frequency due to finite gain.
- The drawback of instability can be eliminated by connected R_f in input path and such differentiator is called practical differentiator.

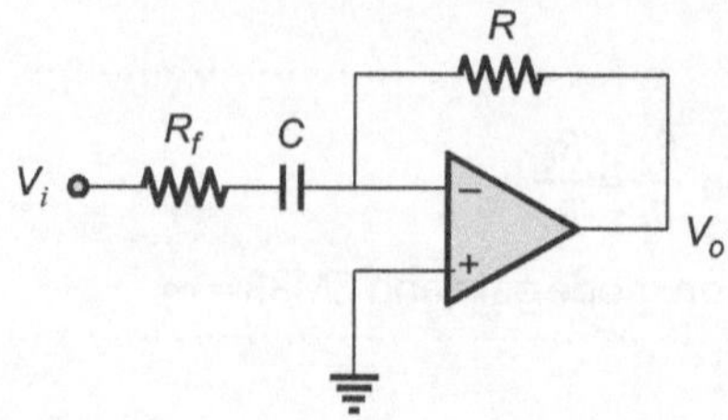

Logarithmic Amplifier

$$V_0 = -\eta V_T ln \frac{V_i}{I_0 R}$$

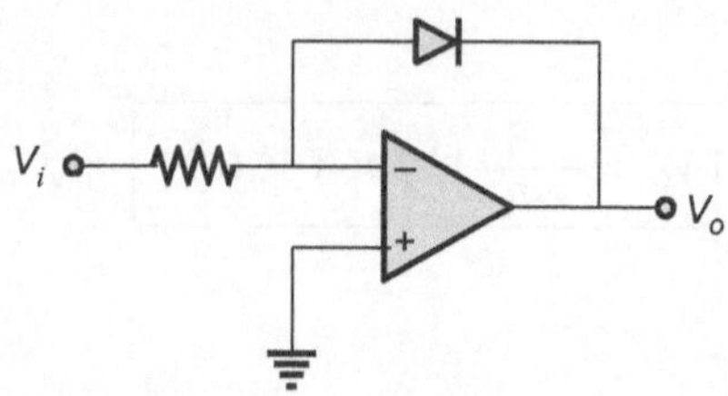

where, $\eta \rightarrow$ Recombination factor
$V_T \rightarrow$ Thermal voltage
$I_o \rightarrow$ Reverse saturation current of diode

Antilog Amplifier

$$V_o = -I_o R \text{ antilog} \frac{V_i}{\eta V_T}$$

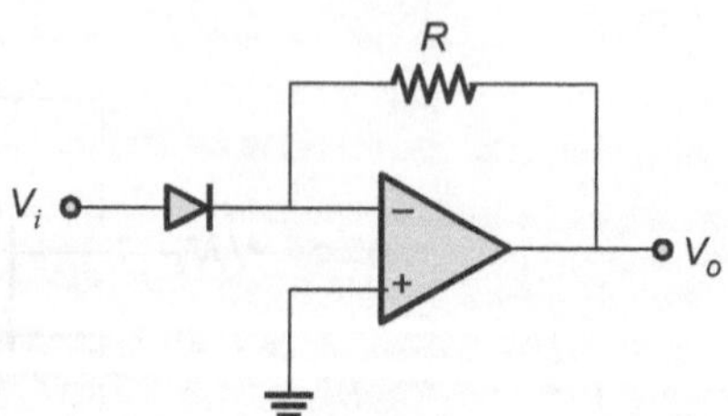

where, $I_0 \rightarrow$ Reverse saturation current of diode
$\eta \rightarrow$ Recombination factor
$V_T \rightarrow$ Thermal voltage

Grounded Load

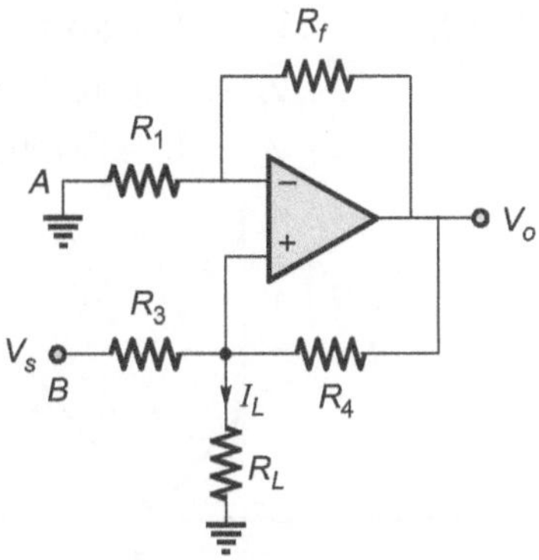

If $\boxed{\dfrac{R_f}{R_1} = \dfrac{R_4}{R_3}}$ then $\boxed{I_L = \dfrac{V_s}{R_3}}$

If V_s is applied at point A and B is grounded and

$\boxed{\dfrac{R_f}{R_1} = \dfrac{R_4}{R_3}}$ then $\boxed{I_L = \dfrac{-V_s}{R_3}}$

Schmitt Trigger

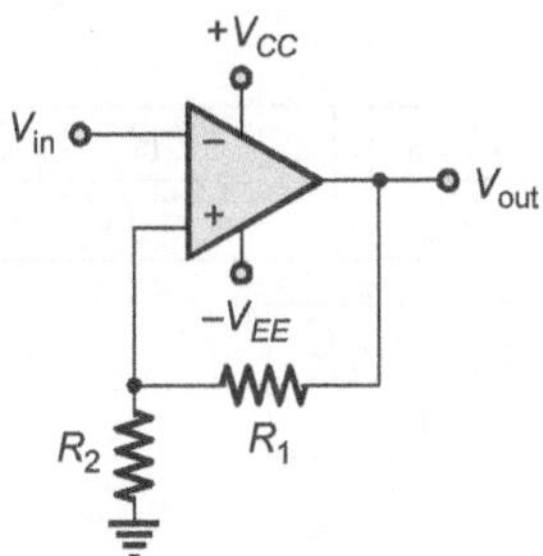

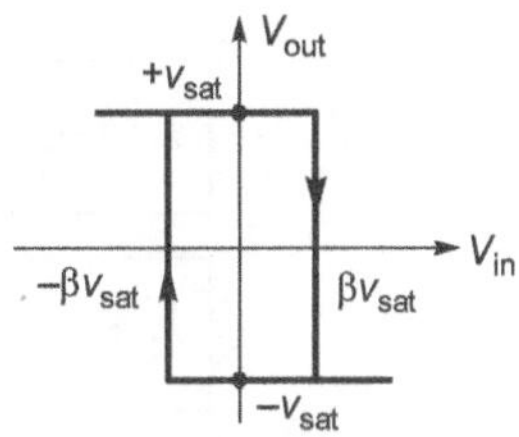

$$\boxed{V_{Th} = \beta V_{sat} = \left(\frac{R_2}{R_1 + R_2}\right) V_{sat}} \quad ; \quad \boxed{V_{TL} = -\beta V_{sat} = \left(\frac{-R_2}{R_2 + R}\right) V_{sat}}$$

Remember: ....

- The complete transfer characteristics of Schmitt trigger shows a hysteresis effect.
- The width of hysteresis is the difference between the two cross-over voltage V_{Th} and V_{TL}.

....

Triangular Wave Generator

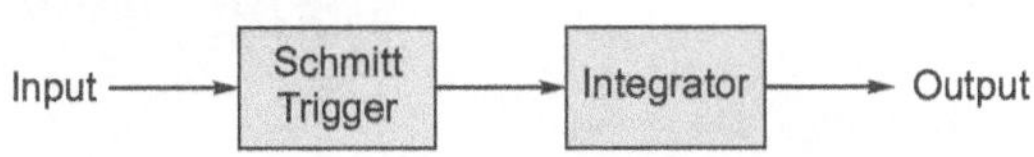

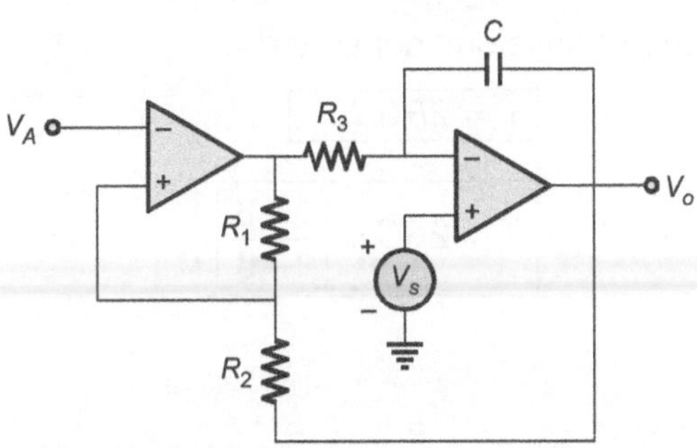

where, $V_s \rightarrow$ Control the slope

$V_A \rightarrow$ Control the amplitude output

$V_s = 0 \rightarrow$ Symmetrical triangular wave is obtained

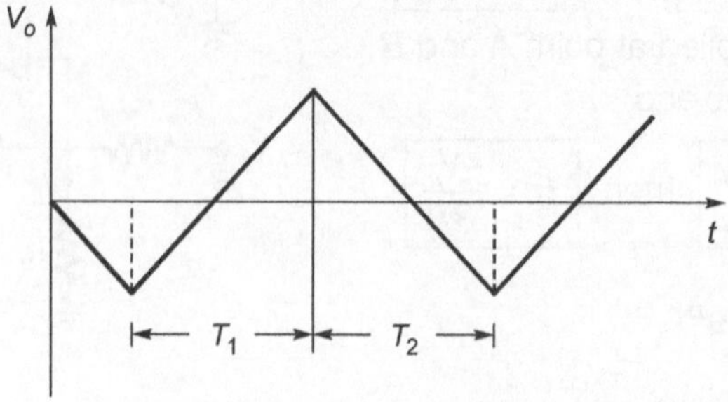

where,

$$T_1 = \frac{2R_2 R_3 C V_{sat}}{R_1(V_{sat} + V_s)} \qquad T_2 = \frac{2R_2 R_3 C V_{sat}}{R_1(V_{sat} - V_s)}$$

$$T_1 + T_2 = T = \frac{4R_2 R_3 C}{R_1}$$

Op-Amp as a Multivibrator

Square Wave Generator or Astable Multivibrator

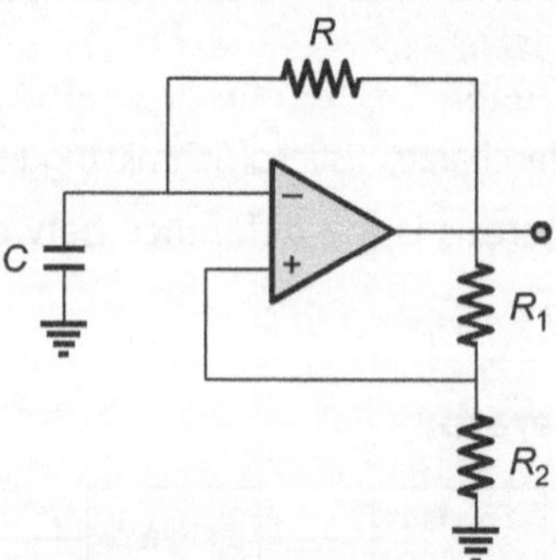

Time period of output waveform generated is

$$T = 2RC\, ln3 \qquad \text{(For } R_1 = R_2\text{)}$$

$$T = 2RC\, ln\left[\frac{1+\beta}{1-\beta}\right] \qquad \left(\text{where } \beta = \frac{R_2}{R_1 + R_2}\right)$$

Monostable Multivibrator

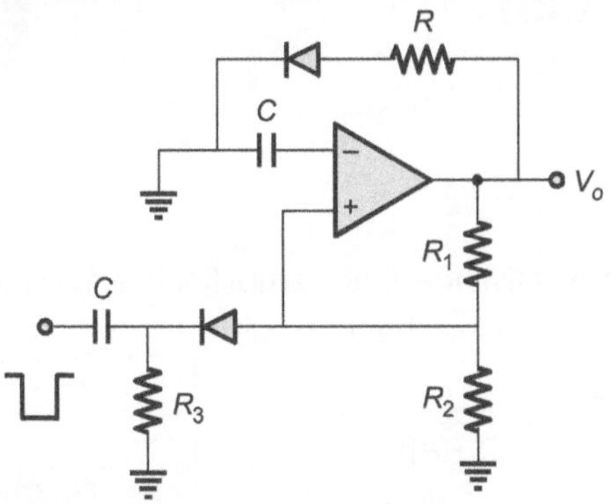

Width of pulse generated at output is

$$T = RC\ ln\left(1+\frac{R_2}{R_1}\right)$$

if $R_2 = R_1$ then $T = RC\ ln\ 2$

Note:

- Astable multivibrator is a square wave generator.
- Monostable multivibrator is used as pulse stretcher and missing pulse detector.

Bistable Multivibrator

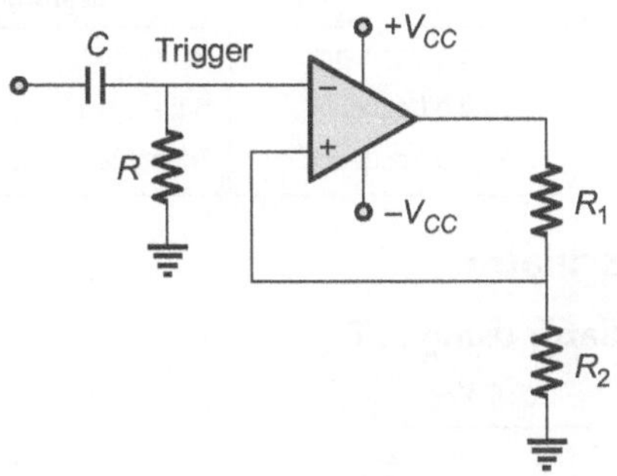

Note:

- Astable multivibrator is a square wave generator.
- Monostable multivibrator is used or pulse stretcher and missing pulse detector.

The Signal Generators and Wave Shaping Circuit

8

Oscillators

The oscillators are positive feedback amplifier in which the part of output is feedback to the input via feedback circuit.

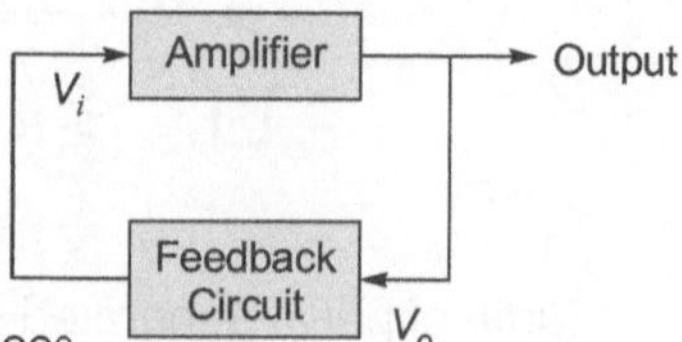

The gain, $\dfrac{V_o}{V_i} = \dfrac{A}{1-A\beta}$

$\because$ $V_i = 0$ (Noise)

$A\beta = 1$ or $1\angle 0° \cong \angle 360°$

Note:

- $|A\beta| = 1$, the amplitude of sinusoidal output remains constant.
- If $|A\beta| < 1$, the amplitude of oscillation will decrease with time and eventually or finally the oscillation will disappear.
- If $|A\beta| > 1$, the amplitude of oscillation will increase with time practically the amplitude becomes constant after reaching maximum value due to supply voltage limitation and non-linearity of transistor.

Oscillator Types

Type of Component	Frequency of oscillation	Waveform generated
RC oscillator	Audio frequency	Sinusoidal
LC oscillator	Radio frequency	Square wave
Crystal oscillator	Radio frequency	Triangular wave sawtooth wave etc.

RC Phase Shift Oscillator

RC phase shift oscillator using BJT

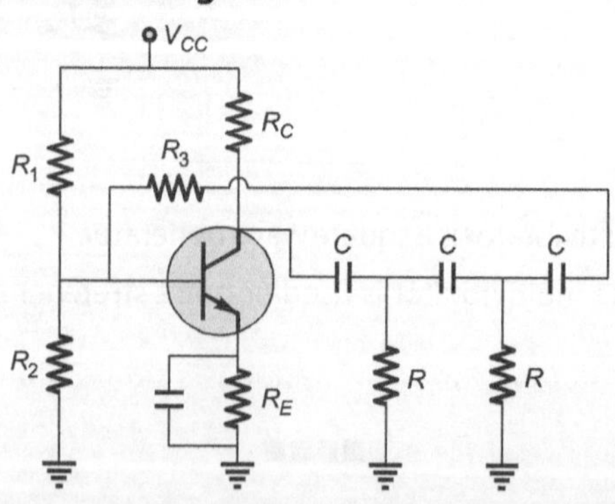

- Frequency of oscillation is

$$\boxed{\omega = \frac{1}{RC\sqrt{4K+6}}} \quad ; \quad \boxed{f = \frac{1}{2\pi RC\sqrt{4K+6}}}$$

where, $K = \frac{R_C}{R}$

Note: ....

- For transistor to work as oscillator

$$h_{fe} \geq \left(4K + 23 + \frac{29}{K}\right)$$

- $h_{fe\,(min)} = 44.5$.

RC Phase Shift Oscillator using FET

- Frequency of Oscillation

$$\boxed{f = \frac{1}{2\pi RC\sqrt{6}}}$$

$$\boxed{\beta = -\frac{1}{29}\text{(feedback factor)}}$$

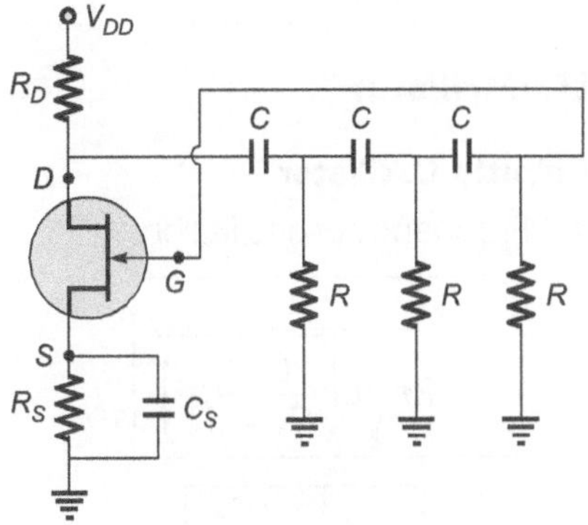

Note: ....

Amplification factor for FET to work as an Oscillator is $\mu \geq 29$.

RC Phase Shift Oscillator using Op-Amp

- Feedback factor

$$\boxed{\beta = \frac{V_f}{V_o} = -\frac{1}{29}\text{(for oscillation)}}$$

- Frequency of oscillation

$$\boxed{f = \frac{1}{2\pi RC\sqrt{6}}}$$

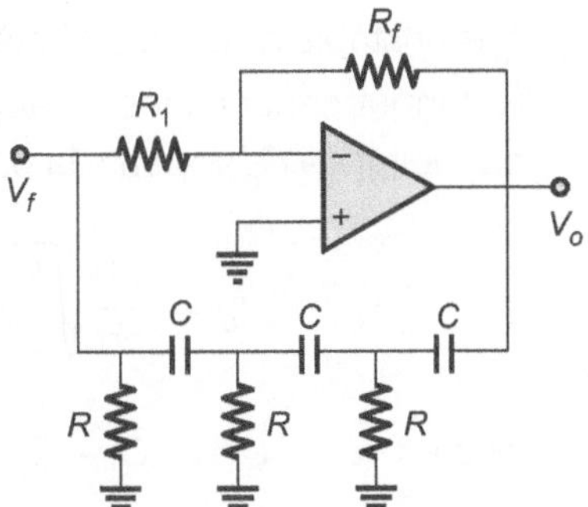

Note: ....

- $R_f \geq 29R_1 \rightarrow$ for oscillation.
- In *RC* phase shift oscillator minimum three *RC* circuits are needed to create a phase shift of 180°.

Wein Bridge Oscillator

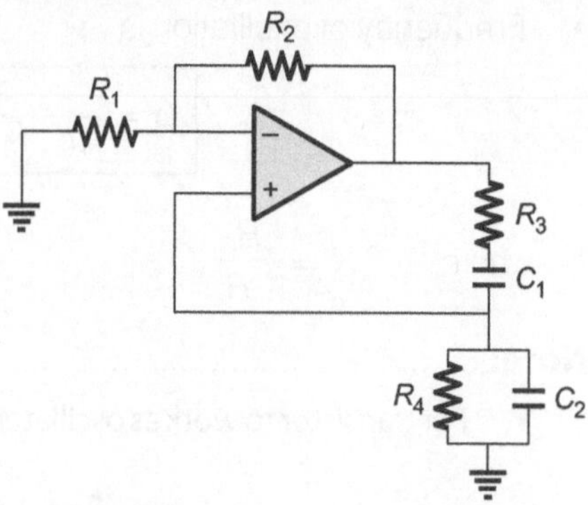

- Frequency of oscillation

$$f = \frac{1}{2\pi\sqrt{R_3 R_4 C_1 C_2}}$$

if $R_3 = R_4 = R$

and $C_1 = C_2 = C$

then $$f = \frac{1}{2\pi R.C}$$

Remember:

- $\beta = \frac{1}{3}$ for oscillation.
- $R_2 \geq 2R_1$ for oscillation.

LC Oscillator

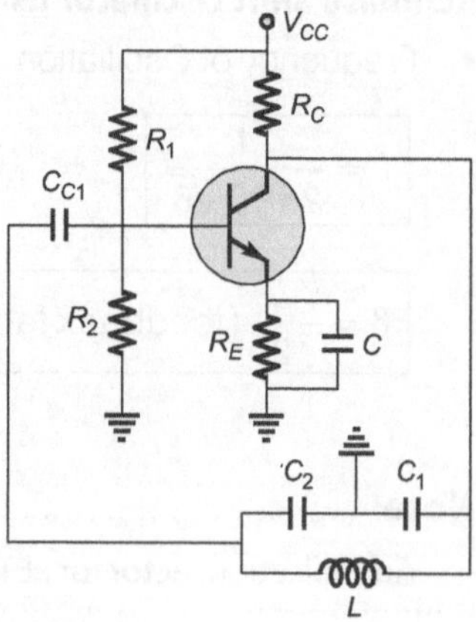

Colpitt's Oscillator

- Frequency of oscillation

$$f = \frac{1}{2\pi\sqrt{L\left(\frac{C_1 C_2}{C_1 + C_2}\right)}}$$

and $$g_m R_C \geq \frac{C_2}{C_1}$$

Drawback of Colpitt's oscillator

f_o is unstable due to the effect of internal capacitances.

Net input capacitance C_i appears in parallel with C_1 and net output capacitance C_o appears in parallel with C_2. As a result

$$f_o = \frac{1}{2\pi}\sqrt{\frac{1}{L}\left(\frac{1}{C_1'} + \frac{1}{C_2'}\right)}$$

where, $C_1' = C_1 + C_i$; $C_2' = C_2 + C_o$.

Since C_i and C_o vary with time and temperature, f_o becomes unstable. To eliminate this drawback an additional capacitance C_3 can be connected in series with inductor. Such circuit is called clapp oscillator.

Clapp Oscillator

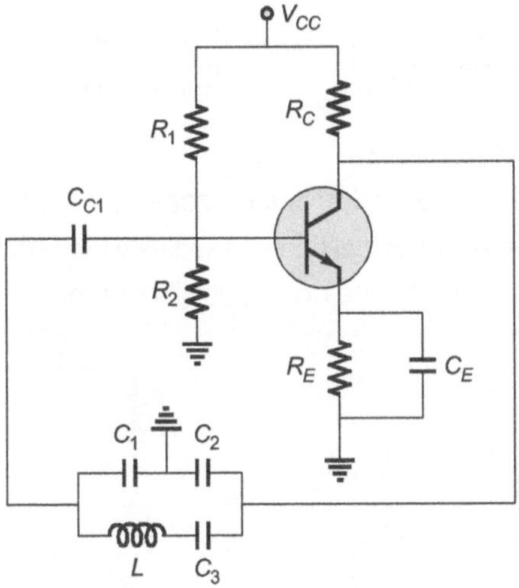

Frequency of oscillation

$$f_o = \frac{1}{2\pi}\sqrt{\frac{1}{L}\left(\frac{1}{C_1}+\frac{1}{C_2}+\frac{1}{C_3}\right)}$$

If C_3 is selected such that it is much smaller than C_1' and C_2' then

$$\frac{1}{C_3} >> \frac{1}{C_1}+\frac{1}{C_2}$$

Thus,
$$f_o = \frac{1}{2\pi}\sqrt{\frac{1}{L}\times\frac{1}{C_3}}$$

f_o becomes independent of C_1 and C_2.

Hartley Oscillator

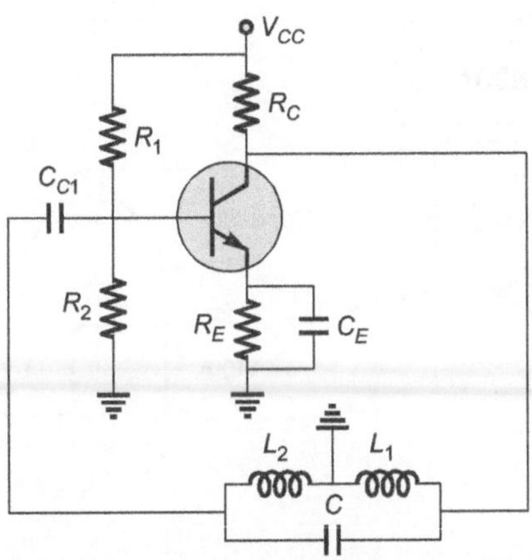

- Frequency of oscillation for Hartley Oscillator

$$f = \frac{1}{2\pi\sqrt{C(L_1 + L_2)}} \quad \text{and} \quad g_m R_C \geq \frac{L_1}{L_2}$$

Crystal Oscillator

- It uses a quartz-crystal which is a piezoelectric material.
- Quartz has excellent physical characteristics which do not get affected by time and temperature, hence crystal oscillator provides highly stable f_o.

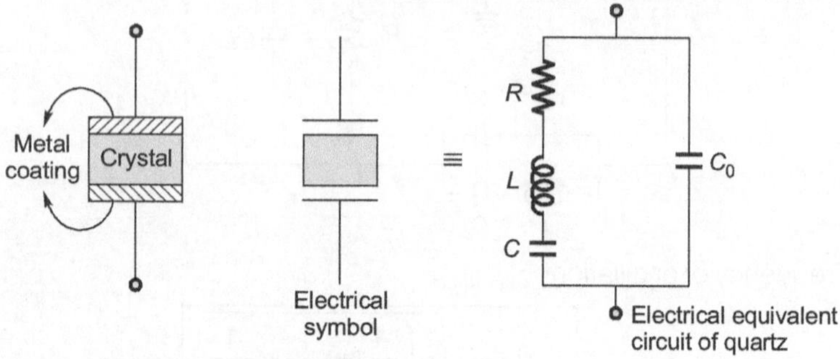

- Series resonance frequency

$$f_s = \frac{1}{2\pi\sqrt{LC}}$$

- Parallel resonance frequency

$$f_p = \frac{1}{2\pi\sqrt{LC_{eq}}} \quad \text{where,} \quad C_{eq} = \frac{CC_0}{(C + C_0)}$$

Note:

Frequency of oscillation is chosen between f_p and f_s.

Relaxation Oscillator

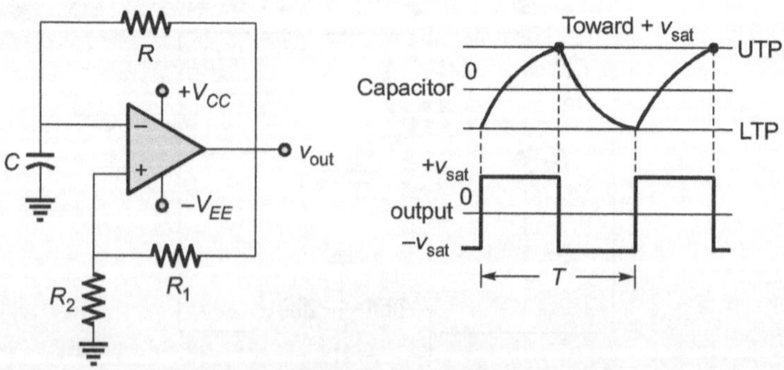

Time period of output signal

$$T = 2RC\, ln\left(\frac{1+B}{1-B}\right)$$

where, $R \rightarrow$ Feedback resistance ; $B \rightarrow$ Feedback fraction

Note: ....

- Linear oscillator generates sinewaves utilizing resonance phenomenon.
- Non linear oscillators or function generators generate square-wave, triangular wave, pulses etc. waveforms. They uses function blocks as multivibrators.
- Relaxation oscillator and UJT oscillator are non-linear oscillator.
- The oscillators which uses a tunnel diode or UJT possesses a current voltage characteristic curve of negative slope within some range of operation are known as negative resistance oscillator.
- Wein's Bridge oscillator is a variable frequency oscillator.

....

Linear Wave Shaping Circuits

High Pass RC Circuit

Pulse response of high pass R.C circuit

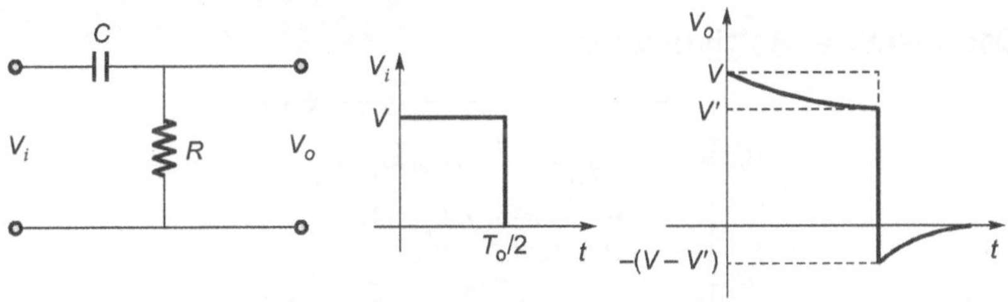

$(V - V')$ is called tilt or sag.

$$\%\,\text{tilt} = \frac{V - V'}{V} \times 100\%$$

Note: ....

Sag or tilt will be obtained if $RC >> T_o/2$.

....

- If input is a square wave of time period T then

$$\%\,\text{tilt} = \frac{\pi f_L}{f} \times 100$$ where, $$f_L = \frac{1}{2\pi RC}$$ and $$f = \frac{1}{T_o}$$

Low Pass RC Circuit

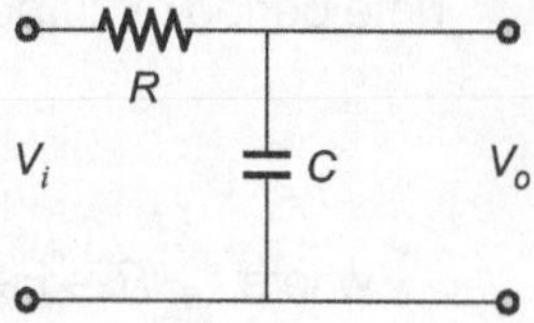

- Rise time

$$t_r = \frac{0.35}{f_H} \text{ (rise from 10\% to 90\%)}$$

where, $$f_H = \frac{1}{2\pi RC}$$

555 Timer

The 555 is a timing circuit that can produce accurate and highly stable time delays or oscillation.

555 Timer Pin Configuration

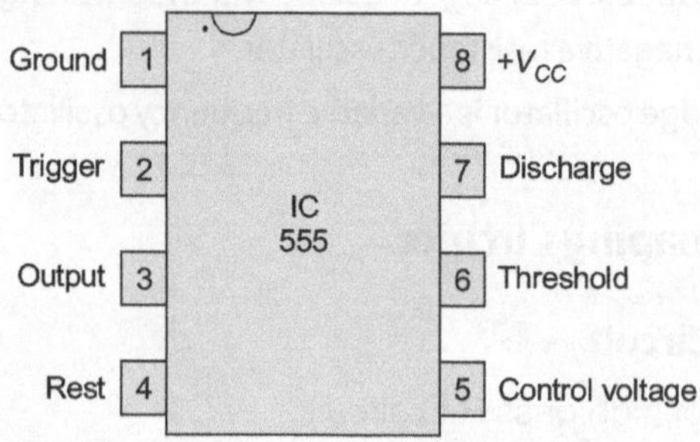

Monostable Multivibrator

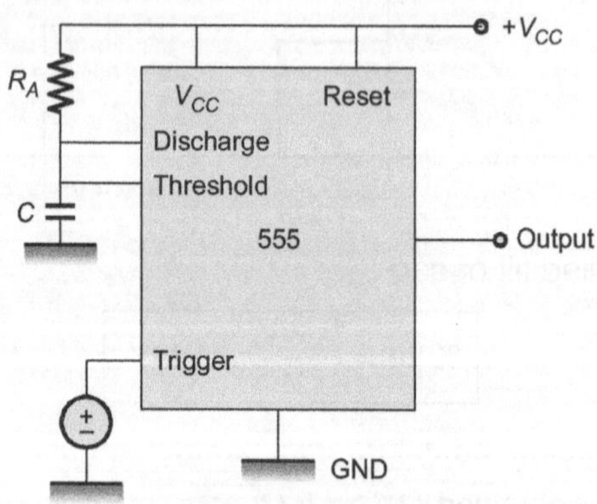

$$T = RC\, ln(3) = 1.1RC$$

Note: The main applications of monostable multivibrator are pulse width modulator, linear ramp generator, frequency divider, pulse stretcher etc.

Astable Multivibrator

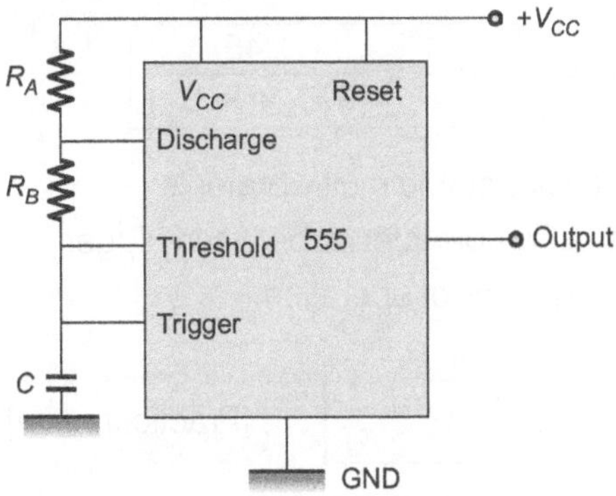

$$\boxed{T = \tau \ln(2)}$$

- Charging time constant, $\boxed{T_C = 0.693(R_A + R_B)C}$
- Discharging time constant, $\boxed{T_D = 0.693\, R_B C}$
- Total time, $\boxed{T = T_C + T_D = 0.693\,(R_A + 2R_B)C}$
- Duty cycle, $\boxed{D = \dfrac{T_C}{T_C + T_D} \times 100\% = \dfrac{R_A + R_B}{R_A + 2R_B} \times 100\%}$

Note:

The main applications of astable are square-wave generator, voltage control oscillator, FSK generator, free running oscillator etc.

Remember:

- Duty cycle (D) > 50%. (Because $T_C > T_D$)
- By placing a diode parallel to R_B and making $R_A + R_f = R_B$, we get square wave output of duty cycle 50%.

Astable multivibrator

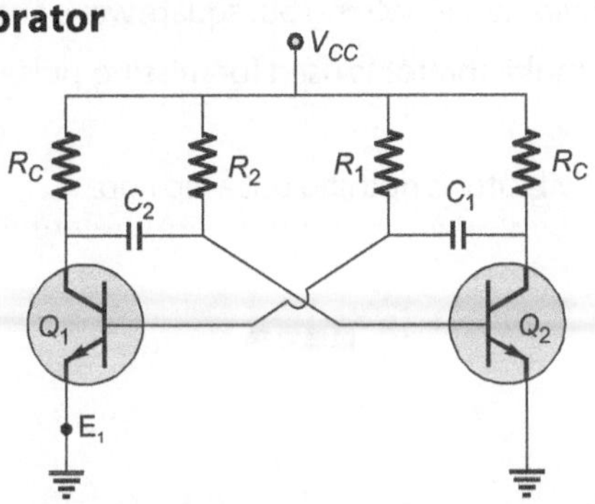

- Time period of output wave form generated is $\boxed{T = 0.69[R_1C_1 + R_2C_2]}$

output frequency $\boxed{f = \frac{1}{T} = \frac{1.45}{(R_1C_1 + R_2C_2)}}$

- Resolution time for bistable multivibrator is
$\tau_s \rightarrow$ Smallest allowable interval between trigger.
So maximum frequency of operation

$$\boxed{f_{max} = \frac{1}{\tau_s}} \quad \text{[Practical value]}$$

Bistable Multivibrator with Commutating Capacitor

- $\boxed{f_{max} = \frac{R_1 + R_2}{2C\,R_1R_2} = \frac{1}{2CR}}$ where, $R = R_1 \parallel R_2$ and $\tau_s = CR$

where, $R_1 \rightarrow$ Resistance between collector of one transistor and base of other transistor.
$R_2 \rightarrow$ Resistance between base and ground of each transistor.
$f_{max} \rightarrow$ maximum frequency of operation

- In case of monostable multivibrator, width of pulse generated is given by

$$\boxed{T = 0.69\ RC}$$

where, $C \rightarrow$ Capacitance between collector of one transistor and base of other transistor.
R $\rightarrow$ Resistance between V_{CC} and base of one transistor.

Note: ....

- **Astable multivibrator is nothing but square wave generator.**
- **Monostable multivibrator is used for missing pulse detection, in PWM, PPM.**
- **Bistable multivibrator is nothing but a flip flop.**

....

■■■■

9 Feedback Amplifiers

Feedback Amplifiers

Basic Feedback Concept

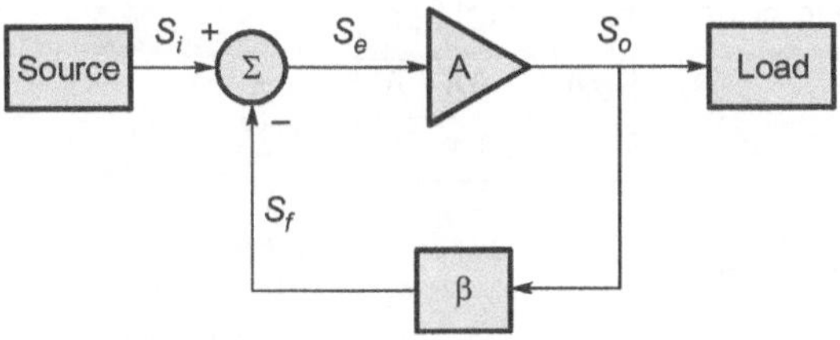

where, $A \rightarrow$ Open loop gain
$\beta \rightarrow$ Feedback factor
$S_i \rightarrow$ Input signal
$S_e \rightarrow$ Error signal
$S_f \rightarrow$ Feedback signal

Gain with feedback

$$A_f = \frac{A}{1 + A\beta}$$ for –ve feedback. $$A_f = \frac{A}{1 - A\beta}$$ for +ve feedback.

Remember:

The amount of feedback introduced into amplifier is often expressed in dB and is given by

$$N(\text{Feedback in dB}) = 20\log\left|\frac{1}{1 + A\beta}\right|$$

Here, $N \rightarrow$ Negative for negative feedback and
$N \rightarrow$ Positive for positive feedback

Loop gain or return ratio

$$\text{Loop gain} = -A\beta$$

Return difference

Difference between unity and loop gain is called return difference (D)

$$D = 1 + A\beta$$

$D \neq$ is also called desensitivity.

Sensitivity

In the case of negative feedback, if

$\frac{\partial A}{A} \rightarrow$ % change without feedback

$\frac{\partial A_f}{A_f} \rightarrow$ % change with feedback

Reduction in sensitivity of amplifier:

$$\boxed{\frac{\partial A_f}{A_f} = \frac{\partial A / A}{1 + A\beta}}$$

Feedback Amplifier

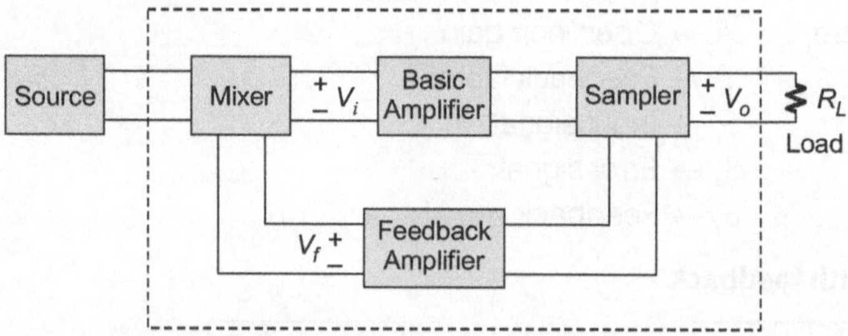

Sample Circuit	Mixer (Comparator)
1. Voltage/Shunt Sampler	**1. Series/Voltage Mixer (Comparator)**
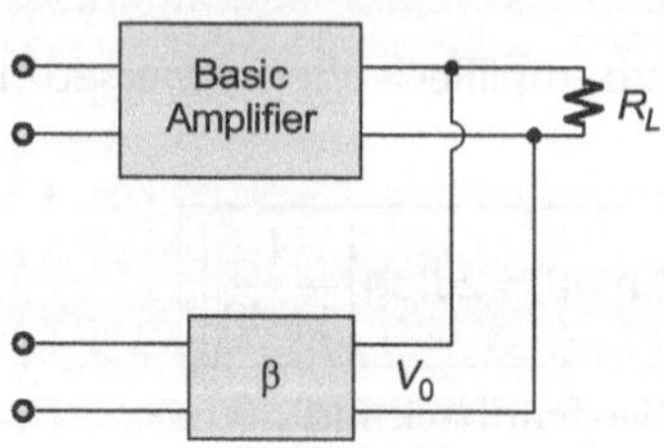	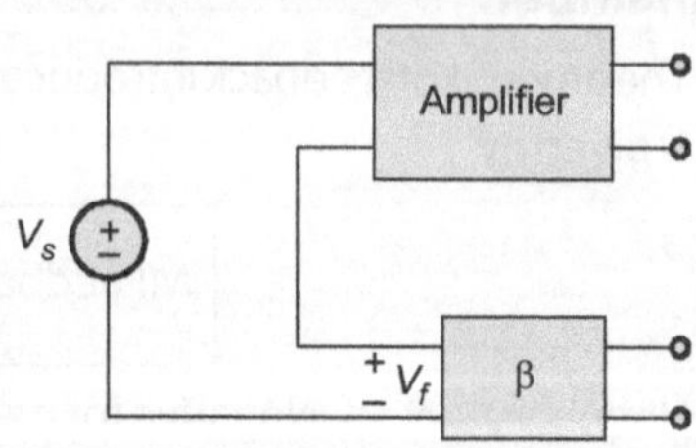
2. Current/Series Sampler	**2. Shunt/Current Mixer (Comparator)**
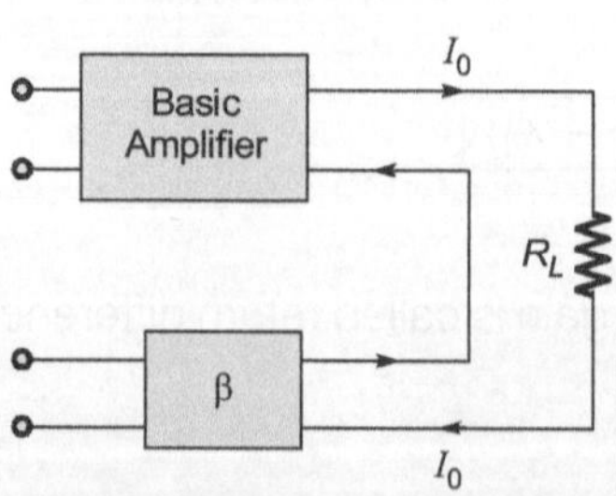	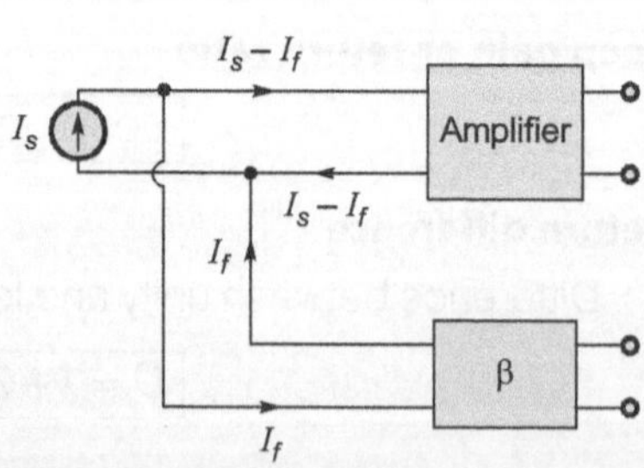

Effect of Feedback

- Reduction in noise

$$N_{of} = \frac{N_o}{1+A\beta}$$

where, $N_o \rightarrow$ Noise without feedback
$N_{of} \rightarrow$ Noise with feedback

- Lower cut-off frequency

$$f_{lf} = \left(\frac{f_l}{1+A\beta}\right)$$

where, $f_{lf} \rightarrow$ Lower 3-dB frequency with feedback
$f_l \rightarrow$ Lower 3-dB frequency without feedback

- Upper cut-off frequency

$$f_{Hf} = f_H(1 + A\beta)$$

where, $f_{Hf} \rightarrow$ Upper 3-dB frequency with feedback
$f_H \rightarrow$ Upper 3-dB frequency without feedback

Note:

Bandwidth increases with negative feedback.

1. Voltage series topology

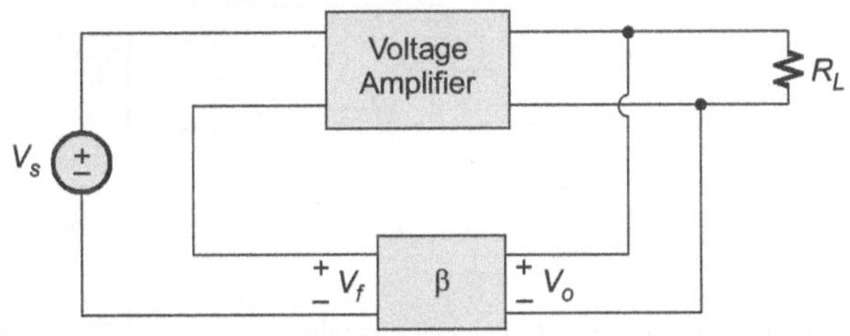

(i) β is dimensionless.

(ii) Voltage series feedback stabilizes the voltage gain.

(iii) Example: Common collector and non-inverting amplifier.

2. Current series topology

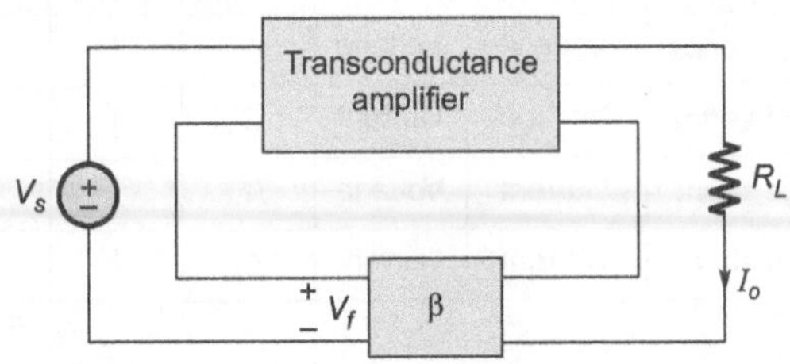

(i) Unit of β is ohm.

(ii) Current series feedback stabilizes the transconductance gain.

(iii) Example: Common emitter.

3. Voltage shunt topology

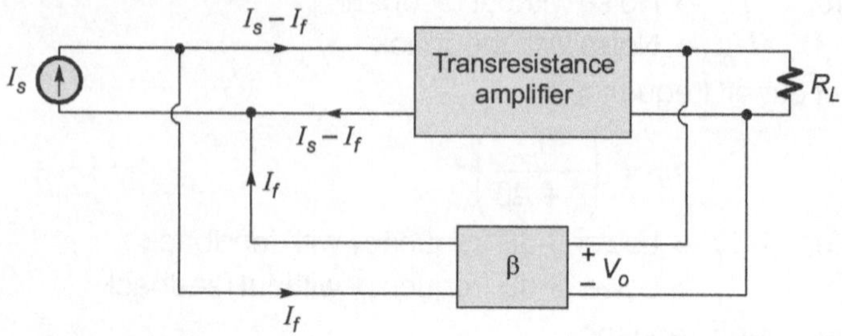

(i) β has unit mho.

(ii) It stabilizes the transresistance gain of the amplifier.

(iii) Example: Collector to base biasing.

4. Current shunt topology

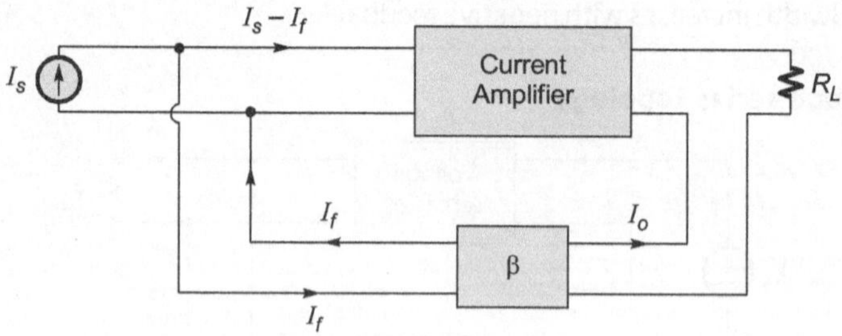

(i) β has no unit.

(ii) It stabilizes the current gain of amplifier.

(iii) Example: Cascading amplifier.

- For different feedback configuration effect on input and output impedance and gain.

Feedback	Input	Output	Gain	R_i	R_o
Voltage series	Voltage	Voltage	A_v	↑	↓
Current series	Voltage	Current	G_m	↑	↑
Voltage shunt	Current	Voltage	R_m	↓	↓
Current shunt	Current	Current	A_I	↓	↑

Note:

- Voltage series feedback is the best arrangement in terms of R_i and R_o.
- Current shunt is the worst for the same.

Remember:

Procedure to identify the type of feedback:

- Identify the element responsible for the feedback.
- If feedback element is directly connected to output node, it indicates voltage sampling otherwise it indicates current sampling.
- If feedback element is directly connected to the input node, it indicates shunt mixing otherwise it indicates series mixing.

■■■■

Power Amplifiers

10

- It is a large signal amplifier which has greater AC output voltage and greater AC output current hence it can provide greater AC output power to load.

Conversion Efficiency

It is the ability of power amplifier to convert DC power into AC power:

$$\eta = \frac{P_{AC}}{P_{DC}} \times 100\%$$

Harmonic Distortion

The harmonic distortion means the presence of the frequency components in the waveform, which are not present in the input signal.

If $I_C = B_0 + B_1 \cos\omega t + B_2 \cos 2\omega t + \cdots + B_n \cos n\omega t$

then $$\% D_n = \frac{|B_n|}{|B_1|} \times 100\% \quad ; \quad \text{where, } n = 1, 2, 3, \text{ and so on}$$

Here, fundamental frequency component has an amplitude B_1 and n^{th} harmonic component has an amplitude of B_n.

Thus, total harmonic distortion is

$$\%D = \sqrt{D_2^2 + D_3^2 + D_4^2 + \cdots + D_n^2} \times 100\%$$

Classification of Power Amplifier

Class-A Operation

- Transistor operates in active region at all times.
- Collector current flows for 360° of the AC cycle.

Unloaded voltage gain of CE amplifier

$$A = -\frac{R_C}{r_e'}$$

Loaded voltage gain

$$A_v = -\frac{r_C}{r_e'}$$

Current gain of the transistor

$$\boxed{A_i = \frac{i_c}{i_b}}$$

where, $A_i \rightarrow$ Current gain
$i_c \rightarrow$ AC collector current
$i_b \rightarrow$ AC base current

Note: ....

$\boxed{A_i \cong \beta}$; In most of the circuits this approximation can be used.

....

Power gain

$$\boxed{A_p = -A_v A_i}$$

Load power

$$\boxed{P_L = \frac{V_L^2}{R_L} = \frac{V_{PP}^2}{8R_L}}$$

where, $P_L \rightarrow$ AC load power
$V_L \rightarrow$ RMS load voltage
$V_{PP} \rightarrow$ Peak-to-peak load voltage
$R_L \rightarrow$ Load resistance

Transistor Power Dissipation

$$\boxed{P_{DQ} = V_{CEQ}\, I_{CQ}}$$

where, $P_{DQ} \rightarrow$ Quiescent power dissipation
$V_{CEQ} \rightarrow$ Quiescent collector-emitter voltage
$I_{CQ} \rightarrow$ Quiescent collector current

Total DC power supplied to an amplifier

$$\boxed{P_S = V_{CC}\, I_S}$$

Maximum AC load power

$$\boxed{P_{L(\max)} = \frac{PP^2}{8R_L}}$$

where, $PP \rightarrow$ Maximum unclipped value of V_{PP}

Stage efficiency

$$\boxed{\eta = \frac{P_{L(\max)}}{P_S} \times 100\%}$$

where, $\eta \rightarrow$ Stage efficiency

$P_S \rightarrow$ DC input power

Remember: ....

- Class-A amplifier produces least distortion in the output among all power amplifier.
- The maximum efficiency of class-A amplifier is 25%.

....

Class-B Operation

- Collector current flows for only 180° of the AC cycle.
- *Q*-point is located approximately at cutoff on both the DC and AC load lines.

Collector-emitter voltage at *Q*-point

$$\boxed{V_{CEQ} = \frac{V_{CC}}{2}}$$

AC load power of a class B push-pull amplifier

$$\boxed{P_L = \frac{V_{PP}^2}{8R_L}}$$

Remember: ....

- Output signal is half sinusoidal.
- Quiescent power dissipation/power drain is almost zero.
- The maximum efficiency of class-B amplifier is 78.5%.

....

Class AB Operation

- Operating point is located between the limits of class-A and class-B.
- Collector current flows for more than half sinusoidal but not fully sinusoidal.
- Distortion in class-AB amplifier is more than class-A but less than class-B.
- Power drain is more than class-B but less than class-A.

Class C Operation

- It is operated either in deep saturation or in deep cut-off region.
- The collector current flows for less than 90° of the AC cycle.

Tuned Amplifier

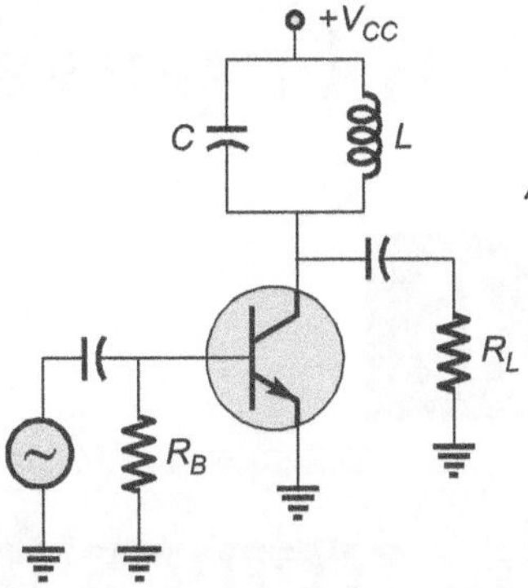

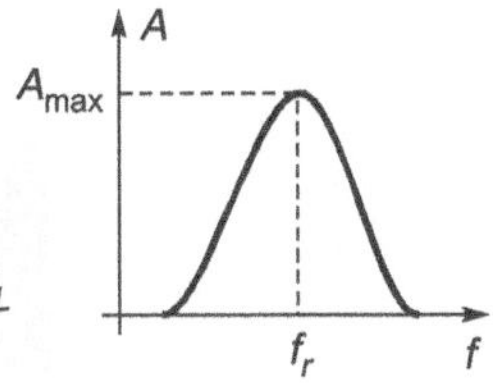

Resonant frequency of tank circuit

$$f_r \cong \frac{1}{2\pi\sqrt{LC}}$$

where, $L \rightarrow$ Inductance ; $C \rightarrow$ Capacitance

AC load power class C amplifier

$$P_L = \frac{V_{PP}^2}{8R_L}$$

Remember:

- It has highest conversion efficiency among all power amplifiers.
- Output is heavily distorted.
- It is used in tuned power amplifiers and radio frequency amplifier.
- The maximum efficiency of class-C amplifier is 87.5%.

Power Amplifier	Conduction Angle	Maximum Efficiency	Figure of Merit
Class A	360°	50%	2
Class B	180°	78.5%	0.4
Class AB	180°-360°	50-78.5%	0.4-2
Class C	< 180°	≥ 90°	< 0.25

■■■■■

A Handbook on
Instrumentation Engineering

6

Digital Electronics & Microprocessors

CONTENTS

Number System and Codes

1

Introduction

A digital system is a combination of devices designed to manipulate physical quantities or information that are represented in digital form, i.e. they can take on only discrete values. Examples of digital system are digital computer, calculator, telephone etc.

Advantages of Digital Techniques

1. Size and cost is less.
2. Power dissipation is less.
3. Digital circuits are less affected by noise.
4. Accuracy and precision are greater.
5. Information storage is easy.

Digital Number System

Many number systems are used in digital technology. The most common are the binary, octal, decimal and hexadecimal system.

Note:

A number system with base 'b' will have 'b' different digits from 0 to (b-1).

Number Representation

$$N_b = \underbrace{d_{n-1}\, d_{n-2} \ldots d_i \ldots d_0}_{\text{Integer Portion}} \underset{\substack{\uparrow \\ \text{Radix} \\ \text{point}}}{\bullet} \underbrace{d_{-1}\, d_2 \ldots d_{-f} \ldots d_{-m}}_{\text{Fraction Portion}}$$

where, $N \rightarrow$ Number ; $b \rightarrow$ Base or radix

Weighted number system

- It is a positional weightage system.
- **Ex:** Binary, octal, hexadecimal, BCD, 2421 etc.

Unweighted number system

- It is non positional weightage system.
- **Ex:** Gray code, Excess 3-code etc.

Number System	Base (b)	Digits
Binary	2	0, 1
Octal	8	0, 1, 2, 3, 4, 5, 6, 7
Decimal	10	0, 1, 2, 3, 4, 5, 6, 7, 8, 9
Hexadecimal	16	0, 1, 2, 3, 4, 5, 6, 7, 8, 9 *A, B, C, D, E, F*

Decimal	Hexadecimal	Binary
0	0	0000
1	1	0001
2	2	0010
3	3	0011
4	4	0100
5	5	0101
6	6	0110
7	7	0111
8	8	1000
9	9	1001
10	*A*	1010
11	*B*	1011
12	*C*	1100
13	*D*	1101
14	*E*	1110
15	*F*	1111

Note:

- In positional weightage system, the position of each digit of a number has some positional weightage.
- In non-positional weightage system, a digit of a number does not indicate any significance in a position and weight.

Codes

Binary Coded Decimal Code (BCD)

Each digit of a decimal number is represented by binary equivalent.

In 4-bit binary formats

Total number of possible representation = $2^4 = 16$

Valid BCD codes = 10

Invalid BCD codes = 6

In 8-bit binary formats

Valid BCD codes = 100

Invalid BCD codes = 256 – 100 = 156

Note: ....

During arithmetic operation if invalid *BCD* is present then add (0110) to get correct result.

....

Excess-3 code

- It is a 4-bit code.
- It can be derived from BCD code by adding "3" to each coded number.
- It is a "self-complementing code".

Gray code

- Also called "minimum change code" or "unit distance code" in which only one bit in the code group changes when going from one step to the next.
- Gray code is a minimum error code.

Binary-to-Gray conversion

- 'MSB' in the gray code is same as corresponding digit in binary number.
- Starting from "Left to Right", add each adjacent pair of binary digits to get next and gray code digit. (Discard the carry if generated).

Gray-to-Binary conversion

- "MSB" of Binary is same as that of gray code .
- Add each binary digit to the generated gray digit in the next adjacent position (discard the carry if generated).

Remember: ....

- **In self-complemented code, the sum of weightage = 9.**
- **The largest number that can be represented by using *N*-bits is $(2^N - 1)_{10}$.**

....

Code Converters

BCD to Excess-3 code

- Let input variables of ***BCD* code** is *A*, *B*, *C*, *D* and output variables of excess-3 is *W*, *X*, *Y*, *Z*.

- Required truth table: (BCD + 0011 = excess-3-code)

Input (BCD)				Output (Excess-3 code)			
A	*B*	*C*	*D*	*W*	*X*	*Y*	*Z*
0	0	0	0	0	0	1	1
0	0	0	1	0	1	0	0
0	0	1	0	0	1	0	1
0	0	1	1	0	1	1	0
0	1	0	0	0	1	1	1
0	1	0	1	1	0	0	0
0	1	1	0	1	0	0	1
0	1	1	1	1	0	1	0
1	0	0	0	1	0	1	1
1	0	0	1	1	1	0	0

Minimized Boolean function:

$$Z = \bar{D}$$
$$Y = CD + \bar{C}\bar{D}$$
$$X = \bar{B}C + \bar{B}D + B\bar{C}\bar{D}$$
$$W = A + BC + BD$$

Binary-to Gray Code Converter

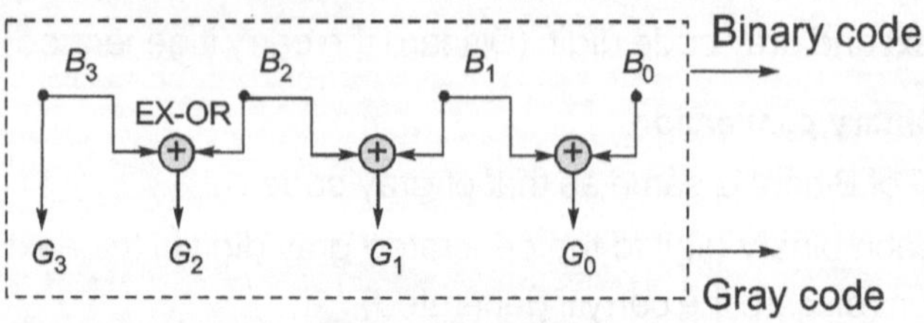

Equivalent logical gate diagram:

$$G_3 = B_3$$
$$G_2 = B_3 \oplus B_2$$
$$G_1 = B_2 \oplus B_1$$
$$G_0 = B_1 \oplus B_0$$

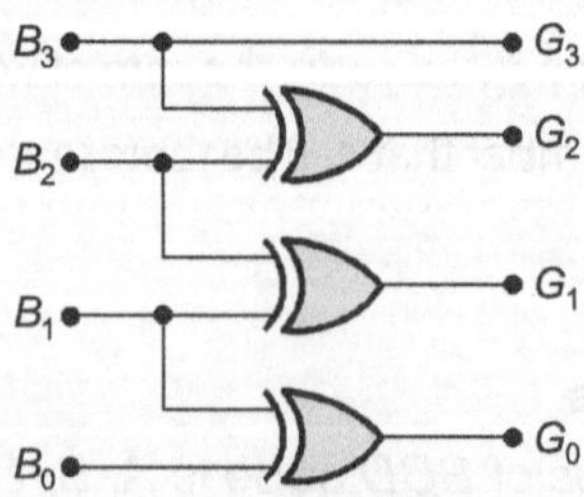

Gray to Binary Converter

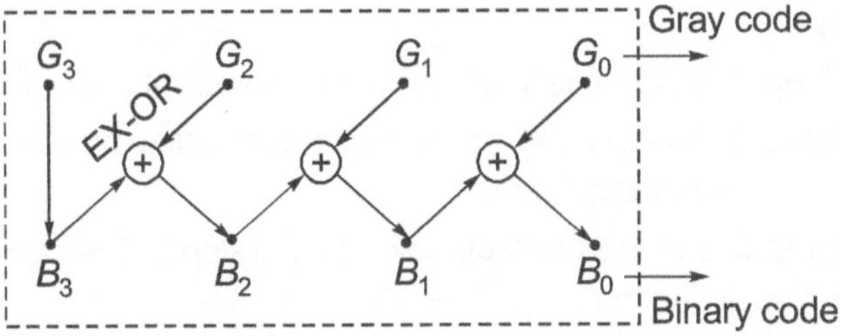

Equivalent Logical Gate Diagram:

$$B_3 = G_3$$
$$B_2 = B_3 \oplus G_2$$
$$B_1 = B_2 \oplus G_1$$
$$B_0 = B_1 \oplus G_0$$

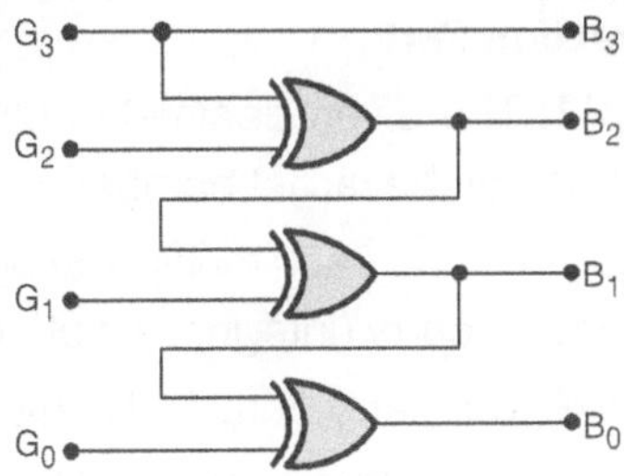

Conversion of Number System

(a) Decimal Number System to Other Number System Conversion

Steps for integer part

- **Step-1:** Divide the decimal number to be converted by the value of new base.
- **Step-2:** Get the remainder from step-1 as the rightmost digit (LSB) of new base number.
- **Step-3:** Divide the quotient of the previous divide by the new base.
- **Step-4:** Record the remainder from step-3 as the next digit (to the left) of the new base number.

Example: $(29)_{10} \rightarrow (\ )_2$

Operation	Result	Remainder
29/2	14	1
14/2	7	0
7/2	3	1
3/2	1	1
1/2	0	1

$\therefore\ (29)_{10} \rightarrow (11101)_2.$

(b) Other Number System to Decimal Number System Conversion

Steps

- **Step-1:** Determine the column (positional value of each digit).
- **Step-2:** Multiply the obtained column value by the digits in the corresponding column.
- **Step-3:** Sum the products calculated in step-2. The total is equivalent value in decimal.

$$(x_2\, x_1\, x_0 \cdot y_{-1}\, y_{-2}) \rightarrow (x_2 \times r^2 + x_1\, r^1 + x_0\, r^0 + y_{-1}\, r^{-1} + y_{-2}\, r^{-2})_{10}$$

Steps for Fraction Part

- **Step-1:** The given decimal fraction is successively multiplied by the base of the target number system.
- **Step-2:** During each multiplication iteration, the product generated will have a carry (integer part of product) and fractional part.
- **Step-3:** The carry obtained at each multiplication step becomes a numeral in the new base system.
- **Step-4:** The fraction part is again multiplied by the new base number and process is repeated until the fractional part becomes zero or the number of multiplication iteration equals the number of digits after the decimal points.
- **Step-5:** Weights are assigned for the carry obtained at each step such that carry obtained from 1st multiplication is most significant digit after the decimal point and last carry is least significant digit.

Example: $(0.125)_{10} \rightarrow (\ )_2$

Operation	Result	Carry
0.125 × 2	0.25	0
0.25 × 2	0.5	0
0.5 × 2	1.0	1

$\therefore\ (0.125)_{10} \rightarrow (0.001)_2.$

(c) Octal to Binary

Each digit is represented with 3 bit binary equivalent.

(d) Hexadecimal to Binary

Each digit is represented with 4 bit binary equivalent.

(e) Octal to Hexadecimal

(i) First convert octal to binary and then binary to Hexadecimal.

(ii) From Hexadecimal to octal conversion first convert Hexadecimal to binary and then binary to octal number.

Digital Number Representation

- In unsigned magnitude representation with 'n' bits, the possible integer values are [0 to $2^n - 1$].
- Extra bit → sign bit → MSB
  - if MSB = 0 → +ve number
  - if MSB = 1 → –ve number
- In a sign magnitude representation the range of number.

$$-(2^{n-1}-1) \text{ to } +(2^{n-1}-1)$$

Compliment: For base 'r' → (r – 1)'s Complement; → r's Complement

Binary		Octal		Decimal		Hexadecimal	
1's	2's	7's	8's	9's	10's	F's	16's

1. To determine (r – 1)'s complement, the given number is subtracted from the maximum possible number in given base.
2. To determine r's complement, first write (r – 1)'s complement then add 1 to the LSB (Least Significant Bit).

- In 1's complement representation, the positive number are represented similar to positive number in sign magnitude, but for representing negative number, first write positive number and then take 1's complement of that.
- **Range of 1's complement number**

$$-(2^{n-1}-1) \text{ to } +(2^{n-1}-1)$$

- **Range of 2's complement representation number**

$$-2^{n-1} \text{ to } +(2^{n-1}-1)$$

Binary Arithmetic

- When both number have same sign then we add only magnitudes and use the sign as MSB.

1's complement addition

- When the numbers have different sign, keep one number as it is and 1's complement the other then add magnitude only.
- If carry is present
  1. add carry to LSB
  2. sign of the result is sign of the uncomplemented number.
- If carry is not present
  1. take 1's complement the result.
  2. sign of the result is sign of the complemented number.

2's complement addition

In 2's complement addition if any carry is present, then discard it.

Note: ....

- In 2's complement there is only one way to represent '0'.
- 2's complement of the 2's complement number is equal to the number itself.
- In 2's complement representation to extend the number of bits MSB bit is is always copied.

....

■■■■

2

Logic Gates

Introduction

- Logic gates are most fundamental digital circuit that can be constructed from diodes, transistors and resistors connected in such a way that the circuit output is the result of a basic logic operation performed on the inputs.
- The Boolean '0' and '1' represents the "logic level".

Logic 0	Logic 1
False	True
OFF	ON
Low	High
No	Yes
Open switch	Closed switch

- A "Truth table" is a means of describing how a logic circuits output depends on the logic levels present at circuits input.

Note:

The number of input combinations will equals to '2^N' for an "N-input" truth table.

- The logic Gates can be classified as
  - (a) **Basic Gate:** NOT, AND, OR.
  - (b) **Universal Gate:** NAND, NOR.
  - (c) **Special Purpose Gates:** EX-OR and EX-NOR. They are used in arithmetic circuit, comparators, code conversion, parity generators and parity checkers etc.

NOT Gate

It is also referred as "inversion" or "complementation".

Symbol and Truth Table

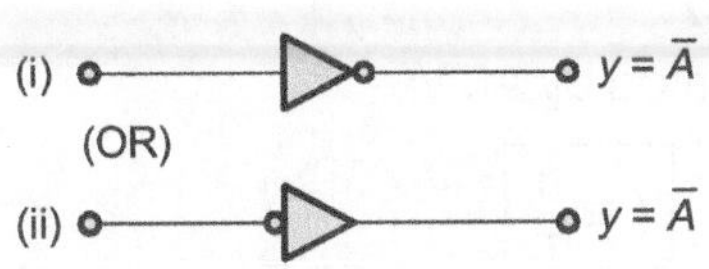

Input	Output
A	$y = \bar{A}$
0	1
1	0

Switching Circuit

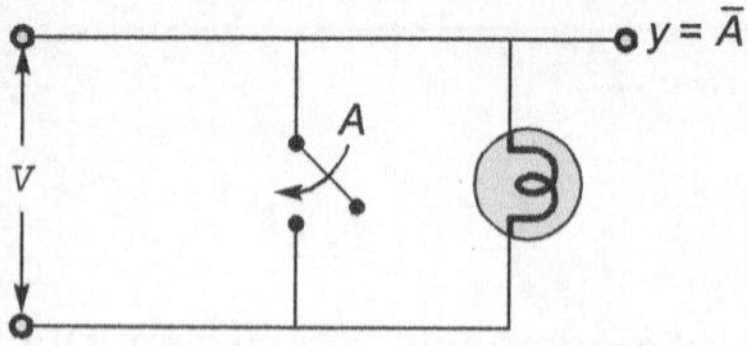

Transistor Circuit

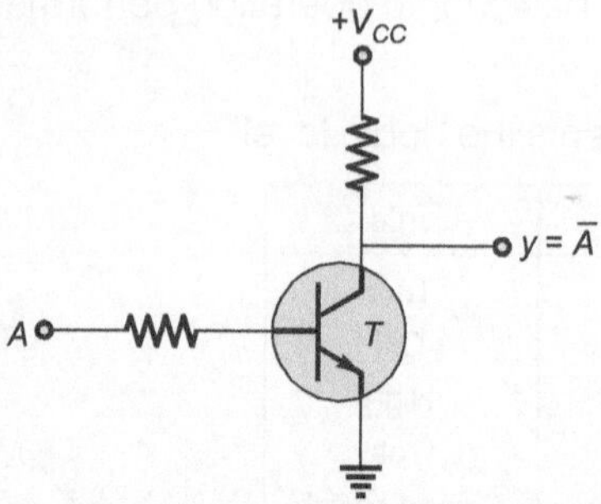

A	T	y
0	OFF	$+V_{CC} \approx 1$
1	0	Ground ≈ 0

- When even number of NOT Gates are connected in series then it acts like Buffer Circuit.

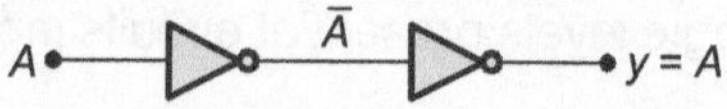

- When even number of NOT Gates are connected with feedback then it acts like a "Bistable multivibrator". It is also a basic memory element.

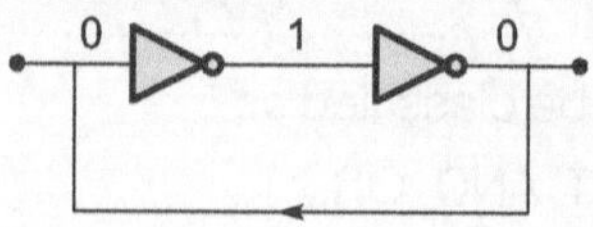

- When odd number of NOT Gates are connected with feedback, then it acts like an astable multivibrator (AMV) or square-wave generator or clock generator or ring oscillator.

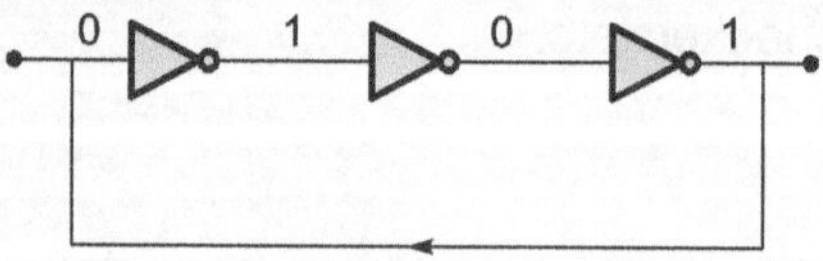

- All inverter take some time to get the response '*Y*', this time is called propagation delay time (t_{pd}).

  For an Astable Multivibrator (AMV)

  Time period of Square Wave Generated by AMV:

$$\boxed{T = 2nt_{pd}} \; ; \; \boxed{f = \frac{1}{T}}$$

where, $n \rightarrow$ Number of inverters (NOT Gates)

$t_{pd} \rightarrow$ Propagation delay time of each inverter

$T \rightarrow$ Time period of a square wave generated by AMV or Ring oscillator

$f \rightarrow$ Frequency of oscillation

AND Gate

Symbol, Truth Table and Switching Circuit

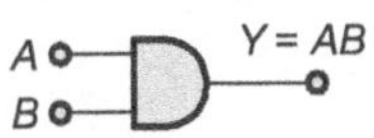

Inputs		Output
A	*B*	*Y* = *AB*
0	0	0
0	1	0
1	0	0
1	1	1

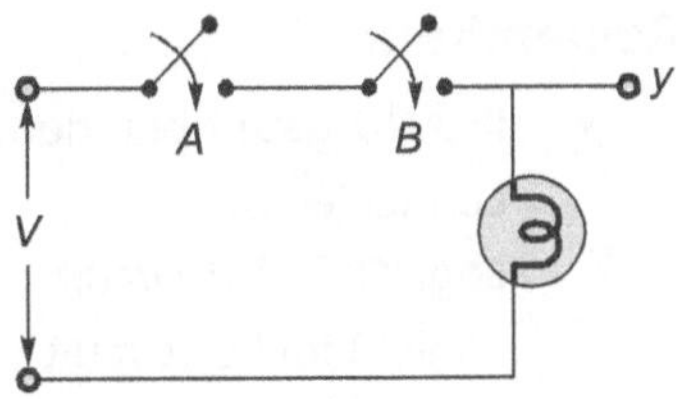

Note:

In AND operation

- ENABLE INPUT ⇒ Logic '1'
- DISABLE INPUT ⇒ Logic '0'

Transistor circuit

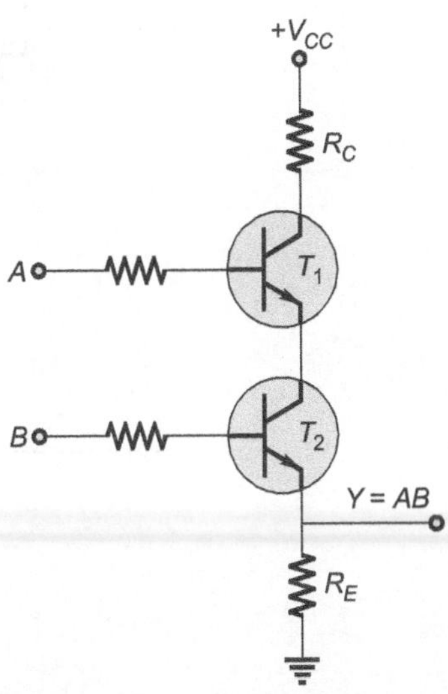

A	B	T_1	T_2	Y
0	0	OFF	OFF	0
0	1	OFF	ON	0
1	0	ON	OFF	0
1	1	ON	ON	1

Diode circuit diagram

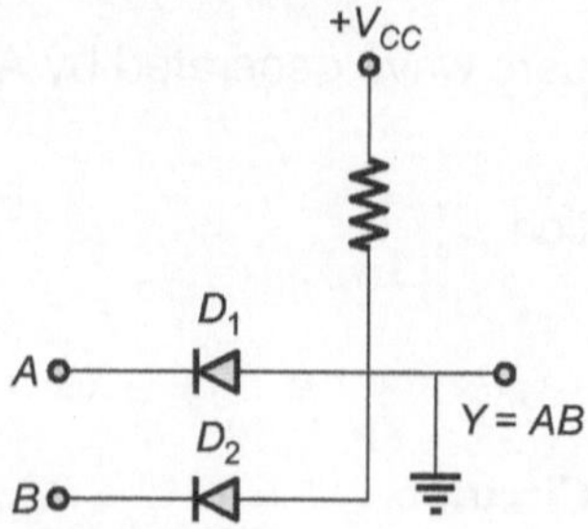

A	*B*	D_1	D_2	*Y*
0	0	ON	ON	0
0	1	ON	OFF	0
1	0	OFF	ON	0
1	1	OFF	OFF	1

Remember:

- In AND gate operation, any unused inputs (Floating inputs) may be connected as:
  Logic '1' for TTL circuit
  Logic '0' for ECL circuit
- AND gate is also known as detector logic

OR Gate

Symbol, Truth Table and Switching Circuit

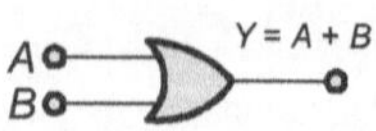

Inputs		Output
A	*B*	*Y* = *A* + *B*
0	0	0
0	1	1
1	0	1
1	1	1

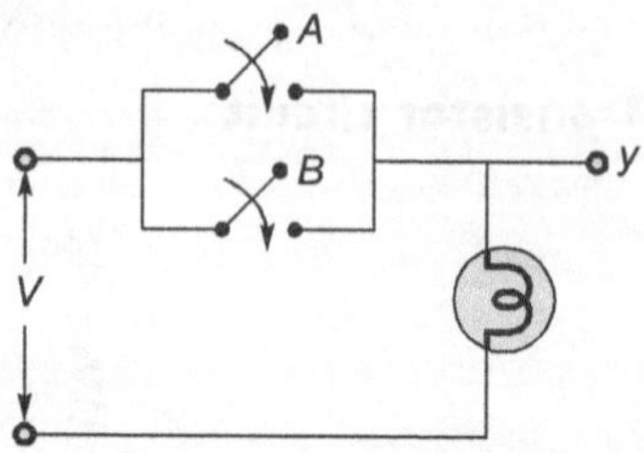

Remember:

In OR operation

- ENABLE INPUT ⇒ Logic '0'
- DISABLE INPUT ⇒ Logic '1'

Transistor Circuit

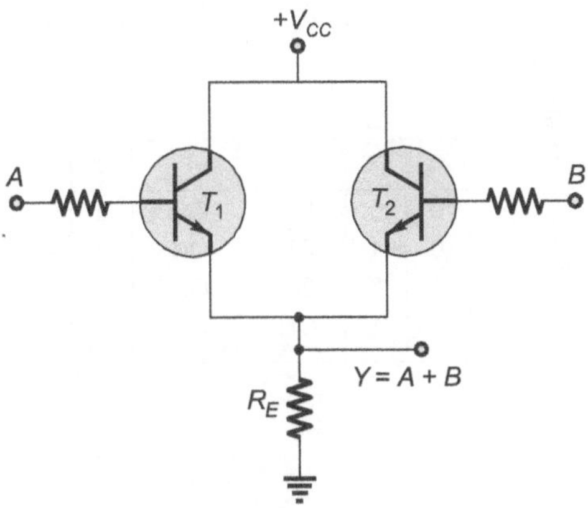

A	B	T_1	T_2	Y
0	0	OFF	OFF	0
0	1	OFF	ON	1
1	0	ON	OFF	1
1	1	ON	ON	1

Diode Circuit Diagram

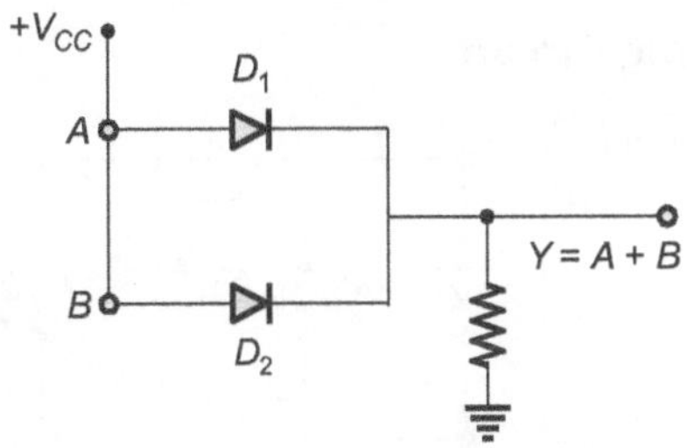

A	B	D_1	D_2	Y
0	0	OFF	OFF	0
0	1	OFF	ON	1
1	0	ON	OFF	1
1	1	ON	ON	1

NAND Gate

Symbol, Truth Table and Switching Circuit

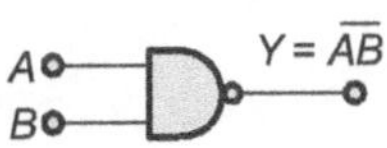

Inputs		Output
A	B	$Y = \overline{AB}$
0	0	1
0	1	1
1	0	1
1	1	0

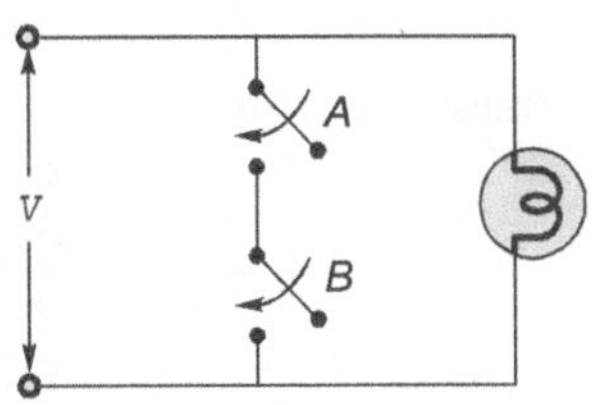

Remember: ....

In NAND operation

- ENABLE INPUT ⇒ Logic '1'
- DISABLE INPUT ⇒ Logic '0'

....

Transistor Circuit

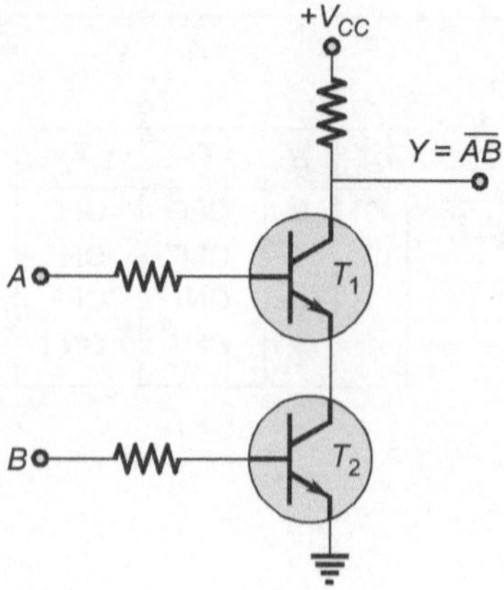

A	B	T_1	T_2	Y
0	0	OFF	OFF	1
0	1	OFF	ON	1
1	0	ON	OFF	1
1	1	ON	ON	0

NOR Gate

Symbol, Truth Table and Switching Circuit

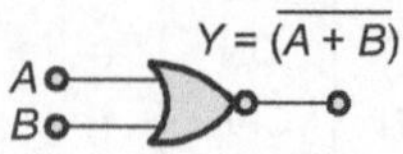

Inputs		Output
A	B	$Y = \overline{(A+B)}$
0	0	1
0	1	0
1	0	0
1	1	0

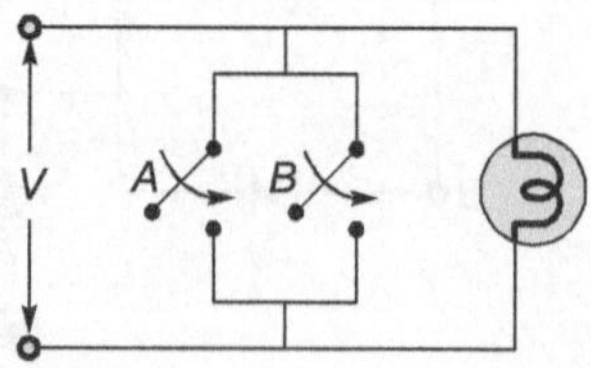

Remember:

In NOR operation

- ENABLE INPUT ⇒ Logic '0'
- DISABLE INPUT ⇒ Logic '1'

Transistor circuit

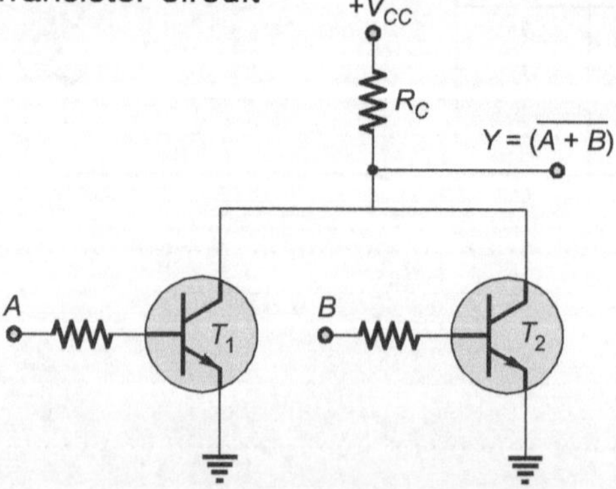

A	B	T_1	T_2	Y
0	0	OFF	OFF	1
0	1	OFF	ON	0
1	0	ON	OFF	0
1	1	ON	ON	0

EX-OR Gate

It is also called "stair case switch".

Symbol and Truth Table

Inputs		Output
A	B	$Y = A \oplus B$
0	0	0
0	1	1
1	0	1
1	1	0

- Boolean function of 2-input EXOR operation

$$Y = A \oplus B = \bar{A}B + A\bar{B}$$

Remember:

- It acts as "odd number of 1's detector in the input".
- It is mostly used in "parity generation and detection".
- When both the inputs are same, then output becomes LOW or Logic '0'.
- When both the inputs are different, then output becomes HIGH or Logic '1'.
- In EXOR operation
  - (i) For BUFFER CIRCUIT ⇒ Logic '0'
  - (ii) For INVERSION CIRCUIT ⇒ Logic '1'

Note:

$$A \oplus A = 0, \quad A \oplus 0 = A$$
$$A \oplus \bar{A} = 1, \quad A \oplus 1 = \bar{A}$$

- $A \oplus A \oplus A \oplus$..........upto n terms = 0, when n = even
  = A, when n = odd

EX-NOR Gate

- It acts as "even number of 1's detector".
- It is also called "Gate of equivalence" or "coincidence logic".

Symbol and Truth Table

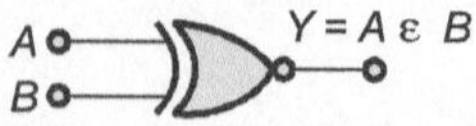

Inputs		Output
A	B	Y = A ε B
0	0	1
0	1	0
1	0	0
1	1	1

- Boolean function of 2-input EXNOR operation

$$Y = A \odot B = \overline{A \oplus B} = \overline{(\bar{A}B + A\bar{B})} = AB + \bar{A}\,\bar{B}$$

Remember: ....

- When both the inputs are same, then output becomes HIGH or Logic '1'.
- When both the inputs are different, then output becomes LOW or Logic '0'.
- In EXNOR operation
  - (i) For BUFFER CIRCUIT ⇒ Logic '1'
  - (ii) For INVERSION CIRCUIT ⇒ Logic '0'

....

Note: ....

$$A \odot A = 1, \quad A \odot 1 = A$$
$$A \odot \bar{A} = 0, \quad A \odot 0 = \bar{A}$$

- $A \odot A \odot A \odot$ .... upto n terms = 1, when n = even
  = A, when n = odd
- $\bar{A} \oplus B = A \odot B$ and $A \oplus \bar{B} = A \odot B$
  $\bar{A} \odot B = A \oplus B$ and $A \odot \bar{B} = A \oplus B$
- $A \oplus B \oplus AB = A + B.$

....

Alternative Symbols of Gates

Bubbled – OR gate ≡ NAND gate

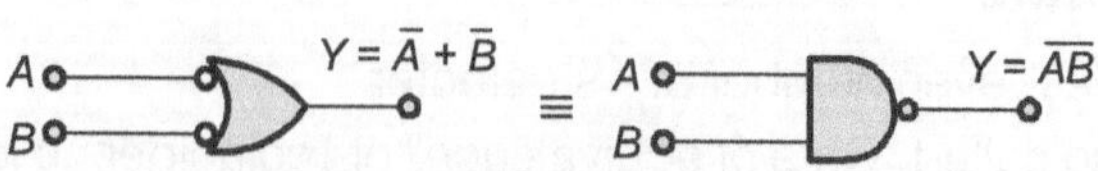

Bubbled – NAND gate ≡ OR gate

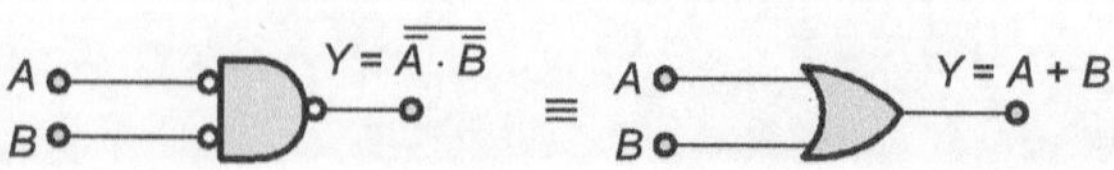

Bubble – NOR gate ≡ AND gate

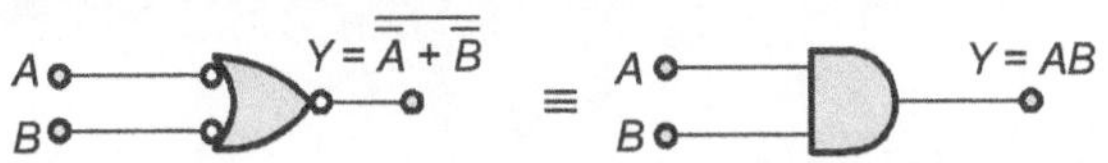

Bubbled – AND gate ≡ NOR gate

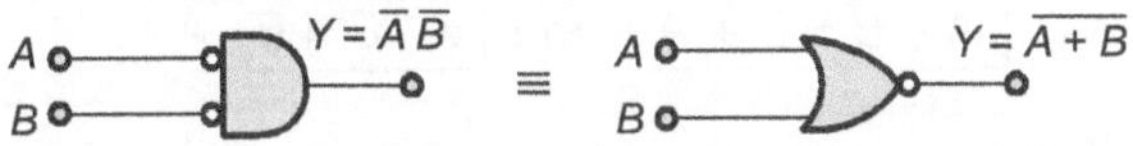

NAND and NOR Gate as Universal Gate

NAND and NOR gates are called universal gates because combinations of them can be used to accomplish any of the basic operations and thus can produce inverter, OR gate or an AND gate.

Logic gates	No. of NAND gate required	No. of NOR gate required
NOT	1	1
AND	2	3
OR	3	2
EX-OR	4	5
EX-NOR	5	4

■■■■

Boolean Algebra and Reduction Techniques

3

Boolean Algebraic Laws

Commutative Law

$$A + B = B + A \quad \text{and} \quad A \cdot B = B \cdot A$$

Associative Law

$$A + (B + C) = (A + B) + C = A + B + C$$

and

$$A \cdot (B \cdot C) = (A \cdot B)\, C = A\, B\, C$$

Distributive Law

$$A(B + C) = AB + AC$$

$$(A + B)\,(C + D) = AC + AD + BC + BD$$

Boolean Algebraic Theorems

AND-Operation Theorem

$$A \cdot A = A \qquad A \cdot 0 = 0$$

$$A \cdot 1 = A \qquad A \cdot \overline{A} = 0$$

Involution Theorem

$$(A')' = \overline{\overline{A}} = A$$

OR-Operation Theorem

$$A + A = A \qquad A + 0 = A$$

$$A + 1 = 1 \qquad A + \overline{A} = 1$$

De Morgan's Theorem

$$\overline{(A_1 \cdot A_2 \cdot A_3 \cdot \ldots . A_n)} = \overline{A}_1 + \overline{A}_2 + \overline{A}_3 + \ldots . + \overline{A}_n$$

$$\overline{(A_1 + A_2 + A_3 + \cdots + A_n)} = \overline{A}_1 \cdot \overline{A}_2 \cdot \overline{A}_3 \cdots \overline{A}_n$$

Transposition Theorem

$$(A + B)(A + C) = A + BC$$

Distribution Theorem

$$A + BC = (A + B)(A + C)$$

Consensus Theorem

- Used to eliminate redundant term.
- It is applicable only when,
  1. A boolean function contains 3 variables.
  2. Each variable is used 2-times
  3. Only one variable is in complemented and uncomplemented form. Then the related terms to that complemented or uncomplemented variable is the answer.

  Ex: $AB + \bar{A}C + BC = AB + \bar{A}C$

Boolean Algebraic Theorems

Theorem No.	Theorem
1.	$(A+B)\cdot(A+\bar{B}) = A$
2.	$AB + \bar{A}C = (A+C)(\bar{A}+B)$
3.	$(A+B)(\bar{A}+C) = AC + \bar{A}B$
4.	$AB + \bar{A}C + BC = AB + \bar{A}C$
5.	$(A+B)(\bar{A}+C)(B+C) = (A+B)(\bar{A}+C)$
6.	$\overline{A\cdot B\cdot C\cdot ...} = \bar{A} + \bar{B} + \bar{C} + ...$
7.	$\overline{A+B+C+...} = \bar{A}\cdot\bar{B}\cdot\bar{C}...$

Duality Theorem

"Dual expression" is equivalent to writing a negative logic of the given boolean relation. The procedure is,

1. Change each OR sign by an AND sign and vice-versa.
2. Complement any '0' or '1' appearing in expression.
3. Keep literals as they are

Note: ....

- If a dual of an expression results the expression itself then it is called "self dual expression".
- For N-variables, maximum possible Self-Dual Function = $(2)^{2^{n-1}} = 2^{(2^n/2)}$.
- With N-variables, maximum possible distinct logic functions = 2^{2^n}.

....

Complementary Theorem

For obtaining complement expression we

1. Change each OR sign by AND sign and vice-versa.
2. Complement any '0' or '1' appearing in expression.
3. Complement the individual literals.

Boolean Function Representation

Canonical Form

All the terms contains each literal.

Ex.: $F(A,B,C) = \overline{A}BC + ABC + \overline{A}B\overline{C}$

Standard Form

All the term do not have each literal.

Ex.: $F(A,B,C) = \overline{A} + BC + A\overline{B}C$

Note:

A binary-variable is called "LITERALS".

Minterms and Maxterms

- n-binary variables have 2^n possible combinations and each of these possible combination is called "Minterm or Standard Product form".
- "Maxterm" is the complement of corresponding "Minterm" i.e. $\boxed{M = \overline{m}}$.

Reduction Techniques

SOP (Sum of Product)

In SOP form each product term is known as min term. SOP form is used when output is logic 1.

e.g. $\underbrace{AB\overline{C}}_{\text{Min term}} + A\overline{B}C + ABC$

POS (Product of SUM)

In POS form each product term is called as max term. POS form is used when output is 0 logic '0'.

e.g. $\underbrace{(A+B+C)}_{\text{Max term}} + (A+\overline{B}+C)\cdot(\overline{A}+B+C)$

Note:

- AND-OR Logic ≡ NAND-NAND Logic and is used in SOP.
- OR-AND Logic ≡ NOR-NOR Logic and is used in POS.

Venn Diagram

It is a diagram that shows all possible relations between a finite collection of differents sets.

Ex:

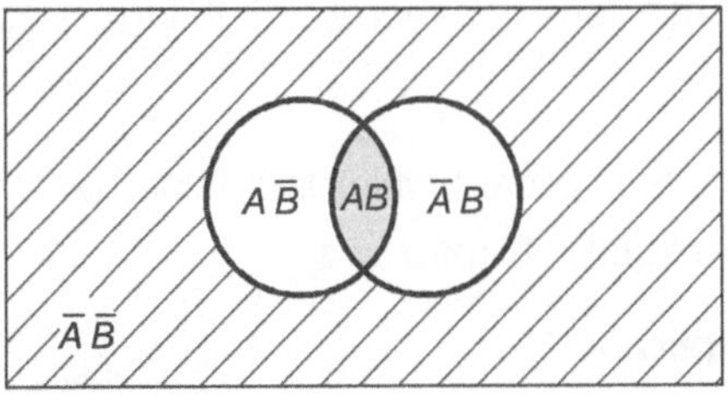

Here, the AND operation is considered as an intersection and the OR operation is considred as a union.

K-Map

Complete Simplification Rules (K-Map)

- Construct the K-map and place 1's in the cells corresponding to the 1's in the truth table. Place 0's in the remaining cells.
- Examine the map for adjacent 1's and loop those 1's which are not adjacent to any other 1's. These are called isolated 1's.
- Next, look for those 1's which are adjacent to only one other 1. Loop any pair containing such a 1.
- Loop any octet even it contains some 1's that have already been looped.
- Loop any quad that contains one or more 1's which have not already been looped, making sure to use the minimum number of loops.
- Loop any pairs necessary to include any 1's that have not yet been looped, making sure to use the minimum number of loops.
- Form the OR sum of all the terms generated by each loop.
- In an n-variable Karnaugh-map there are 2^n cells.

Don't Care Condition

- Some logic circuit can be designed so that there are certain input condition for which there is no specified output levels, usually because these conditions will never occur.
- So, a circuit designer is free to make the output for any "don't care" condition either a '0' or '1' in order to produce the simplest output expression.

Implicant

Each individual min-term in canonical SOP form is called implicant.

Prime Implicant

Prime implicant is an implicant, which is obtained by combining maximum possible adjacent cell in K-map.

Essential Prime Implicant

An essential prime implicant is a prime implicant in which one or more min term are unique i.e. which is obtain by combining only one way in K-map.

Ex:

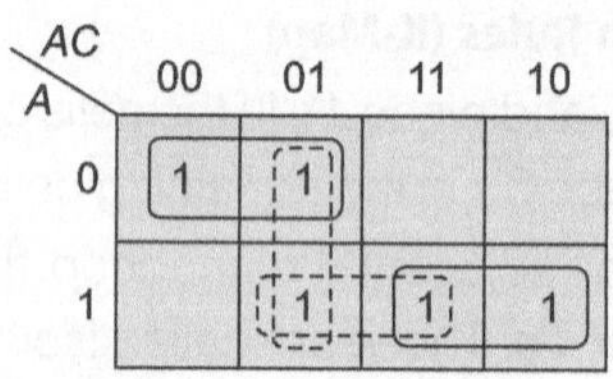

Here, Implicants: $\bar{A}\bar{B}\bar{C}, \bar{A}\bar{B}C, A\bar{B}C, ABC, AB\bar{C}$.

Prime implicants: $\bar{A}\bar{B}, AB, AC, \bar{B}C$.

Essential prime implicants: $\bar{A}\bar{B}, AB$.

Queen-McClusky Minimization Procedure

- **Step-1:** Describe individual minterms of the given expression by their equivalent binary number.
- **Step-2:** Form a table by grouping numbers with equivalent number of 1's in them i.e. first numbers with no 1's then numbers with one 1, and then numbers with two 1's etc.

- **Step-3**: Compare each number in the top group with each minterm in the next lower group. If the two numbers are the same in every position but one, place a check sign (✓) to the right of both numbers to show that they have been paired and covered. Then enter the newly formed number in the next column (a new table). The new number is the old numbers but where the literal differs, an "*X*" is placed in the position of that literal.
- **Step-4**: Using step-3, form a second table and repeat the process until no further pairing is possible on second repeat, compare numbers to numbers in the next group that have the same "*X*" position.
- **Step-5**: Term which were not covered are the prime implicants and are ORed and ANDed together to form final function.

Note:

The above *QM* method gives the prime implicants but not the essential prime implicants.

■■■■

Digital Logic Circuits

4

Introduction

Digital logic circuits are classified as:

(a) Combinational circuit

(b) Sequential circuit

Combinational Circuit	Sequential Circuit
1. Present output depends on present input only.	1. Present output depends on present input as well as previous output
2. No feedback is present.	2. Feedback is present.
3. No memory is present.	3. Due to feedback, memory is present
4. Example: Comparator, half adder, full adder, half subtractor, full-subtractor, multiplexer, demultiplexer,encoder, decoder.	4. Example: Flip-flop, counter, register.

Combinational Circuits

(a) Arithmetic Circuit

Half Adder (H.A)

A logic circuit for the addition of two one-bit numbers is referred to as an "HALF ADDER (H.A)".

Logic circuit and Truth table:

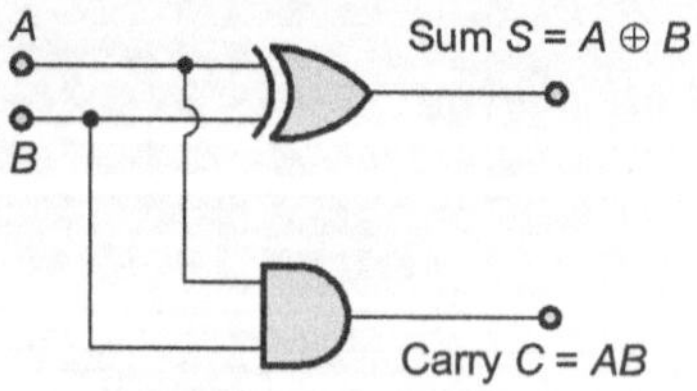

Inputs		Outputs	
A	B	Sum(S)	Carry(C)
0	0	0	0
0	1	1	0
1	0	1	0
1	1	0	1

Logical expression:

Sum $S = \bar{A}B + A\bar{B} = A \oplus B$

Carry $C = AB$

Remember: ....

- Total number of NAND-gates required to implement half adder = 5
- Total number of NOR-gates required to implement half adder = 5
- To implement the half adder circuit by minimum number of logic gates (if we have all gates except EXOR and EXNOR) is "3".
- Total number of MUX required to implement half adder = 3.

....

Full Adder (F.A.)

It performs the arithmetic sum of the three input bits i.e. addend bit, augend bit and carry bit.

Logical expression:

Sum, $S = A \oplus B \oplus C$

Carry, $C = AB + BC + CA = AB + C\,(A \oplus B)$

Remember: ....

- A F.A. can be implemented by two H.A. and one OR-gate
- Total number of NAND-gate/NOR-gate required to implement a F.A is equals to "9".
- Total number of MUX required to implement full adder = 7.

....

Half Subtractor (H.S)

Logic Circuit

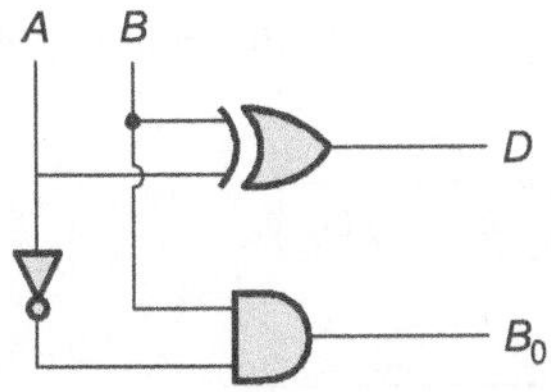

Logical expression:

When B is subtracted from A.

Difference, $D = \bar{A}\,B + A\,\bar{B} = A \oplus B$

Borrow, $B_0 = \bar{A}\,B$

Remember: ....

- Total number of NAND/NOR gates required to implement the H.S is equal to "5".
- Total number of MUX required to implement half subtractor = 3.

....

Full Subtractor (F.S)

It is a circuit which performs subtraction between two bits taking into account that a '1' may have been borrowed by a lower significant stage.

Logical expression:

Difference, $\boxed{D = A \oplus B \oplus C}$

Borrow, $\boxed{B_o = \bar{A}B + \bar{A}C + BC = \bar{A}B + \left(\overline{A \oplus B}\right) \cdot C}$

Remember: ....

- A full subtractor can be implemented with two half subtractor and one OR Gate.
- Number of NAND/NOR gates required to implement the full subtractor is equal to '9'
- In parallel adder n full adder or {(n – 1) F.A. and 1 H.A.} or {(2n – 1) H.A. and (n – 1) OR Gate} are required to add two n-bit numbers.
- Total number of MUX required to implement full subtractor = 7.

....

Look Ahead Carry Adder

- Look ahead carry adder makes the addition faster.
- It is based on the two additional signal that must be computed.

  Generator signal

$$\boxed{G_i = A_i B_i}$$

  Propagation signal

$$\boxed{P_i = A_i \oplus B_i}$$

- The carry signal generated is

$$\boxed{C_{i+1} = P_i C_i + G_i}$$

- Number of AND gates required in carry circuit for n-bit adder is

$$\boxed{N = \frac{n(n+1)}{2}}$$

- Number of OR gates required in carry circuit fo n-bit adder is $\boxed{N = n}$.

(b) Non Arithmetic Circuits

Multiplexer

Multiplexer is a combinational circuit which have many data input and single output depending on select or control input, one of the input line is transferred to the output, hence it is known as "many to one circuit" or "universal logic circuit" or "data selector circuit".

- The selection of a particular input line is controlled by a set of select lines.
- There are 2^n input lines where 'n' is the number select line.

4 : 1 MUX

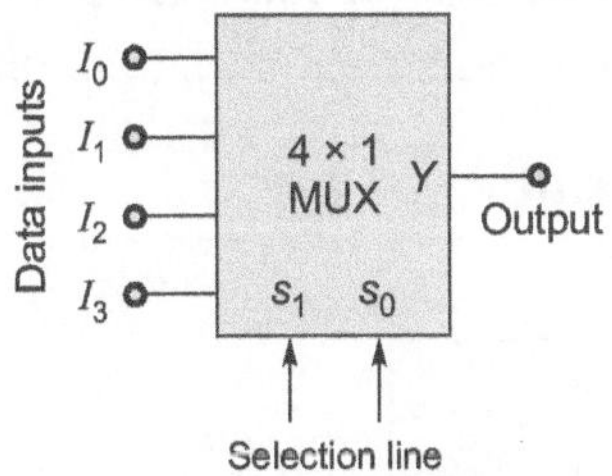

S_1	S_0	Y
0	0	I_0
0	1	I_1
1	0	I_2
1	1	I_3

$$Y = output = \bar{S}_1\,\bar{S}_0\,I_0 + \bar{S}_1\,S_0\,I_1 + S_1\,\bar{S}_0\,I_2 + S_1\,S_0\,I_3$$

Note: ....

- The size of the MUX is specified by the 2^n of its input line and the single output line.
- MUX contains AND gate followed by an OR-gate.

2 : 1 MUX

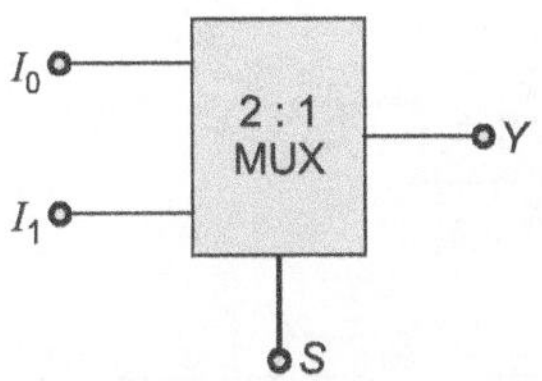

$$Y = \bar{S}I_0 + SI_1$$

Higher order MUX using Lower order MUX:

Given MUX	To be Implemented MUX	Required number of MUX
2 : 1	4 : 1	3
4 : 1	16 : 1	4 + 1 = 5
4 : 1	64 : 1	16 + 4 + 1 = 21
8 : 1	64 : 1	8 + 1 = 9
8 : 1	256 : 1	32 + 4 + 1 = 37

Remember:

- To implement $2^n : 1$ MUX by using 2 : 1 MUX, the total number of 2 : 1 MUX required is $(2^n - 1)$.
- MUX is a Universal Logic gate.
- Using one $2^n : 1$ MUX
  - → All 'n' variable logic functions can be implemented.
  - → Some '$n + 1$' variable loigc function can be implemented.

  Using one $2^n : 1$ MUX and 1 NOT gate.
  - → All 'n' and '$n + 1$' variable logic functions can be implemented.

MUX as Universal Logic Circuit

- AND gate using MUX

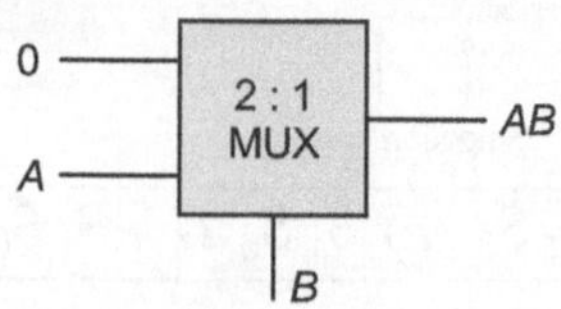

- OR gate using MUX

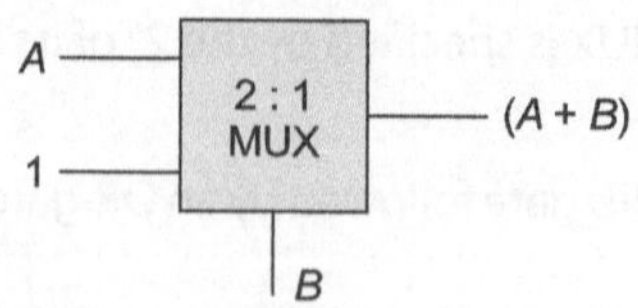

- NOT gate using MUX

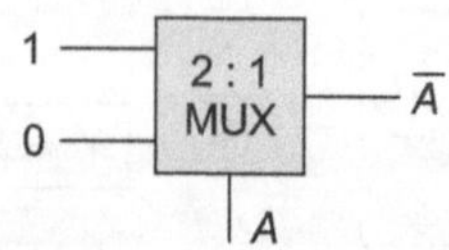

- NAND gate using MUX

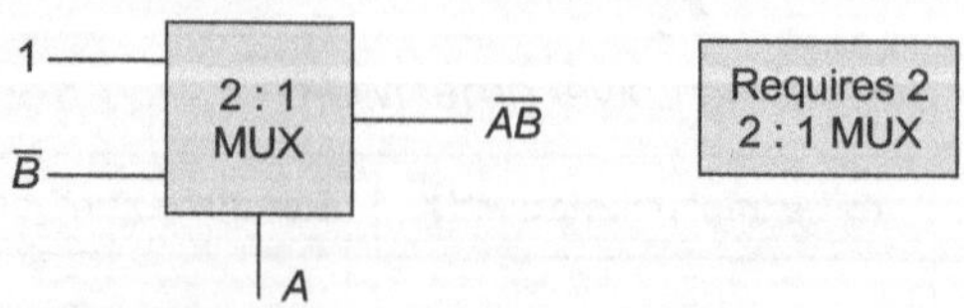

- NOR gate using MUX

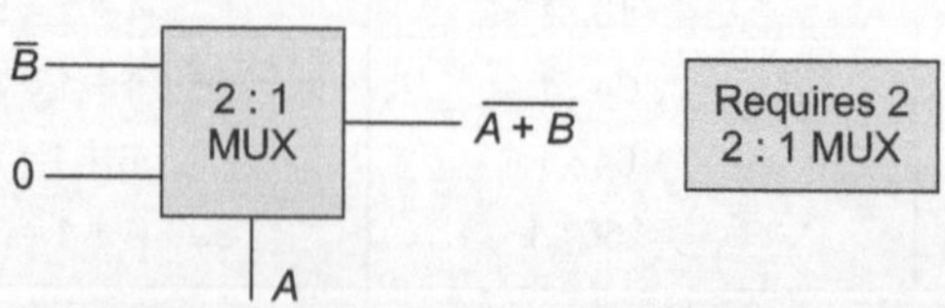

- EXOR gate using MUX

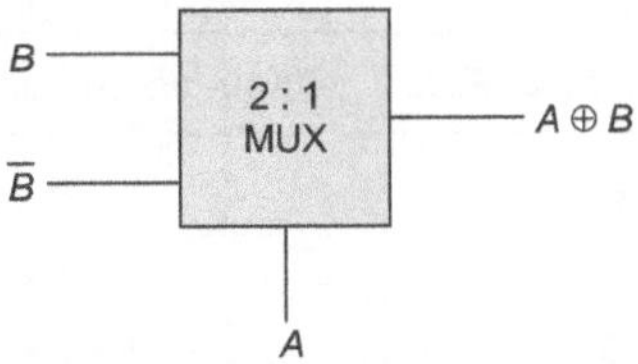

It requires 2, 2 : 1 MUX

- EXNOR gate using MUX

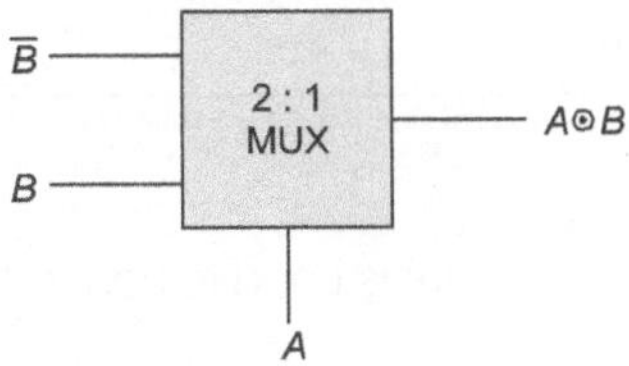

Requires 2 2 : 1 MUX

Demultiplexer (DEMUX)

- It receives information on a single line and transmits it on one of 2^n possible output lines.
- A "Decoder" with an enable input can function as a "Demultiplexer".

1 : 2 Demux

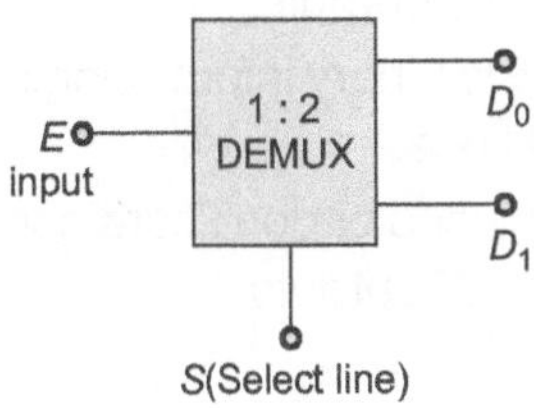

$D_0 = \overline{S}\,E$ and $D_1 = SE$

1 : 4 DEMUX

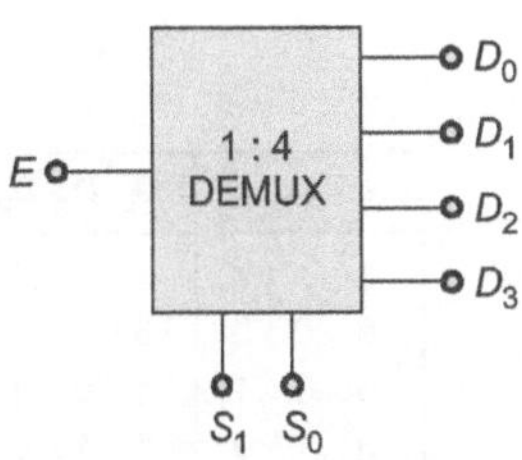

Select lines		Outputs			
S_1	S_0	D_3	D_2	D_1	D_0
0	0	0	0	0	E
0	1	0	0	E	0
1	0	0	E	0	0
1	1	E	0	0	0

$D_0 = \overline{S}_1\overline{S}_0 E,\ D_1 = \overline{S}_1 S_0 E,\ D_2 = S_1\overline{S}_0 E,\ D_3 = S_1 S_0 E$

Remember:

- Other name of demultiplexer are "one to many circuit" or "data distributor circuit".
- Number of select line is $\log_2 n$ (where n is number of output line).

Higher Order DEMUX Using Lower Order DEMUX:

Given DEMUX	To be Implemented MUX	Required number of MUX
1 : 2	1 : 4	1 + 2 = 3
1 : 2	1 : 8	1 + 2 + 4 = 7
1 : 2	1 : 16	1 + 2 + 4 + 8 = 15
1 : 2	1 : 64	1 + 2 + 4 + 8 + 16 + 32 = 63
1 : 4	1 : 16	1 + 4 = 5

Decoder

- Decoder is a combinational circuit that is used to convert binary to other codes such as
  (a) Binary to octal
  (b) Binary to decimal
  (c) Binary to hexadecimal
- In decoder, depending on binary data applied, one of the output line will become logic 1.
- A "Decoder" has many inputs and many outputs line.
- It is a combinational circuit that converts binary information from n input lines to a maximum 2^n unique output lines.
- If the n-bit decoded information has unused or don't care combinations, the decoder output will have less than 2^n outputs.
- **Total number of output lines**

$m \leq 2^n$ where, $n \rightarrow$ Total number of input lines

2 × 4 Decoder

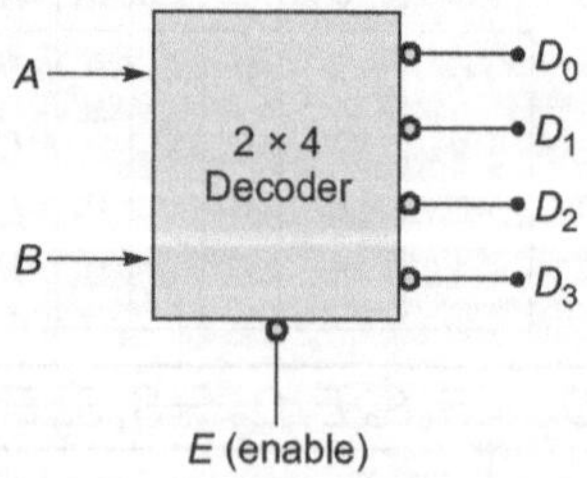

E	A	B	D_0	D_1	D_2	D_3
1	X	X	1	1	1	1
0	0	0	0	1	1	1
0	0	1	1	0	1	1
0	1	0	1	1	0	1
0	1	1	1	1	1	0

Note:

- 2 × 4 - Decoder may act like 1 : 4 DE MUX and vice-versa.
- Decoder and Demux circuits are almost same.
- Decoder contains AND-gates or NAND-gates.

Higher Order Decoder Using Lower Order Decoder:

Given Decoder	To be Implemented Decoder	Required number of Decoder
2 : 4	4 : 16	1 + 4 = 5
2 : 4	3 : 8	2 + a NOT Gate
4 : 16	8 : 256	1 + 16 = 17

Encoder

- Encoder is used to convert other codes to binary such as
  - (a) Octal to binary
  - (b) Hexadecimal to binary
  - (c) Decimal to binary
  - (d) Decimal to *BCD*
- In encoder one of the input line is high and corresponding binary is available at the output.
- In priority encoder any number of inputs can be high but based on priority highest priority input corresponding binary is available at the output.

Magnitude Comparator

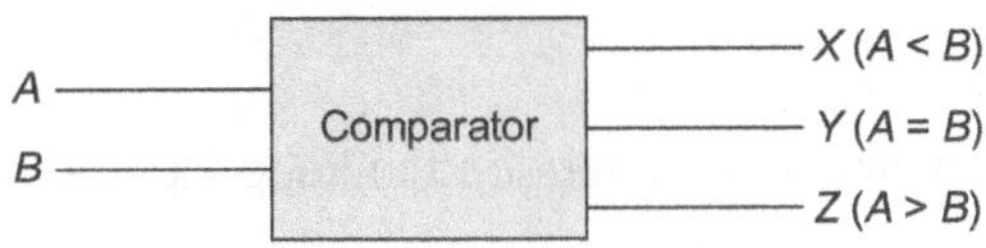

Truth table:

A	*B*	*X*	*Y*	*Z*
0	0	0	1	0
0	1	1	0	0
1	0	0	0	1
1	1	0	1	0
Output expression		$\bar{A}B$	$AB + AB = A \varepsilon B$	$A\bar{B}$

■■■■

5

Sequential Circuits

Flip-Flops

- Flip-flop is a basic memory element.
- It can store "one bit of information".
- A flip-flop has 2 outputs, which are always complementary to each other.
- A flip-flop has 2 stable state hence it is known as bi-stable multivibrator.
- A simplest form of flip-flop is called "latch". It can be constructed with two cross coupled NAND gates or NOR gates.

Triggering

Triggering is used to initiate the operation of latches or flip-flops.

(i) **Level trigger:** Input signal affects the flip-flop only when the clock is at logic '1' or logic '0'.

(ii) **Edge trigger:** Input signal affects the flip-flop only if they are present at the positive or negative going edge of the clock pulse.

Setup Time

It is minimum time required to keep input at proper level before applying clock.

Hold Time

Hold time is minimum time required to keep input at same level after applying clock.

Note:

- **If set-up time and hold time is not properly given, the output may go to meta stable state.**

Clocked S-R Flip-Flop

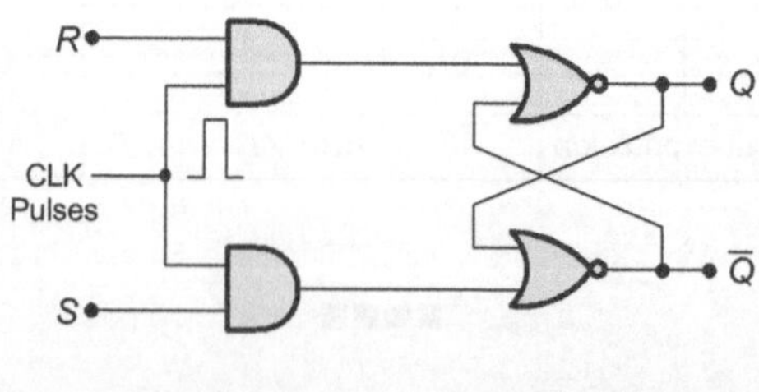

Truth table

Clock	S_n	R_n	Q_{n+1}	
0	X	X	Q_n	
1	0	0	Q_n	→ HOLD state
1	0	1	0	→ RESET state
1	1	0	1	→ SET state
1	1	1	Invalid	→ FORBIDDEN state

- S_n and R_n denotes the inputs and Q_n the output during the bit time 'n'.
- 'Q_{n+1}' denotes the output Q after CLK passes, i.e. in bit time $(n + 1)$.

Characteristic table

S	R	Q_n	Q_{n+1}	
0	0	0	0	→ Q_n
0	0	1	1	
0	1	0	0	→ 0
0	1	1	0	
1	0	0	1	→ 1
1	0	1	1	
1	1	0	X	→ Invalid
1	1	1	X	

Characteristic Equation

$$Q_{n+1} = S + \bar{R}Q_n$$

Remember:

Disadvantage of SR flip-flop is invalid state will be present when both inputs are 1 or high. To avoid this J-K flip-flop is used.

J-K Flip-Flop

JK-FF is called universal FF because the FFs like *D*-FFs, *SR*-FF and *T*-FF can be derived from it.

$$S = J\bar{Q} \text{ and } R = KQ$$

Logic diagram

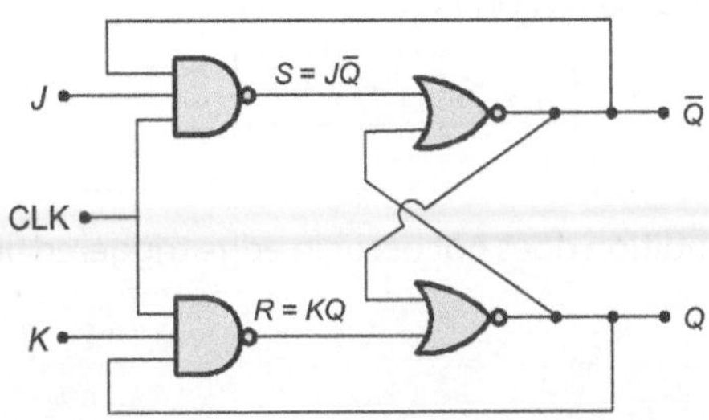

Truth table

Clock	J	K	Q_{n+1}	
0	X	X	Q_n	
1	0	0	Q_n	→ HOLD state
1	0	1	0	→ RESET state
1	1	0	1	→ SET state
1	1	1	$\bar{Q}_n$	→ TOGGLE state

Characteristic table

J	K	Q_n	Q_{n+1}	
0	0	0	0	→ Q_n
0	0	1	1	
0	1	0	0	→ 0
0	1	1	0	
1	0	0	1	→ 1
1	0	1	1	
1	1	0 →	1	→ $\bar{Q}_n$
1	1	1 →	0	

Characteristic equation

$$Q_{n+1} = J\bar{Q}_n + \bar{K}Q_n$$

Race Around Condition

- The Race around condition will occur when
  - (i) $J = K = 1$
  - (ii) $t_{pd(FF)} < t_{pw}$
  - (iii) Flip-flop is level triggered.
- For the duration "t_{pw}" of the clock pulse, the output Q will oscillate back and forth between 0 and 1. At the end of the clock pulse, the value "Q" is uncertain. This situation is referred as "race around condition".
- To avoid this, the condition is

$$t_{pw} < t_{pd\,(FF)} < T$$

Remember:

Race around condition does not occur in edge triggered flip-flop.

Master-Slave J-K Flip-Flop

To avoid race around condition master slave FF is used.

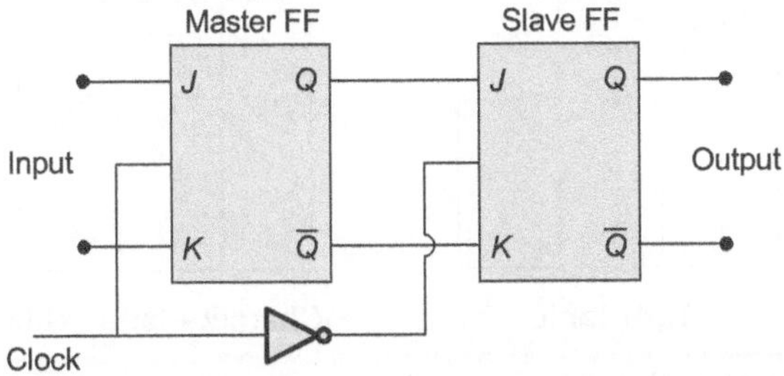

- In master slave flip-flop master is applied with input clock and slave is applied with inverted clock. Due to this when master output is changing slave output remains in previous state.
- In master slave flip-flop, output will change only when slave output changes.
- Output of master can change many times but, slave output can change only 1 time so master FF act as level triggered and slave FF act as edge triggered. Therefore there is no "race around condition" at the output of the slave.

D-Flip-Flop

It is a FF with a delay equal to exactly one cycle of CLK.

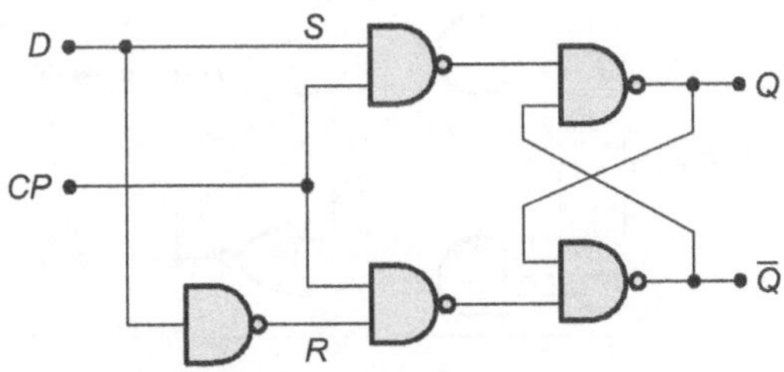

Graphical diagram

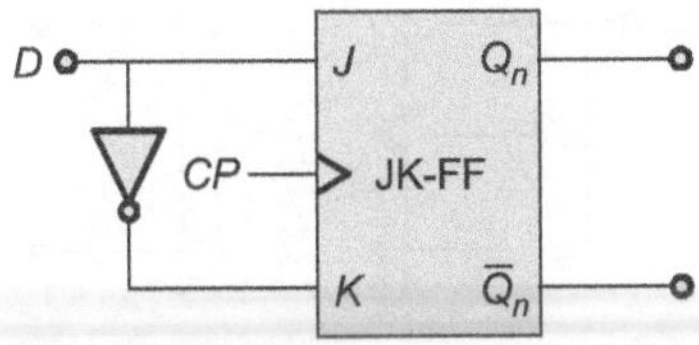

$$J = D \text{ and } K = \overline{D}$$

Truth table and characteristic table

CLK	D	Q_{n+1}
0	X	Q_n
1	0	0
1	1	1

Truth Table

D	Q_n	Q_{n+1}
0	0	0
0	1	0
1	0	1
1	1	1

Characteristic Table

Characteristic equation

$$Q_{n+1} = D$$

Note: It is also known as "Transparent Latch" because $Q_{n+1} = D$.

Toggle Flip-Flop (T-FF)

- The T-FF is a single input version of the J-K FF.
- T-FF can be obtained from a J-K FF if J and K are tied together as in figure shown below.
- The designation T, comes from the ability of the FF to "toggle" or "change state".

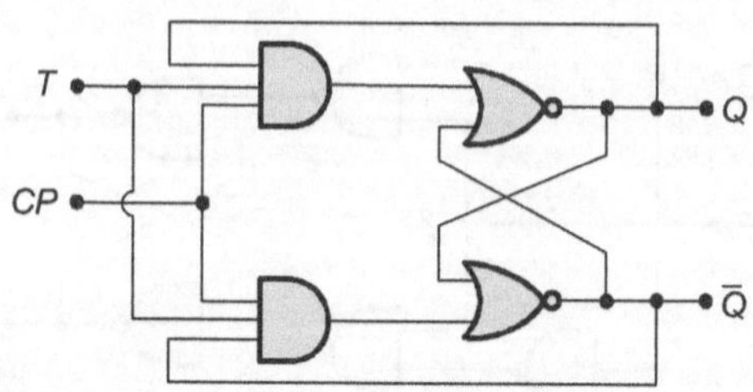

Graphical symbol of T-FF

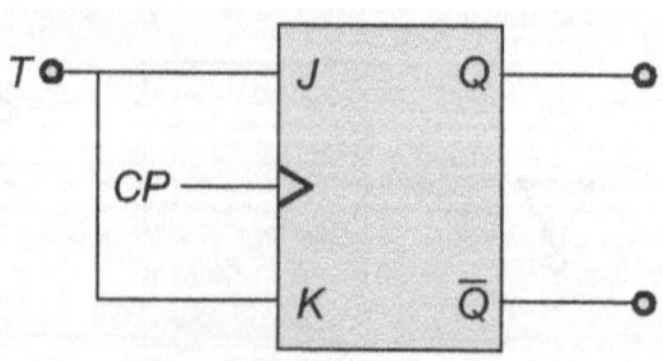

$$J = K = T$$

Truth table and characteristic table

CLK	T	Q_{n+1}	
0	X	Q_n	
1	0	Q_n	→ HOLD state
1	1	$\overline{Q}_n$	→ Toggle state

Truth Table

T	Q_n	Q_{n+1}
0	0	0
0	1	1
1	0	1
1	1	0

Characteristic Table

Characteristic equation

$$Q_{n+1} = \overline{T}\, Q_n + T\, \overline{Q}_n = T \oplus Q_n$$

Conversion of Flip-Flops

SR FF to other FFs	JK FF to other FFs	D-FF to other FFs	T-FF to other FFS
$S = J\overline{Q}_n$ & $R = KQ_n$	$J = S$ & $K = R$	$D = S + \overline{R}Q_n$	$T = S\overline{Q}_n + RQ_n$
$S = D$ & $R = \overline{D}$	$J = D$ & $K = \overline{D}$	$D = J\overline{Q}_n + \overline{K}Q_n$	$T = J\overline{Q}_n + KQ_n$
$S = T\overline{Q}_n$ & $R = TQ_n$	$J = K = T$	$D = T \oplus Q_n$	$T = D \oplus Q_n$

Excitation Table of Flip-Flops

Q_n	Q_{n+1}	S	R	J	K	D	T
0	0	0	X	0	X	0	0
0	1	1	0	1	X	1	1
1	0	0	1	X	1	0	1
1	1	X	0	X	0	1	0

■■■■

Registers

6

Shift Registers (SRs)

- Registers are used to store group of bits.
- In register to store *N*-bit, it requires N flip-flop.
- The "serial input" determines what goes into the left most FF during the shift.
- Depending upon input and output, registers can be classified into 4 types.:
  (a) SISO: Serial In Serial Out
  (b) SIPO: Serial In Parallel Out
  (c) PISO: Parallel In Serial Out
  (d) PIPO: Parallel In Parallel Out

SISO (Serial In Serial Out)

4-bit Right-Shift SISO Register

- In right shift SISO register, LSB data is applied at the MSB FF′(D-FF).
- In '*n*' bit register, to enter '*n*' bit data, it requires 'n' clock pulses in serial form.
- If '*n*' bit data is stored in SISO register then output is taken serially for this $(n-1)$ clock pulses are required.
- SISO register is used to provide '*n*' clock pulse delay to the input data.
- If '*T*' is the time period of clock pulse, then delay provided by SISO is *nT*.

4-bit Left-Shift SISO Register

- In this above SISO register MSB data is applied to the LSB FF(D-FF).
- To enter the '*n*' bit data in serial form we require '*n*' clock pulse.
- To exit or getting output of '*n*' bit data as serially we require $(n-1)$ clock pulse.

SIPO (Serial In Parallel Out)

- For '*n*' bit- serial input data to be stored the number of CLK pulse required = *n*.
- For '*n*' bit-parallel output data to be stored the number of CLK pulse required = 0 (there is no need of CLK pulse).

PISO (Parallel In Serial Out)

- To store parallel in data, if we store 'n' bit then the number of CLK pulse required = 1 CLK pulse.
- To store serial out data if we store 'n' bit then the number of CLK pulse required = $(n - 1)$.

Note:

To convert temporal code into spatial code, we use SIPO register, whereas to convert spatial code into temporal code we use PISO register.

PIPO (Parallel In Parallel Out)

- For parallel in data, the number of CLK pulse required = 1 CLK pulse.
- For parallel out data, the number of CLK pulse required = 0 CLK pulse.

Remember:

- All shift registers are made of JK flip-flops.
- "PIPO" register is a storage register made up with D-FFs.
- Clock pulses required by various shift registers.

	Input	Output
SISO	n	$n-1$
SIPO	n	0
PISO	1	$n-1$
PIPO	1	0

- If 'n' shift left operations are performed then data will be multiplied 2^n times.
- If 'n' shift right operation are performed then data will be divided 2^n times.

Time delay

A SISO SRs may be used to introduce time delay "Δt" in digital signals

$$\Delta t = N \times T = N \times \frac{1}{f_c}$$

where, $N \rightarrow$ Number of FFs

$T \rightarrow$ Time period of CLK pulse

$f_C \rightarrow$ CLK frequency

- The amount of delay can be controlled by the "f_c" or number of FFs in the SR.

■■■■

Counters

7

Introduction

- It is a sequential circuit formed by the cascading of FFs.
- Counter are basically used for
  - (i) Counting of the number of clock pulses applied
  - (ii) Frequency division
  - (iii) Timers
  - (iv) Frequency measurement
  - (v) Waveform generation
- Counters are classifed as:
  - (i) Asynchronous counter
  - (ii) Synchronous counter

Remember:

- If N = total number of states and 'n' = number of FFs then

$$\boxed{N \leq 2^n}$$

$$\boxed{n \geq 3.32 \log_{10} N} \text{ or } \boxed{n \geq \log_2 N}$$

- If $N = 2^n$, then we get BINARY COUNTER.
- If $N < 2^n$, then we get NON-BINARY COUNTER.

MOD number

- The "MOD-number" indicates the number of states in counting sequence.
- For n-FFs, counter will have 2^n different states and then this Counter is said to be "MOD-2^n Counter".
- MOD number indicates the frequency division factor obtained from the Last FF.
- It would be capable of counting upto $(2^n - 1)$ before returning to zero state.

Note:

- In "MOD-N Counter", if applied input frequency is "*f*", then output frequency is f/N.
- If two counters are cascaded with MOD-M followed by MOD-*N*, then number of overall states of combined counter is ($M \times N$) and counter is called "MOD-*MN*" counter.

Asynchronous (Series) Counter

Binary Ripple Counter

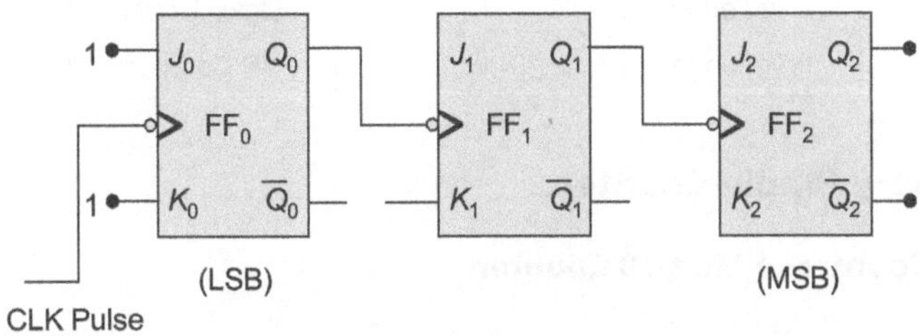

- In ripple counter with *n*-FFs there are 2^n possible states.
- With n-FFs the maximum count that can be counted by this counter is $(2^n - 1)$.
- If input frequency is '*f*' then output frequency is $\frac{f}{2^n}$.
- It is also known as $2^n : 1$ scalar divider.

For proper operation of the ripple counter:

$$\boxed{T_{CLK} \geq n t_{pd(FF)}}$$

$$\boxed{f_{CLK} \leq \frac{1}{n t_{pd(FF)}}}$$

Maximum CLK frequency:

$$\boxed{f_{CLK,m} = \frac{1}{n t_{pd(FF)}}}$$

Disadvantage of Ripple Counter:

Decoding error is present due to propagation delay of FFs i.e. $t_{pd(FF)}$.

Up/Down Counter

An "Up/Down Counter" can count in any direction depending upon the control input.

For determination of Up/Down Counter:

Triggering with	CLK connection in	Access as
(–ve) edge	Q	UP Counter
(–ve) edge	$\overline{Q}$	Down Counter
(+ve) edge	Q	Down Counter
(+ve) edge	$\overline{Q}$	UP Counter

Non-Binary Ripple Counter

Decade Counter or Mod-10 Counter

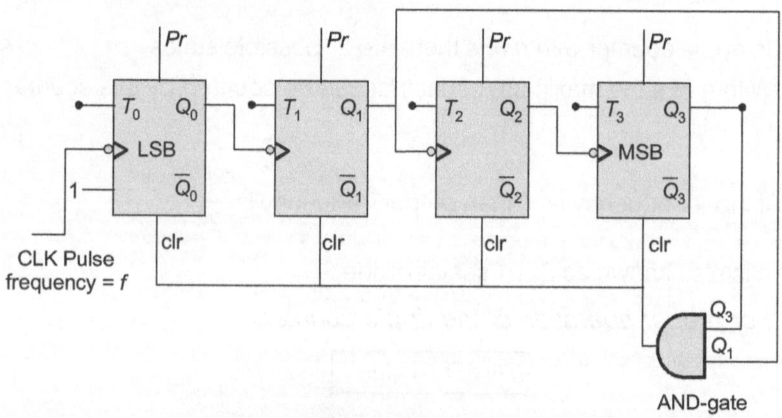

- Used states = 10 and unused states = 6
- Output frequency of MOD-10 counter = $f/10$.
- Clear and preset is used in non binary ripple counter.
- Clear is used to reset counter without clock.
- Preset is used to set counter.
- When decade counter counts from 0 to 9 then it is known as BCD counter.

Synchronous (Parallel) Counters

Ring Counter

It is SISO shift register.

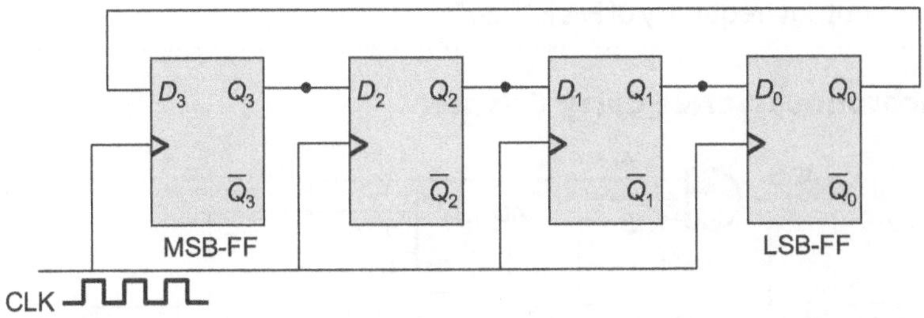

Remember: ....

- Ring counter is a non-self starting counter.
- With 'n' FFs, there are n-states present in ring counter.
- With 'n' FFs, maximum count possible in ring counter is (2^{n-1}).
- Decoding is very easy in Ring counter, because there is no aid of extra circuit.

....

In a 4-bit ring counter:

Used states = 4

Unused states = $2^4 - 4 = 12$

- In a ring counter if CLK frequency is "f" the FFs output frequency is "f/N" (where N = Number of states = modulus of the counter).

Twisted-Ring Counter

Also known as Johnson Counter or Switch Tail Ring Counter

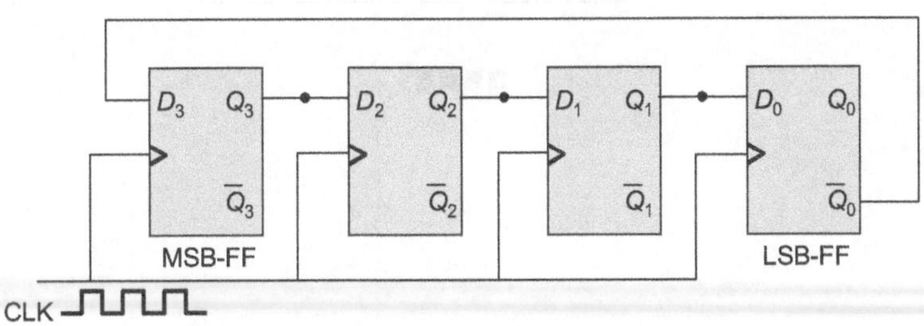

Remember: ....

- With 'n' FFs there are '$2n$' states in this counter.
- With 'n' FFs the maximum count by this counter is $(2^n - 1)$.
- In normal "Johnson Counter" with 'n' FFs and the input frequency is 'f' then output frequency of FFs is "$f/2n$"

Synchronous-Series Carry Counter

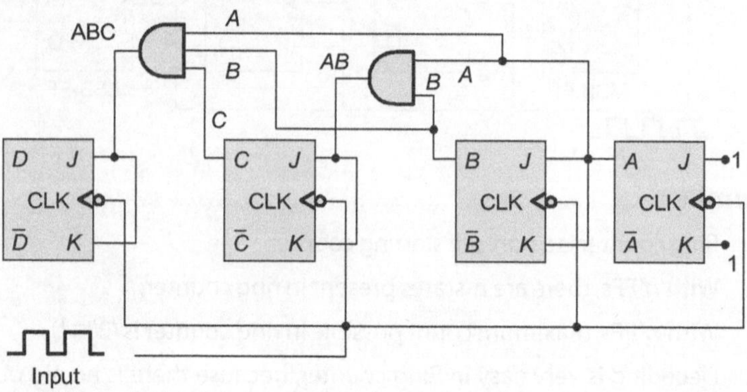

Clock frequency

$$f_{CLK} \le \frac{1}{t_{pd(FF)} + (n-2)\, t_{pd(\text{AND gate})}}$$

Synchronous-Parallel Carry Counter

It is the "Fastest Counter".

Clock frequency:

$$f_{CLK} = \frac{1}{t_{pd(FF)} + t_{pd(\text{AND gate})}}$$

■■■■

Digital ICs Family

8

Logic Family

A group of compatible ICs with the same logic levels and supply voltages for performing various logic functions fabricated using a specific circuit configuration is called "logic family".

Characteristics of Logic Families

1. Propagation Time Delay (t_{pd})

This is the average transition delay time for the signal to propagate from input to output when signal changes its value. This determines how fast the logic system can operate.

Delay time:

$$\boxed{t_{pd} = \frac{t_{PHL} + t_{PLH}}{2}} \text{ ns}$$

where, $t_{PLH} \rightarrow$ Delay time in going form LOW logic to HIGH logic

$t_{PHL} \rightarrow$ Delay time in going form HIGH logic to LOW logic

Remember:

- The delay times are measured between the 50% points on the input and output transitions.
- In BJT, $\boxed{t_{PLH} > t_{PHL}}$ due to reverse recovery time.
- In FET, $\boxed{t_{PLH} < t_{PHL}}$ due to large capacitance formed.

2. Power Dissipation (P_D)

- It is the amount of power dissipated in an IC.
- It is determined by the current "I_{CC}" which draws from the V_{CC} supply.

Collector current:

$$\boxed{I_{CC} = \frac{I_{CCH} + I_{CCL}}{2}}$$

where, $I_{CCH} \rightarrow$ current value when all inputs are HIGH.

$I_{CCL} \rightarrow$ current value when all inputs are LOW.

Power dissipation:

$$\boxed{P_{D(avg)} = I_{CC} \times V_{CC}} \text{ mW}$$

3. Figure of Merit (FOM)

$$\boxed{\text{FOM} = t_{pd} \times P_{D(avg)}} \text{ pico Joules}$$

For the best operation of ICs, figure of merit (FOM) should be as small as possible.

4. Fan-in and Fan-out

The maximum number of inputs which can be applied to a logic gate is known as fan-in. On the other hand, fan out of a logic gate is the number of gates that can be driven by it.

$$\boxed{\text{Fan out} = \left(\frac{I_{OH}}{I_{IH}}\right) \text{ or } \left(\frac{I_{OL}}{I_{IL}}\right)} \text{ ; whichever is minimum}$$

5. Noise-Margin

It is the property of a logic circuit to withstand unwanted noise voltage at input or the maximum value of noise signal that a system can reject with performance unaffected.

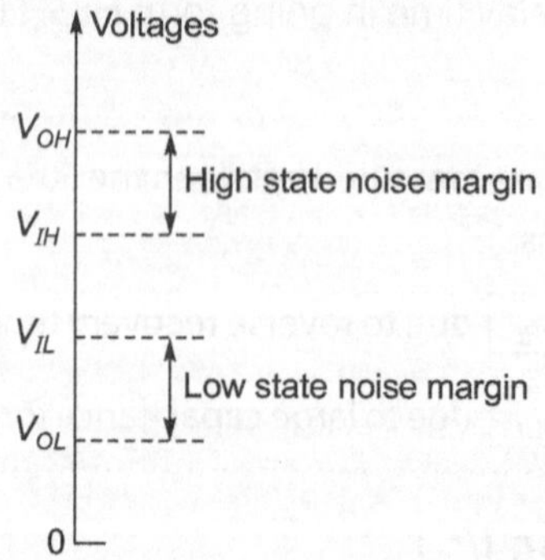

$$\boxed{V_{OH} > V_{IH} > V_{IL} > V_{OL}}$$

$$\boxed{\text{High state N.M.} = V_{NH} = V_{OH(min)} - V_{IH(min)}}$$

$$\boxed{\text{Low state N.M.} = V_{NL} = V_{IL(max)} - V_{OL(max)}}$$

$$\boxed{\text{Noise margins} = (V_{NH}, V_{NL})_{min}}$$

Bipolar Logic Families

Resistor Transistor Logic (*RTL*)

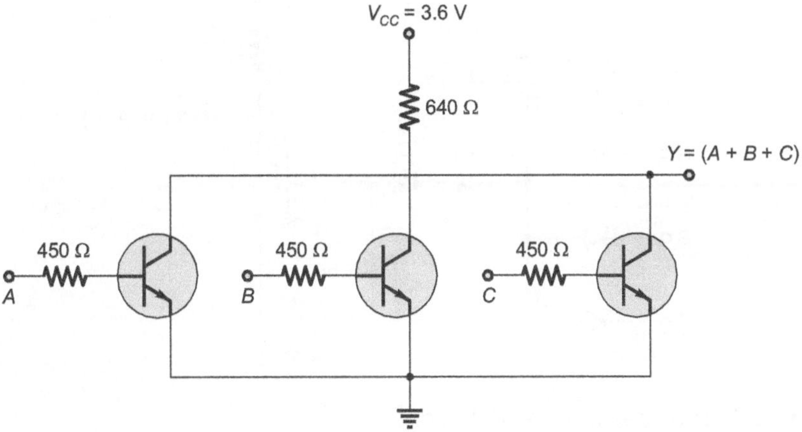

- RTL provides wired AND logic.

Diode Coupled Transistor Logic (DCTL)

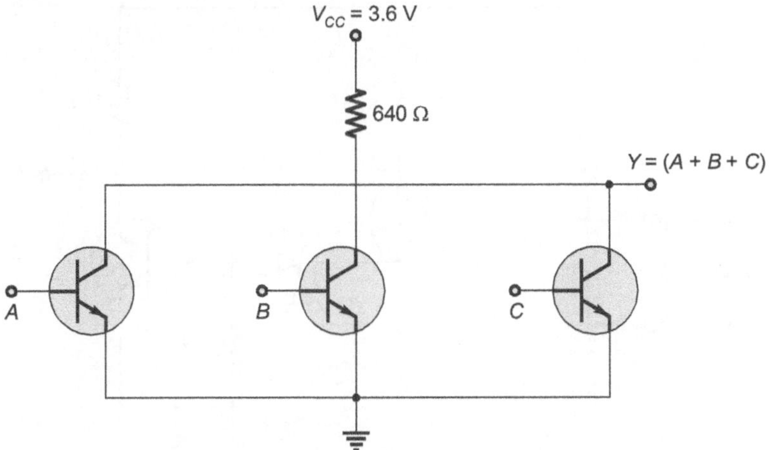

- DCTL provides wired AND logic.
- It suffers from "current Hogging".

Diode Transistor Logic (DTL)

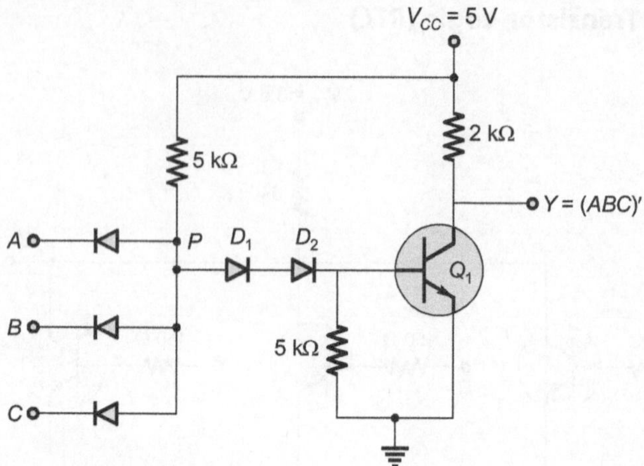

- DTL provides wired AND logic.

High Threshold Logic (HTL)

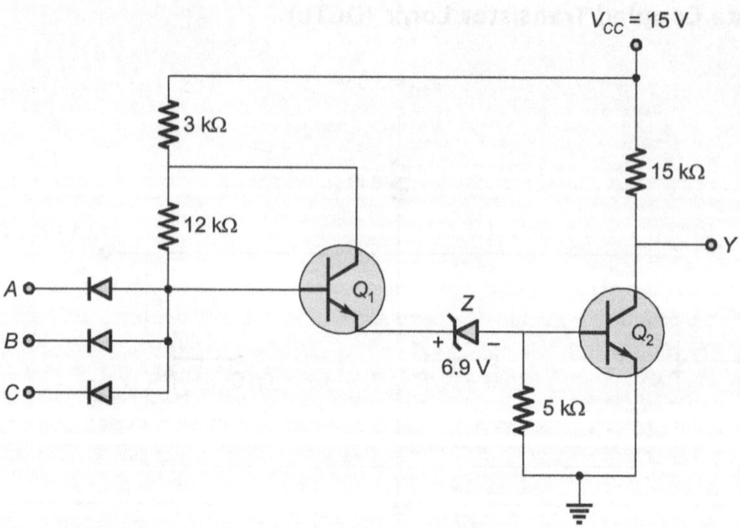

- HTL circuits has excellent noise margin and largest voltage swing.
- It provides wired AND operation.

Transistor Transistor Logic (TTL)

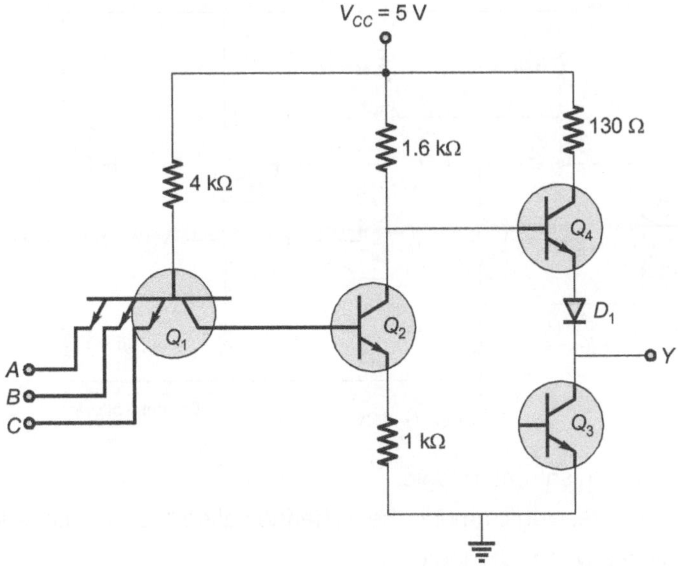

TTL gate with totem-pole output

- TTL circuits are classified as:
  - (i) Tristate logic (high impedance logic)
  - (ii) Totem pole logic (active pull-up)
  - (iii) Open collector logic (passive pull-up)
- Tristate logic is used in bus oriented system.
- Totem pole logic does not provide wired logic.
- The advatange of active pull-up over passive pull-up is, reduced power dissipation and increased speed of operation.
- Any floating input TTL is considered as logic '1'.

Emitter Coupled Logic (ECL)

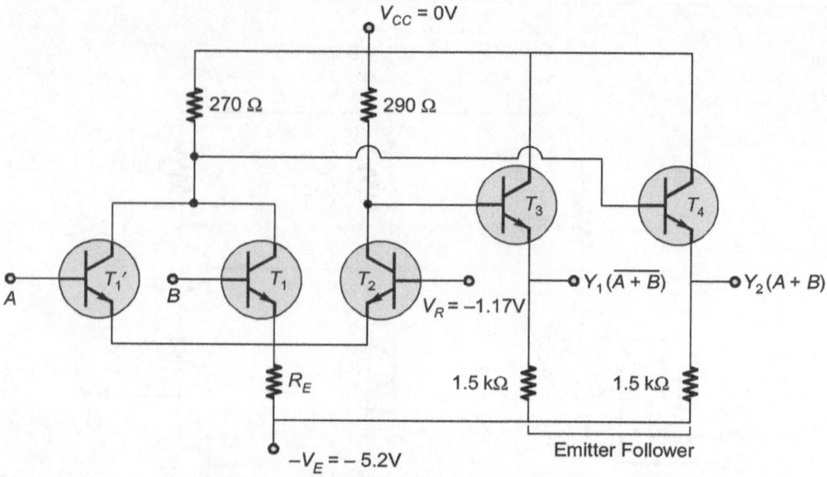

- ECL is a non saturated logic.
- It is the fastest logic among the all hence called "current mode logic".
- It provides wired 'OR' logic.
- It uses negative power supply to avoid noise and glitches.
- Any floating input in ECL is considered as logic '0'.

Integrated Injection Logic (IIL or I^2L)

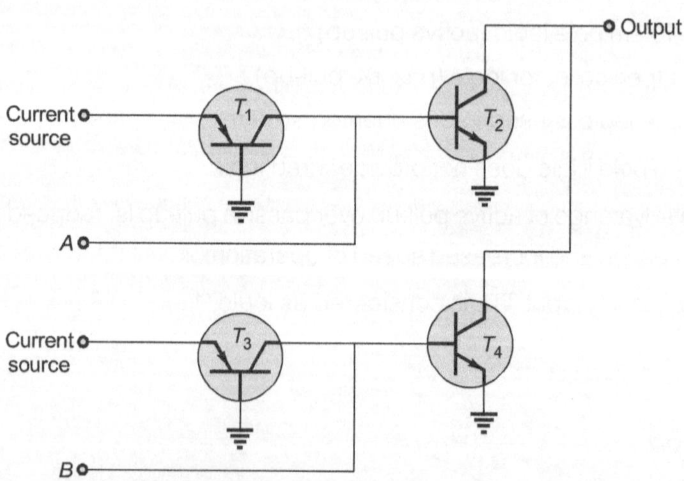

- I^2L logic has best figure of merit.

MOS Logic Family

- MOS digital ICs use enhancement MOSFETs exclusively.
- Because of the symmetrical construction of source (*S*) and drain (*D*), the MOS transistor can be operated as a bilateral device.

N-Channel MOS (NMOS)

- It is faster than PMOS.
- NMOS conducts whenever input is HIGH.

Symbol

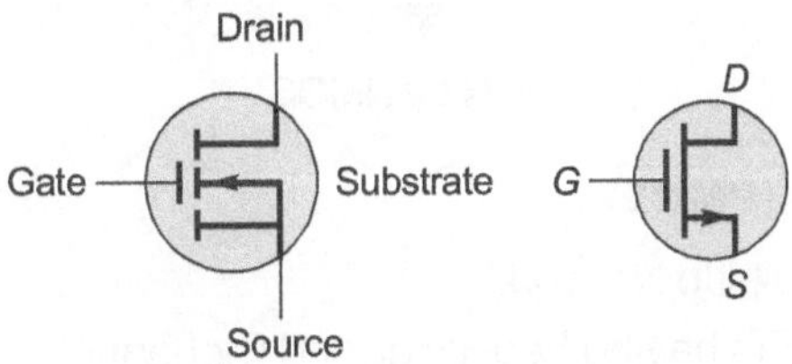

Different Logic Gates by NMOS

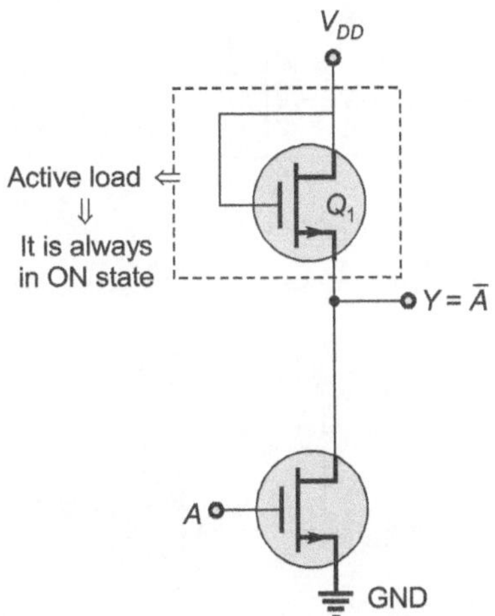

Inverter by NMOS

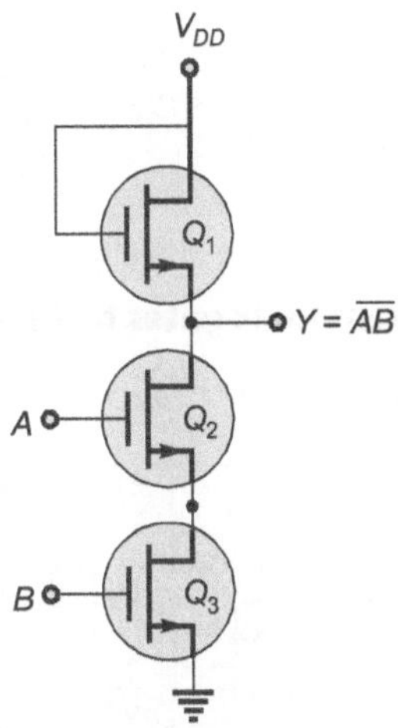

NAND by NMOS

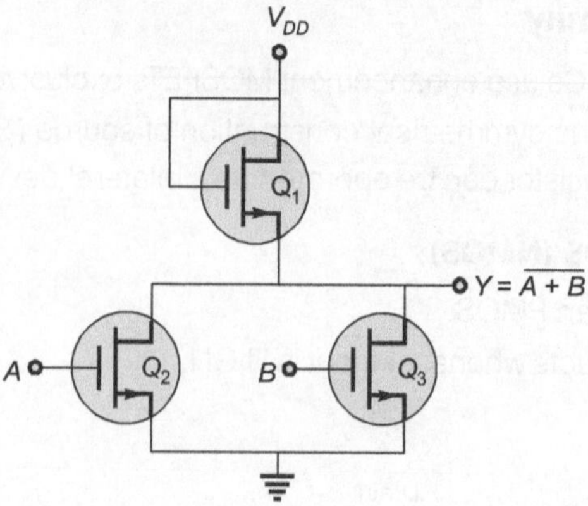

NOR by NMOS

P-channel MOS (PMOS)

- Also called "Pull-Up Network".
- PMOS uses FETs having heavily doped P-channel.
- PMOS conducts whenever input is LOW.

Symbol

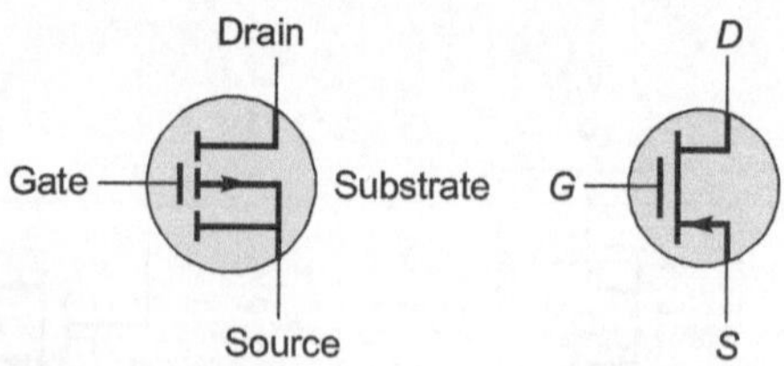

Different logic gates by PMOS

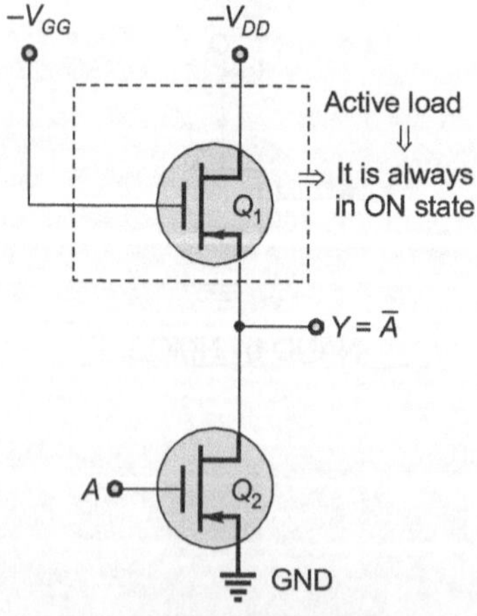

Inverter by PMOS

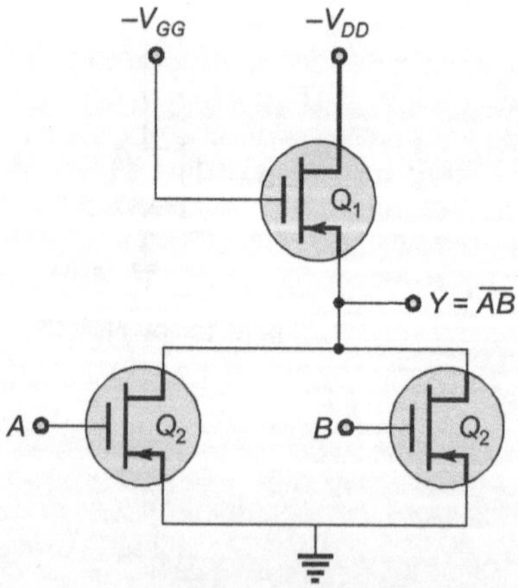

NAND by NMOS

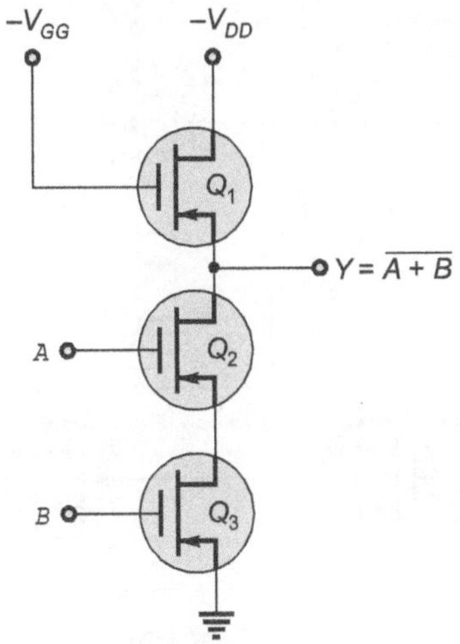

NOR by PMOS

Complementary MOS (CMOS)

CMOS is a combination of NMOS and PMOS and is fabricated as twin tub. Various logic functions can be realized using CMOS. They are characterized by low power dissipation and excellent noise immunity.

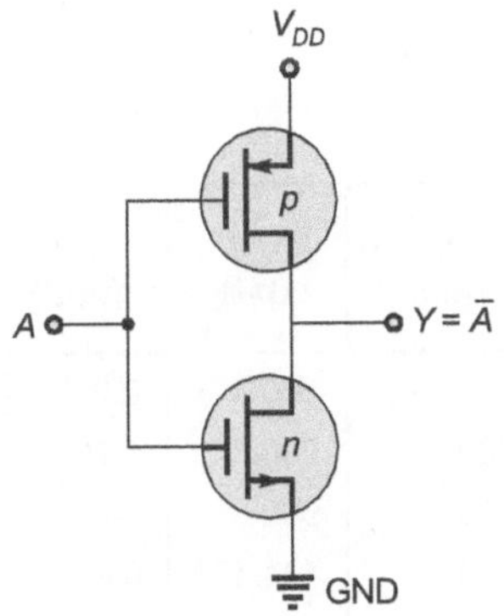

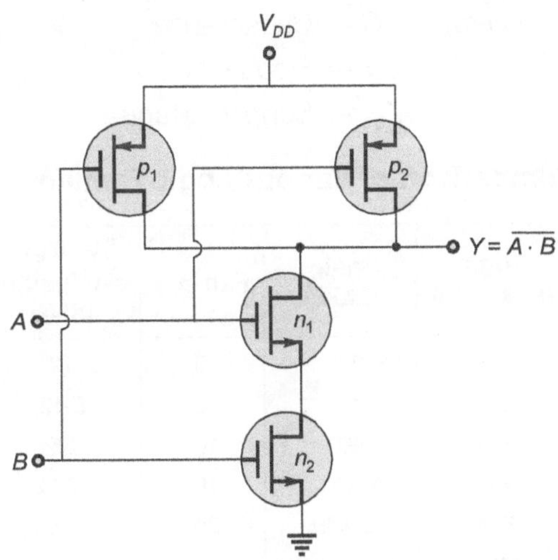

NOR by CMOS

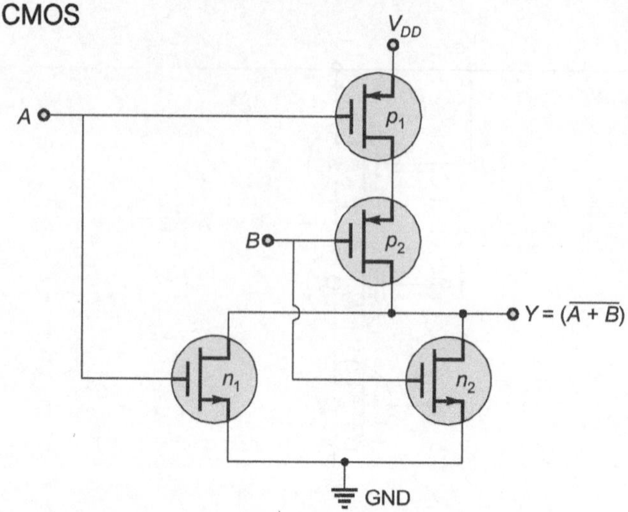

Power Dissipation

Power dissipation is CMOS is of two types:

(a) Static: It will be present during logic '0' or logic '1' input.

(b) Dynamic: It will be present during logic level transition

$$P_d(\text{dynamic}) = Cf\, V_{DD}^2$$

where, $C \rightarrow$ Capacitance

$f \rightarrow$ Frequency

$V_{DD} \rightarrow$ Supply voltage

Comparison of Various Logic Families

Logic parameter	Basic GATE	Fan-out	Power dissipation (mW)	t_{PD} (nsec)	F.O.M. (pJ)	N.M. (volt)
RTL	NOR	5	12	50	500	0.2
DTL	NAND	8	8-12	30	240	0.75
HTL	NAND	10	55	90	5000	4-5
TTL	NAND	10	12-22	12-6	144-132	0.4
ECL	OR/NOR	25	40-55	4-1	55-160	0.3
MOS	NAND/NOR	20	0.2-10	300	60	1.5
CMOS	NAND/NOR	>50	0-0.01	70	0.7	$\frac{V_{DD}}{2}$

■■■■

DACs and ADCs

9

Digital to Analog Converter (DAC)

Specification of DAC

1. Resolution

Resolution of DAC is change in analog voltage corresponding to LSB bit increment at the input.

$$\text{Resolution} = \frac{V_r}{2^n - 1}$$

where, $V_r \rightarrow$ Reference voltage corresponding to logic 1

$n \rightarrow$ Number of bit

2. Analog Output Voltage

- Resolution × Decimal equivalent of binary data = Output voltage
- Analog output voltage of an N-bit straight binary DAC

$$\boxed{V_0 = K\left[2^{N-1} b_{N-1} + 2^{N-2} b_{N-2} + \ldots\ldots\ldots + 2^2 b_2 + 2^1 b_1 + b_0\right]}$$

where, $K \rightarrow$ Proportionality factor.

$b_n \rightarrow$ 1; if the nth bit of digital input is '1'.

$\rightarrow$ 0; if the nth bit of digital input is '0'.

- Full scale output (V_{FS}) voltage is maximum output of DAC

$$V_{FS} = \text{Resolution} * \text{Maximum decimal}$$

$$= \frac{V_r}{2^n - 1} * (2^n - 1)$$

$$V_{FS} = V_r$$

3. % Resolution

$$\%\ \text{Resolution} = \frac{\text{Resolution}}{V_{FS}} \times 100$$

$$\%\ \text{Resolution} = \frac{1}{2^n - 1} \times 100$$

4. Error/Accuracy

Maximum error acceptable in ADC/DAC is 1 LSB bit which is equal to resolution.

Weighted-Resistor DAC

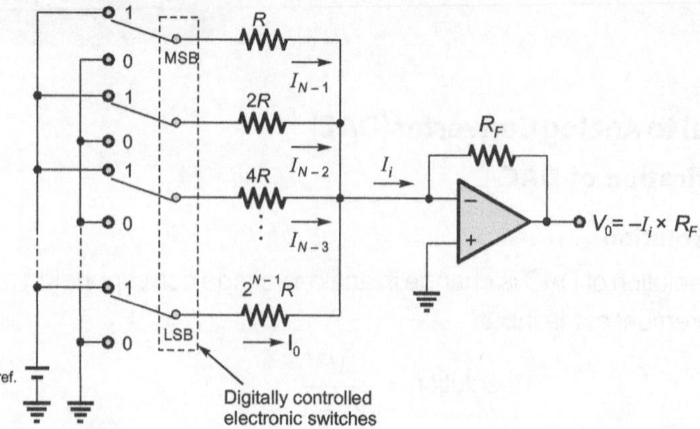

Output voltage:

$$V_0 = \frac{R_F}{2^{N-1} \cdot R}(2^{N-1}V_{N-1} + 2^{N-2}V_{N-2} + \ldots\ldots + 2^1V_1 + 2^0V_2)$$

Proportionality factor:

$$K = \frac{R_F}{2^{N-1} \cdot R} \cdot V_R$$

Input current to OP-AMP:

$$\boldsymbol{I_i} = \frac{V_{ref.}}{2^{N-1} \cdot R}\sum_{i=0}^{N-1} 2^i b_i$$

Maximum output current:

$$\boldsymbol{I}_{max} = \frac{V_{ref.}}{2^{N-1} \cdot R}(2^N - 1)$$

Resistance:

$$\text{LSB Resistance} = (2^{N-1}) \times \text{MSB Resistance}$$

Remember:

- The resistance values are weighted in accordance with the binary weight.
- Here "OP-AMP" is employed as a "summing amplifier".
- OP-AMP is used in negative feedback mode to work as a "current to voltage converter".
- For N-bit DAC
  (i) Number of different levels = 2^N
  (ii) Number of steps = $2^N - 1$

R-2R Ladder DAC by Using Non-inverting OP-AMP

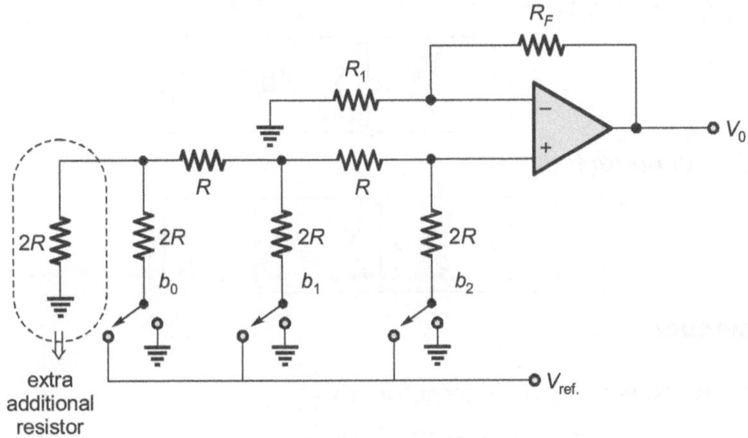

3-bit (b_2 b_1 b_0) R-$2R$ ladder DAC

Gain of OP-AMP :

$$\boxed{\frac{V_0}{V_{ref.}} = \left(1 + \frac{R_F}{R_1}\right)}$$

where, $V_{ref.} \rightarrow$ Reference voltage or Input voltage

$V_0 \rightarrow$ Output voltage

Output analog voltage:

$$\boxed{V_0 = \frac{V_{ref.}}{2^N}\left[\sum_{i=0}^{N-1} 2^i b_i\right] \times \left[1 + \frac{R_F}{R_1}\right]}$$

R-2R Ladder DAC by Using Inverting OP-AMP

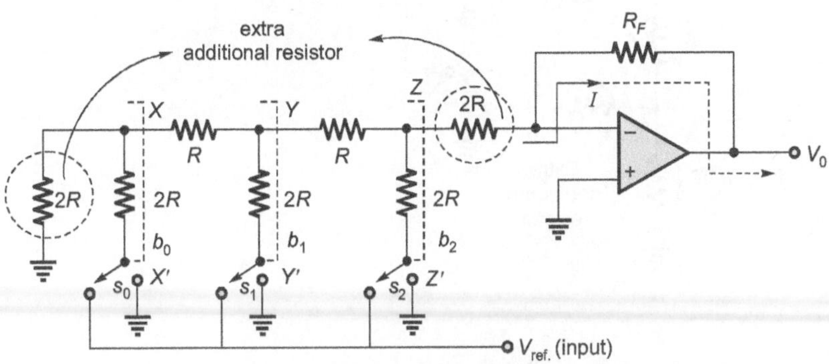

3-bit (b_2 b_1 b_0) R-$2R$ ladder DAC

Output voltage:

For N-bit DAC

$$V_0 = \frac{V_{ref.}}{2^N} \times \left[\sum_{i=0}^{N-1} 2^i b_i\right] \times \left[\frac{-R_F}{3R}\right]$$

Output current:

$$I = \frac{V_{ref.}}{2^N} \times \left[\sum_{i=0}^{N-1} 2^i b_i\right] \times \left[\frac{1}{3R}\right]$$

Remember:

Resolution of R-2R ladder network is $\frac{V_r}{2^n}$.

Inverted Ladder (R-2R)type DAC

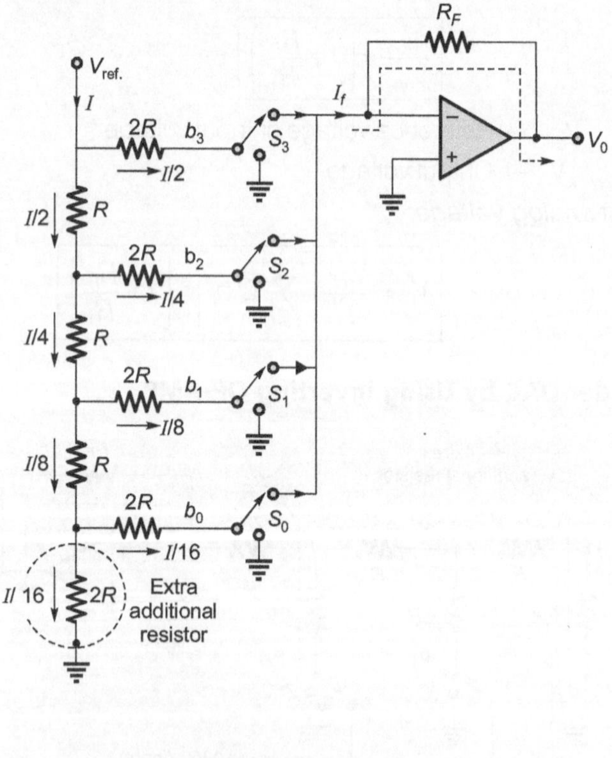

Output voltage:

For N-bit DAC

$$V_0 = \frac{V_{ref.}}{2^N} \times \left[\sum_{i=0}^{N-1} 2^i b_i\right] \times \left(\frac{-R_F}{R}\right)$$

Forward current :

$$I_f = \frac{V_{ref.}}{2^N} \times \left[\sum_{i=0}^{N-1} 2^i b_i\right] \times \frac{1}{R}$$

Note:

- ON-OFF Switches (S_0, S_1, S_2, S_3) are at the same potential.
- We always consider the bit as MSB, where the input reference voltage or supply to be given.
- The bit stream $(b_3\, b_2\, b_1\, b_0)_2$ has MSB = b_3 and LSB = b_0.

Analog to Digital Converter

Specifications of ADC

1. Voltage Range

$$\text{Voltage range} = V_{max} - V_{min}$$

2. Resolution

$$\text{Resolution} = \frac{V_{range}}{2^n - 1}$$

$$\%\ \text{Resolution} = \frac{1}{2^n - 1} \times 100$$

3. Dynamic range

$$\text{Dynamic range} = (1.8 + 6n)\ \text{dB}$$

Types of ADCs are

(a) Counter type

(b) Successive Approximation Resistor type (SAR)

(c) Flash type

(d) Dual slope type

(a) Counter type ADC

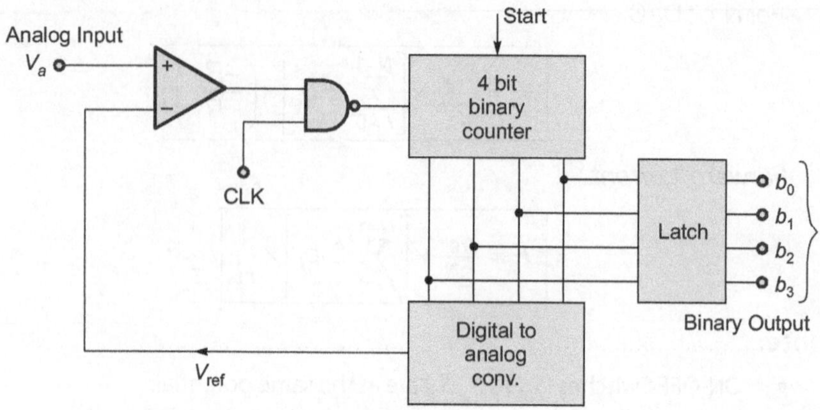

- Number of clock pulse required for n-bit conversion = $2^n - 1$
- Maximum conversion time = $(2^N - 1)\ T_{clk}$.
- Counter type ADC also known as 'ramp type' ADC.
- In this ADC, conversion time depends on input analog voltage.

(b) SAR type ADC

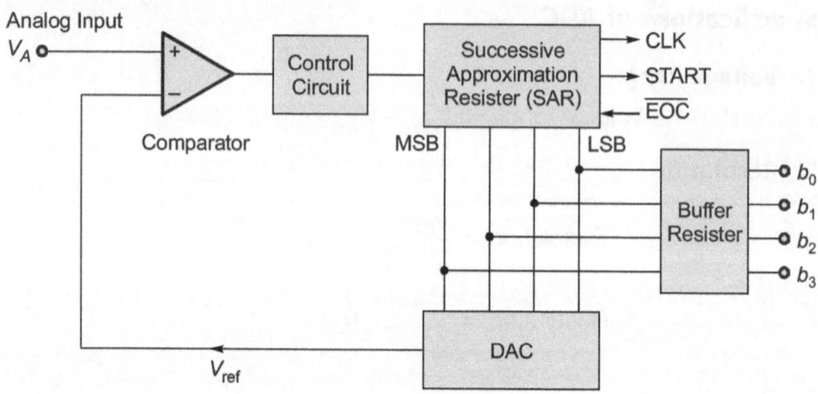

- Number of clock pulse required for n-bit conversion = n.
- Maximum conversion time = $n * T_{clk}$.
- In SAR ADC, conversion time is independent of input analog voltage.

(c) Flash type ADC

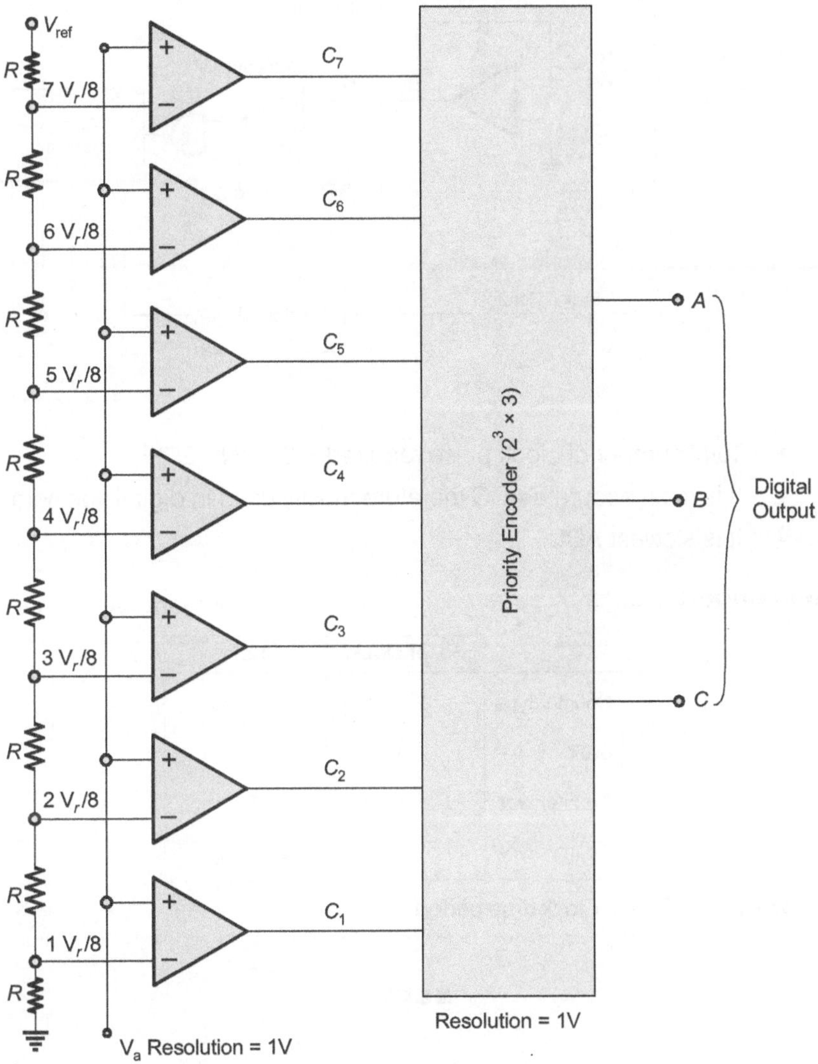

- For n bit conversion flash type ADC requires
  - (i) $2^n - 1$ comparators
  - (ii) 2^n resistors
  - (iii) One $2^n \times n$ priority encoder
- It is the fastest ADC among all.
- It is also known as parallel comparator type ADC.

(d) Dual slope integrating type ADC

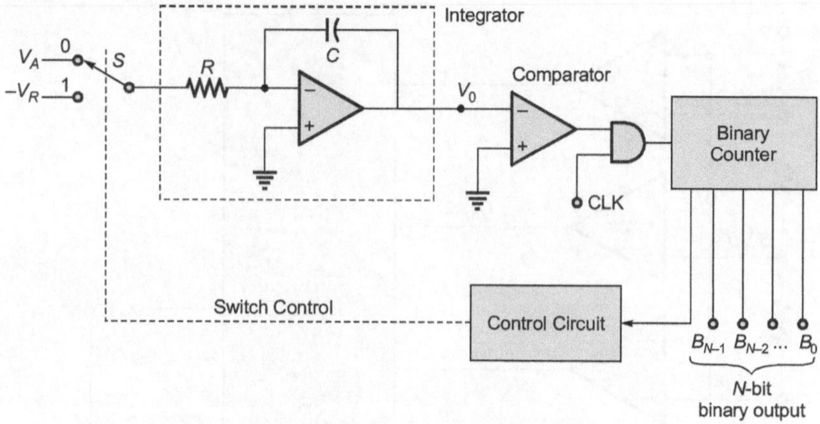

- Total number of clock pulse required = $2^n + N \approx 2^{n+1}$.
- It is most accurate ADC therefore mostly used in digital voltmeter.
- It is slowest ADC.

Remember:

	No. of clocks	Time
Counter type	2^{n-1}	$(2^{n-1})\,T_{clk}$
SAR	n	nT_{clk}
Counter type	1	T_{clk}
Counter type	2^{n+1}	$2^{n+1}\,T_{clk}$

where, $T_{clk} \rightarrow$ Clock time period

■■■■

Microprocessors

CONTENTS

Introduction

1

Definition

Microprocessor is a programmable integrating device having computing, storing, retrieving and decision making capability.

Note:

- Computer with microprocessor as its CPU is known as microcomputer.
- Microcomputer on a single chip is known as microcontroller.

History of Microprocessor

Microprocessor	Word length	Memory capacity
Intel4004 (PMOS)	4-bit	640 B
Intel 8008	8-bit	16 kB
Intel 8080 (NMOS)	8-bit	64 kB
Intel 8085 (NMOS)	8-bit	64 kB
Intel 8086 (HMOS)	16-bit	1 MB
Intel 8088	8/16-bit	1 MB
Intel 80186	16-bit	1 MB
Intel 80286	16-bit	16 MB real, 4 GB virtual
Intel 80386	32-bit	4 GB real, 4 GB virtual
Intel 80486	32-bit	4 GB real, 64 TB virtual
Pentium-II	64-bit	64 GB real
Z-80	8-bit	64 kB
Z-800	8-bit	500 kB

Table : *A brief review of various microprocessors*

Computer language

1. Mnemonic

A combination of letters to suggest the operation of an instruction.

2. Program

A set of instruction written in a specific sequence for the computer to accomplish a given task.

3. Machine language

A set of instructions executed directly by computer's CPU.

4. Assembly language

It is a low level programming language in which there is a strong correspondence between the language and architecture's machine code instructions.

5. Low level language

Machine specific languages are known as low level language Ex: Machine language, Assembly language.

6. High level language

A programming language that enables a programmer to write programs that are independent of particular type of computer.

7. Assembler

It is a computer program that translates an assembly language program from mnemonics to the binary machine code of a computer.

8. Compiler

A program that translates high level language program into machine level language by translating the entire program at a time and executing it.

9. Interpreter

It translates high level language program in object codes statementwise.

Remember:

High level language programming requires a translator like a compiler or an interpreter to translate the program written in a high level language to binary form.

Bit: A binary digit, 0 or 1

Nibble: A group of four bits

Byte: A group of eight bits

Word: A group of byte the computer recognizes and processor at a time.

■■■■

Architecture of 8085

2

Introduction

There are two types of architecture depending upon storage of program and data in memory

(i) **Von Neumann architecture** : Both data and program are in the same memory. Intel 8085 Intel 8086

(ii) **Harvard architecture:** Program and data are in different memories. TMS 32010, Intel 8051, e.g. Intel's Pentium etc.

Pin Configuration

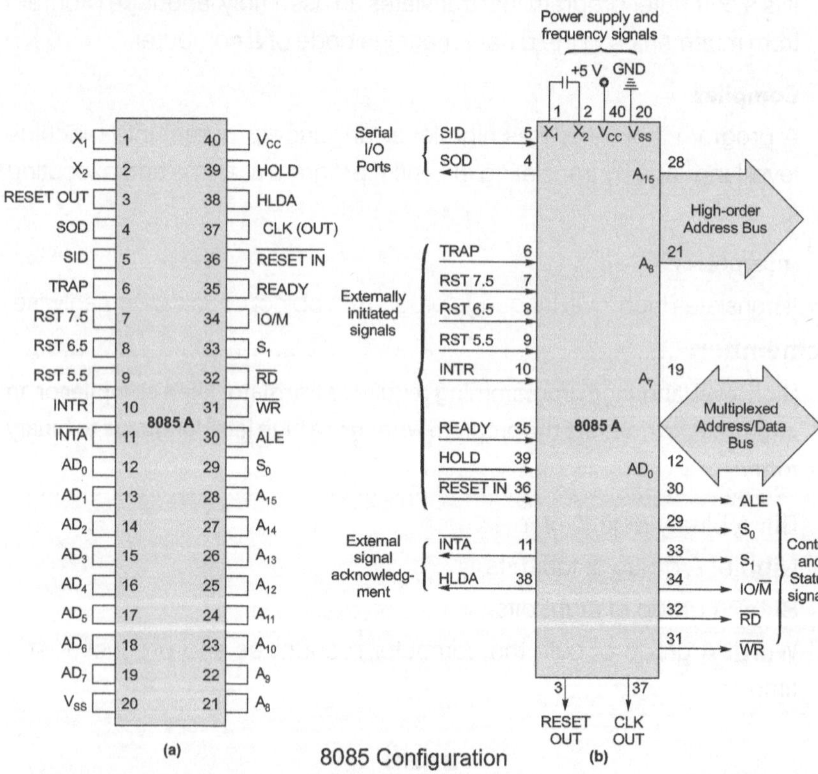

8085 Configuration

Key features of 8085

(i) It is a 8-bit processor.

(ii) It has 8 data bus lines, which is the bit capacity of the microprocessor.

(iii) It has total 16 address lines with addressing capacity of 64 kB.

(iv) The crystal frequency of processor is 6 MHz and the clock frequency is 3.07 MHz (= 3 MHz), which is approximately half the crystal frequency.

(v) Low order address bus ($AD_0 - AD_7$) is multiplexed with data bus in order to reduce the number of pins.

(vi) To de-multiplex address from data, ALE (Address Latch Enable) signal is used.

ALE = 1, Address transfer to bus

ALE = 0, Data transfer to bus.

(vii) Disadvantage of multiplexing is that speed will be reduced.

(viii) There are 5 hardware interrupts available for 8085.

(ix) 8085 has 74 basic instructions with 246 Opcodes.

Sr. No.	Signals	Pins
1.	Power supply signals	V_{CC}, V_{SS}
2.	Clock signals	X_1 X_2, CLK OUT
3.	Reset signals	RESETIN, RESET OUT
4.	Interrupt signals	TRAP, Restart interrupts (RST 7.5, REST 6.5 and RST 5.5), INTR, INTA
5.	Address bus and data bus	Address bus ($A_8 - A_{15}$) Multiplexed address/data bus ($AD_0 - AD_7$)
6.	Status signals and control signals	Address latch enable (ALE), Input output/ memory ($IO/\overline{M}$), Status signals (S_1 and S_0), Read ($\overline{RD}$), Write ($\overline{WR}$), READY
7.	Serial input/output signals	HOLD, HLDA
8.	DMA request signals	SID, (Serial input data), SOD (Serial output data)

System Bus

Bus is a group of wires or conductors used for communication between processor, memory and I/O device.

(a) Address Bus

(i) It is used to transfer address of memory or I/O from the processor. It is unidirectional.

(ii) Address Bus defines maximum memory that can be connected to a processor related by $2^n = N$.

where, $n \rightarrow$ Number of address lines

$N \rightarrow$ Number of address/memory location

(b) Data Bus

(i) It is used to transfer data between processor, memory or peripheral devices. It is bidirectional.

(ii) In 8085, it is multiplexed with lower order address bus.

ALE = 1 ; All lines are address bus

ALE = 0 ; A_{15} - A_8 → Address bus

AD_7 - AD_0 → Data bus

(c) Control Bus

(i) It is a group of control signals required for various operations of the processor.

(ii) It is considered to the unidirectional if signals from processor are considered. It is partially unidirectional and bidirectional if the control signals are considered.

Control and status signal

(i) $\overline{RD}$ (Read) : When the signal is low on this pin, the microprocessor performs memory reading or I/O reading operation.

(ii) $\overline{WR}$ (Write) : When the signal is low on this pin, the microprocessor performs memory writing or I/O writing operation.

(iii) $IO/\overline{M}$: This status signal is used to give information of operation to be performed with memory or I/O device.

$IO/\overline{M}$	$\overline{RD}$	$\overline{WR}$	Description
0	0	1	Memory read ($\overline{MEMR}$)
0	1	0	Memory write ($\overline{MEMW}$)
1	0	1	IO Read ($\overline{IOR}$)
1	1	0	IO Write ($\overline{IOW}$)

S_1 **and** S_0 : These two status signal are used to indicate the status of the operation.

S_1	S_0	Microprocessor operation
0	0	Halt (No operation)
0	1	Write operation
1	0	Read operation
1	1	Opcode fetch (reading instruction)

Power supply and frequency signals

V_{CC} : Power supply pin of +5V

X_1 and X_2 : A crystal is connected across these pins. The frequency is internally divided by two.

CLK (out) : This pin, there will be synchronization between the different peripherals and microprocessor.

Serial I/O Port

(i) **SID (Serial Input Data)** : This pin is used for receiving the data into microprocessor serially.

(ii) **SOD (Serial output data)** : This pin is used for sending the data from the microprocessor serially.

Externally initiated signals

1. Hardware interrupts

TRAP, RST-7.5, RST-6.5, RST-5.5, INTR are called Hardware interrupts.

It is also used to accept external interrupts to provide acknowledgment (Ack) to the external device.

Remember:

In priority order, interrupts are as TRAP > RST-7.5 > RST-6.5 > RST-5.5 > INTR.

2. READY

It is used to interface with the slow peripheral devices (Memory or I/O devices) to the microprocessor.

3. Reset

(a) $\overline{\text{RESET IN}}$: It is an active low signal, when this signal gets activated, microprocessor is reset.

(b) RESET OUT : This signal is used to reset the other peripheral devices.

4. HOLD and HLDA

(a) HOLD : A peripheral like DMA controller sends the HOLD request to the microprocessor through this pin to leave the data bus which is being used by microprocessor.

(b) HLDA : HOLD acknowledge

Internal Architecture of 8085 Microprocessor

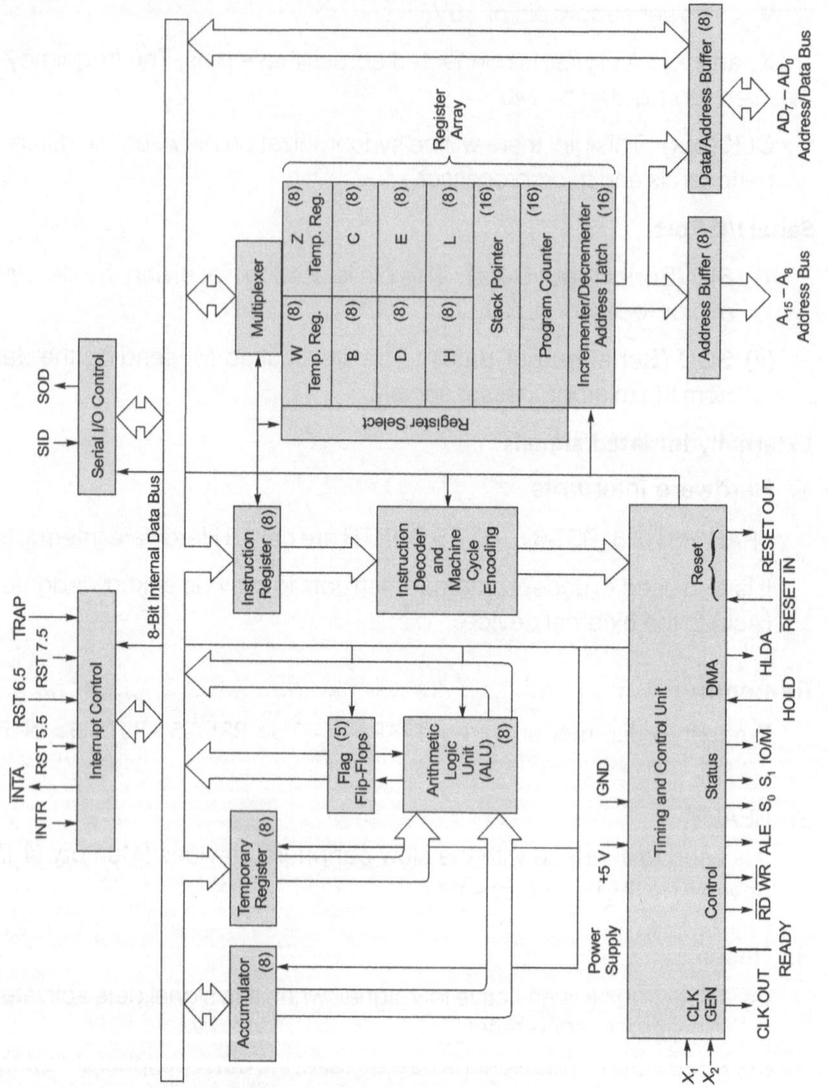

The architecture is divided into following functional units:

1. Arithmetic and Logical Unit (ALU)

The ALU performs arithmetical and logical operations. It includes

(a) Accumulator : It is a 8 bit programmable register. All the arithmetic and logical operations performed with contents of accumulator and the results are stored in accumulator only.

(b) Temporary Register : It is an 8 bit non-programmable register used to hold data during an arithmetic and logic operation.

(c) Arithmetic and logic circuits : This unit performs the actual numerical and logical operation.

(d) Flags : The flags generally reflect the status of arithmetic or logical operation.

D_7	D_6	D_5	D_4	D_3	D_2	D_1	D_0
S	Z	×	AC	×	P	×	CY

- **Carry flag (CY):** If an arithmetic operation results in a carry, the CY flag is **set** otherwise it is **reset.**
- Parity flag (P):
  ⇒ If the result has an even number of 1's, the flag is **set.**
  ⇒ If the result has an odd number of 1's the flag is **reset.**
- **Auxiliary Carry (AC):** In an arithmetic operation.
  ⇒ If carry is generated by bit D_3 and passed to D_4, flag is **set.**
  ⇒ Otherwise it is **reset.**
- Zero Flag (Z):
  ⇒ Zero flag is **set** to 1, when the result is zero.
  ⇒ Otherwise it is **reset.**
- Sign Flag (S):
  ⇒ Sign flag is **set** if bit D_7 of the result is 1.
  ⇒ Otherwise it is **reset.**

Remember: ....

- **Among the five flags, the AC flags is used internally for BCD arithmetic; the instruction set does not include any conditional Jump instruction based on the AC flag.**
- **'x' in the flag register indicate the unused flip-flops.**
- **The values of D_1, D_3 and D_5 bits should be taken as '0' in programs while using PSW instruction.**

....

2. Register array

The architecture of 8085 consist of following registers.

(a) Temporary registers :

(i) Temporary data register : It is also called as operand register (8-bit). It provides operands to the ALU.

(ii) Temporary registers (W and Z) : This registers are not available to the user.

Remember: ....

W and Z registers are used by 8085 for swap instruction.

....

(b) General purpose registers :

The 8085 microprocessor consists of 6 general purpose registers i.e. B, C, D, E, H, and L of 8 bits each. These registers are available to the user.

Remember:

- **The general purpose registers put together is called scratch and memory.**
- **The valid register pairs available are BC, DE and HL to store 16-bit of data.**
- **The HL register pair functions as default data pointer. If used as memory pointer it holds the address of a 16 bit address at a memory location.**
- **Accumulator can also be used along with status register to form a 16 bit programmable register called program status word (PSW).**

(c) Special purpose registers

There are two 16-bit special purpose register i.e. program counter (PC) and stack pointer (SP).

(i) Program counter : It is a 16 bit register used to hold memory addresses.

(ii) Stack pointer : It is a 16 bit register used as a memory pointer.

Remember:

- **When the microprocessor is reset, the PC sets to zero.**
- **Stack grows from bottom to top following last in first out (LIFO) structure and the SP contents keep decreasing as stack grows.**

3. Timing and control unit

It controls all internal and external circuits in the microprocessor system.

Remember:

The microprocessor uses a quartz crystal (LC or RC circuits) to determine the clock frequency, so that other timing and control signals are developed.

4. Instruction Register and Decoder

It is the part of ALU.

(a) **Instruction register** : The instruction register holds the opcode of the instruction that is decoded and executed.

(b) **Instruction decoder** : The output of instruction register connected to the decoder. The decoder decodes the instruction and establishes the sequence of events to follow.

5. Interrupt Control Unit

The interrupt control units job is to service the interrupt and after completing the interrupt service routine return back the control to the main program where it was interrupted.

■■■■

Instruction Set and Data Formats

3

Timing Diagram

Instruction cycle

The CPU fetches one instruction from the memory at a time and executes it. One instruction cycle can consists of 1-6 machine cycle.

Note:

An instruction cycle consists of a fetch cycle and execute cycle.

Machine Cycle

The time required by the microprocessor to complete the operation of accessing memory or I/O device is called as a machine cycle. One machine cycle can consists of 3-6 T-states.

Note:

- An instruction cycle consists of several machine cycles.
- Different types of machine cycles are opcode fetch (4T-6T), memory read (3T), memory write (3T), I/O read(3T), I/O write (3T), INTR Acknowledge and BUS Idle(3T).

T-State

Microprocessor perform an operation in a specific time period i.e. specific clock cycles. Each clock cycle is called as T-state.

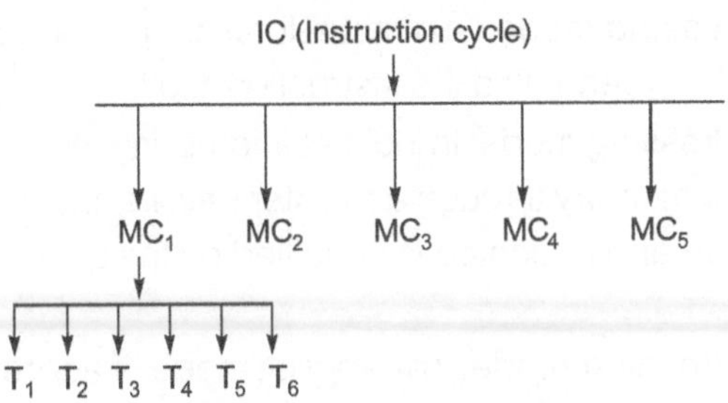

Classification of Instructions Set of 8085 Microprocessor

Instructions set

- Based on Length of wordsize
  - One-byte
  - Two-byte
  - Three-byte
- Based on addressing modes
  - Direct
  - Register
  - Register indirect
  - Immediate
  - Implicit/Implide
- Based on operations
  - Data transfer
  - Arithmetic
  - Logical
  - Branch control
  - Stack, I/O and machine control

An instruction is a binary pattern designed inside a microprocessor to perform a specific function.

Based on Length of Word Size

1. **One-Byte Instructions:** The one-byte instructions specify the operation to be performed and who is going to perform it. These instructions require single memory location.
2. **Two-Byte Instructions:** The Two-byte instruction uses first byte to specify the operation i.e. opcode and second byte to specify the operand. These instructions required two successive memory locations in the memory.
3. **Three-Byte instructions:** First byte stores opcode, second byte stores lower order 8-bit of 16-bit operand or address and third byte stores higher order 8-bit of 16-bit operand or address.

Based on Addressing Modes

1. **Register addressing mode:** In register addressing mode the source and the destination are general purpose registers.
2. **Immediate addressing mode:** In immediate addressing mode the data (8/16 bit) is specified in the instruction itself.
3. **Direct addressing mode:** In direct addressing mode 16 bit address of the operand is given within the instruction itself.
4. **Indirect addressing mode:** In indirect addressing mode the instructions reference the memory through a register pair, i.e. the memory address where the operand is located is specified by the contents of a register pair.
5. **Implicit addressing mode:** The implicit mode of addressing does not require any operand. The data is specified within the opcode itself.

Based on Operations

Abbreviations used in the description of the instruction set:

R = Register	Rs = Register source
Rd = Register destination	M = Memory
Rp = Register pair	( ) = Contents of
XX = Random information	S = Sign flag
Z = Zero flag	AC = Auxiliary carry flag
P = Parity flag	CY = Carry flag

Data Transfer Instruction

MOV : Move – Copy from Source to Destination

Opcode	Operand	Bytes	M-Cycles	T-States
MOV	Rd/M, Rs/M	1	1/2	4T/7T

Description:

This instruction copies the contents of the source register into the destination register; the contents of the source register are not altered. MOV Rd, Rs requires 1 M-cycle and 4T. MOV R_d, M and MOV M, Rs requires 2M-cycle and 7T.

MVI : Move Immediate 8-Bit

Opcode	Operand	Bytes	M-Cycles	T-States
MVI	R/M, 8-bit data	2	2/3	7T/10T

Description

The 8-bit data are stored in the destination register or memory.

LXI : Load Register Pair Immediate

Opcode	Operand	Bytes	M-Cycles	T-States
LXI	Rp, 16-bit data	3	3	10T

Description:

The instruction loads 16-bit data in the register pair designated in the operand.

Comments: ....

This is the only instruction that can directly load a 16-bit address in the stack pointer register.

....

LDA : Load Accumulator Direct

Opcode	Operand	Bytes	M-Cycles	T-States
LDA	16-bit address	3	4	13T

Description:

The contents of a memory location, specified by a 16-bit address in the operand, are copied to the accumulator. The contents of the source are not altered.

STA : Store Accumulator Direct

Opcode	Operand	Bytes	M-Cycles	T-States
STA	16-bit address	3	4	13T

Description:

The contents of the accumulator are copied to a memory location specified by the operand.

LHLD : Load H and L Registers Direct

Opcode	Operand	Bytes	M-Cycles	T-States
LHLD	16-bit address	3	5	16T

Description:

The instruction copies the contents of the memory location pointed out by the 16-bit address in register L and copies the contents of the next memory location in register H.

SHLD : Store H and L Registers Direct

Opcode	Operand	Bytes	M-Cycles	T-States
SHLD	16-bit address	3	5	16T

Description:

The contents of register L are stored in the memory location specified by the 16-bit address in the operand, and the contents of H register are stored in the next memory location by incrementing the operand. The contents of registers HL are not altered.

LDAX : Load Accumulator Indirect

Opcode	Operand	Bytes	M-Cycles	T-States
LDAX	Rp	1	2	7T

Description:

The contents of the designated register pair point to a memory location. This instruction copies the contents of that memory location into the accumulator.

Flags:

No flags are affected.

STAX : Store Accumulator Indirect

Opcode	Operand	Bytes	M-Cycles	T-States
STAX	Rp	1	2	7T

Description:

The contents of the accumulator are copied into the memory location specified by the contents of the operand (register pair).

Flags:

No flags are affected.

XCHG : Exchange H and L with D and E

Opcode	Operand	Bytes	M-Cycles	T-States
XCHG	None	1	1	4T

Description:

The contents of register H are exchanged with the contents of register D, and the contents of register L are exchanged with the contents of register E.

Arithmetic Instructions

ADD : Add Register to Accumulator

Opcode	Operand	Bytes	M-Cycles	T-States
ADD	R/M	1	1/2	4T/7T

Description:

The contents of the operand (register or memory) are added to the contents of the accumulator and the result is stored in the accumulator.

ADC : Add Register to Accumulator with Carry

Opcode	Operand	Bytes	M-Cycles	T-States
ADC	R/M	1	1/2	4T/7T

Description:

The contents of the operand (register or memory) and the Carry flag are added to the contents of the accumulator and the result is placed in the accumulator.

ACI : Add Immediate to Accumulator with Carry

Opcode	Operand	Bytes	M-Cycles	T-States
ACI	8-bit data	2	2	7

Description:

The 8-bit data (operand) and the Carry flag are added to the contents of the accumulator, and the result is stored in the accumulator.

ADI : Add Immediate to Accumulator

Opcode	Operand	Bytes	M-Cycles	T-States
ADI	8-bit data	2	2	7T

Description:

The 8-bit data (operand) is added to the contents of the accumulator, and the result is placed in the accumulator.

Flags:

All flags are modified to reflect the result of the addition.

SBB : Subtract Source and Borrow from Accumulator

Opcode	Operand	Bytes	M-Cycles	T-States
SBB	R/M	1	1/2	4T/7T

Description:

The contents of the operand (register or memory) and the Borrow flag are subtracted from the contents of the accumulator and the results are placed in the accumulator.

SUB : Subtract Register or Memory from Accumulator

Opcode	Operand	Bytes	M-Cycles	T-States
SUB	R/M	1	1/2	4T/7T

Description:

The contents of the register or the memory location specified by the operand are subtracted from the contents of the accumulator, and the result is placed in the accumulator.

SBI : Subtract Immediate with Borrow

Opcode	Operand	Bytes	M-Cycles	T-States
SBI	8-bit data	2	2	7T

Description:

The 8-bit data (operand) and the borrow are subtracted from the contents of the accumulator, and the result is placed in the accumulator.

SUI : Subtract Immediate from Accumulator

Opcode	Operand	Bytes	M-Cycles	T-States
SUI	8-bit data	2	2	7T

Description:

The 8-bit data (the operand) is subtracted from the contents of the accumulator, and the result is placed in the accumulator.

INR : Increment Contents of Register/Memory by 1

Opcode	Operand	Bytes	M-Cycles	T-States
INR	R/M	1	1/3	4T/10T

Description:

The contents of the designated register/memory is incremented by 1 and the result is stored in the same place.

DCR : Decrement Source by 1

Opcode	Operand	Bytes	M-Cycles	T-States
DCR	R/M	1	1/3	4T/10T

Description:

The contents of the designated register/memory is decremented by 1 and the result is stored in the same place.

Note:

In INR and DCR operation except carry, all flags are affected.

INX : Increment Register Pair by 1

Opcode	Operand	Bytes	M-Cycles	T-States
INX	Rp	1	1	6T

Description:

The contents of the specified register pair are incremented by 1. The instruction views the contents of the two registers as a 16-bit number.

DCX : Decrement Register Pair by 1

Opcode	Operand	Bytes	M-Cycles	T-States
DCX	Rp	1	1	6T

Description:

The contents of the specified register pair are decremented by 1. This instruction views the contents of the two registers as a 16-bit number.

Flags:

No flags are affected for INX and DCX.

DAA : Decimal Adjust Accumulator

Opcode	Operand	Bytes	M-Cycles	T-States
DAA	None	1	1	4T

Description:

1. If the value of the low-order four bits ($D_3 - D_0$) in the accumulator is greater than 9 or if AC flag is set, the instruction adds 6(06) to the low-order four bits.
2. If the value of the high-order four bits ($D_7 - D_4$) in the accumulator is greater than 9 or if the Carry flag is set, the instruction adds 6(60) to the high-order four bits.

Flags:

S, Z, AC, P, CY flags are altered to reflect the results of the operation.

DAD : Add Register Pair to H and L Registers

Opcode	Operand	Bytes	M-Cycles	T-States
DAD	Rp	1	3	10T

Description:

The 16-bit contents of the specified register pair are added to the contents of the HL register and the sum is saved in the HL register. The contents of the source register pair are not altered.

Flags:

If the result is larger than 16-bits the CY flag is set. No other flags are affected.

Note: ....

After the execution of the instruction, the contents of the stack pointer register are not altered.

....

Logical Instructions

ANA : Logical AND with Accumulator

Opcode	Operand	Bytes	M-Cycles	T-States
ANA	R/M	1	1/2	4T/7T

Description:

The contents of the accumulator are logically ANDed with the contents of the operand (register or memory), and the result is placed in the accumulator.

ANI : AND Immediate with Accumulator

Opcode	Operand	Bytes	M-Cycles	T-States
ANI	8-bit data	2	2	7T

Description:

The contents of the accumulator are logically ANDed with the 8-bit data (operand) and the result is placed in the accumulator.

Flags:

AC = 1, CY = 0 after execution

ORA : Logically OR with Accumulator

Opcode	Operand	Bytes	M-Cycles	T-States
ORA	R/M	1	1/2	4T/7T

Description:

The contents of the accumulator are logically ORed with the contents of the operand (register of memory).

ORI : Logically OR Immediate

Opcode	Operand	Bytes	M-Cycles	T-States
ORI	8-bit data	2	2	7T

Description:

The contents of the accumulator are logically ORed with the 8-bit data in the operand and the result is placed in the accumulator.

Flags: AC = 0, CY = 0 after execution.

XRA : Exclusive OR with Accumulator

Opcode	Operand	Bytes	M-Cycles	T-States
XRA	R/M	1	1/2	4T/7T

Description:

The contents of the operand (register or memory) are Exclusive ORed with the contents of the accumulator.

XRI : Exclusive OR Immediate with Accumulator

Opcode	Operand	Bytes	M-Cycles	T-States
XRI	8-bit data	2	2	7T

Description:

The 8-bit data (operand) are Exclusive ORed with the contents of the accumulator, and the result is placed in the accumulator.

Flags:

Z, S, P are altered to reflect the results of the operation. CY and AC are reset.

CMA : Complement Accumulator

Opcode	Operand	Bytes	M-Cycles	T-States
CMA	None	1	1	4T

Description:

The contents of the accumulator are complemented.

Flags:

No flags are affected.

CMP : Compare with Accumulator

Opcode	Operand	Bytes	M-Cycles	T-States
CMP	R/M	1	1/2	4T/7T

Description:

- The contents of the operand (register or memory) are compared with the contents of the accumulator. Both contents are preserved and the comparison is shown by setting the flags as follows:
  - ⇒ If (A) < (R/M) : Carry flag is set and Zero flag is reset.
  - ⇒ If (A) = (R/M) : Zero flag is set and Carry flag is reset.
  - ⇒ If (A) > (R/M) : Carry and Zero flags are reset.
- The comparison of two bytes is performed by subtracting the contents of the operand from the contents of the accumulator; however, neither contents are modified.

Flags:

S, P, AC are also modified in addition to Z and CY to reflect the results of the operation.

CPI : Compare Immediate with Accumulator

Opcode	Operand	Bytes	M-Cycles	T-States
CPI	8-bit	2	2	7T

Description:

- The second byte (8-bit data) is compared with the contents of the accumulator.
  - ⇒ If (A) < Data : Carry flag is set and Zero flag is reset.
  - ⇒ If (A) = Data : Zero flag is set and Carry flag is reset.
  - ⇒ If (A) > Data : Carry and Zero flags are reset.
- The comparison of two bytes is performed by subtracting the data byte from the contents of the accumulator; however, neither contents are modified.

Flags:

S, P, AC are also modified in addition to Z and CY to reflect the result of the operation.

CMC : Complement Carry

Opcode	Operand	Bytes	M-Cycles	T-States
CMC	None	1	1	4T

Description:

The Carry flag is complemented.

Flags:

The Carry flag is modified, no other flags are affected.

STC : Set Carry

Opcode	Operand	Bytes	M-Cycles	T-States
STC	None	1	1	4T

Description:

The Carry flag is set.

Flags:

No other flags are affected.

RLC : Rotate Accumulator Left

Opcode	Operand	Bytes	M-Cycles	T-States
RLC	None	1	1	4T

Description:

Each binary bit of the accumulator is rotated left by one position. Bit D_7 is placed in the position of D_0 as well as in the Carry flag.

Flags:

CY is modified according to bit D_7. S, Z, P, AC are not affected.

RAL : Rotate Accumulator Left through Carry

Opcode	Operand	Bytes	M-Cycles	T-States
RAL	None	1	1	4T

Description:

Each binary bit of the accumulator is rotated left by one position through the Carry flag. Bit D_7 is placed in the bit in the Carry flag and the Carry flag is placed in the least significant position D_0.

Flags:

CY is modified according to bit D_7. S, Z, AC, P are not affected.

RRC : Rotate Accumulator Right

Opcode	Operand	Bytes	M-Cycles	T-States
RRC	None	1	1	4T

Description:

Each binary bit of the accumulator is rotated right by one position. Bit D_0 is placed in the position of D_7 as well as in the Carry flag.

Flags:

CY is modified according to bit D_0. S, Z, P, AC are not affected.

RAR : Rotate Accumulator Right through Carry

Opcode	Operand	Bytes	M-Cycles	T-States
RAR	None	1	1	4T

Description:

Each binary bit of the accumulator is rotated right by one position through the Carry flag. Bit D_0 is placed in the Carry flag and the bit in the Carry flag is placed in the most significant position D_7.

Flags:

CY is modified according to bit D_0. S, Z, P, AC are not affected.

Branch Control Instructions

JMP : Jump Unconditionally

Opcode	Operand	Bytes	M-Cycles	T-States
JMP	16-bit	3	3	10T

Description:

This instruction is equivalent to a 1-byte unconditional Jump instruction. The program sequence is transferred to the memory location specified by the 16-bit address.

Conditional Jump

Op Code		Operand	Bytes	M-cycles	T-States
Carry =	1 JC	16-bit	3	2/3	7/10
	0 JNC	16-bit	3	2/3	7/10
Zero =	1 JZ	16-bit	3	2/3	7/10
	0 JNZ	16-bit	3	2/3	7/10
Sign =	1 JM	16-bit	3	2/3	7/10
	0 JP	16-bit	3	2/3	7/10
Parity =	1 JPE	16-bit	3	2/3	7/10
	0 JPO	16-bit	3	2/3	7/10

Description:

Jump execution of program immediately at the given address if condition is true. Thre machine cycles and 10 T-states are consumed if condition is true while two machine cycles and 7 T-states are consumed if condition is false.

PCHL : Load Program Counter with HL Contents

Opcode	Operand	Bytes	M-Cycles	T-States
PCHL	None	1	1	6T

Description:

The contents of registers H and L are copied into the program counter.

Flags:

No flags are affected.

CALL : Unconditional Subroutine Call

Opcode	Operand	Bytes	M-Cycles	T-States
CALL	16-bit address	3	5	18T

Description:

The program sequence is transferred to the address specified by the operand. Before the transfer, the address of the next instruction to CALL (the contents of the program counter) is pushed on the stack.

Flags:

No flags are affected.

Conditional Call

Op Code		Operand	Bytes	M-cycles	T-States
Carry =	1 CC	16-bit	3	2/5	9/18
	0 CNC	16-bit	3	2/5	9/18
Zero =	1 CZ	16-bit	3	2/5	9/18
	0 CNZ	16-bit	3	2/5	9/18
Sign =	1 CM	16-bit	3	2/5	9/18
	0 CP	16-bit	3	2/5	9/18
Parity =	1 CPE	16-bit	3	2/5	9/18
	0 CPO	16-bit	3	2/5	9/18

Description:

Call immediately the subroutine stored at given 16-bit address if condition is true. Five machine cycles and 18 T-states are consumed if condition is true while two machine cycles and 9-T states are consumed if condition is false.

RET : Return from Subroutine Unconditionally

Opcode	Operand	Bytes	M-Cycles	T-States
RET	None	1	3	10T

Description:

The program sequence is transferred from the subroutine to the calling program. The instruction is equivalent to POP program counter.

Flags:

No flags are affected.

Conditional Return

Op Code		Operand	Bytes	M-cycles	T-States
Carry =	1 RC	—	1	1/3	6/12
	0 RNC	—	1	1/3	6/12
Zero =	1 RZ	—	1	1/3	6/12
	0 RNZ	—	1	1/3	6/12
Sign =	1 RM	—	1	1/3	6/12
	0 RP	—	1	1/3	6/12
Parity =	1 RPE	—	1	1/3	6/12
	0 RPO	—	1	1/3	6/12

Description:

Return to main program if condition is true. Three machine cycles and 12 T-states are consumed if condition is true while one machine and 6T-states are consumed if condition is false.

XTHL : Exchange H and L with Top of Stack

Opcode	Operand	Bytes	M-Cycles	T-States
XTHL	None	1	5	16T

Description:

The contents of the L register are exchanged with the stack location pointed out by the contents of the stack pointer register. The contents of the H register are exchanged with the next stack location (SP + 1).

Flags:

No flags are affected.

Stack, I/O and Machine Control Instructions

NOP : No Operation

Opcode	Operand	Bytes	M-Cycles	T-States
NOP	None	1	1	4T

Description:

No operation is performed.

Comments:

The instruction is used to fill in time delays or to delete and insert instructions while troubleshooting.

HLT : Halt and Enter Wait State

Opcode	Operand	Bytes	M-Cycles	T-States
HLT	None	1	2 or more	5T or more

Description:

The microprocessor finishes executing the current instruction and halts any further execution.

Note:

NOP and HLT bot are same as no operation. However, when NOP is executed the PC is increased by 1 time where as for HLT PC remains the same.

PUSH : Push Register Pair onto Stack

Opcode	Operand	Bytes	M-Cycles	T-States
PUSH	Rp	1	3	12T

Description:

The contents of the register pair designated in the operand are copied into the stack in the following sequence. The stack pointer register is decremented and the contents of the high-order register (B, D, H, A) are copied into that location. The stack pointer register is decremented again and the contents of the low-order register (C, E, L, flags) are copied to that location.

Flags:

No flags are modified.

POP : Pop off Stack to Register Pair

Opcode	Operand	Bytes	M-Cycles	T-States
POP	Rp	1	3	10T

Description:

The contents of the memory location pointed out by the stack pointer register are copies to the low-order register (such as C, E, L and flags) of the operand. The stack pointer is incremented by 1 and the contents of that memory location are copied to the high-order register (B, D, H, A) of the operand. The stack pointer register is again incremented by 1.

Flags:

No flags are modified.

Note:

The contents of the source, stack locations, are not altered after the POP instruction.

RSTn : Restart

Opcode	Operand	Bytes	M-Cycles	T-States
RSTn	None	1	3	12T

Description:

This instruction is equivalent to 1-byte CALL instruction. Here $n = 0$ to 7 are 8 software interrupts.

Flags:

No flags are affected.

SPHL : Copy H and L Registers to the Stack Pointer

Opcode	Operand	Bytes	M-Cycles	T-States
SPHL	None	1	1	6T

Description:

The instruction loads the contents of the H and L registers into the stack pointer register.

Flags:

No flags are affected.

Note:

- This instruction performs the same function as MOV A, M except this instruction uses the contents of BC or DE as memory pointers.
- This instruction is used in conjunction with CALL or conditional call instructions.

IN : Input Data to Accumulator from a Port with 8-bit Address

Opcode	Operand	Bytes	M-Cycles	T-States
IN	8-bit port address	2	3	10T

Description:

The contents of the input port designated in the operand are read and loaded into the accumulator.

Flags:

No flags are affected.

OUT : Output Data from Accumulator to a Port with 8-Bit Address

Opcode	Operand	Bytes	M-Cycles	T-States
OUT	8-bit port address	2	3	10T

Description:

The contents of the accumulator are copied into the output port specified by the operand.

Flags:

No flags are affected.

EI : Enable Interrupts

Opcode	Operand	Bytes	M-Cycles	T-States
EI	None	1	1	4T

Description:

The Interrupt Enable flip-flop is set and all interrupts are enabled.

Flags:

No flags are affected.

Comments:

After a system reset or the acknowledgment of an interrupt, the Interrupt Enable flip-flop is reset, thus disabling the interrupts. This instruction is necessary to enable the interrupts (except TRAP).

DI : Disable Interrupts

Opcode	Operand	Bytes	M-Cycles	T-States
DI	None	1	1	4T

Description:

The Interrupt Enable flip-flop is reset and all the interrupts except the TRAP (8085) are disabled.

Flags:

No flags are affected.

Comments:

This instruction is commonly used when the execution of a code sequence cannot be interrupted.

RIM : Read Interrupt Mask

Opcode	Operand	Bytes	M-Cycles	T-States
RIM	None	1	1	4T

Description:

This is a multipurpose instruction used to read the status of interrupts 7.5, 6.5, 5.5 and to read serial data input bit.

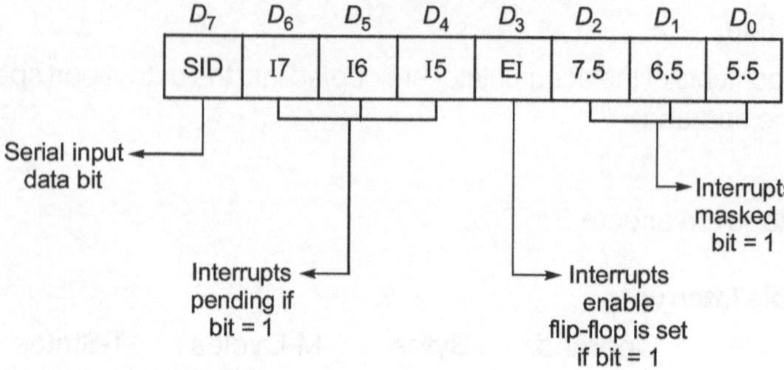

Flags

No flags are affected.

SIM : Set Interrupt Mask

Opcode	Operand	Bytes	M-Cycles	T-States
SIM	None	1	1	4T

Description:

This is a multipurpose instruction and used to implement the 8085 interrupts (RST 7.5, 6.5 and 5.5) and serial data output.

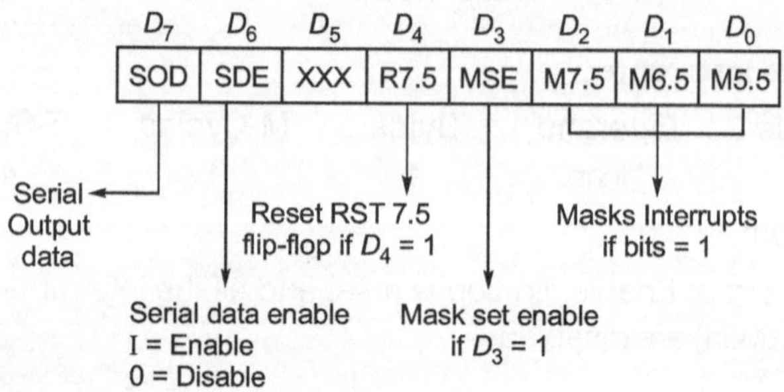

⇒ **SOD (Serial Output Data):** Bit D_7 of the accumulator is latched into the SOD output line and made available to a serial peripheral if bit $D_6 = 1$.

⇒ **SDE (Serial Data Enable):** If this bit = 1, enables the serial output. To implement serial output, this bit needs to be enabled.

⇒ **XXX:** Don't care

⇒ **R7.5 (Reset RST 7.5):** If this bit = 1, RST 7.5 flip-flop is reset. This is an additional control to reset RST 7.5.

⇒ **MSE (Mask Set Enable):** If this bit is high, it enables the functions of bits D_2, D_1, D_0. This is a master control over all the interrupt masking bits. If this bit is low, bits D_2, D_1 and D_0 do not have any effect on the masks.

⇒ **M7.5:** D_2 = 0, RST 7.5 is enabled.
= 1, RST 7.5 is masked or disabled

⇒ **M6.5:** D_1 = 0, RST 6.5 is correct.
= 1, RST 6.5 is masked or disabled.

⇒ **M6.5:** D_0 = 0, RST 5.5 is enabled.
= 1, RST 5.5 is masked or disabled.

Comments: ....

This instruction does not affect TRAP interrupt.

....

Counter and Time Delays

- Counters are used primarily to keep track of events.
- Time delays are important in setting up reasonably accurate timing between two events.
- Time delay can be introduced using a loop and total delay = time to execute instructions outside loop + time to execute loop instructions.

$$\boxed{T_D = T_O + T_{LA}}$$

where T_D = Total delay
T_O = Time to execute instructions outside loop.
T_{LA} = Time to execute loop instructions.

- Normally T_O is very small and neglected in most of the cases.
- The accuracy of time delay depends on the accuracy of the system's clock.
- Intel 8254 is a programmable timer chip that can be interfaced with microprocessor and programmed to provide with considerable accuracy.

Remember: ....

- Data copy instructions do not affect the flags.
- Operand PSW (Program Status Word) represents the contents of the accumulator and the flag register; the accumulator is high-order register and the flags are low-order register.
- Add and subtract are performed in relation to contents of the accumulator, however the increment or decrement operation can be performed in any register.

- Out of all instructions PUSH , CALL, RET, RSTn, INX, DCX, SPHL and PCHL uses 6 T-states for fetch machine cycle.
- DAD instruction uses bus idle machine cycle.
- NOT operation does not affect any flags.
- JMP: Instruction is a type of immediate addressing.
- Conditional jump instructions allow the microprocessor to make decision, based on certain conditions indicated by the flags.
- INX and DCX does not affect any of the flags.
- Compare instruction work like subtraction but content of accumulator and register does not change.
- Stack is a set of memory locations used to store binary information (byte) temporarily during execution of a program.
- Stack and stack pointer are two different things.
- Subroutine is also a program written outside main program.
- CALL and RET instructions are used for execution of subroutine and after execution, return to main program.

■■■■

Interrupts

4

Introduction

- When an interrupt is acknowledged the microprocessor perform following tasks:
  1. Microprocessor jumps to vectored location at page 00H where a subroutine (interrupt service routine) is written.
  2. Before jumping microprocessor saves the content of the program counter on the stack.
  3. It automatically resets the interrupt enable flip-flop.
- Interrupt which can be masked or stopped are maskable interrupt otherwise non-maskable interrupt. To mask and demask maskable interrupt of 8085 has two instruction i.e. EI and DI.
- Interrupt whose vectored location is fixed are known as vectored interrupt otherwise non-vectored interrupt.

Hardware Interrupts

The 8085 microprocessor has five interrupt signals that can be used to interrupt a program execution.

Interrupt	Triggering	Vectored Address	Maskable/ Non-maskable	Instructions
TRAP	Edge and level	0024H	Non-maskable	Independent of EI and DI
RST 7.5	Edge	003CH	Maskable	Controlled by EI and DI unmasked by SIM
RST 6.5	Level	0034H	Maskable	Controlled by EI and DI unmasked by SIM
RST 5.5	Level	002CH	Maskable	Controlled by EI and DI unmasked by SIM
INTR	Level	Non vectored	Maskable	Controlled by EI and DI

$\overline{\text{INTA}}$:

It is the active low interrupt acknowledgment signal which is only used with INTR.

Note: ....

To calculate vector address of RSTn

$$n \times 8 = (X)_{10} = (Y)_{16}$$

Example : For RST 4.5,

So, $(4.5 \times 8)_{10} = (36)_{10} \xrightarrow{\text{Heaxadecimal}} (24)_H \approx (0024)_H$

....

Software Interrupt

There are 8 software interrupts which are used either in instructions or alongwith INTR interrupt. They are defined as RSTn where n → 0 to 7.

Software interrupt (RSTn)

	Vectored address
RST 0	0000 H
RST 1	0008 H
RST 2	0010 H
RST 3	0018 H
RST 4	0020 H
RST 5	0028 H
RST 6	0030 H
RST 7	0038 H

Remember: ....

- TRAP is also called RST 4.5 interrupt.
- In interrupts priority order is
  TRAP > RST 7.5 > RST 6.5 > RST 5.5 > INTR
- Hold has higher priority than TRAP.
- RIM and SIM instructions are not only used for interrupt process but also used for serial I/O process.
- 8259A is a programmable interrupt controller and is used to implement and extend the capability of the 8085 interrupt. It manages 8 interrupt requests.

....

■■■■

Interfacing with Microprocessor

5

Interfacing

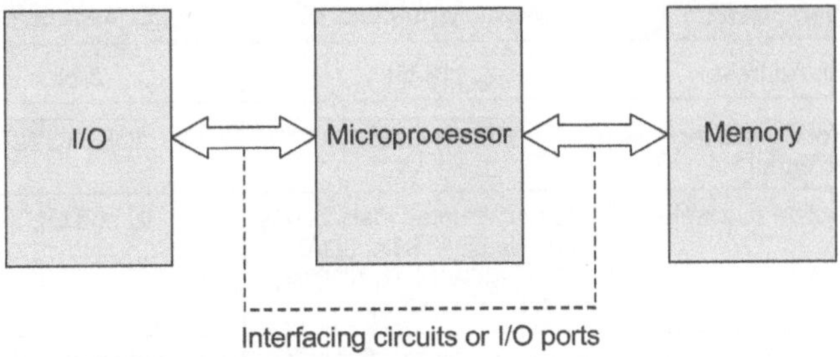

- Designing logic circuits and writing program to make the processor communicate either with memory or I/O is known as interfacing.
- The logic circuits used are known as interfacing circuits or I/O ports.

Characteristics of Memory

- Capacity

Memory capacity depends upon the amount of data that can be stored.

- **Memory size** $= 2^A \times D$

  where, $A \rightarrow$ Address lines

  $D \rightarrow$ Data lines

- **Number of chips required**

$$\boxed{n = \frac{\text{Size of } M_2}{\text{Size of } M_1}}$$

  where, $n \rightarrow$ Number of chips required

  $M_1 \rightarrow$ Available capacity

  $M_2 \rightarrow$ Memory to be designed

- If initial address and memory size is given then formulae for last address

  last address = [Initial address in hexadecimal + hexadecimal equivalent of memory size $- (1)_H$]

- If memory range is given then formulae for calculating memory size

  Memory size = [(last address)$_H$ − (Initial address)$_H$ + $(1)_H$] byte

Interfacing with I/O Ports

There are two ways by which I/O port can be connected to the microprocessor:

1. Memory mapped I/O scheme
2. I/O mapped I/O scheme

Comparison between Memory Mapped I/O and I/O Mapped I/O Scheme

Characteristic	Memory-mapped I/O	I/O Mapped I/O
Device Address	16-bit	8-bit
Control signals for Input/Output	$\overline{MEMR}$ / $\overline{MEMW}$	$\overline{IOR}$ / $\overline{IOW}$
Instruction available	Memory-related instructions such as LDA; STA; LDAX; STAX; MOV M, R; ADD M, SUB M; ANA M; etc.	IN and OUT
Data Transfer	Between any register and I/O	Only between I/O and the accumulator.
Maximum number of input/output devices possible	The memory map (64 K) is shared between I/O devices and system memory.	The I/O map is independent of the memory map; 256 input devices and 256 output devices can be connected.
Execution speed	13 T-states (STA, LDA) 7 T-states (MOV M.R)	10 T-states
Hardware requirements	More hardware is needed to decode 16-bit address.	Less-hardware is needed to decode 8-bit address.
Other features	Arithmetic or logical operations can be directly performed with I/O data.	Not available

Interfacing Devices

Intel 8155 : Programmable Peripheral Interface

The 8155 includes 256 bytes of RD/WR memory i.e. RAM, 3 I/O ports and a 16-bit timer.

Format of 8155 Timer

LSB Timer =	T_7	T_6	T_5	T_4	T_3	T_2	T_1	T_0

MSB Timer =	M_2	M_1	T_{13}	T_{12}	T_{11}	T_{10}	T_9	T_8

Where, M_2 M_1 Modes

0 0 Mode 0 $\Rightarrow$ Single square wave cycle

0 1 Mode 1 $\Rightarrow$ Square wave

1	0	Mode 2	⇒	Single pulse after terminal count
1	1	Mode 3	⇒	Pulse every terminal count

Intel 8255 : Programmable Peripheral Interface/Adopter

- The 8255 is widely used programmable parallel I/O device. It can be programmed to transfer data under various conditions from I/O to interrupt I/O.
- Operating modes of 8255
    1. Mode 0 ⇒ Simple input/output
    2. Mode 1 ⇒ Strobed input/output
    3. Mode 2 ⇒ Bidirectional Strobed

Structure of Control Word

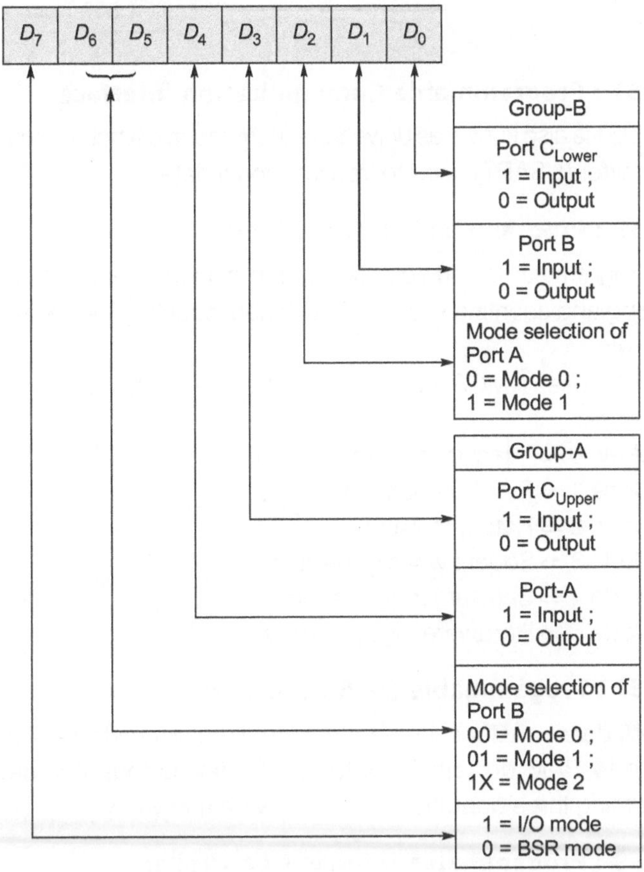

Control Word for BSR Mode

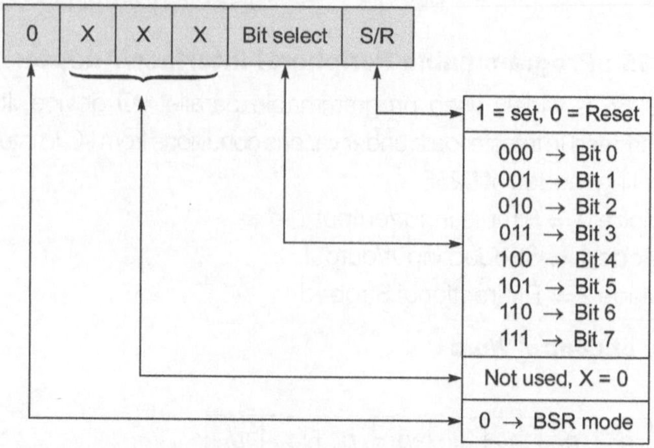

Intel 8251 : Programmable Communication Interface

Intel 8251 is also known as universal synchronous/asynchronous receiver/ transmitter (USART) used to transmit serial data.

Intel 8253 : Programmable Interval Timer

- The programmable counter/interval time is used in real time application of timing and counting such as BCD/binary counting, generation of time delay etc.
- The 8253 uses NMOS technology and operates in any of the following six modes:
  1. Mode 0 ⇒ Interrupt on terminal count
  2. Mode 1 ⇒ Programmable one shot
  3. Mode 2 ⇒ Rate generator
  4. Mode 3 ⇒ Square wave generator
  5. Mode 4 ⇒ Software trigger strobe
  6. Mode 5 ⇒ Hardware trigger strobe

Intel 8257 : Progammable DMA Controller

In DMA data transfer scheme, data are directly transferred from an I/O device to RAM or from RAM to an I/O device thus it is capable of performing three operations i.e. read, write and verify.

Intel 8259 : Progammable Interrupt Controller

The programmable interrupt controller is used when several I/O devices transfer data using interrupt and they are to be connected to the same interrupt level of the microprocessor.

Intel 8272 : Floppy Disk Controller

The function of floppy disk controller is to interface a floppy disk system to a microprocessor.

Intel 8275 : Progammable CRT Controller

The Intel 8275 is a single chip device. Its function is to interface CRT raster scan display with the microcomputer.

Intel 8279 : Progammable Keyboard/Display Interface

The Intel 8279 is a programmable keyboard interfacing device. It has two sections namely keyboard section and display section. The function of keyboard section is to interface the keyboard and the function of the display section is to drive alphanumeric displays or indicator lights.

■■■■

Introduction to 8086

6

Important Features of 8086

- 8086 is a 16 bit processor having 20 address lines.

  ∴ a maximum of 2^{20} = 1 Mega Byte can be connected.
- All registers are 16 bit in length.
- There are 8 8-bit general data registers.

AX		BX		CX		DX	
AH	AL	BH	BL	CH	CL	DH	DL

- It can be operated in 2 modes with the help of control signal

  $MN/\overline{MX}$ = 0; maximum/multi processor mode

  = 1; minimum/single processor mode.
- When processor is reset, control of the program is transferred to FFFFOH.
- It contains 6 bytes of pre-decoding instruction byte queue by which 6 bytes of instruction can also be prefetched into the processor.

  At the same time interval more than one instruction can be at different stages of execution i.e. parallel execution of instruction. This process is known as PIPELINING.

  Technique involved in 6 bytes of queue is First In First Out.
- There are 9 flags in 8086

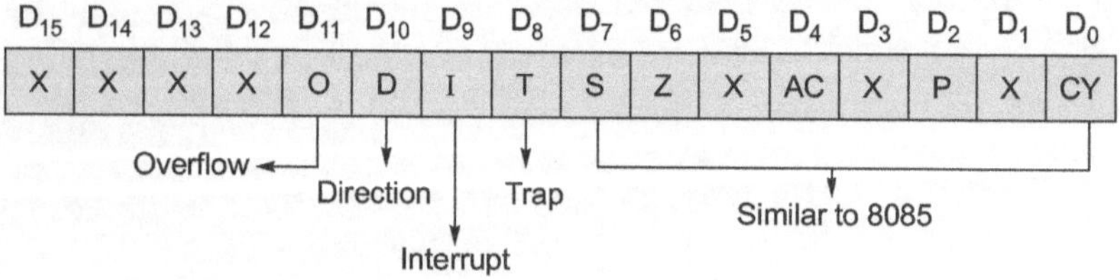

  (i) Status flags → S, Z, A, C, P, CY, O.

  (ii) Control flags → D, I, T.
- It follows VON NEUMAN architecture for memory but it is segmented memory.

  1 MB is segmented into 16 logical segment.

  1 MB → $2^4 \times 2^{16} = 2^{20}$

  16 (64 kB)

- Interrupts:

	Hardware		Software
(i)	NMI	(i)	INTn n = 0 to 255
(ii)	INTR		Type 0 to type 255.

Architecture of 8086

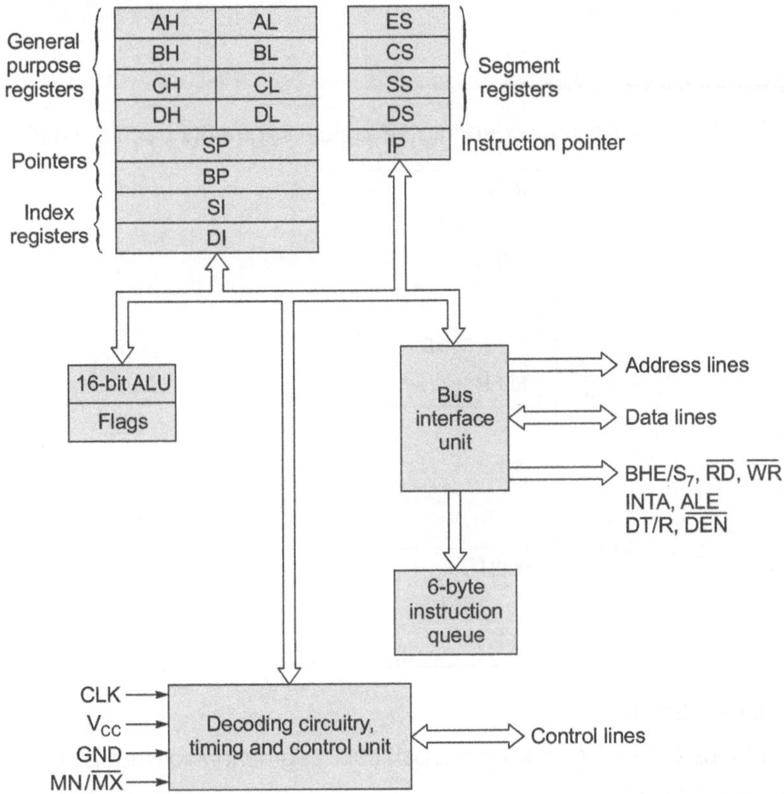

Functional Units

1. Bus Interface Unit

It contains address conversion mechanism in which 2 16-bit registers are used to find out or calculate 20-bit effective or physical address. It contains 6-bytes of instruction byte queue used for pre fetching the instructions for parallel execution known as pipelining technique. It has different segment registers to hold base address of **code, data, stack** and **extra segments**. It makes the processor to communicate with external devices including memory.

2. Execution Unit

It is responsible for decoding the prefetched instruction in BIU. It generates various control signals for BIU and also for operations of processor.

It contains different registers like

(i) general data registers,

(ii) pointer and index registers,

(iii) flag register

Memory Segmentation

1 MB of memory is divided into 16 logical segments each of 64 kB.

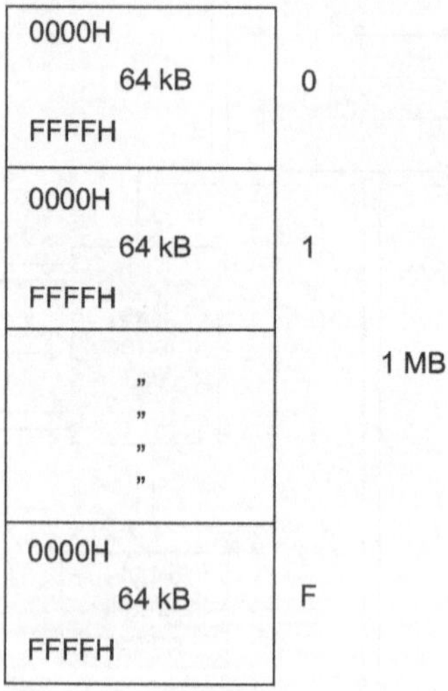

(a) Base Address

It indicates starting location of a particular segment which may be present in segment register.

(b) Offset Address

It is the distance of actual memory location from base address in a particular segment. It is present in pointer or index registers or BX register.

(c) Physical Address

It is a position of actual memory location from the beginning of memory.

(d) Segments

There are 4 types of segments:

(i) Code Segment :

Contains code/program,

Code Segment Register CS → Base Address.

IP → Offset Address.

Instruction pointer is similar to program counter which contains offset address of next instruction to be executed.

(ii) Data Segment :

Contains data

DS → Base Address

BP/SI/DI/BX → Offset address

(iii) Stack Segment:

Contains temporary data

SS → Base Address

SP → Offset address

(iv) Extra Segment:

Contains data

ES → Base Address

DI → Offset address

Remember:

- **Range of Segment Address** can be from 000H-F000H
- **Range of Offset Address** can be from 0000H-FFFFH

Register Organisation

(i) General Data Register

AX	AH	AL
BX	BH	BL
CX	CH	CL
DX	DH	DL

(ii) Special Purpose Registers

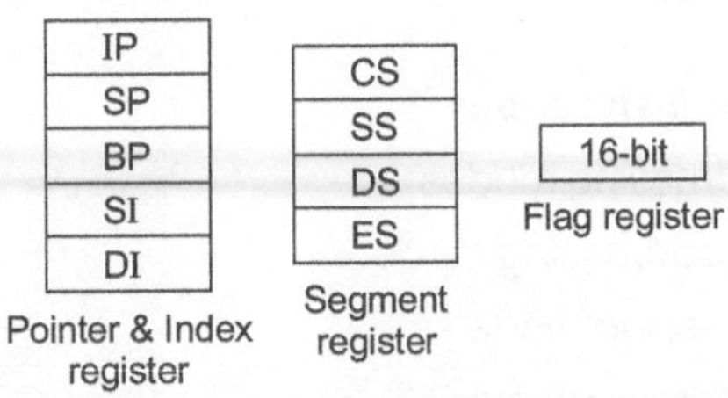

General Data Registers

(a) AX:

It is also called as Accumulator.

One of the operands in multiplication & division operation must be present in accumulator.

(b) BX:

It is used to hold offset address in Data Segment.

(c) CX:

Used as default counter in instruction like LOOP, LOOPZ, LOOPNZ.

(d) DX:

Used in multiplication and division operations and also as implicit register.

Flag Registers

(a) TRAP:

TF = 1; processor enters into single step execution mode in which execution steps after every instruction such that contents of registers can be verified.

(b) Interrupt Flag:

IF = 1; maskable interrupts are recognized by processor.

(c) Direction flag:

It is used in string manipulation instructions

D = 0; Auto increment mode

= 1; Auto decrement mode.

(d) Overflow flag:

It is SET (1) if result of a signed operation (8-bit on 16-bit) is not possible to be accommodated in destination register.

It is 0 otherwise.

Addressing Modes

1. Sequential flow instructions

There are 8 addressing modes

(i) Immediate addressing mode

(ii) Direct addressing mode

(iii) Register addressing mode

(iv) Register indirect addressing mode

(v) Register Relative addressing mode

(vi) Indexed Relative addressing mode

(vii) Based Indexed addressing mode

(viii) Relative based addressing mode

2. Control transfer instructions

(i) Intra Segment
- Direct
- Indirect

(ii) Inter Segment
- Direct
- Indirect

Intra Segment

If control of the program is transferred within the same segment, it is known as intra segment which can be **direct** or **indirect addressing** modes.

- If an 8-bit signed displacement is used, there can be +127 to –128 locations possible known as short jump.
- If a 16-bit signed displacement is used, there can be +32,767 to –32,768 locations possible known as long jump.

Inter Segment

If control of the program is transferred from one segment to another, it is known as intersegment which can be direct or indirect addressing modes. Unconditional jump instructions are used in this type.

■■■■

A Handbook on Instrumentation Engineering

7

Measurements

CONTENTS

Characteristics of Instruments and Measurement Systems

1

Measurements

Measurement is a process by which one can convert physical parameters to meaningful numbers. The measuring process is one in which the property of an object or system under consideration is compared to an accepted standard unit, a standard defined for that particular property.

Static Characteristics

1. Accuracy

It is the closeness with which an instrument reading approaches the true value of the quantity being measured.

2. Precision

It is a measure of the reproducibility of the measurements. It is a measure of degree of agreement within a group of measurements.

Remember:

- Precision is not the guarantee of accuracy.
- An instrument with more significant figure has more precision.

3. Sensitivity

It is the ratio of the magnitude of output signal to the magnitude of input signal applied to the instrument.

$$\text{Sensitivity} = \frac{\text{Output}}{\text{Input}}$$

Note:

- An instrument requires high degree of sensitivity.
- $\text{Sensitivity} \propto \frac{1}{\text{Deflection factor}}$

4. Resolution

The smallest change in input which can be detected with certainity by an instrument is its resolution.

5. Linearity

The output is linearly proportional to the input. For a linear instrument the sensitivity is constant for the entire range of instrument. Linearity is the most important parameter compared to all other parameters.

Remember:

- Linearity is more important than the sensitivity.
- Accuracy is more important than resolution.

6. Dead Zone

It is the largest change of input quantity for which there is no output of the instrument.

7. Dead time

Time required by an instrument to begin to respond to the change in a measurand.

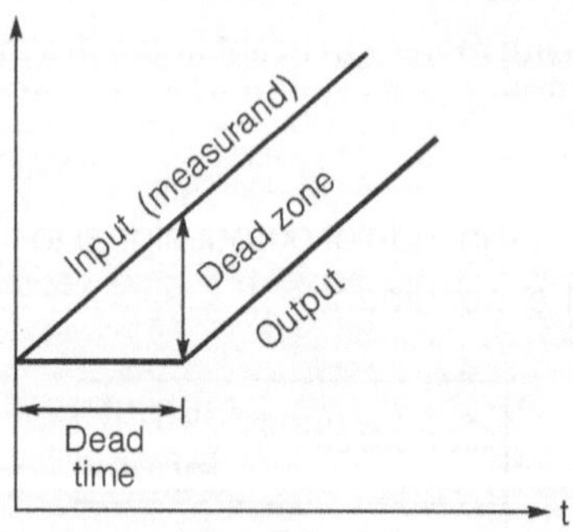

8. Range and Span

The difference between the maximum and minimum values of the scale is called range. The maximum value of the scale is called span.

Errors

$$\boxed{\text{Error} = \text{Measured value} - \text{True value}}$$

$$\boxed{\text{Error} = -\text{Accuracy}}$$

- **Static Error**

$$\boxed{\delta A = A_m - A_t}$$

where, A_m = Measured value of quantity or Actual value

A_t = True value of quantity or Nominal value

- **Relative static error**

$$\boxed{\epsilon_r = \frac{\delta A}{A_t}}$$

- **Static correction**

$$\boxed{\delta C = A_t - A_m = -\delta A}$$

- **Static sensitivity**

$$\boxed{\text{Static sensitivity} = \frac{\Delta q_0}{\Delta q_i}}$$

where, Δq_0 = Infinitesimal change in output

Δq_i = Infinitesimal change in input

- **Non-linearity (N.L.)**

$$\boxed{\text{N.L.} = \frac{\text{(Max. deviation of output from the idealized straight line)}}{\text{Full scale deflection}} \times 100}$$

$$\boxed{\text{Error at desired value} = \frac{\text{Full scale value} \times \text{Error at full scale}}{\text{Desired value}}}$$

Combination of Quantities with Limiting Errors

Sum or Difference of Two or More than Two Quantities

Let $X = \pm x_1 \pm x_2 \pm x_3 \pm x_4$

$$\boxed{\frac{\delta X}{X} = \pm\left(\frac{x_1}{X}\frac{\delta x_1}{x_1} + \frac{x_2}{X}\frac{\delta x_2}{x_2} + \frac{x_3}{X}\frac{\delta x_3}{x_3} + \frac{x_4}{X}\frac{\delta x_4}{x_4}\right)}$$

where $\pm\delta x_1$ = Relative increment in quantity x_1

$\pm\delta x_2$ = Relative increment in quantity x_2

$\pm\delta X$ = Relative increment in X

$\frac{\delta x_1}{x_1}$ = Relative limiting error in quantity x_1

$\frac{\delta x_2}{x_2}$ = Relative limiting error in quantity x_2

$\frac{\delta X}{X}$ = Relative limiting error in X

Product or Quotient of Two or More than Two Quantities

Let $$X = x_1 x_2 x_3 \text{ or } X = \frac{x_1}{x_2 x_3} \text{ or } X = \frac{1}{x_1 x_2 x_3}$$

$$\boxed{\frac{\delta X}{X} = \pm\left(\frac{\delta x_1}{x_1} + \frac{\delta x_2}{x_2} + \frac{\delta x_3}{x_3}\right)}$$

Composite Factors

Let $$X = x_1^n \cdot x_2^m$$

$$\boxed{\frac{\delta X}{X} = \pm\left(n\frac{\delta x_1}{x_1} + m\frac{\delta x_2}{x_2}\right)}$$

- Arithmetic Mean

$$\boxed{\bar{X} = \frac{\Sigma x}{n} = \frac{x_1 + x_2 + \cdots + x_n}{n}}$$

where, $x_1, x_2, ..., x_n$ = Readings or samples

n = Number of readings

Deviation

$$\boxed{d_n = x_n - \bar{x}}$$

Note: ....

Algebraic sum of deviation is zero.

....

- Average deviation

$$\boxed{\bar{D} = \frac{\Sigma |d|}{n} = \frac{|d_1| + |d_2| + \cdots |d_n|}{n}}$$

- **Standard deviation**

  For $n > 20$

$$\text{S.D.} = \sigma = \sqrt{\frac{\Sigma d^2}{n}} = \sqrt{\frac{d_1^2 + d_2^2 + \cdots + d_n^2}{n}}$$

  For $n < 20$

$$\text{S.D.} = s = \sqrt{\frac{\Sigma d^2}{n-1}}$$

- **Variance**

  For $n > 20$

$$V = \sigma^2 = \frac{\Sigma d^2}{n}$$

  For $n < 20$

$$V = s^2 = \frac{\Sigma d^2}{n-1}$$

Normal or Gausian Curve of Errors

1. For Infinite Numbers of Reading

$$y = \frac{1}{\sigma\sqrt{2\pi}} \exp(-x^2 / 2\sigma^2)$$

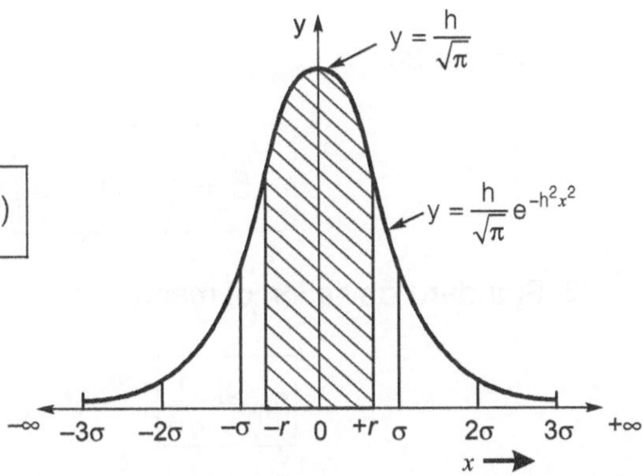

where, x = magnitude of deviation from mean,

y = number of readings at any deviation x, (the probability of occurance of deviation x)

σ = standard deviation

- Precision Index

$$h = \frac{1}{\sigma\sqrt{2}}$$

- Probable error (P.E.)

$$r = \frac{0.4769}{h}$$

- Average deviation

$$\bar{D} = \frac{r}{0.8453} = \frac{1}{\pi h^2}$$

- Standard deviation

$$\sigma = \frac{r}{0.6745} = \frac{1}{h\sqrt{2}}$$

$$\text{P.E.} = r = 0.8453\,\bar{D} = 0.6745\,\sigma$$

2. For Finite Numbers of Reading

- For $n > 20$

$$\text{P.E.} = r = 0.6745\sqrt{\frac{\sum |d|^2}{n}}$$

- For $n < 20$

$$\text{P.E.} = r = 0.6745\sqrt{\frac{\sum |d|^2}{n-1}}$$

- Standard deviation of mean

$$\sigma_m = \frac{\sigma}{\sqrt{n}}$$

- Standard deviation of standard deviation

$$\sigma_\sigma = \frac{\sigma_m}{\sqrt{2}}$$

- **Variance of combination of components**

Let $X = f(x_1, x_2, \cdots, n)$

$$V_x = \left(\frac{\partial X}{\partial x_1}\right)^2 V_{x_1} + \left(\frac{\partial X}{\partial x_2}\right)^2 V_{x_2} + \cdots + \left(\frac{\partial X}{\partial x_n}\right)^2 V_{x_n}$$

where, $V_{x1}, V_{x2}, \cdots, V_{xn}$ = Variance of $x_1, x_2, \cdots x_n$

- **Standard Deviation of Combination of Components**

Let $X = f(x_1, x_2, , \ldots\ldots, x_n)$

$$\sigma_x = \sqrt{\left(\frac{\partial X}{\partial x_1}\right)^2 \sigma_{x_1}^2 + \left(\frac{\partial X}{\partial x_2}\right)^2 \sigma_{x_2}^2 + \cdots + \left(\frac{\partial X}{\partial x_n}\right)^2 \sigma_{x_n}^2}$$

where, $\sigma_{x1}, \sigma_{x2}, \cdots, \sigma_{xn}$ = Standard deviation of $x_1, x_2, \cdots, x_n$

- **Probable Error of Combination of Components**

Let $X = f(x_1, x_2, \ldots\ldots, x_n)$

$$r_x = \sqrt{\left(\frac{\partial X}{\partial x_1}\right)^2 r_{x_1}^2 + \left(\frac{\partial X}{\partial x_2}\right)^2 r_{x_2}^2 + \cdots + \left(\frac{\partial X}{\partial x_n}\right)^2 r_{x_n}^2}$$

where, $r_{x1}, r_{x2}, \cdots, r_{xn}$ = Probable error of $x_1, x_2, \cdots, x_n$

- **Uncertainty of Combination of Components**

Let $X = f(x_1, x_2, \ldots\ldots, x_n)$

$$w_x = \sqrt{\left(\frac{\partial X}{\partial x_1}\right)^2 w_{x_1}^2 + \left(\frac{\partial X}{\partial x_2}\right)^2 w_{x_2}^2 + \cdots + \left(\frac{\partial X}{\partial x_n}\right)^2 w_{x_n}^2}$$

where, $w_{x1}, w_{x2}, \cdots, w_{xn}$ = Uncertainties of $x_1, x_2, \cdots, x_n$

Order of Instrument

1. Zero Order System

As input changes, output also changes immediately called zero order system. Example : Resistor.

2. First Order System

As input changes, output also changes but not immediately, it takes some delay but without oscillation. Example : heater.

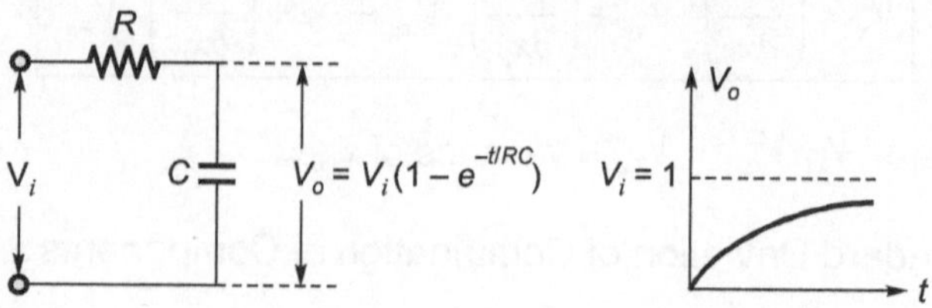

3. Second Order System

As input changes, output also changes, with some delay and oscillation.

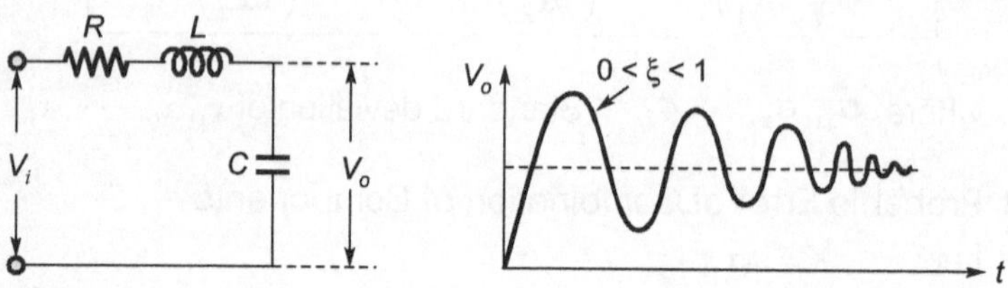

Remember: ....

- **The analog instruments are of second order instrument which has damping factor (ξ) between 0.6 to 0.8. It is an underdamped system.**

Standards

Quantity	Unit	Definition
Length	Metre	The length of path travelled by light in an interval of $\frac{1}{299792458}$ sec.
Time	second	9.192631770×10^9 cycles of radiation from vapourised cesium-133 atom.
Temp.	Kelvin	The temperature difference between the absolute and the triple point of water is defined as 273.16°K.
Voltage	Volt	Standard cell voltage of weston cell i.e. 1.0186 V.
Current	Ampere	One ampere is the current flowing through two infinite long parallel conductor of negligible cross section placed 1 meter apart produced a force of 2×10^{-7} N/m.

■■■■

Circuit Components (Resistors, Inductors, Capacitors)

2

Frequency Errors in Resistors

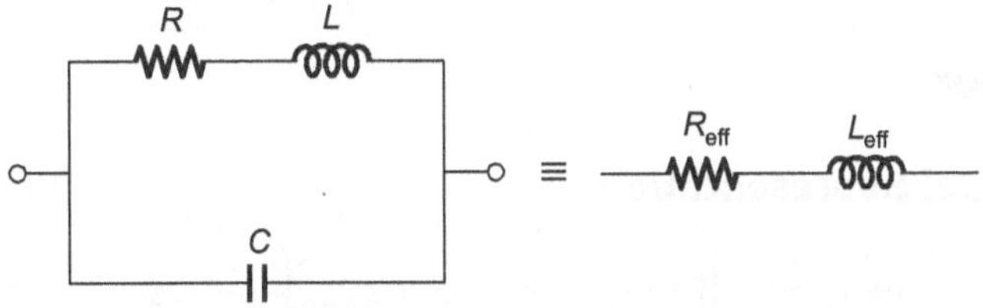

(Equivalent circuit of a resistor at low and medium frequencies)

- Effective resistance

$$R_{eff} = \frac{R}{1 + \omega^2 C(CR^2 - 2L)}$$

- Effective inductance or residual inductance

$$L_{eff} = \frac{L - CR^2}{1 + \omega^2 C(CR^2 - 2L)}$$

$$\tan\phi = \frac{X_{eff}}{R_{eff}} = \frac{\omega L_{eff}}{R_{eff}} = \frac{\omega(L - CR^2)}{R} = \omega\left(\frac{L}{R} - CR\right)$$

where, ϕ = Phase deflection angle

- Time constant

$$\tau = \frac{L_{eff}}{R_{eff}} = \frac{L - CR^2}{R} = \frac{L}{R} - CR$$

- Condition for resistance to remain independent of frequency $\quad CR^2 = 2L$

- Condition for resistance to show no inductive effect $\quad CR^2 = L$

- Effective resistance for zero effective inductance $\quad R_{eff} = \dfrac{R}{1 - \omega^2 LC}$

- Quality factor $\quad Q = \dfrac{\omega L}{R}$

Frequency Errors in Inductors

- Effective resistance $$R_{eff} = \frac{R}{(1-\omega^2 LC)^2}$$
- Effective inductance $$L_{eff} = L(1+\omega^2 LC)$$

Capacitor

1. Parallel Representation

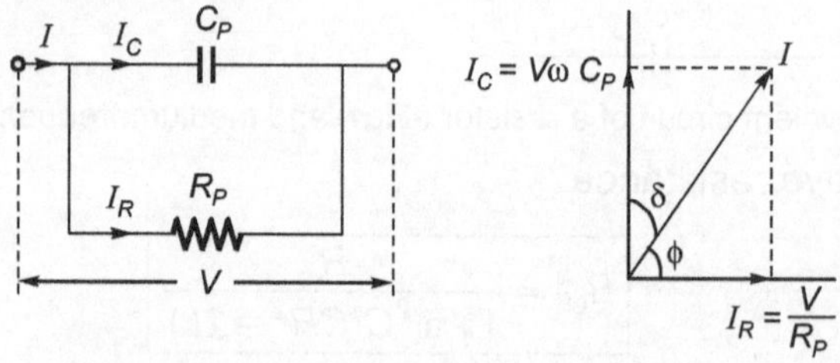

- Dielectric loss

$$P_L = \omega C_P V^2 \tan\delta$$

- Dissipation factor

$$D = \tan\delta = \frac{1}{\omega C_P R_P}$$

where, δ = loss angle of the capacitor.

2. Series Representation

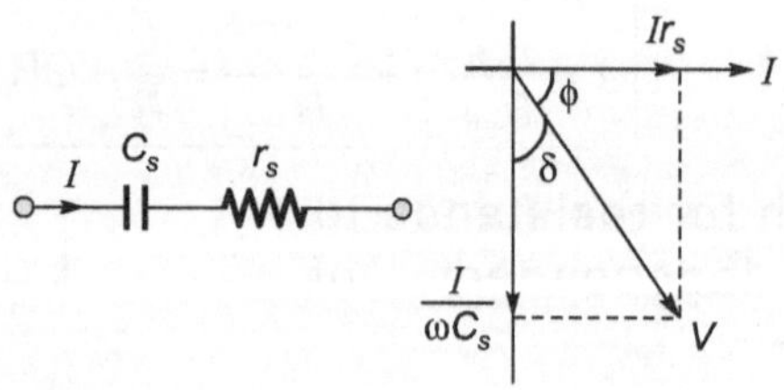

- Dielectric loss

$$P_L = \frac{I^2}{\omega C_s} \tan\delta$$

- Dissipation factor

$$D = \tan\delta = \omega C_s r_s$$

Frequency Errors in Capacitors

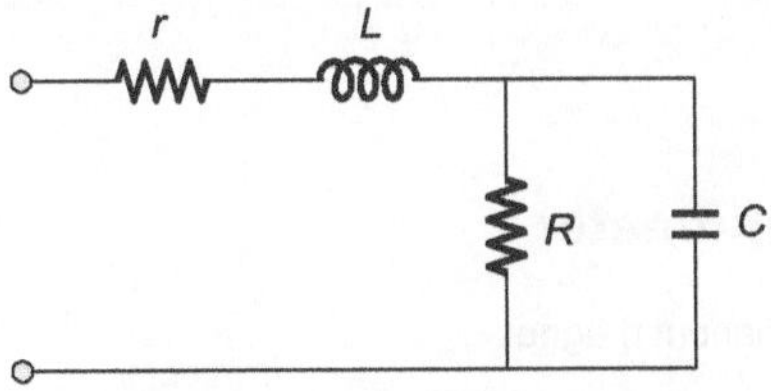

Equivalent Circuit of a Capacitor

- Effective capacitance

$$C_{eff} = \frac{C}{1-\omega^2 LC}$$

1. For Medium Frequency

- Effective capacitance

$$C_{eff} = C(1+\omega^2 LC)$$

- Effective series resistance

$$R_{eff} = r + \frac{R}{1+\omega^2 R^2 C^2}$$

where, r = resistance of lead

- Loss angle

$$\tan\delta = \frac{1-\omega^2 LC}{\omega r + \frac{1}{\omega CR}}$$

2. For Low Frequency

- Effective capacitance

$$C_{eff} = C + \frac{1}{\omega^2 CR^2}$$

- Effective series resistance

$$R_{eff} = \frac{R}{1+\omega^2 C^2 R^2}$$

- Loss angle

$$\tan\delta = \omega C r + \frac{1}{\omega CR}$$

■■■■

Galvanometers

3

D' Arsonval Galvanometer

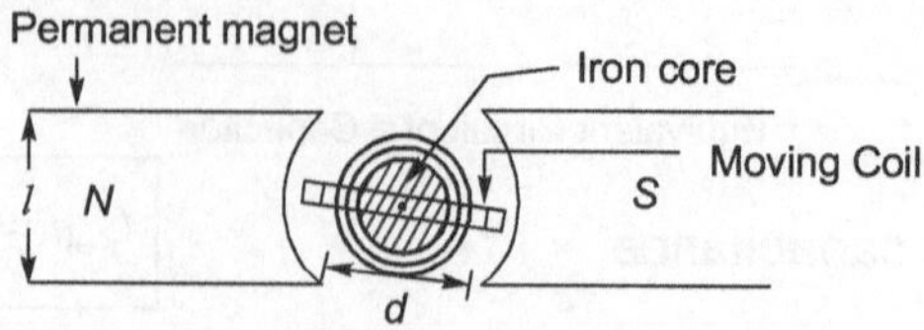

❑ Deflecting torque

$$T_d = B\,i\,N\,A = G\,i$$

where, B = Flux density in air gap; Wb/m^2

i = Current through moving coil; A

N = Number of turns in coil

A = ld = Area of coil; m^2

l, d = Length of vertical and horizontal side (width) of coil respectively; m

G = Displacement constant of galvanometer

❑ Controlling torque

$$T_c = K\theta_f$$

where, K = Spring constant of suspension; Nm/rad

θ_f = Final steady deflection of moving coil; rad

❑ Final steady deflection

$$\theta_f = \left(\frac{NBA}{K}\right) i = \left(\frac{G}{K}\right) i$$

Dynamic behaviour of Galvonometers

Torques in Galvonometers

❑ Inertia torque

$$T_j = J\frac{d^2\theta}{dt^2}$$

where, J = moment of inertia of moving system about the axis of rotation; kg-m^2,

θ = deflection at any time t, rad.

- Damping torque

$$T_D = D\frac{d\theta}{dt}$$

where, D = damping constant

- Controlling torque

$$T_C = K\theta$$

where, K = control constant

- Deflecting Torque

$$T_d = G\,i$$

- Equation of motion

$$T_j + T_D + T_C = T_d$$

$$J\frac{d^2\theta}{dt^2} + \frac{Dd\theta}{dt} + K\theta = Gi$$

Note:

- If $D^2 < 4$ kJ, galvanometer is underdamped.
- If $D^2 = 4$ kJ, galvanometer is critically damped.
- If $D^2 > 4$ kJ, galvanometer is overdamped.

- Total resistance of galvanometer circuit for critical damping

$$R = \frac{G^2}{2\sqrt{KJ}}$$

- External series resistance required for critical damping

$$R_e = R - R_g = \frac{G^2}{2\sqrt{KJ}} - R_g$$

where, R_g = Resistance of galvanometer

Sensitivity

- Current sensitivity

$$\boxed{S_i = \frac{\theta_f}{i} = \frac{G}{K}} \text{ rad/A}$$

$$\boxed{S_i = \frac{d}{i \times 10^6}} \text{ scale divisions/}\mu\text{A}$$

$$\boxed{S_i = \frac{2000\,G}{K \times 10^6}} \text{ mm/}\mu\text{A}$$

- Voltage sensitivity

$$\boxed{S_v = \frac{d}{i R_g \times 10^6}} \text{ scale division/}\mu\text{V}$$

- Megohm sensitivity

$$\boxed{S_o = \frac{d}{i \times 10^{-6}}} \text{ M}\Omega\text{/scale division}$$

Remember:

Sensitive galvanometer is one which produces a large deflection for a small current.

■■■■

4

Analog Meters

Classification of Analog Meters

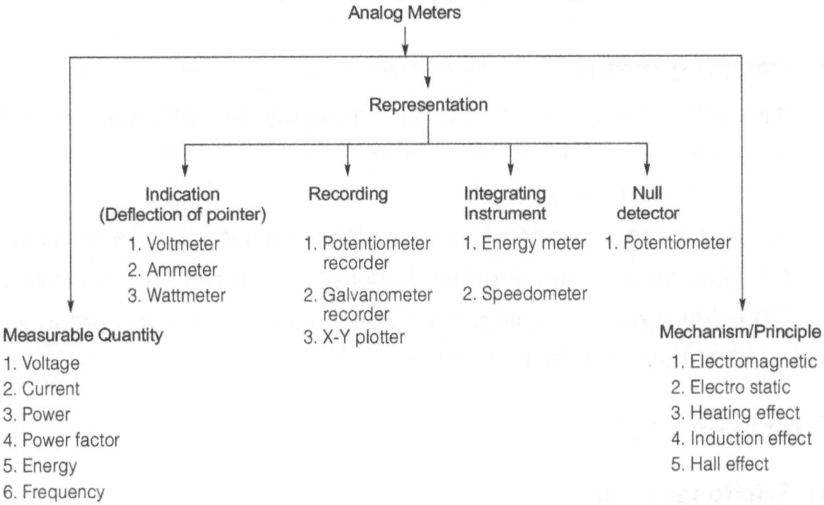

Torque in Analog Meter

1. Deflecting Torque (T_D)

Deflecting torque is proportional to quantity under measurement. This torque deflect the pointer away from initial or zero position.

$$\boxed{T_D \propto \text{Measurable quantity}}$$

2. Controlling Torque (T_C)

The controlling torque is opposite to deflecting torque. When, deflecting torque equals to controlling torque, pointer comes to final steady state position.

At equilibrium, $\boxed{T_C = T_D}$

Note: ....

- Control torque is also used to bring the pointer in zero initial position, if there is no deflecting torque.
- Except in PMMC, in all other instruments if the control spring is failed or broken then pointer moves to the maximum position of scale.
- Control torque is provided by

  (i) Spring control (ii) Gravity control

....

3. Damping Torque

It is used to damp out oscillation at final steady state position. The time response of the instrument depends on damping torque.

Damping torque provided by:

(i) **Air friction damping:** Used where low magnetic fields are produced

(ii) **Fluid friction damping:** Used where deflecting torque is minimum.

(iii) **Eddy current damping:** Used where permanent magnet produces the required deflecting torque.

Error in Analog Meters

1. Frictional Error

To reduce the frictional error, the torque to weight ratio of the instrument should be high.

2. Temperature Error

Due to change in temperature, change in resistance of meters and shunts and series multiplier occurs. To reduce this effect, resistances are made up of manganin material.

3. Frequency Error

Due to change in frequency, error produce in instrument because change in frequency cause change in reactance. To reduce this error, a capacitance is used in case of voltmeter and for ammeter, the time constant and shunt impedances are maintained at same value.

Permanent Magnet Moving Coil (PMMC)

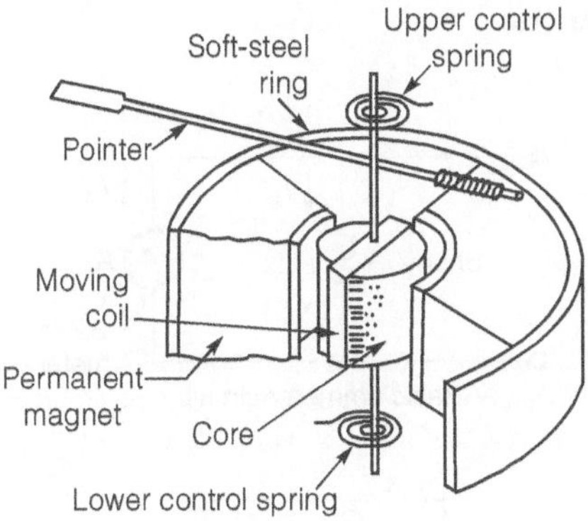

❑ Deflection torque

$$\boxed{T_D = nBAI}$$

$$\boxed{T_D = GI}$$

where, $G = nBA$

n = Number of turns

B = Flux density

A = Area of core

I = Current to be measured

❑ Final steady state deflection

$$\boxed{\theta = \left(\frac{G}{K}\right) I}$$

where, K = Spring constant

Note:

- PMMC instrument measures only DC or average values.
- Scale is linear.
- Spring is used for controlling torque.
- Damping torque provided by eddy current damping.
- It has more, torque to weight ratio so accuracy and sensitivity is higher compare to other instrument.
- In direct measurement, the PMMC measures upto a current of 50 mA or a voltage of 100 mV, without any external device.

Enhancement of Ammeters and Voltmeters

1. Ammeter Shunts

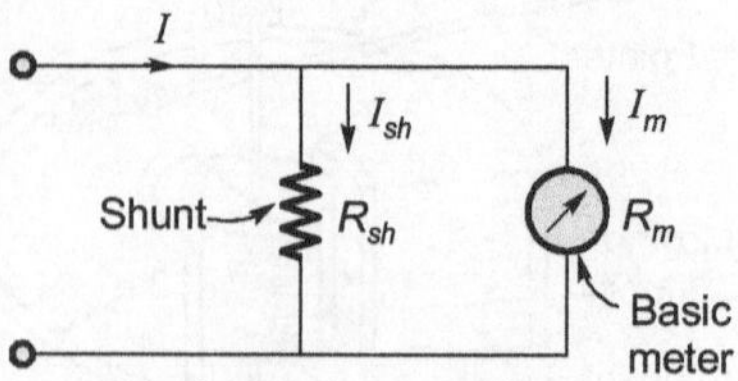

Basic ammeter circuit

$$I_{sh} R_{sh} = I_m R_m$$

$$I = \left(1 + \frac{R_m}{R_{sh}}\right) I_m$$

where, I = Current to be measured ;

$I_m = I_{fs}$ = Full scale deflection current ; A

R_m = Internal resistance of meter ; Ω

R_{sh} = Resistance of the shunt ; Ω

❑ Shunt resistance

$$R_{sh} = \frac{R_m}{m-1}$$

where, $m = \frac{I}{I_m} = 1 + \frac{R_m}{R_{sh}}$

m = Multiplying factor for shunt

Note:

To reduce the temperature effect, swamp resistance made up of manganin is added in series with ammeter.

2. Universal or Ayrton Shunt

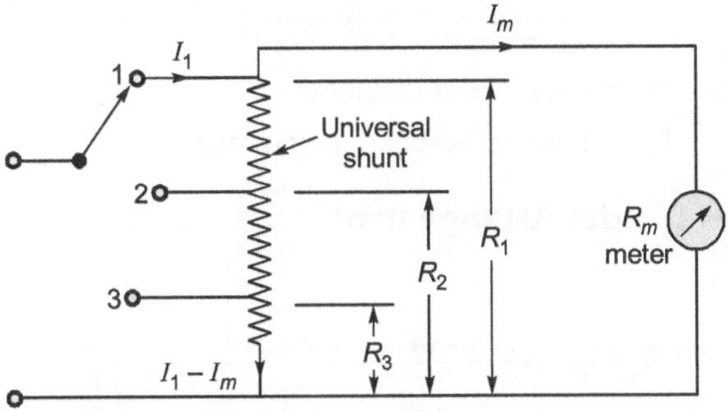

(Multi-range ammeter using universal shunt)

- For switch at a position 1

$$R_1 = \frac{R_m}{(m-1)}$$

- For switch at a position 2

$$R_2 = \frac{(R_1 + R_m)}{m_2}$$

- For switch at a position 3

$$R_3 = \frac{(R_1 + R_m)}{m_3}$$

where, $m_1 = \frac{I_1}{I_m}$, $m_2 = \frac{I_2}{I_m}$, $m_3 = \frac{I_3}{I_m}$

3. Voltmeter Multipliers

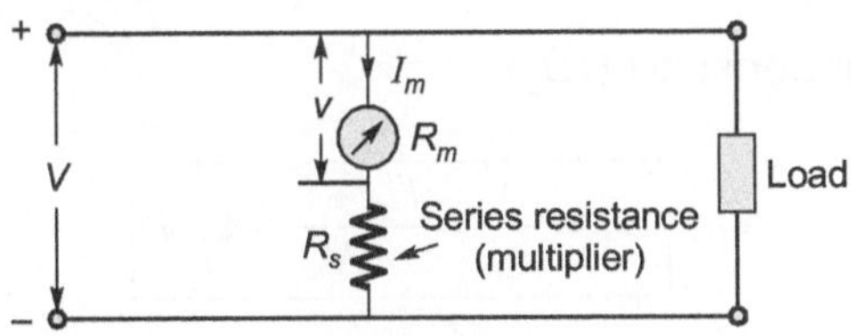

- Multiplying factor for multiplier

$$m = \frac{V}{v} = 1 + \frac{R_s}{R_m}$$

❑ Resistance of multiplier

$$R_s = (m-1)R_m$$

where, R_s = Multiplier resistance

R_m = Internal resistance of meter

4. Potential Divider Arrangement

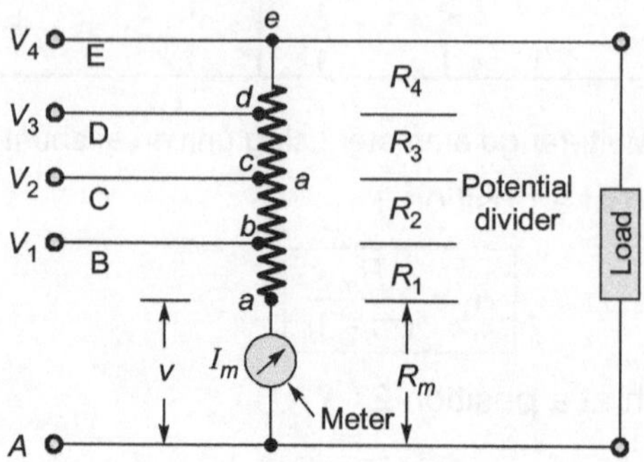

$$R_1 = (m_1 - 1)\,R_m$$
$$R_2 = (m_2 - m_1)\,R_m$$
$$R_3 = (m_3 - m_2)\,R_m$$
$$R_4 = (m_4 - m_3)\,R_m$$

where, R_1 = Resistance between point a and b

R_2 = Resistance between point b and c

R_3 = Resistance between point c and d

R_4 = Resistance between point d and e

❑ Voltmeter Sensitivity (S_v)

$$S_V = \frac{1}{I_{fs}} = \frac{R_s + R_m}{V}\ \Omega/\text{V}$$

Remember: ....

To reduce loading effect, a voltmeter with higher value of sensitivity is preferred.

Moving Iron Instruments

- Deflecting torque

$$T_d = \frac{1}{2} I^2 \frac{dL}{d\theta}$$

- Deflection

$$\theta = \frac{1}{2} \frac{I^2}{K} \frac{dL}{d\theta}$$

- For linear scale

$$\theta \cdot \frac{dL}{d\theta} = \text{constant}$$

Scale is cramped at lower and higher end.

Note:

- Moving iron instrument measure both A.C. and D.C. quantities.
- In case of A.C., it measure RMS value.
- Scale is non linear.
- Controlling torque is provided by spring and air friction damping is used.
- Curve between $\frac{dL}{d\theta}$ and θ is rectangular hyperbola.

Shunts for Moving Iron Instruments

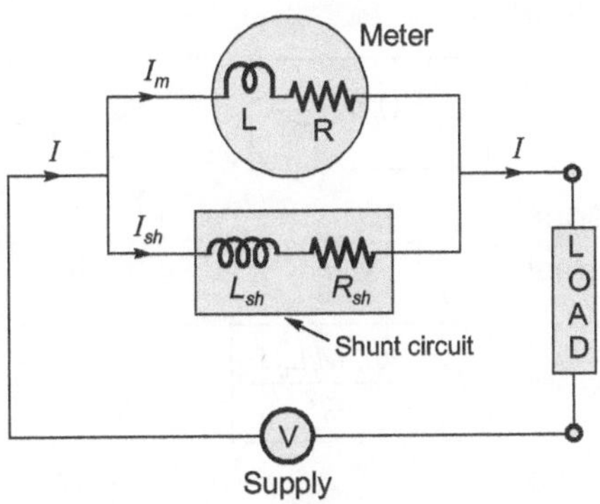

$$\frac{I_{sh}}{I_m} = \frac{R}{R_{sh}} \frac{\sqrt{1+(\omega L/R)^2}}{\sqrt{1+(\omega L_{sh}/R_{sh})^2}}$$

Multipliers for Moving Iron Instruments

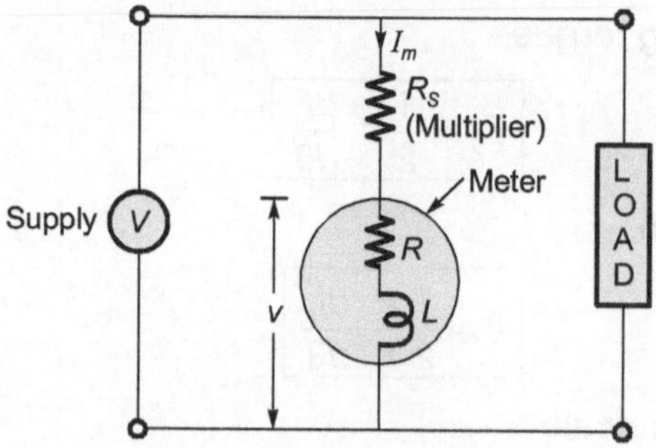

- Voltage multiplying factor

$$m = \frac{V}{v} = \frac{\sqrt{(R+R_s)^2 + \omega^2 L^2}}{\sqrt{R^2 + \omega^2 L^2}}$$

Errors in Moving Iron Instruments

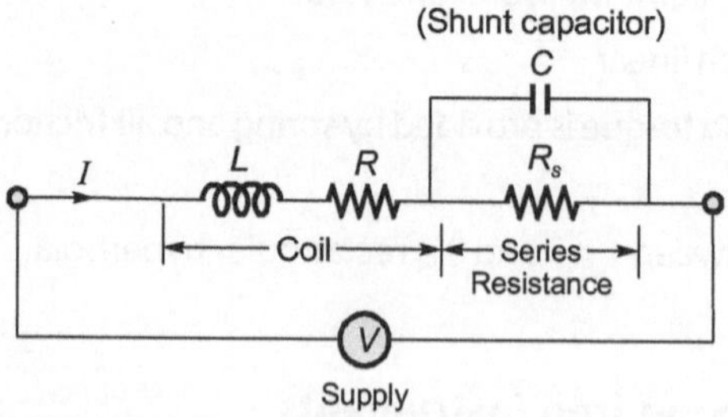

- Shunt capacitance

$$C = 0.41 \frac{L}{R_s^2}$$

- Eddy currents

When ω is small

$$I_e' = \frac{\omega^2 M L_e I}{R_e^2}$$

When ω is large

$$I_e' = \frac{MI}{L_e} = \text{constant}$$

where, R_e, L_e = resistance and inductance of eddy current path

Note:

- Moving iron instrument is not suitable for measurement of current or voltage for frequency above 125 Hz because eddy current is constant at higher frequency.
- If meter time constant is equal to shunt time constant then ammeter is made independent of input supply frequency.
- The voltmeter is made independent of input supply frequency by connecting a capacitor in parallel to the series multiplier resistance R_s.
- To reduce hysteresis error, the iron part of moving iron is made up of Nickel iron alloy.
- To reduce the external stray magnetic field, the instrument is kept inside the iron case or iron shielding is done.

Electrodynamometer

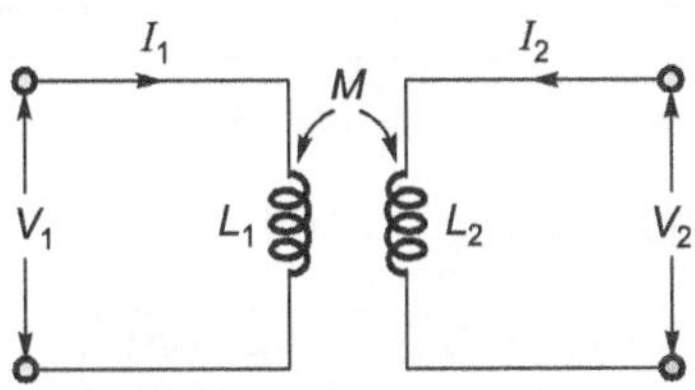

(a) If i_1 and i_2 are D.C. current i.e. $i_1 = i_2 = I$

$$\boxed{T_d = I^2 \frac{dM}{d\theta}}$$ (Measure average value)

(b) If i_1 and i_2 are A.C. current and no phase shift

$$i_1 = i_2 = I$$

$$\boxed{T_d = I^2 \frac{dM}{d\theta}}$$ (Measure RMS value)

(c) If $i_1 = i_{m_1} \sin \omega t$ and $i_2 = i_{m_2} \sin (\omega t - \phi)$

$$\boxed{T_d = I_1 I_2 \cos\phi \frac{dM}{d\theta}}$$ (Measure RMS value)

Where, $I_1 = \frac{I_{m_1}}{\sqrt{2}}$ and $I_2 = \frac{I_{m_2}}{\sqrt{2}}$

Note:

- Electrodynamometer instrument is a transfer instrument.
- It measures both A.C. and D.C.
- Scale is nonlinear.
- Its sensitivity is lesser than PMMC and M.I. type instruments.

Ratiometer

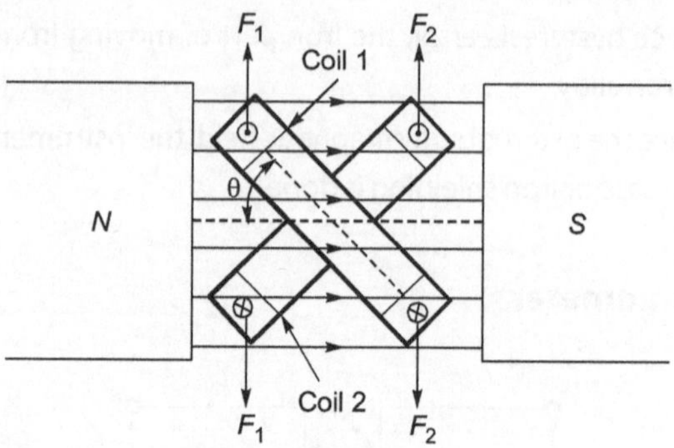

- Deflecting torque acting on coil 1

$$T_{d1} = N_1 B l_1 d_1 I_1 \cos\theta$$

- Deflecting torque acting on coil 2

$$T_{d2} = N_2 B l_2 d_2 I_2 \cos\theta$$

where, I_1, I_2 = current in coil 1 and 2

N_1, N_2 = number of turns in coil 1 and 2

l_1, l_2 = length of coil 1 and 2

d_1, d_2 = width of coil 1 and 2

B = flux density of magnetic field

- Deflection at equilibrium

$$\theta = k\left(\frac{I_1}{I_2}\right)$$

■■■■

Instrument Transformers

5

Ratios of Instrument Transformers

1. Transformation Ratio (*R*)

It is the ratio of the magnitude of the primary phasor to the secondary phasor.

$$R = \frac{|\text{primary phasor}|}{|\text{secondary phasor}|}$$

- For current transformer (C.T.)

$$R = \frac{\text{primary winding current}}{\text{secondary winding current}}$$

- For potential transformer (P.T.)

$$R = \frac{\text{primary winding voltage}}{\text{secondary winding voltage}}$$

2. Nominal Ratio (K_n)

It is the ratio of rated primary winding current (or voltage) to the rated secondary winding current (or voltage).

- For C.T.

$$K_n = \frac{\text{rated primary winding current}}{\text{rated secondary winding current}}$$

- For P.T.

$$K_n = \frac{\text{rated primary winding voltage}}{\text{rated secondary winding voltage}}$$

3. Turns Ratio (n)

- For C.T.

$$n = \frac{\text{number of turns of secondary winding}}{\text{number of turns of primary winding}}$$

❑ For P.T.

$$n = \frac{\text{number of turns of primary winding}}{\text{number of turns of secondary winding}}$$

4. Ratio Correction Factor

$$\text{RCF} = \frac{R}{K_n}$$

Remember: ....

The ratio marked on the transformers is their nominal ratio.

....

Current Transformer

Equivalent Circuit

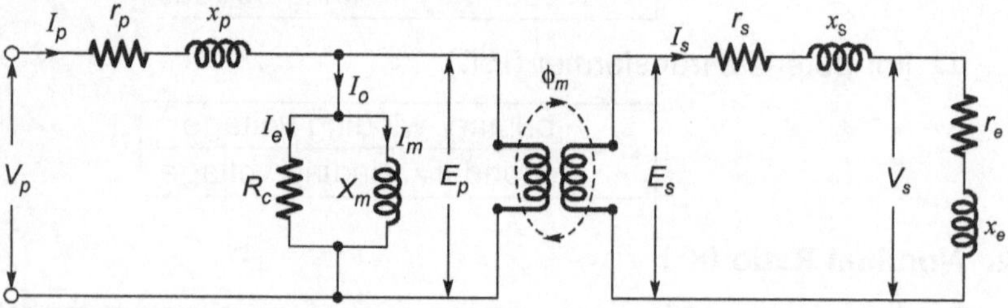

where, r_s, x_s = resistance, reactance of secondary winding

r_e, x_e = resistance, reactance of external burden

E_p, E_s = primary and secondary winding induced voltage

N_p, N_s = number of primary and secondary winding turns

I_p, I_s = primary and secondary winding current

ϕ = working flux of transformer

θ = phase angle of transformer

δ = angle between secondary winding induced voltage and secondary winding current

Δ = phase angle of secondary winding load circuit

I_0 = exciting current

I_m = magnetizing component of exciting current

I_e = loss component of exciting current

α = angle between exciting current and flux

Phasor Diagram

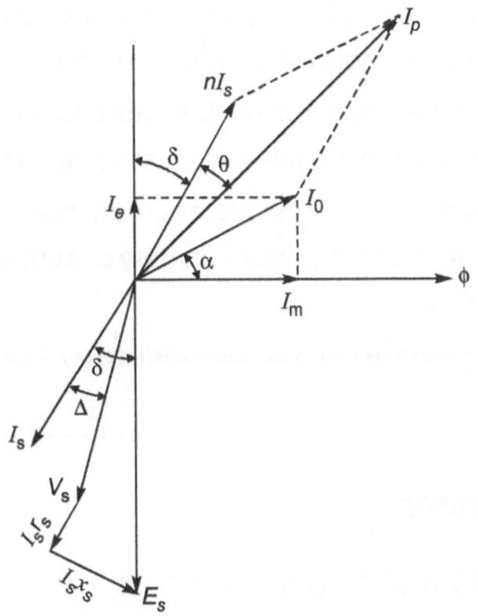

Phasor diagram of C.T.

- Transformation ratio

$$R = \frac{I_p}{I_s} \simeq n + \frac{I_o}{I_s}\sin(\delta + \alpha)$$

$$R \simeq n + \frac{(I_m \sin\delta + I_e \cos\delta)}{I_s}$$

$$R \simeq n + \frac{I_e}{I_s} \simeq n\left(1 + \frac{I_e}{I_p}\right)$$

where, $I_m = I_o \cos\alpha$
$I_e = I_o \sin\alpha$

- Phase angle

$$\theta \simeq \frac{180}{\pi}\left(\frac{I_m \cos\delta - I_e \sin\delta}{nI_s}\right) \text{ degree}$$

$$\theta \simeq \frac{180}{\pi}\frac{I_m}{I_p} \text{ degree}$$

- Ratio error

$$\text{Ratio error} = \frac{\text{nominal ratio}(K_n) - \text{actual ratio}(R)}{\text{actual ratio}(R)}$$

Remember:

- The primary current of C.T. is depending on load connected to system but it is not depending secondary winding burden.
- Primary winding is single turn or bar winding and secondary has more number of turns to reduce the current at secondary.
- If primary current is very high, it causes reduction in ratio error and phase angle error. So to increase value of primary current the primary is maintain with single turn.
- The secondary number of turns are reduce by 1 or 2 turns then the ratio error reduces.

Potential transformer

- Actual transformation (voltage) ratio

$$R = n + \frac{\frac{I_s}{n}\left[R_p \cos\Delta + X_p \sin\Delta\right] + I_e r_p + I_m x_p}{V_s}$$

- Phase angle

$$\theta = \frac{I_s}{V_s}(X_s \cos\Delta - R_s \sin\Delta) + \frac{I_e x_p - I_m r_p}{nV_s} \text{ rad.}$$

Note:

- C.T. never operates with secondary winding open but P.T. can be operated with secondary winding open.
- Strip wound core is used to reduce ratio error and phase angle error.

Application of C.T. and P.T.

- Multiple operation with a single device.
- Higher current and higher voltages are step down to lower current and lower voltage so that metering is easier.
- Measuring circuit is isolated from the power circuit.
- Low power consumption.
- Replacement is easier.

■■■■

Measurement of Power and Wattmeters

6

Measurement of Power

1. D.C. Circuits

- Ammeter connected between load and voltmeter

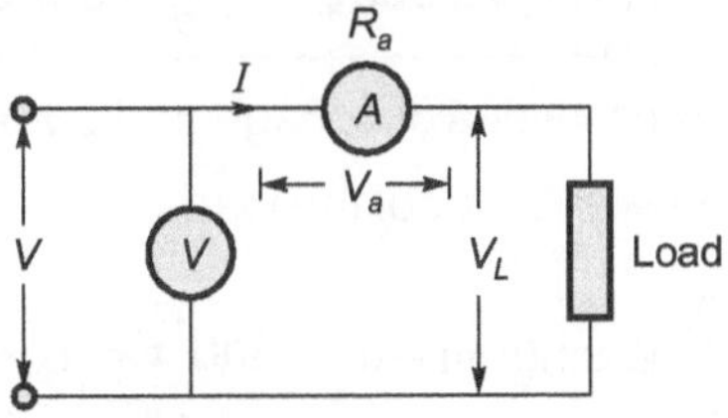

Power consumed by load:

$$P = VI - I^2 R_a$$

where, V = Voltage across voltmeter
I = Current through ammeter
R_a = Resistance of ammeter

- Voltmeter connected between load and ammeter

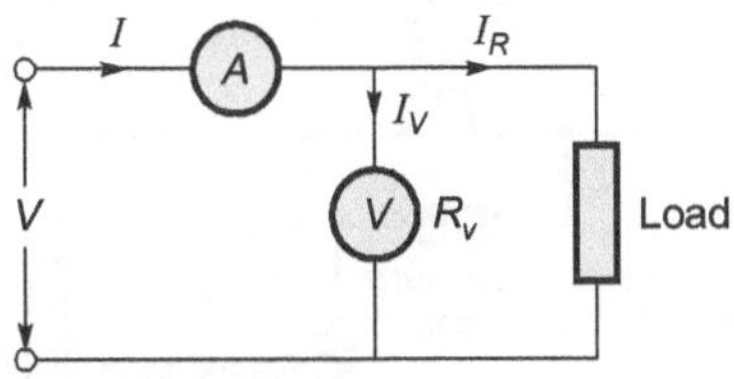

Power consumed by load:

$$P = VI - \frac{V^2}{R_V}$$

where, V = Voltage across voltmeter
I = Current through ammeter
R_V = Resistance of voltmeter

2. A.C. Circuits

- Instantaneous power

$$p = v\,i = V_m I_m \sin\omega t \cdot \sin(\omega t - \phi)$$

where, $v = V_m \sin\omega t$

$i = I_m \sin(\omega t - \phi)$

- Average power

$$P = VI\cos\phi = \frac{V_m I_m}{2}\cos\phi$$

where, V, I = Rms values of voltage and current

$\cos\phi$ = Power factor of the load

- Let $v = V_o + \sum_{n=1}^{m} V_n \sin(n\omega t + \theta_n)$ and $i = I_o + \sum_{n=1}^{m} I_n \sin(n\omega t + \phi_n)$

then
$$P_{avg.} = V_o I_o + \frac{1}{2}\sum_{n=1}^{m} V_n I_n \cos[\theta_n - \phi_n]$$

Remember: ....

Wattmeter reads average active power.

....

electrodynamometer wattmeter

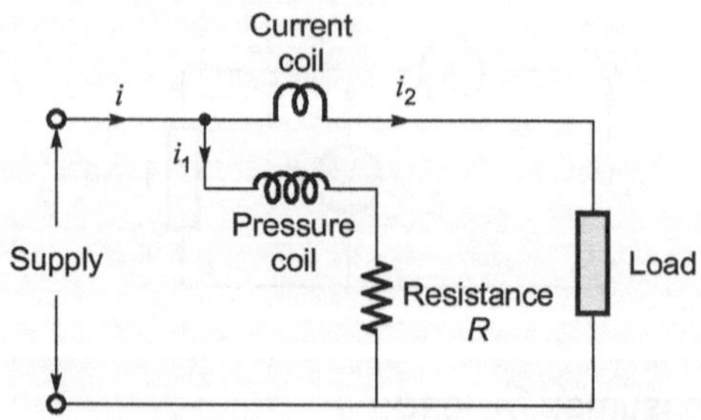

- Instantaneous torque

$$T_i = i_1 i_2 \left(\frac{dM}{d\theta}\right)$$

where i_1, i_2 = instantaneous value of current in pressure and current coils

❑ **Deflecting torque**

$$T_d = \frac{VI}{R_p}\cos\phi \cdot \frac{dM}{d\theta}$$

where, R_p = resistance of pressure coil circuit

❑ **Controlling torque**

$$T_c = K\theta$$

where K = spring constant

θ = final steady deflection

❑ **Deflection**

$$\theta = \left(K_1 \frac{dM}{d\theta}\right) P$$

where, P = power being measured = $VI\cos\phi$

$$K_1 = \frac{1}{R_p K}$$

Note:

Scale is linear in terms of power as $\theta \propto P$.

Errors in Electrodynamometer Wattmeters

Correction Factor (*K*)

The correction factor is a factor by which the actual wattmeter reading is multiplied to get the true power.

❑ **For lagging power factor**

$$K = \frac{\cos\phi}{\cos\beta\,\cos(\phi - \beta)}$$

❑ **For leading power factor**

$$K = \frac{\cos\phi}{\cos\beta\,\cos(\phi + \beta)}$$

where, ϕ = Angle between current in the current coil and voltage of pressure coil

β = Angle between current and voltage of pressure coil

$$\text{True power} = \text{Correction factor} \times \text{actual wattmeter reading}$$

- For β very small

$$\text{Actual wattmeter reading} = \text{true power } (1 + \tan\phi \tan\beta)$$

$$\text{Error} = \tan\phi \tan\beta \times \text{true power} = VI \sin\phi \tan\beta$$

$$\%\text{error} = \tan\phi \tan\beta \times 100$$

$$\text{True power} = VI \cos\phi$$

where, V = Voltage applied to pressure coil
I = Current in current coil

Power in Poly-Phase Systems

Blondel's Theorem

If a network is supplied through n conductors, the total power is measured by summing the reading of n wattmeters so arranged that a current element of a wattmeter is in each line and the corresponding voltage element is connected between that line and a common point, if the common point is located on one of the lines, then the power may be measured by $(n-1)$ wattmeters.

Two wattmeter method

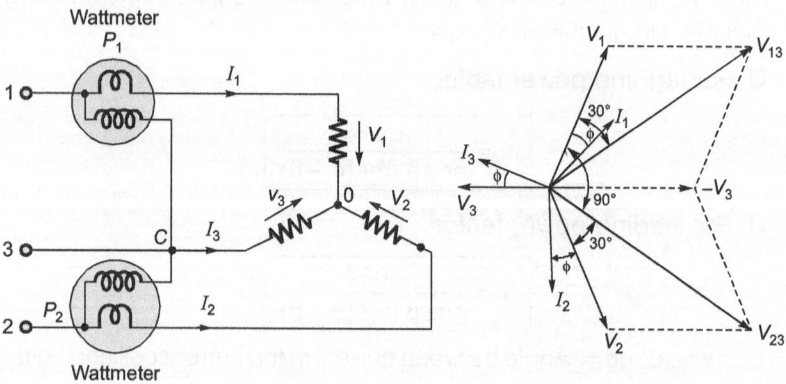

- Reading of P_1 wattmeter

$$P_1 = \sqrt{3}\, V I \cos(30° - \phi)$$

- Reading of P_2 wattmeter

$$P_2 = \sqrt{3}\, V I \cos(30° + \phi)$$

- Total power consumed by load

$$P = P_1 + P_2$$

- Power factor

$$\cos\phi = \cos\left[\tan^{-1}\left(\sqrt{3}\,\frac{P_1 - P_2}{P_1 + P_2}\right)\right]$$

$$\tan\phi = \sqrt{3}\,\frac{P_1 - P_2}{P_1 + P_2}$$

where, V = Phase voltage
I = Phase current
ϕ = Angle between phase current and phase voltage

Reading of Wattmeter at Different Power Factor

S.No	ϕ	$\cos\phi$	P_1	P_2	$P = P_1 + P_2$	Comment
1.	0	1	$\frac{\sqrt{3}}{2} V_L I_L$	$\frac{\sqrt{3}}{2} V_L I_L$	$\sqrt{3} V_L I_L$	$P_1 = P_2$ (eqaul reading)
2.	30°	0.866	$V_L I_L$	$\frac{V_L I_L}{2}$	$1.5 V_L I_L$	$P_1 = 2P_2$
3.	60°	0.5	$\frac{\sqrt{3}}{2} V_L I_L$	0	$\frac{\sqrt{3}}{2} V_L I_L$	$P_2 = 0, P_2 = P$
4.	90°	0	$+\frac{V_L I_L}{2}$	$-\frac{V_L I_L}{2}$	0	$P_1 = -P_2$

Note: When wattmeter reading comes into negative, reverse either P.C. or C.C. terminal and then take the reading of negative wattmeter.

Measurement of Energy

For the measurement of energy, we use energy meter. Energy meter is an integrating instrument which adds the energy cumulatively over a period of time.

$$\boxed{\text{Energy} = \text{Power} \times \text{time}}$$

$$\boxed{\text{Energy} = \int_0^t P \cdot dt} \text{ kWhr}$$

Note:

- Energy meter works on principle of induction motor.
- The meter which measure A.C. energy is called watt hour meter.
- The meter which measure D.C. energy is called amp-hour meter.

❑ **Deflection torque**

$$\boxed{T_D \propto P}$$

❑ **Breaking torque**

$$\boxed{T_B \propto N}$$

where, N = Speed of disc in rps

❑ **At balance**

$$\boxed{T_d = T_B}$$

$$\boxed{\int P \cdot dt = K\int N \cdot dt}$$

$$\boxed{\text{Energy} \propto \int N \cdot dt}$$

❑ **Energy meter constant (EMC)**

$$\boxed{\text{EMC} = \frac{\text{Number of revolution made by disc}}{\text{Energy recorded in kWhr.}}}$$

$$\boxed{K = \frac{N}{P \times t}}$$

where, P = Power in kW

t = Time in hrs.

❑ $$\% \text{ Creeping Error} = \frac{\text{Revolution of disc due to creeping per hour}}{\text{Revolution of disc due to total load per hour}} \times 100$$

Remember: ....

Potential coil of energy meter should be highly inductive so that it measures true energy.

....

Compensation in Energy Meter

1. **Lag compensation :** Through lag coil or shading coil.
2. **Low load or friction adjustment :** By using shading loop.
3. **Over friction or creeping :** By providing holes or slots on rotating disc.
4. **Over load compensation :** By keeping saturable shunt magnet in series magnet or current coil.
5. **Over voltage compensation :** By keeping saturable shunt magnet in shunt magnet.
6. **Temperature compensation :** By making permanent magnet of "mutemp" material.
7. **Speed adjustment :** By adjusting position of break magnet.

Remember: ....

- Creeping error is always positive.
- If either potential coil or current coil is wrongly connected then the disc rotates in opposite direction.

....

■■■■

Measurement of Resistance

7

Classification of Resistance

1. **Low resistance:** All resistance of the order of 1 Ω and below.
   Example: Winding coils of electrical motors, generators and transformers.
2. **Medium resistance:** Resistances from 1 Ω upwards to about 0.1 MΩ.
   Example: Resistance of heaters, potentiometers.
3. **High resistance:** All resistances of the order of 0.1 MΩ and above.
   Example: Insulation of electrical cable and windings, insulation of motors, generators and transformers.

Measurement of Medium Resistance

The different methods employed are:

(i) Ammeter-voltmeter method
(ii) Wheatstone bridge method
(iii) Ohmmeter method
(iv) Substitution method

1. Ammeter Voltmeter Method

$$R_m = \frac{\text{voltmeter reading}}{\text{ammeter reading}} = \frac{V}{I}$$

where, R_m = measured value of resistance

(a) Circuit for higher resistance

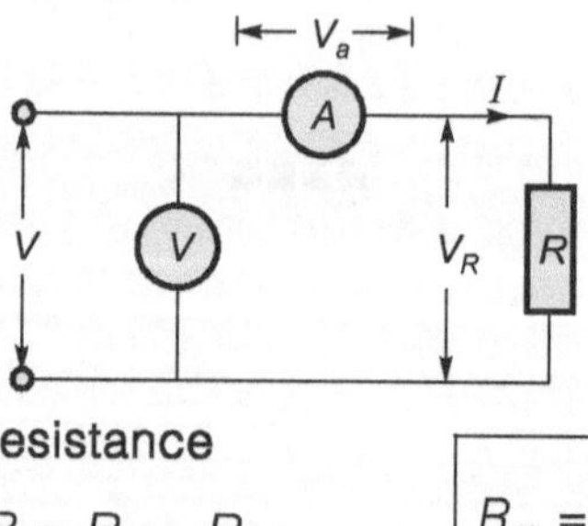

❑ True value of resistance

$$R = R_{m_1} - R_a$$

$$R_m = R_{m_1}\left(1 - \frac{R_a}{R_{m_1}}\right)$$

where, R_{m_1} = Measured value of resistance

R_a = Resistance of ammeter

❑ Relative error

$$\epsilon_r = \frac{R_{m_1} - R}{R} = \frac{R_a}{R}$$

- To get minimum error, the test resistance should be more than the ammeter resistance so that this adjustment is suitable for measurement of high resistance.

(b) Circuit for lower resistance

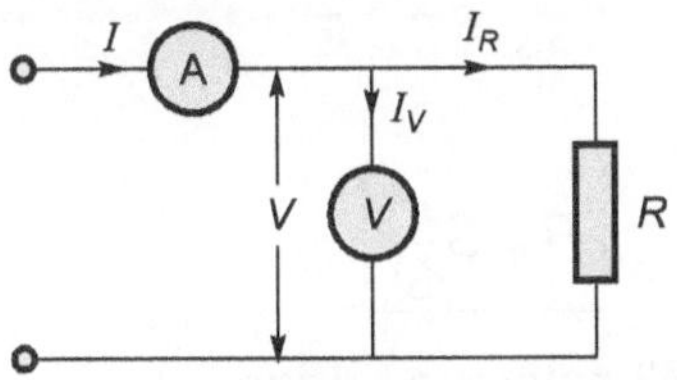

❑ True value of resistance

$$R = \frac{R_{m_2} R_V}{R_V - R_{m_2}}$$

where, R_{m_2} = Measured value of resistance
R_v = Resistance of voltmeter

For $R_v >> R_{m_2}$

$$R = R_{m_2}\left(1 + \frac{R_{m_2}}{R_v}\right)$$

❑ Relative error

$$\epsilon_r = \frac{R_{m_2} - R}{R} = \frac{-R_{m_2}^2}{R_V R} \qquad \left[\text{For } R_v >> R_{m_2}\right]$$

❑ Approximate relative error

$$\epsilon = -\frac{R}{R_v} \qquad \left[\text{For } R_{m_2} \approx R\right]$$

- This circuit is suitable for measurement of low resistance under medium scale range to get minimum error.

Note: ....

Relative errors for above two cases are equal when true value of resistance, $R = \sqrt{R_a R_v}$

....

2. Wheatstone Bridge

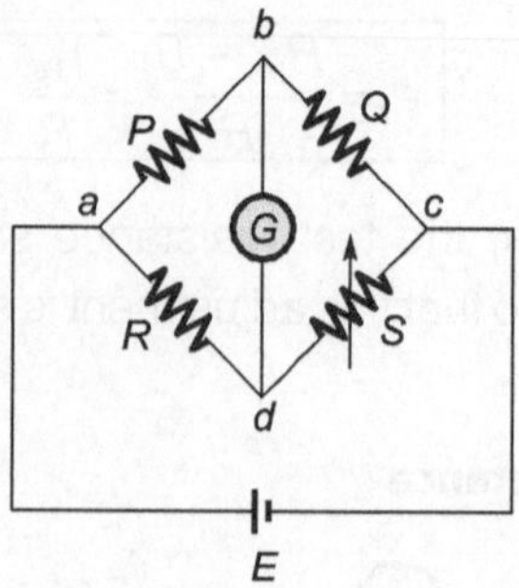

- At balance

$$R = S\frac{P}{Q}$$

- Sensitivity of Wheatstone bridge

$$S_B = \frac{\theta}{\Delta R / R} = \frac{S_v ESR}{(R+S)^2}$$; mm where, S_v = Voltage sensitivity of galvanometer, mm/volt

$$S_B = \frac{S_v E}{\frac{P}{Q} + 2 + \frac{Q}{P}}$$

E = Bridge voltage

P, Q = Branch resistances

θ = Deflection of galvanometer,mm

- For a bridge with equal arms

$$S_B = \frac{S_v E}{4}$$

Note: ....

- For maximum bridge sensitivity

$$\frac{P}{Q} = \frac{R}{S} = 1$$

- Sensitivity of bridge is most important parameter as compared to accuracy, precision and resolution.

....

Equivalent circuit of Wheatstone bridge

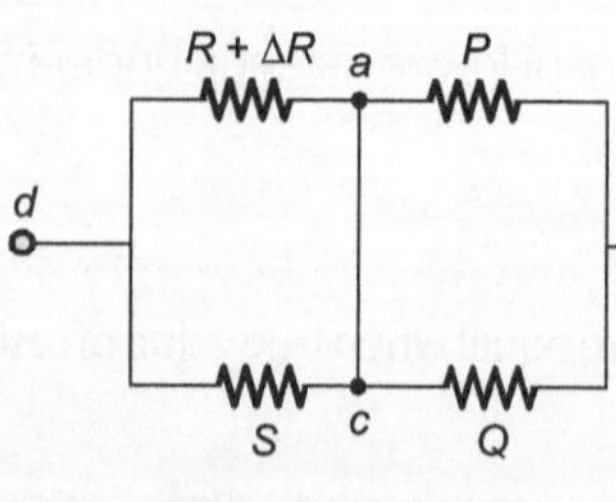

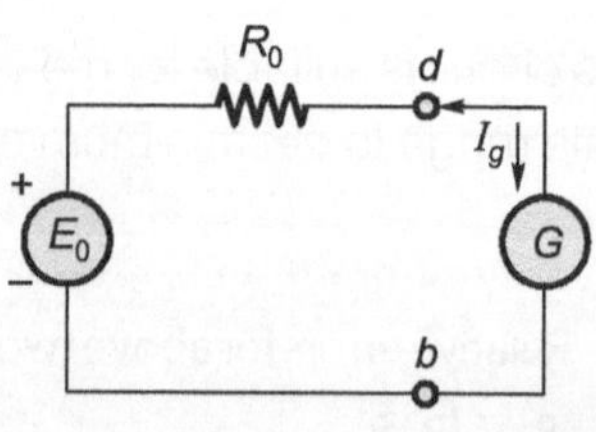

- **Galvanometer current**

$$I_g = \frac{E_0}{R_0 + G}$$

where E_0 = Thevenin's or open circuit voltage appearing between terminals b and d with galvanometer circuit open circuited.

G = Resistance of the galvanometer circuit

$$E_0 = E\left[\frac{R+\Delta R}{2R+\Delta R} - \frac{1}{2}\right] \approx E\left(\frac{\Delta R}{4R}\right) \quad \text{as } \Delta R << R$$

ΔR = Change in resistance R

- **Thevenin equivalent resistance of bridge**

$$R_0 = \frac{RS}{R+S} + \frac{PQ}{P+Q}$$

- **Galvanometer deflection**

$$\theta = \frac{S_v ES\,\Delta R}{(R+S)^2} = \frac{S_i\,ES\Delta R}{(R_0+G)\,(R+S)^2}$$

where, S_i = Current sensitivity of galvanometer

- **Bridge sensitivity**

$$S_B = \frac{\theta}{\Delta R/R} = \frac{S_i\,E\,SR}{(R_0+G)\,(R+S)^2}$$

- **Current Sensitivity**

$$S_i = \frac{\theta}{I_g};\ \text{mm/}\mu\text{A}$$

θ = Deflection in galvanometer

I_g = Current in galvanometer

- **Voltage Sensitivity**

$$S_v = \frac{\theta}{V_{Th}};\ \text{mm/V}; \quad V_{Th} = \text{Voltage across galvanometer}$$

Note: ....

In Wheatstone bridge method, the effect of lead resistance is not eliminated hence it is not suitable for measurement of low resistance.

....

3. Ohmmeters

(a) Series Type Ohmmeter

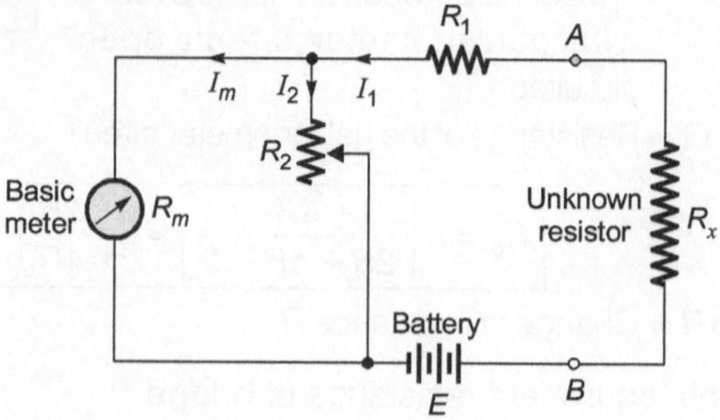

❑ Half scale resistance

$$R_h = R_1 + \frac{R_2 R_m}{R_2 + R_m}$$

❑ Meter current

$$I_m = \frac{ER_2}{(R_h + R_x)(R_2 + R_m)}$$

❑ Full scale deflection current

$$I_{fs} = \frac{ER_2}{R_h(R_2 + R_m)}$$

❑ Friction of full scale reading

$$S = \frac{I_m}{I_{fs}} = \frac{R_h}{R_x + R_h}$$

(b) Shunt Type Ohmmeter

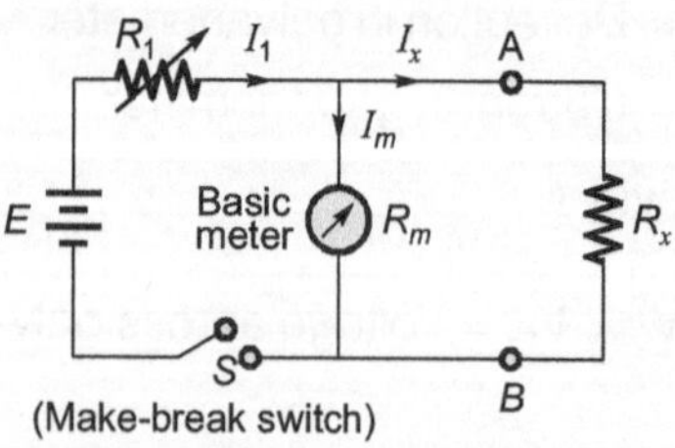

❑ Half scale reading of unknown resistance R_x is

$$R_h = \frac{R_1 R_m}{R_1 + R_m}$$

- Half scale reading of the meter

$$I_h = 0.5\, I_{fs} = \frac{ER_h}{R_1 R_m + R_h(R_1 + R_m)}$$

where, R_m = Internal resistance of meter
R_1 = Adjustable resistor (as shown in figure)
E = Supply voltage

Measurement of low Resistance

The different methods employed are:

(i) Kelvin's double bridge method
(ii) Ammeter voltmeter method
(iii) Potentiometer method

Kelvin's Double Bridge Method

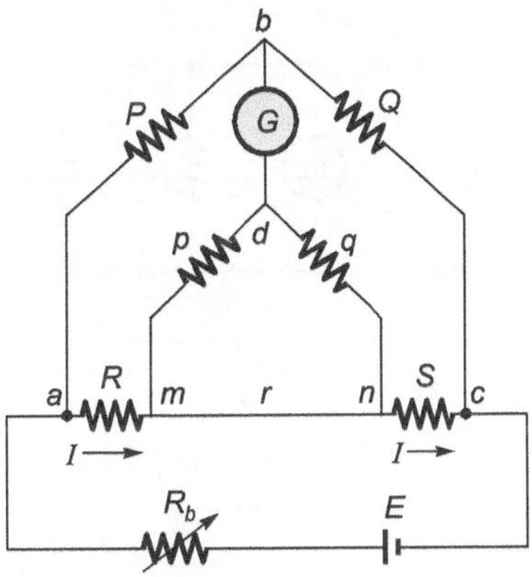

- For zero galvanometer deflection

$$E_{ab} = E_{amd}$$

$$R = \frac{P}{Q}\cdot S + \frac{qr}{p+q+r}\left[\frac{P}{Q} - \frac{p}{q}\right]$$

- If $\frac{P}{Q} = \frac{p}{q}$

then $$R = \frac{P}{Q}\cdot S$$

Note: ....

Accuracies by Kelvin double bridge method

(i) From 1000 μΩ to 1.0 Ω : 0.05%

(ii) From 100 μΩ to 1000 μΩ : 0.5% to 0.05%

(iii) From 10 μΩ to 100 μΩ : 0.5% to 0.2%

....

Measurement of High Resistance

The differenece methods employed are:

(i) Loss of charge method

(ii) Meggar

(iii) Direct deflection method

(iv) Megohm bridge

1. Loss of Charge Method

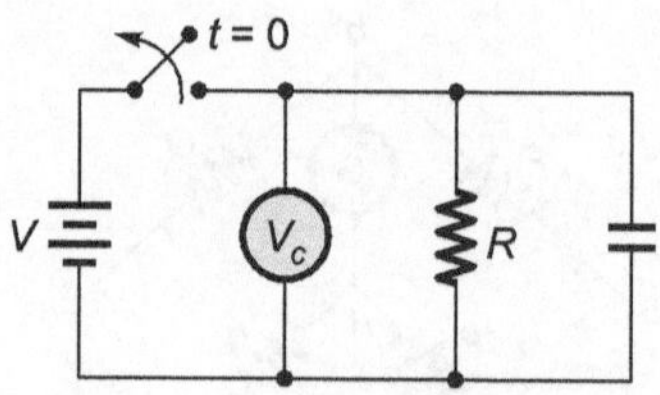

$$R = \frac{0.4343\,t}{C \log_{10}\left(\frac{V}{V_c}\right)}$$

2. Meggar

- Meggar works on the principle of electrodynamometer.
- Meggar is used to measure the insulation resistance of cable, motor and generator, etc.
- Deflecting torque angle is proportional to the resistance of the insulator, which is under test.
- It is also used to check the continuity of cable.
- No external control torque provided.
- Air friction damping is used.
- No need of external power supply.

Note: ....

High resistance have a guard terminal which is used to avoid leakage current.

....

Carry Foster Slide Wire Bridge

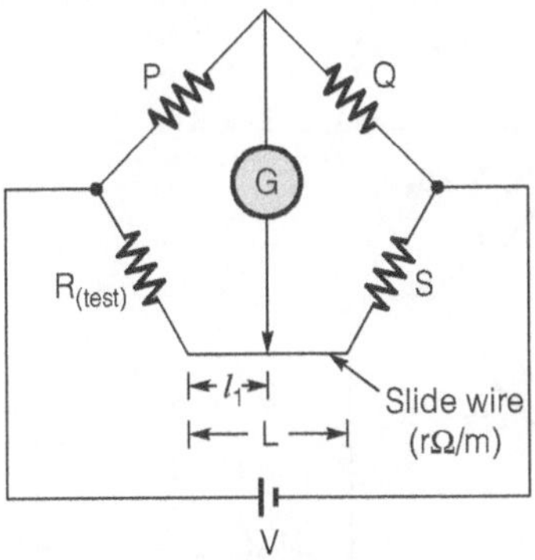

(Bridge-1)

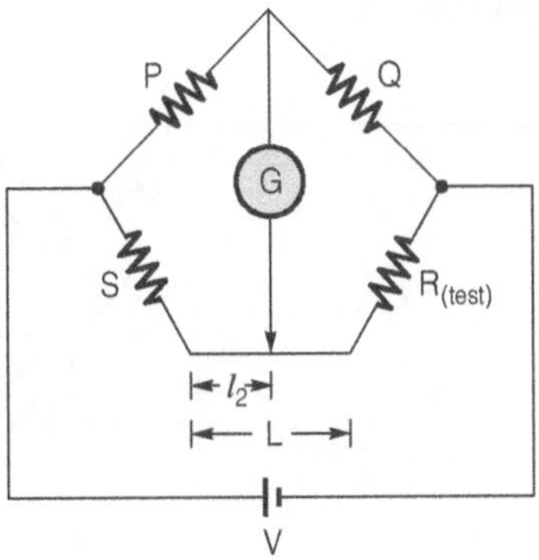

(Bridge-2)

From bridge (1)

$$\frac{P}{Q}+1=\frac{R+S+Lr}{S+(L-l_1)r} \quad \cdots\text{(i)}$$

From bridge (2)

$$\frac{P}{Q}+1=\frac{R+S+Lr}{R+(L-l_2)r} \quad \cdots\text{(ii)}$$

Equating equation (i) and (ii)

$$\boxed{R-S=(l_2-l_1)r}$$

Note: ....

Carry Foster bridge method is used for medium resistance measurement by comparing with standard resistance.

....

■■■■

A.C. Bridges

8

Introduction

Used to measure self inductance, Mutual inductance, capacitance, and frequency.

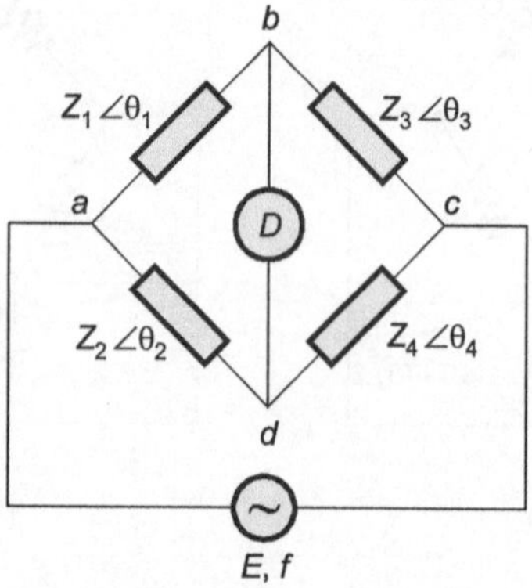

- **General equation for bridge balance**

$$\boxed{\bar{Z}_1 \bar{Z}_4 = \bar{Z}_2 \bar{Z}_3}$$

- **Magnitude condition**

$$\boxed{|Z_1||Z_4| = |Z_2||Z_3|}$$

- **Angle condition**

$$\boxed{\angle\theta_1 + \angle\theta_4 = \angle\theta_2 + \angle\theta_3}$$

Note:

Magnitude condition and angle condition both must be satisfied for the bridge to be balanced.

- **Depending upon the frequency, different null detectors are used**

  Vibration galvanometer – 5 Hz to 1 kHz

  Head phones – 250 Hz to 4 kHz

  Tuned amplifier detector – 10 Hz to 100 kHz

  D'Arsonval Galvanometer – DC frequency = 0 Hz

- Depending upon phase angle θ, elements are

θ	Elements
0°	R
90°	L_1
–90°	C_1
$0° < \theta_1 < 90°$	R_1, L_1
$-90 < \theta_1 < 0°$	R_1, C_1

- **Convergence to balance point:**
  If the variables are in the same arm of bridge then minimum time is required for balancing of bridge. This is called convergence to balance point.
- **Quality factor (Q.F.)**

$$\text{Q.F.} = \frac{\text{Energy stored}}{\text{Energy dissipated}}$$

Measurement of Self Inductance

1. Maxwell's Inductance Bridge

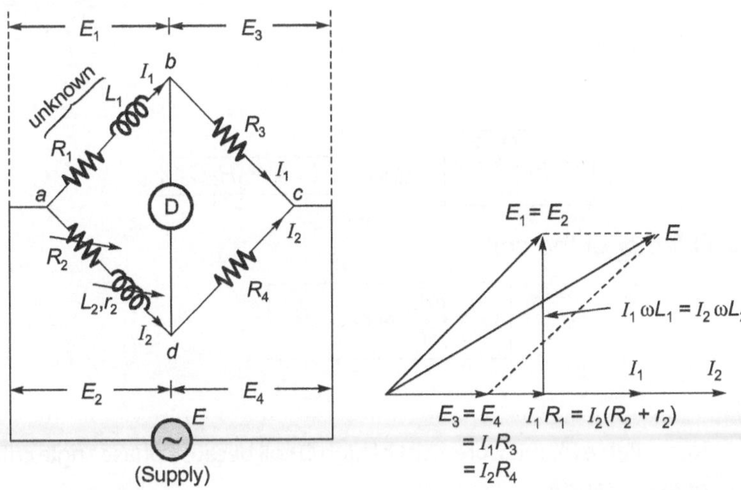

$$\boxed{R_1 = \frac{R_3}{R_4}(R_2 + r_2)}$$

$$\boxed{L_1 = \frac{R_3}{R_4} \cdot L_4}$$

where, L_1 = Unknown inductance of resistance R_1

L_2 = Variable inductance of fixed resistance r_2

R_2 = Variable resistance connected in series with L_2

R_3, R_4 = Known non-inductive resistances

2. Maxwell's Inductance-Capacitance Bridge

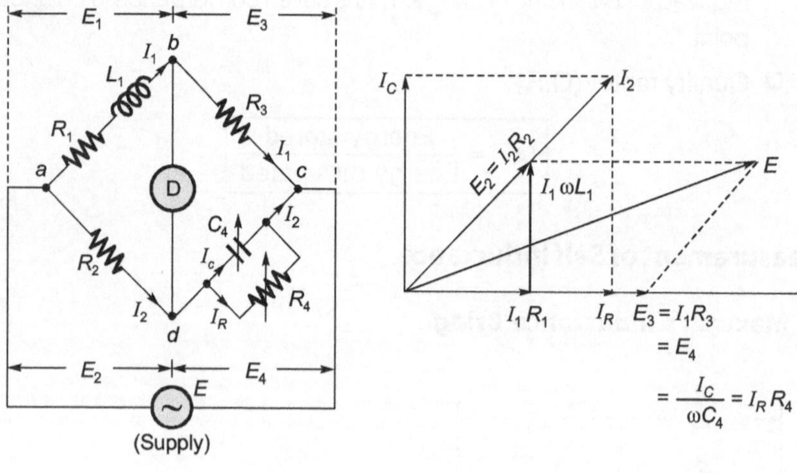

$$\boxed{R_1 = \frac{R_2 R_3}{R_4}} \quad \text{and} \quad \boxed{L_1 = R_2 R_3 C_4}$$

❑ Q factor of the coil

$$\boxed{Q = \frac{\omega L_1}{R_1} = \omega C_4 R_4}$$

Note: ....

- Not suitable for measurement of high Q coil because phase angle criteria does not satisfy.

- Not suitable for measurement of low Q-coil because of sliding balance problem.
- Suitable for measurement of medium Q coil i.e. $(1 < Q < 10)$.

3. Hay's Bridge

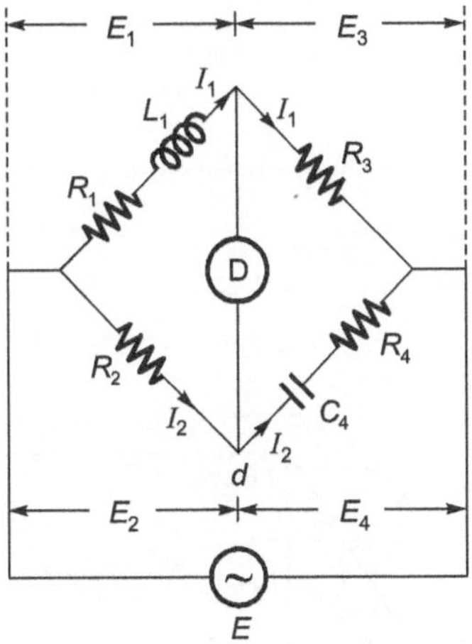

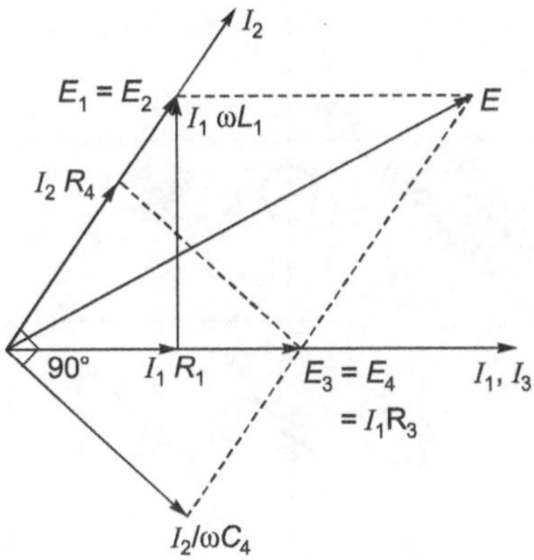

$$R_1 = \frac{\omega^2 R_2 R_3 R_4 C_4^2}{1+\omega^2 C_4^2 R_4^2} = \frac{R_2 R_3}{R_4}\left(\frac{1}{1+Q^2}\right)$$

$$L_1 = \frac{R_2 R_3 C_4}{1+\omega^2 C_4^2 R_4^2} = \frac{R_2 R_3 C_4}{1+\left(\frac{1}{Q}\right)^2}$$

❑ For Q > 10

$$L_1 = R_2 R_3 C_4$$

$$Q = \frac{\omega L_1}{R_1} = \frac{1}{\omega C_4 R_4}$$

where, L_1 = Unknown-inductance having a resistance R_1

R_2, R_3, R_4 = known non-inductive resistance

C_4 = Standard capacitor

Note: ....

The Hay's bridge is suited for the measurement of high Q inductors.

....

4. Anderson's Bridge

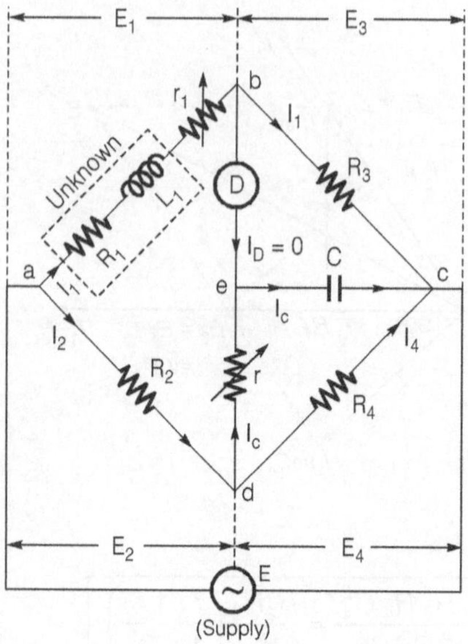

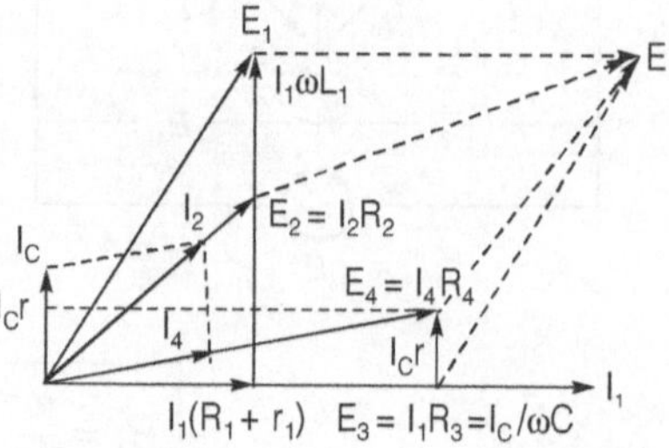

$$R_1 = \frac{R_2 R_3}{R_4} - r_1$$

$$L_1 = \frac{CR_3}{R_4}\left[r(R_4 + R_2) + R_2 R_4\right]$$

where, L_1 = Self-inductance to be measured

R_1 = Resistance of self-inductor

r_1 = Resistance connected in series with self-inductor

r, R_2, R_3, R_4 = Known non-inductive resistances

C = Fixed standard capacitor

5. Owen's Bridge

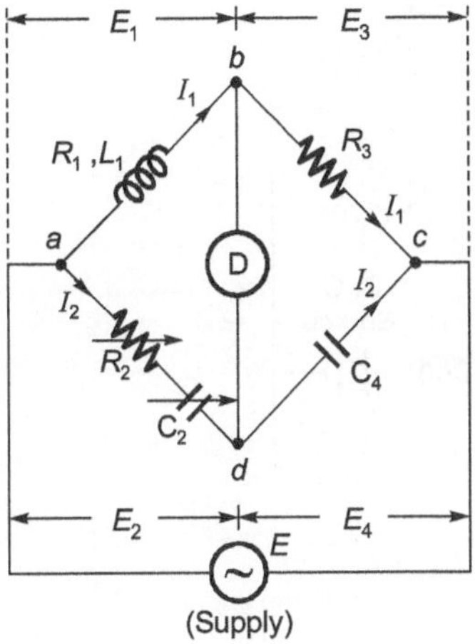

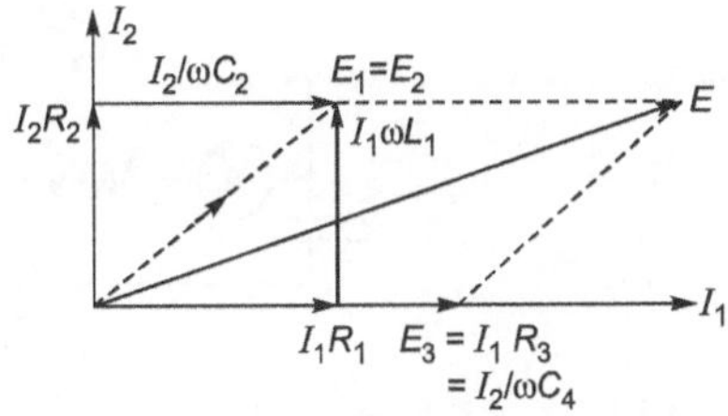

$$\boxed{L_1 = R_2 R_3 C_4}$$

$$\boxed{R_1 = R_3 \frac{C_4}{C_2}}$$

where, L_1 = Unknown self inductance of resistance R_1

R_2 = Variable non-inductive resistance

R_3 = Fixed non-inductive resistance

C_2 = Variable standard capacitor

C_4 = Fixed standard capacitor

Note: Owen's bridge is used for measurement of unknown inductance and incremental inductance and incremental permeability (μ).

Measurement of Incremental Inductance

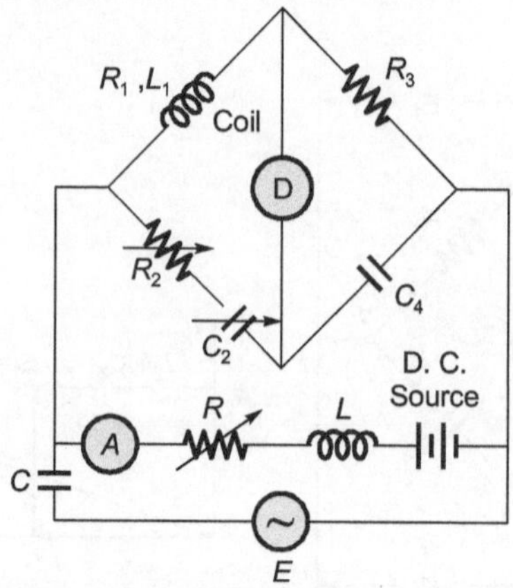

- Incremental inductance

$$L_1 = R_2 R_3 C_4$$

- Incremental permeability

$$\mu = \frac{L_1 l}{N^2 A}$$

where, N = Number of turns
A = Area of flux path
l = Length of flux path
R_2 = Variable non-inductive resistance
R_3 = Fixed non-inductive resistance
C_4 = Fixed standard capacitor

Note:

- External D.C. source is used to compensate residual magnetism.
- Capacitor, C is to block D.C. to enter into A.C. and inductor, L is to block A.C to enter into D.C.

Measurement of Capacitance

1. De Sauty's Bridge

(a) For lossless capacitor

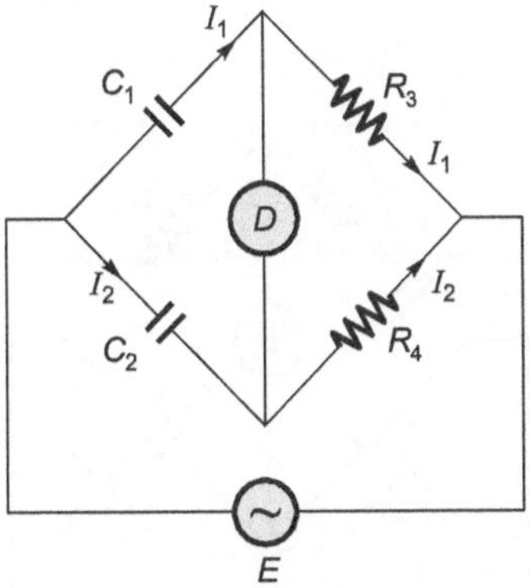

$$C_1 = \frac{C_2 R_4}{R_3}$$

where, C_1 = Capacitor whose capacitance to be measured

C_2 = A standard capacitor

R_3, R_4 = Non-inductive resistors

(b) For imperfect capacitor having dielectric loss

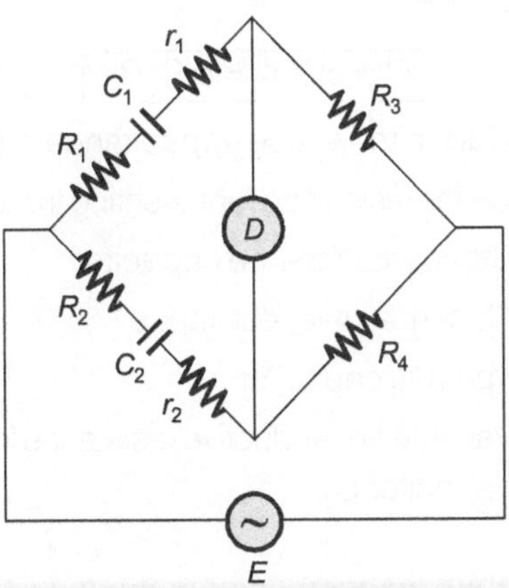

$$\frac{C_1}{C_2} = \frac{R_4}{R_3} = \frac{R_2 + r_2}{R_1 + r_1}$$

where, r_1, r_2 = Resistance representing the loss component of the two capacitor.

❑ Dissipation factor

$$D = \tan\delta = \omega C_1 r_1 = \omega C_2 r_2$$

2. Schering Bridge

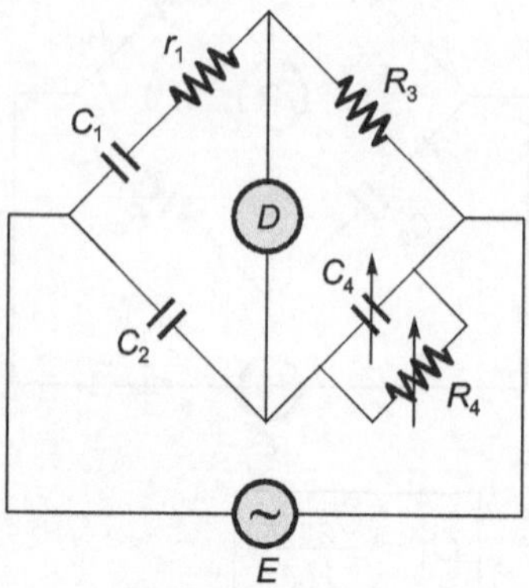

$$r_1 = \frac{R_3 C_4}{C_2}$$

$$C_1 = C_2\left(\frac{R_4}{R_3}\right)$$

❑ Dissipation factor

$$D = \omega C_1 r_1 = \omega C_4 R_4$$

where, C_1 = Capacitor whose capacitance is to be determined

r_1 = Series resistance representing the loss in the capacitor C_1

C_2 = Standard loss-free capacitor

R_3 = Non-inductive resistance

C_4 = Variable capacitor

R_4 = Variable non-inductive resistance in parallel with variable capacitor C_4

Note: Schering bridge is shielded with metal screen to reduce the stray capacitance exists between the arms and arms to the earth.

Measurement of Frequency

Wien's Bridge

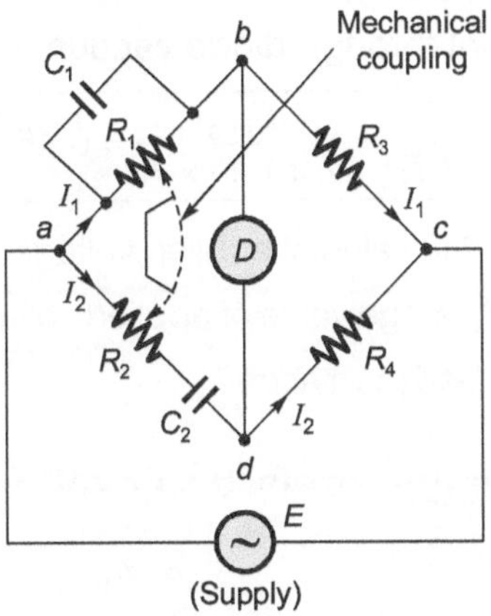

- Frequency

$$\frac{R_4}{R_3} = \frac{R_2}{R_1} + \frac{C_1}{C_2}$$

$$f = \frac{1}{2\pi\sqrt{R_1 R_2 C_1 C_2}} \text{Hz}$$

- For $R_1 = R_2 = R$ and $C_1 = C_2 = C$

$$f = \frac{1}{2\pi RC}$$

$$\frac{R_4}{R_3} = 2$$

Limitation of Wein's Bridge

If the input signal is not a sinusoidal or signal containing harmonics then balancing of bridge is not possible because null detector is sensitive to the frequencies.

■■■■

Electronic Instruments

9

- Average current through diode vacuum tube voltmeters

$$I_{av} = \frac{E_{av}}{2R} = \frac{E_{rms}}{2 \times 1.11 \times R} = 0.45\,\frac{E_{rms}}{R}$$

where, E_{rms} = RMS value of applied voltage
E_{av} = Average value of applied voltage
R = Load resistance

Difference amplifier type of electronic voltmeter

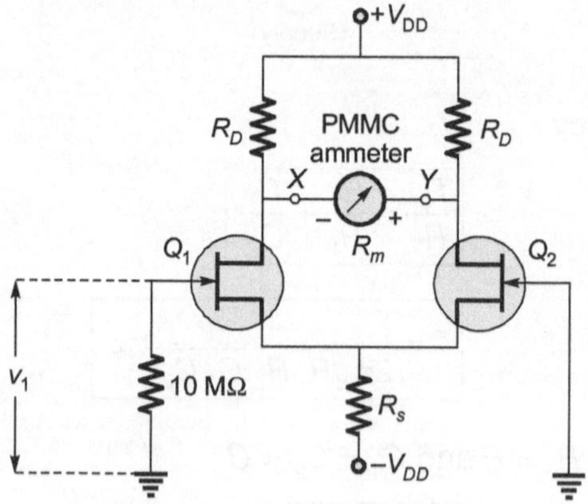

- Thevenin's voltage across terminal X-Y

$$V_{Th} = g_m\left(\frac{r_d R_D}{r_d + R_D}\right)V_1$$

- Thevenin's resistance looking into terminals X-Y

$$R_{Th} = \frac{2 r_d R_D}{r_d + R_D}$$

where, R_d = A.C. drain ressitance in Ω
g_m = Transconductance in mho

❑ Current through ammeter

$$i = \frac{V_{Th}}{R_{Th} + R_m} = \frac{g_m r_d R_D /(r_d + R_D)}{2r_d R_D /(r_d + R_D) + R_m} \cdot v_1$$

when $R_D << r_d$

$$i = \frac{g_m R_D}{2R_D + R_m} \cdot v_1$$

Digital Meters

Basic measurable quantity in digital meter is D.C.

1. Resolution (*R*) of Digital Meter

The smallest change in the input which a digital meter can be able to detect is called resolution.

$$R = \frac{1}{10^n}$$

where, n = Number of full digit.

2. Sensitivity (*S*)

The smallest change in input that can be displayed within given range.

$$S = \text{Resolution} \times \text{Range of meter}$$

3. Over ranging

Switch on the extra half (1/2) is called over ranging. Due to this over ranging the range of the instrument increases.

■■■■

Cathode Ray Oscilloscope

10

CRO is a digital instrument, which works on the principle of thermionic emission i.e. emission of electron from a heated surface. It is a linear device. With the use of CRO one can measure peak to peak, rms, peak or average value of voltage and current.

- **Calibration of CRO**

  Calibration of CRO is done by applying a known quality of square signal having a frequency of 1 kHz and peak to peak magnitude of 1 mV.

- **The rise time (t_r), of signal applied to CRO and bandwidth of CRO are related as**

$$\boxed{t_r \times \text{B.W.} = 0.35}$$

  If this condition fails then the signal is distorted at the output of CRO.

- **Electrostatic Deflection**

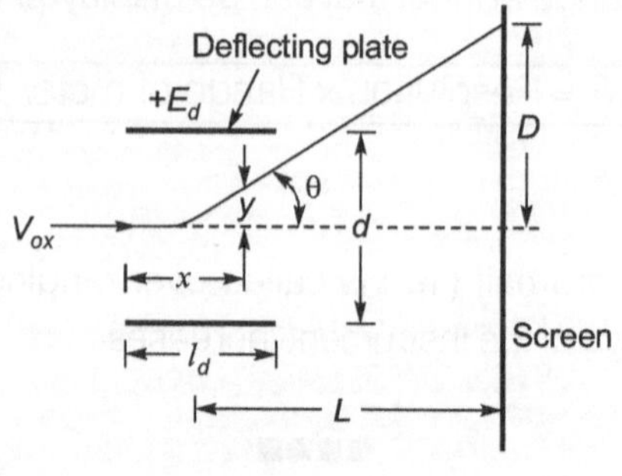

$$\boxed{y = \frac{1}{2}\frac{e\,E_y}{mv_{ox}^2}x^2}$$

where, y = Displacement in y-direction; m

e = Charge of an electron; Coulomb

E_y = Electric field intensity in Y-direction; V/m

m = Mass of electron; kg

V_{ox} = Velocity of electron when entering the fields of deflecting plates; m/s

x = Displacement in x-direction ; m

- Deflection

$$\boxed{D = \frac{L l_d E_d}{2d E_a}}$$

where, L = Distance between screen and the centre of deflecting plates; m

l_d = Length of deflecting plates; m

E_d = Potential between deflecting plates; V

d = Distance between deflecting plates; m

E_a = Voltage of pre-accelerating anode; V

- Deflection sensitivity

$$\boxed{S = \frac{D}{E_d} = \frac{L l_d}{2d E_a} \text{ m/V}}$$

- Deflection factor

$$\boxed{G = \frac{1}{S} = \frac{2d E_a}{L l_d} \text{ V/m}}$$

Sawtooth Generator

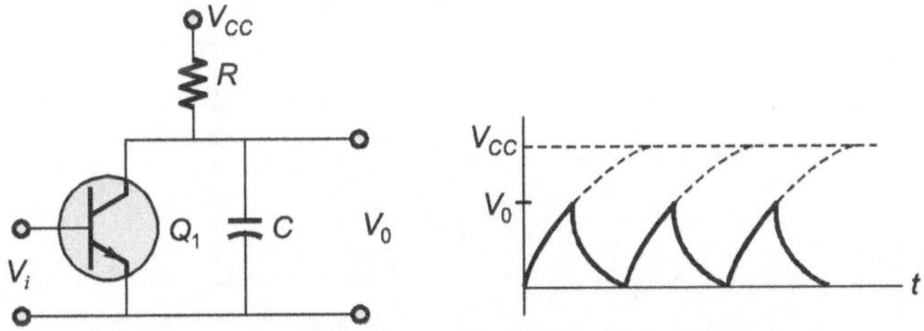

$$\boxed{V_0 = V_{cc}\left[1 - \exp(-t/RC)\right]}$$

where, V_0 = Instantaneous voltage across the capacitor at time t; V

V_{cc} = Supply voltage

Lissajous patterns

If horizontal and vertical deflecting plate are applied with sinusoidal signal, the waveform pattern appearing on the screen is called Lissajous pattern.

Application

- Used for finding the phase angle difference between the signal applied to vertical and horizontal plate.
- Used for finding the frequency ratio between vertical and horizontal plates voltage.

Phase angle (ϕ) between V_x and V_y	Lissajous pattern
0° or 360°	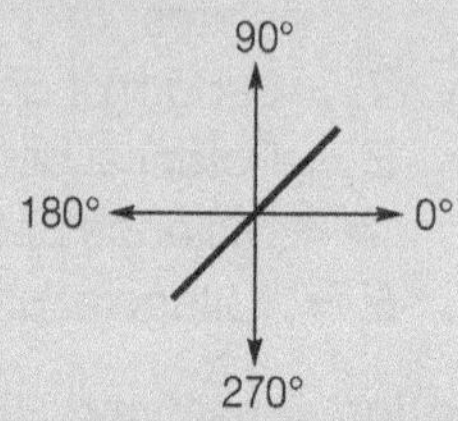
0° < ϕ < 90° (or) 270° < ϕ < 360°	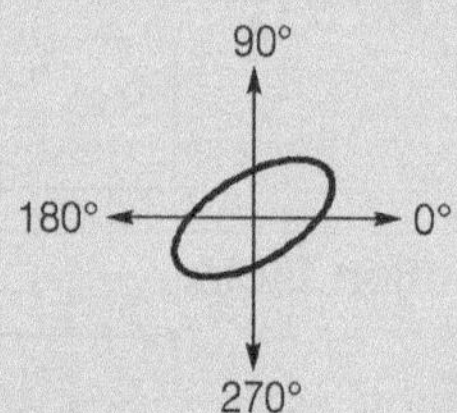
ϕ = 90° or 270°	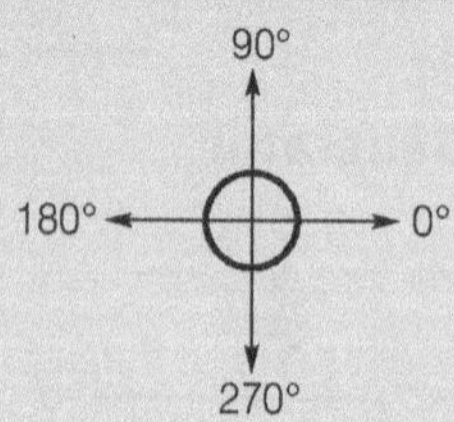
90° < ϕ < 180° (or) 180° < ϕ < 270°	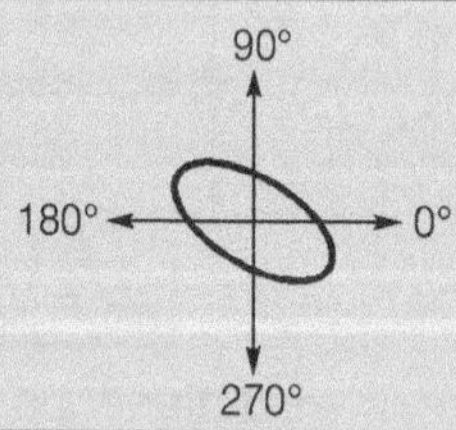
ϕ = 180°	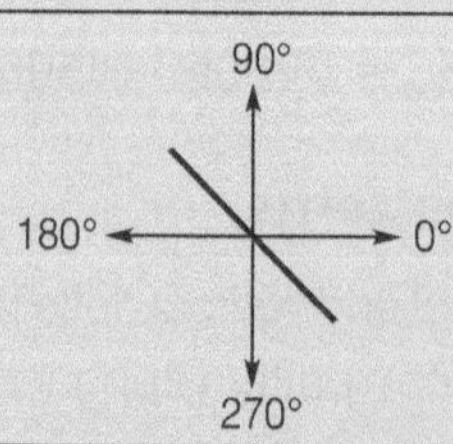

Finding the phase angle φ from given Lissajous pattern

(a) When Lissajous pattern is in first and third quadrant

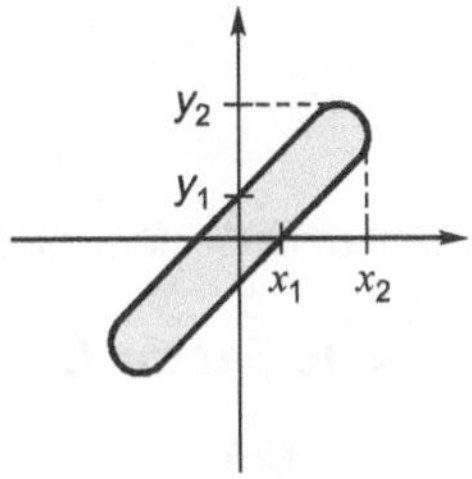

- First possibility

$$\phi = \sin^{-1}\left(\frac{x_1}{x_2}\right) = \sin^{-1}\left(\frac{y_1}{y_2}\right)$$

- Second possibility = 360° – φ

(b) When Lissajous pattern is in second and fourth quadrant

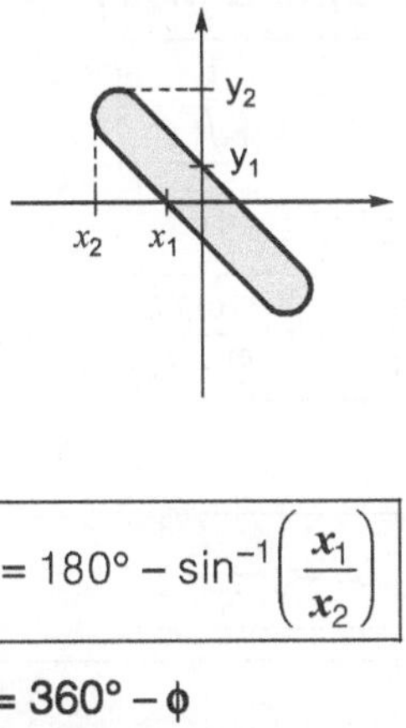

- First possibility

$$\phi = 180° - \sin^{-1}\left(\frac{x_1}{x_2}\right)$$

- Second possibility = 360° – φ

Measurement of Frequency Using Lissajous Pattern

$$\frac{f_y}{f_x} = \frac{\text{(number of intersections of the horizontal line with the curve)}}{\text{(number of intersections of the vertical line with the curve)}}$$

where, f_y = Frequency of signal applied to Y plates

f_x = Frequency of signal applied to X plates

■■■■

High Frequency Measurements

11

Q-meter

It works on the principle of series resonance.

Measurement of the Storage Factor Q

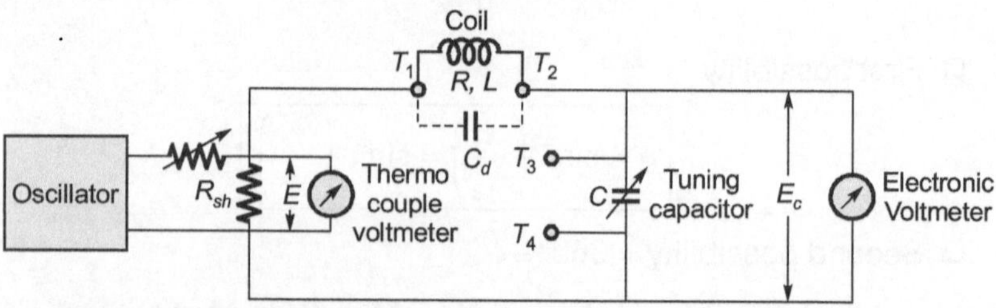

- Resonant frequency of Q-Meter

$$f_0 = \frac{1}{2\pi\sqrt{LC}}$$

- Measured value of Q

$$Q_m = \frac{\omega_0 L}{R + R_{sh}}$$

- True value of Q

$$Q_t = \frac{\omega_0 L}{R}$$

$$Q_t = Q_m\left(1+\frac{R_{sh}}{R}\right) = Q_m\left(1+\frac{C_d}{C}\right)$$

where, R = Resistance of coil
L = Inductance of coil
R_{sh} = Shunt resistance
C = Tuning capacitance
C_d = Distributed or self-capacitance

- Measurement of inductance

$$L = \frac{1}{4\pi^2 f_o^2 C}$$

❑ Measurement of effective resistance

$$\boxed{R = \frac{\omega_o L}{Q_t}}$$

Measurement of Distributed or self-capacitance

❑ Resonance frequency

$$\boxed{f_1 = \frac{1}{2\pi\sqrt{L(C_1 + C_d)}}} \; ; \; \boxed{f_2 = \frac{1}{2\pi\sqrt{L(C_2 + C_d)}}}$$

when, $f_2 = nf_1$

then, $$\boxed{C_d = \frac{C_1 - n^2 C_2}{n^2 - 1}}$$

where, C_1 = Tuning capacitance at frequency f_1
C_2 = Tuning capacitance at frequency f_2

Measurement of Unknown Capacitance C_x

Adjust capacitor $C = C_1$ to get resonance frequency f_1 with unknown capacitance C_x in parallel.

$$f_1 = \frac{1}{2\pi\sqrt{L(C_x + C_1)}} \quad \text{...(i)}$$

Now remove C_x and again adjust $C = C_2$ to get same resonance frequency f_1

$$f_1 = \frac{1}{2\pi\sqrt{LC_2}} \quad \text{...(ii)}$$

By equating equation (i) and (ii),

$$\boxed{C_x = C_2 - C_1}$$

■■■■

A Handbook on

Instrumentation Engineering

8

Sensors and Industrial Instrumentation

CONTENTS

Introduction

1

Transducers

- A transducer is a device which converts one kind of energy to the other.
- The conversion can be to or from electrical, electromechanical, photonic, photo-voltaic or any other form of energy.

Sensors:

A sensor is a type of transducer which detects or measures a physical property and records, indicates or otherwise responds to it.

Classification of Transducers

Based on Principle of Transduction

- The transduction depends on how they transduce the input quantity into the resistance, capacitance or inductance which can be measured by bridge or any other electronic circuitry.

Electrical parameter and class of transducer	Principle of Operation and nature of device	Typical Application
	Passive transducers (externally powered)	
1. Resistance		
Potentiometer device	Positioning of the slider by an external force varies the resistance in a potentiometer or a bridge circuit.	Pressure, displacement
Resistance strain gauge	Resistance of a wire or semiconductor is changed by elongation or compression due to externally applied stress.	Force, torque, displacement
Pirani gauge or hot wire meter	Resistance of a heating element is varied by convection cooling of a steam of gas.	Gas flow, gas pressure

Resistance thermometer	Resistance of pure metal wire with a large positive temperature co-efficient of resistance varies with temperature.	Temperature, radiant heat
Thermistor	Resistance of certain metal oxides with negative temperature coefficient of resistance varies with temperature.	Temperature, flow
Resistance hygrometer	Resistance of a conductive strip changes with moisture content.	Relative humidity
Photoconductive cell	Resistance of the cell as a circuit element varies with incident light.	Photosensitive relay
2. Capacitance		
Variable capacitance pressure gauge	Distance between two parallel plate is varied by an externally applied force.	Displacement, pressure
Capacitor microphone	Sound presssure varies the capacitance between a fixed plate and a movable diaphram.	Speech, music, noise
3. Inductance		
Magnetic circuit transducter	Self-inductance or mutual inductance of a.c. excited coil is varied by changes in the magnetic circuit.	Pressure, displacement
Reluctance pick up	Reluctance of the magnetic circuits is varied by changing the position of the iron core of coil.	Pressure, displacement, vibration, position
Differential transformer	The differential voltage of two secondary windings of a transformer is varied by positioning the magnetic core thorugh an externally applied force.	Pressure, force, displacement, position

Eddy current gauge	Inductance of a coil is varied by the proximity of an eddy current plate.	Displacement, thickness
Magnetostriction gauge	Magnetic properties are varied by pressure and stress.	Force, pressure, sound
4. Voltage and Current		
Hall effect pickup	A potential difference is generated across a semiconductor plate (germanium) when magnetic flux interacts with an applied current.	Magnetic flux, current power
Ionization chamber	Electron flow induced by ionization of gas due to radioactive radiation.	Particle counting radiation
Photoemissive cell	Electron emission due to incident radiation upon photoemissive surface.	Light and radiation
Photomultiplier tube	Secondary electron emission due to incident radiation on photosensitive cathode.	Light & radiation, photosensitive relays
Self-generating transducers (no external power) active transducer		
Thermocouple and thermopile	An emf is generated across the junction of two dissimilar metals or semiconductors when that junction is heated.	Temperature, heat, flow, radiation
Moving coil generator	Motion of a coil in a magnetic field generates a voltage.	Velocity, vibrations
Piezoelectric pick-up	An emf is generated when an external force is applied to certain crystalline materials, such as quartz.	Sound, vibrations, acceleration, pressure changes.
Photovoltaic	A voltage is generated in a semiconductor junction device when radiant energy stimulates the cell.	Light meter, solar cell

Primary and Secondary Transducers

- The transducers which convert the physical quantity into the mechanical displacement are called as the Primary transducers. eg. Bourdon Tube, Diaphragm, Bellows, etc.
- The transducers which convert the mechanical displacement into the electrical output are called as the Secondary transducers. eg. LVDT.

Active and Passive Transducers

- Active transducers are those transducers which do no need any external power source to produce output.
- They produce the electrical power output themselves.
  eg. Piezoelectric transducers, thermocouple, photovoltaic cell
- Passive transducers are those transducers which draws power from an external source to produce their output.
  eg. Resistive, Capacitive, Inductive transducer.

Analog and Digital Transducers

- The transducers which convert the input quantity into an analog output (i.e. the output which is continuous with time) are called analog transducers.
  eg. LVDT, Strain gauge etc.
- The transducers which convert the input quantity into an output which is in the form of pulses are called as the digital transducers.
  eg. Shaft encoders.

Inverse Transducers

- Inverse Transducers are the device which convert the electric quantity into mechanical quantity.
  eg. Piezoelectric crystal.

Characteristics of Transducers

Transfer Function

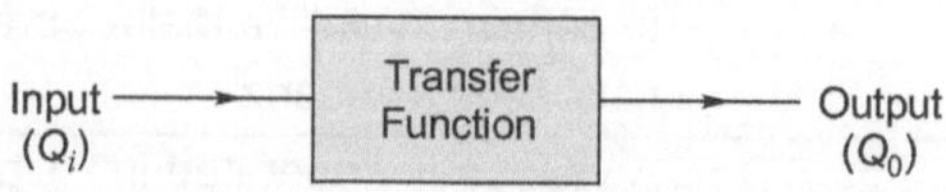

- The transfer function is the ratio of Laplace transform of output to the Laplace transform of input.

Sensitivity

- Sensitivity is defined as the ratio of the change in the output to the change in the input for a small time.

$$\text{Sensitivity} = \frac{\Delta Q_0}{\Delta Q_i}$$

Accuracy

- Accuracy of a measuring system is defined as the closeness of the instrument output to the true value of the measured quantity.
- The accuracy of the instruments (which represents really its inaccuracy) can be specified in either of the following forms :
  1. Percentage of true value

$$\frac{M.V - T.V}{T.V} \times 100$$

where
M.V. = measured value
T.V = true value

  2. Percentageof full-scale deflection

$$\frac{M.V - T.V}{\text{FSD}} \times 100$$

FSD = full-scale deflection

Precision

- Precision is defined as the ability of the instrument to reproduce a certain set of readings within a given accuracy.

Resolution

- It is defined as the smallest increment in the measured value that can be detected with certainty by the instrument.
- The least count of any instruement is taken as the resolution of the instrument.

Threshold

- It is defined as the minimum value of input below which no output can be detected.

Range and Span

- The range of the instrument is specified by the lower and upper limits in which it is designed to operate for measuring, indicating or recording the measured variable.
- The algebraic difference between the upper and lower range values is termed as the **span** of the instrument.

■■■■

Resistive Transducers

2

Resistive Transducers

- Resistive transducers are used for the measurement of a physical quantity in terms of the change in resistance produced by them. Expression for the resistance of a conductor is given as:

$$R = \frac{\rho l}{A}\ \Omega$$

where, ρ = Resistivity of the conductor (Ω-m)

l = Length of the conductor (m)

A = Area of the cross-section of the conductor (m^2)

- Resistance of the conductor can be changed by change in either of the given three parameters (i.e. resistivity, length or the area of cross-section).

Potentiometer Transducers

- These are the passive transducers used for the measurement of translatory and the rotational motions.
  Depending on the type of motion, they can be classified as:

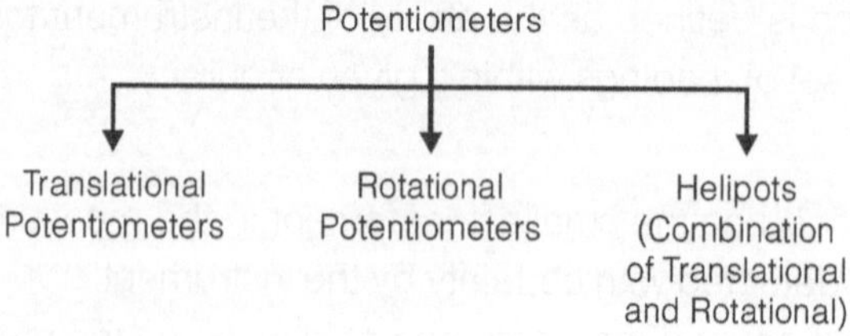

Translational potentiometer Transducers

- The output signal of these transducers is a DC or AC voltage which is proportional to the potentiometer resistance that is proportional to the slider linear displacement

$$V_O = \frac{x_i}{x_t} \times V_i$$

$$V_0 = \frac{R_x}{R_T} \times V_i$$

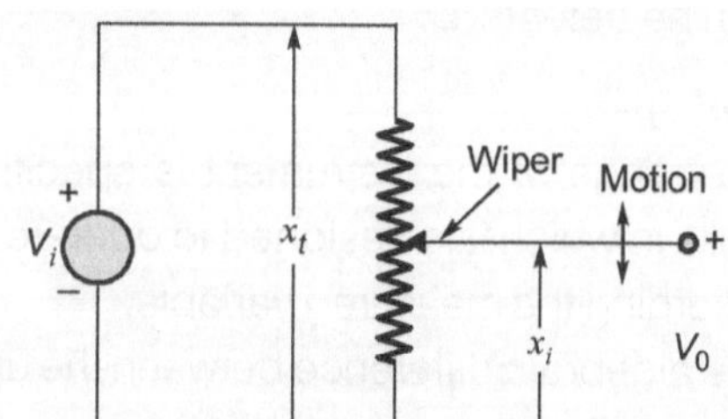

$$\text{Sensitivity (S)} = \frac{V_0}{x_i} = \frac{V_i}{x_t} \text{ V/m}$$

where, V_i and V_0 = input and output voltages respectively (V).
x_t = total length of the translational potentiometer (m).
x_i = displacement of the wiper from its zero position (m).
R_x = resistance at output terminal of the potentiometer (Ω).
R_t = resistance of the potentiometer (Ω).

Rotational Potentiometer Transducers

- The output signal of these transducers is a DC or AC voltage which is proportional to the potentiometer resistance that is proportional to the slider angular displacement

$$V_O = \frac{\theta_i}{\theta_t} \times V_i$$

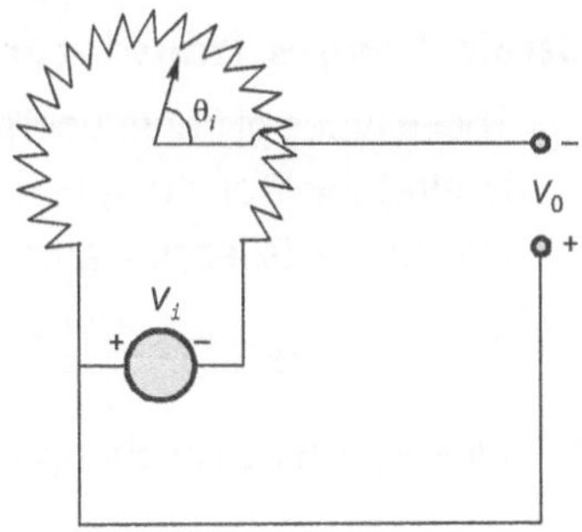

Where, θ_i = Input Angular displacement of wiper in degrees
θ_t = Total wiper angular displacement in degrees

$$\text{Sensitivity } (S) = \frac{V_0}{\theta_i} = \frac{V_i}{\theta_t} \text{ V/degree}$$

Note: ....

- Pot is a zero order instrument.
- For measurement of angular and translational displacement helipot is used.
- Material used for pot are platinum, Nickel chromium, Nickel Copper etc.

....

Loading Effect in Potentiometer

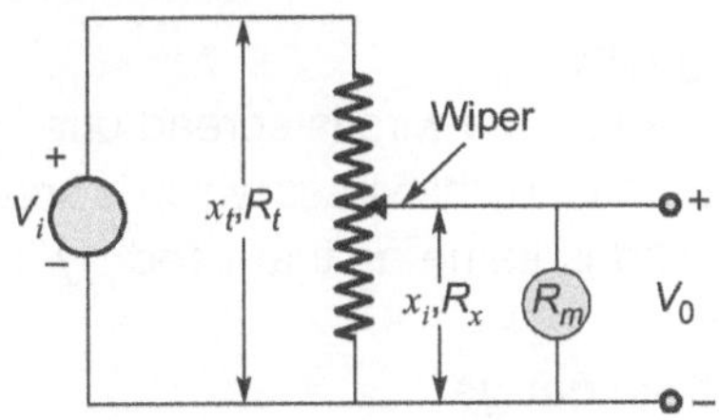

Case I $\quad R_m = \infty \Rightarrow V_0 = \frac{x_i}{x_t} V_i = kV_i$

Case II $\quad R_m = \infty \Rightarrow V_0 = \frac{k}{k(1-k)\left(\frac{R_t}{R_m}\right)+1} V_i$

Note: ....

- For higher linearity of the potentiometer, the potentiometer resistance (Rp) must be as low as possible, and meter resistance (R_m) must be high (ideally infinite).
- For higher sensitivity of the potentiometer, the potentiometer resistance (R_p) must be high.

....

Strain Gauges Transducers

These types of transducers are based on the principle that if a conductor is stretched or compressed, its resistance will change, because of change in its length, area and resistivity.

$$\text{Gauge factor}(G_F) = \frac{\text{Per unit change in Resistance}}{\text{Per unit change in Length}} = \frac{\Delta R / R}{\Delta L / L} = \frac{\Delta R / R}{\varepsilon}$$

Where, ΔR being change in resistance R due to axial strain(ε)

$$G_F = 1 + 2\nu + \frac{\Delta\rho/\rho}{\varepsilon}$$

Where ε = Axial Strain = $\frac{\Delta l}{L}$

ν = Poisson ratio

$$= \frac{\text{Lateral strain}}{\text{longitudinal strain}} = \frac{\text{Transverse strain}}{\text{Axial strain}}$$

$$V = \frac{-\partial D / D}{\partial L / L} = -\frac{\varepsilon_t}{\varepsilon}$$

Types of Strain Gauges

Unbonded metal wire strain gauges

- In such strain gauges a resistance wire is stretched between the two frames, one being the moving frame and the other one is fixed.
- They are made up of various alloys like copper-nickel, chrome-nickel or nickel-iron.

Bonded wire strain gauge

- In such strain gauge, the wire is spread uniformly and hence they can be used to measure the stress which is spread uniformly over it.
- The material used is same as that used by unbonded metal wire strain gauge.

Bonded metal foil strain gauge

- These type of strain gauges are the extension to the metal wire strain gauges.
- They have large surface area hence they have large heat dissipation capacity. So they are used at the higher operating range of temperature.

Semiconductor strain gauge

- These type of strain gauge have higher sensitivity and higher gauge factor, also they have higher accuracy.
- For P type semiconductor strain gauge G_F is positive and for N type semiconductor strain gauge G_F is negative

Note:

- Gauge factor for metals is nearby 2.
- Poisson ratio of metals lies between 0 to 0.5.
- For metalic and P type strain gauges G_F is positive that is on tensile stress resistance increases $(R+\Delta R)$ where as on the compressive stress resistance decreases $(R-\Delta R)$.
- For N type strain gauges G_F is negative that is on tensile stress resistance decreases $(R-\Delta R)$ where as on the compressive stress resistance increases $(R+\Delta R)$.

Signal Conditioning for Strain Gauges

- The signal conditioning of strain gauge involves the use of the Wheatstone Bridge Circuit.
- The Wheatstone Bridge can operate in two modes.

Null Mode (No deflection)	Deflection Mode
• More accurate	• Accuracy is low
• Used for static input only	• Used for both static and dynamic inputs
• Less used	• More used
• Slow in operation	• Fast in operation

NULL MODE

- In the null-mode the strain gauge behaves as one of the arm of the Wheatstone bridge circuit.
- The resistance R_2 (kept variable) is varied until the galvanometer (G) shows a zero or null deflection in the voltmeter.
- This change in the value of R_2 (i.e. ΔR_2) is calibrated in term of the applied force.

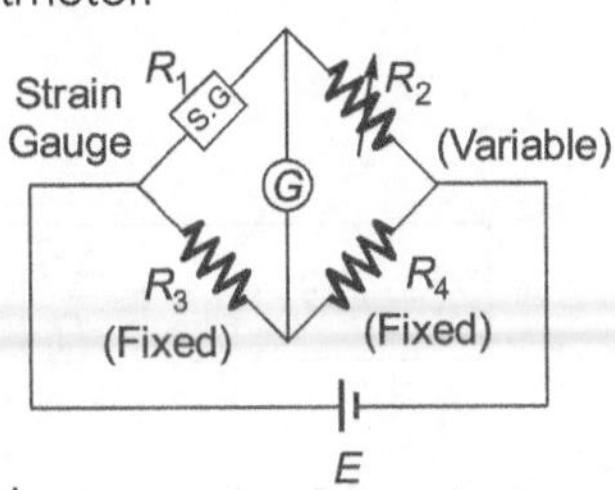

$$F \propto \varepsilon_a \propto \Delta R_1 \propto \Delta R_2$$

$$\therefore \quad \Delta R_2 \propto \varepsilon_a \propto F$$

where,

F = applied force; ε_a = applied axial strain

ΔR_2 = change in resistance R_2, to produce null deflection

DEFLECTION MODE

Quarter Bridge

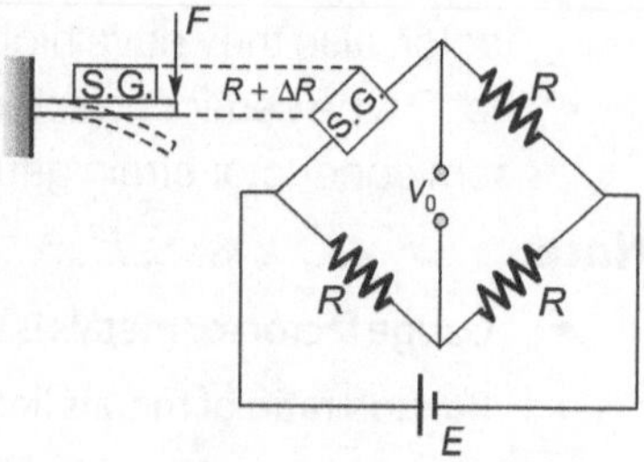

- One strain gauge is used
- This configuration has temperature related error which is cancel out by the connecting dummy strain gauge in the adjacent arm of the active strain gauge in the quarter bridge arrangement.

$$V_0 = \frac{E}{4} G_F . \varepsilon_a = \frac{E}{4} \frac{\Delta R}{R}$$

Half Bridge

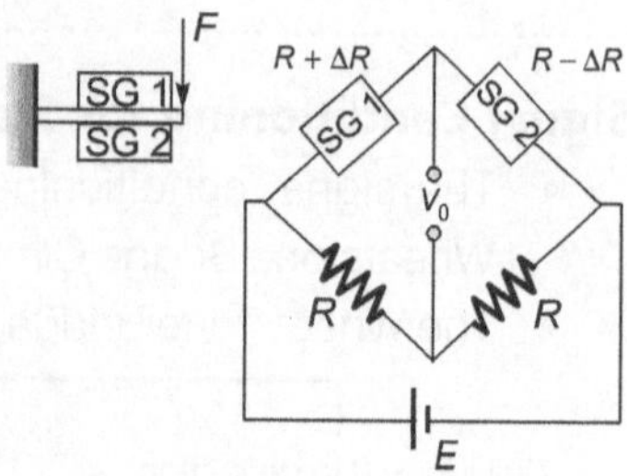

- Two strain gauges are used
- Sensitivity is improved than the quarter

$$S_{\text{Half Bridge}} = 2\, S_{\text{Quarter Bridge}}$$

- Temperature error is removed

$$V_0 = \frac{E}{2} G_F . \varepsilon_a = \frac{E}{2} \frac{\Delta R}{R}$$

SG1 ⇒ Tenslie Force ⇒ Positively strained → R + ΔR
SG2 ⇒ Compressive Force ⇒ Negatively strained → R–ΔR

Full Bridge

- Four strain gauges are used
- Sensitivity of full bridge is more than the half bridge.

$$S_F : S_H : S_Q = 4 : 2 : 1$$

- Temperature error is removed.
- This method is always preferred for the stress and force measurement.

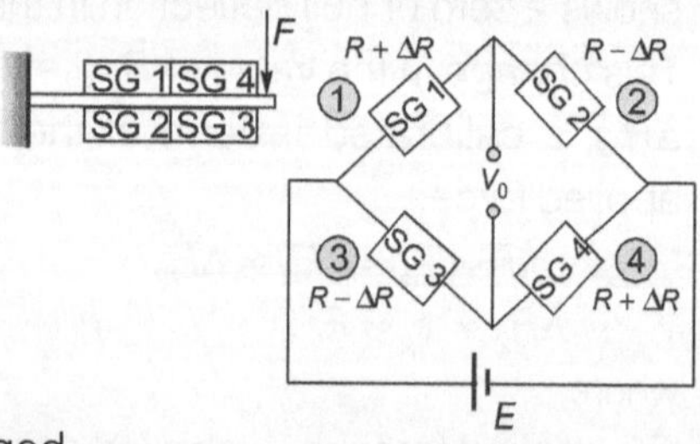

$$V_0 = E G_F . \varepsilon_a = E \frac{\Delta R}{R}$$

- 1 and 2 can be interchanged.
- 2 and 3 can be interchanged.

Note: ....

- For getting maximum sensitivity connect the strain gauges with opposite strain in the adjacent arm (i.e. in 1, 2 or 3, 4) and with same strain in opposite arm (i.e. in 1, 4 or 2, 3).

....

Strain Measurement By Poissions Method

- In this method two strain gauges are bounded at right angles to each other (i.e. at 90°).

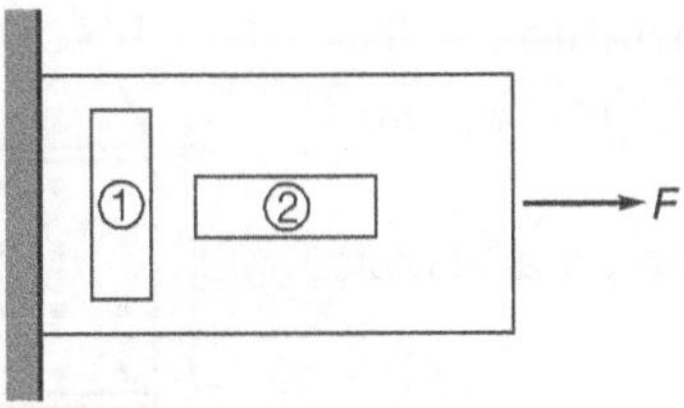

(i) if force is away ($\longrightarrow F$)

① ⇒ Compressive

② ⇒ Tensile

(ii) if force is towards ($\longleftarrow F$)

① ⇒ Tensile

② ⇒ Compressive

The output voltage is given as

$$V_0 = \frac{E}{4} G_f \cdot \varepsilon (1 + \mu)$$

$\mu \Rightarrow$ Poissions constant

The output voltage is increased $(1 + \mu)$ times.

■■■■

Capacitive Transducers

3

Capacitive Transduces

- Capacitive transducers are the transducers whose capacitance is changed by the change in the physical variables like displacement, force, and pressure.
- Capacitance of a parallel plate capacitor.

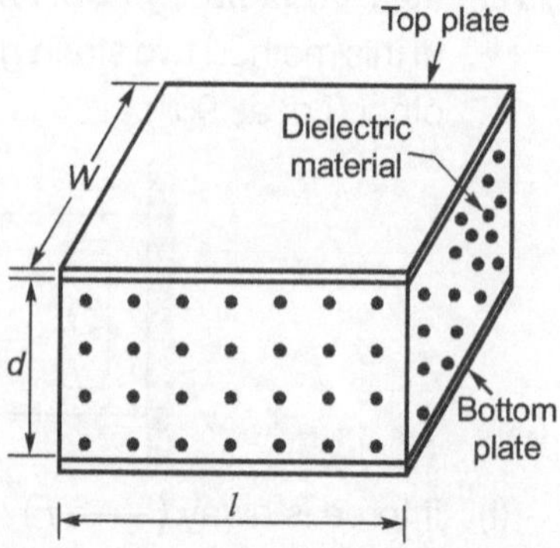

$$C = \frac{A\varepsilon_r\,\varepsilon_0}{d}$$

A = overlapping area of the plates (m^2).
d = distance between two plates (m).
ε_0 = permittivity of free space, (8.85×10^{-12}) F/m.
ε_r = relative permittivity

Transducers Using Change in Area of Plates

Sheet type capacitor

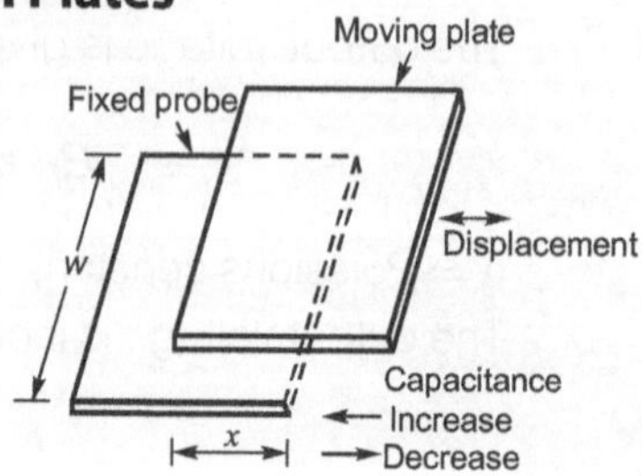

$$C = \frac{A\varepsilon}{d} = \frac{(wx)\times\varepsilon}{d}\ F$$

x = length of overlapping part of plates (m).
w = width of overlapping part of plates (m).

$$\text{Sensitivity}(S) = \frac{\partial C}{\partial x} = \frac{\varepsilon w}{d}\ \text{F/m}$$

Note:

- This type of capacitive transducer is suitable for measurement of linear displacement ranging 1 mm to 10 mm.

Cylindrical Capacitor

$$C = \frac{2\pi\varepsilon_0 x}{\text{Log}_e\left[\frac{D_2}{D_1}\right]}$$

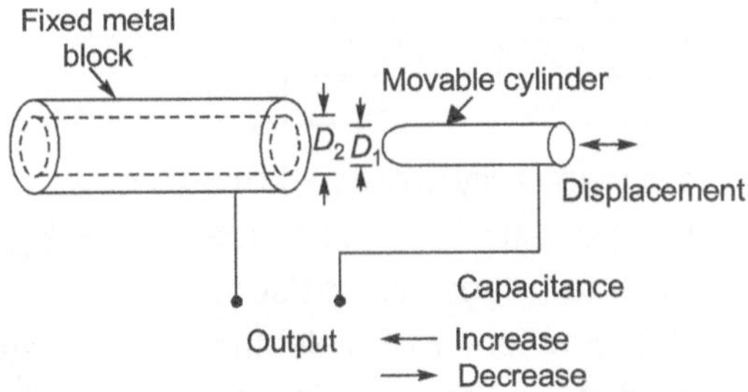

x = Length of overlapping part of cylinder (m)

D_2 = inner diameter of outer cylindrical electrode (m)

D_1 = Outer diameter of inner cylindrical electrode (m)

$$\text{Sensitivity } (S) = \frac{\partial C}{\partial x} = \frac{2\pi\varepsilon}{\text{Log}_e\left[\frac{D_2}{D_1}\right]} \text{ F/m}$$

Measurement of Angular Displacement

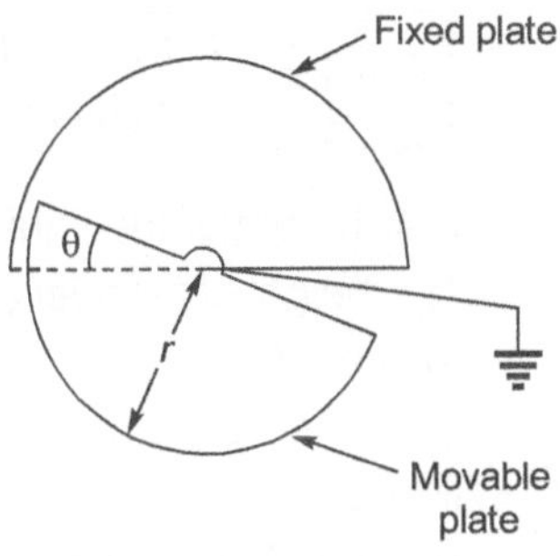

$$C = \frac{\varepsilon\theta r^2}{2d} \text{F}$$

θ = Angular displacement in rad.

r = radius of the movable plate

$$\text{Sensitivity } (S) = \frac{\partial C}{\partial \theta} = \frac{\varepsilon r^2}{2d} \text{F/rad}$$

Transducer Using Change in Distance Between Plates

- The capacitance can be varied by the change of distance 'd' between the plates.

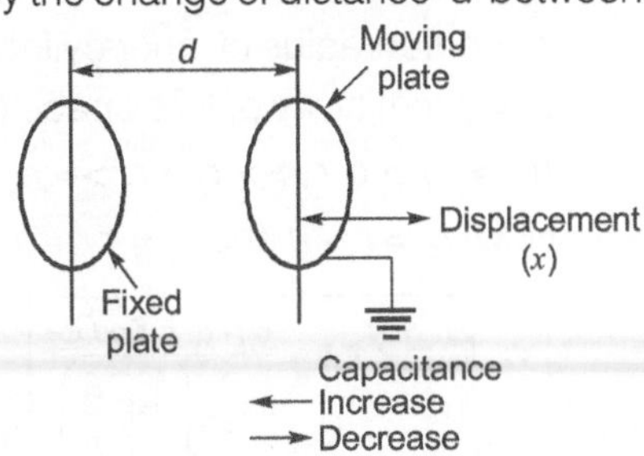

$$C = \frac{\varepsilon A}{d}$$

$$\text{Sensitivity } (S) = \frac{\partial C}{\partial d} = \frac{-\varepsilon A}{d^2}$$

Variation of Dielectric Constant Between Plates

$$C = \frac{\varepsilon_0 w l_1}{d} + \frac{\varepsilon_0 \varepsilon_r w l_2}{d}$$

ε_r = permittivity of dielectric medium
ε_0 = permittivity of free space
Change in capacitance due to displacement x of dielectric medium in the direction as shown in the figure

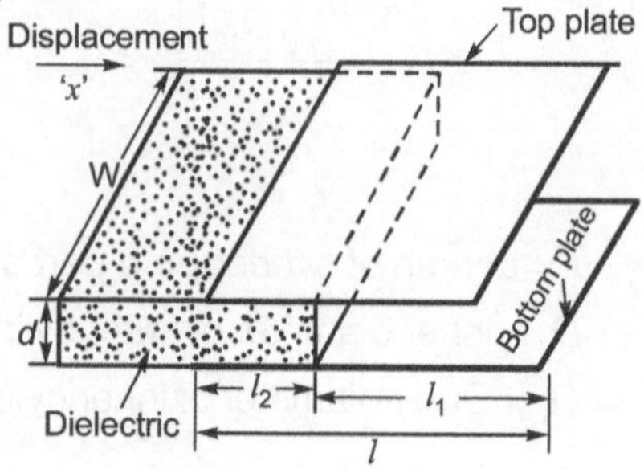

$$\Delta C = \varepsilon_0 \frac{w x}{d}(\varepsilon_r - 1)$$

$$\text{Sensitivity}(S) = \frac{\partial C}{\partial x} = \varepsilon_0 \frac{w}{d}(\varepsilon_r - 1)$$

Measurement of liquid level by varying dielectric constant

$$C = 2\pi\varepsilon_0 \frac{\varepsilon_1 h_1 + \varepsilon_2 h_2}{\log_e(r_2 / r_1)}$$

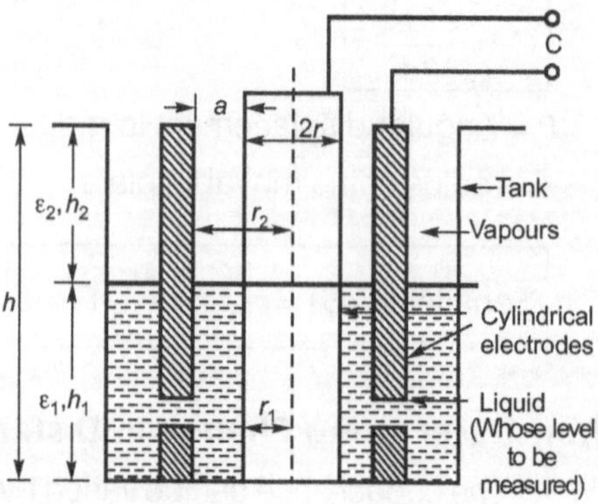

h_1 = height of the liquid (m).
h_2 = height of cylinder above the liquid (m).
ε_1 = relative permittivity of liquid.
ε_2 = relative permittivity of vapour above liquid.
r_2 = inner radius of outer cylinder (m).
r_1 = outer radius of inner cylinder (m).
ε_0 = permittivity of free space (F/m).
$h >> r_2$ and $r_2 >> r_2 - r_1 >> a$
Now, $r_2 = r + a$ and $r_1 = r$

$$\therefore\ C = 2\pi\varepsilon_0 \frac{\varepsilon_1 h_1 + \varepsilon_2 h_2}{\log_e\left(1 + \frac{a}{r}\right)}$$

Differential Arrangement

- This arrangement is used to make linear relationship between change in capacitors (ΔC) and displacement x applied to the moving plate.

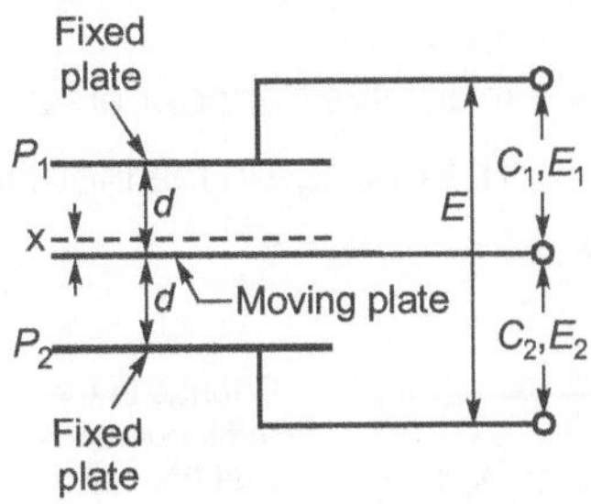

$$\Delta C = C_1 - C_2 = C_0 \frac{x}{d}$$

C_0 = nominal capacitance value (no displacement is applied)
x = displacement of moving plate

$$S = \frac{\partial C}{\partial x} = \frac{C_0}{d} \text{ pF/m}$$

Differential output voltage $\Delta E = E_1 - E_2 = E \frac{x}{d}$

Frequency response on Capacitance Transducers

$$\frac{E_0(s)}{X_i(s)} = \frac{\tau s}{1 + \tau s}$$

$$\text{Magnitude}(M) = \left|\frac{E_0(J\omega)}{X_i(J\omega)}\right| = \frac{\omega\tau}{\sqrt{1+(\omega\tau)^2}}$$

$$\phi = \frac{\pi}{2} - \tan^{-1}(\omega\tau) \text{ rad}$$

Note:

- Under static inputs $\omega = 0$

$$\therefore \quad M = \frac{0}{\sqrt{1+0}} = 0$$

  Thus, capacitive transducers are not used under static conditions.
- Capacitive transducers have high input impedance, Thus less loading.
- Capacitive transducers are used to measure dynamic inputs.
- Capacitive inputs transducers have frequency response same, as of high pass filter.
- Capacitive transducers have high sensitivity

■■■■

4 Inductive Transducers

- Inductive transducers are used for the measurement of displacement.
- The change in the inductance can be caused by three factors.
  - Change in the Self inductance of the coil
  - Change in the Mutual inductance
  - Production of eddy current which vary inductance

Inductive Transducers based on the change of self inductive

Self inductance of the coil (L) is given by

$$L = N^2\mu\left(\frac{A}{l}\right)$$

where,
N = number of turns in the coil
A = area of cross-section of the coil (m^2)
l = length of the coil (m)
μ = permeability of the medium in and around the coil. (H/m)

The change in inductance can be brought by bringing change in any of the above physical factor (i.e., N, A, l or μ).

Inductive Transducers based on the change of Mutual Inductance

The mutual inductance between the two coils L_1 and L_2 is given by

$$M = k\sqrt{L_1 L_2}$$

where,
L_1 = inductance of coil 1
L_2 = inductance of coil 2
k = coefficient of coupling between the coils

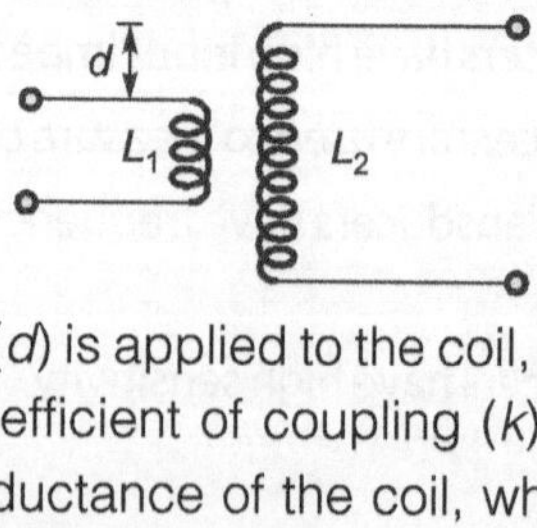

When the displacement (d) is applied to the coil, the inductance L_1 and L_2 remain fixed, but the coefficient of coupling (k) changes leading to the change in the mutual inductance of the coil, which can be calibrated in terms of the magnitude of displacement.

Inductive Transducers based on the change of inductance due to production of eddy currents

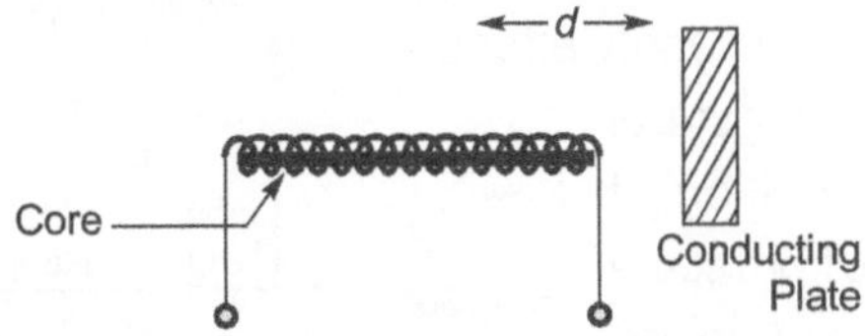

- In this method when the conducting plate is nearer to the coil, the lesser will be the inductance of the coil, as the plate is taken away from the coil, the inductance of the coil increases.

Linear Variable Different Transformer (LVDT)

- When the primary winding is excited by the alternating source, it produces an alternating magnetic field, which induces an alternating voltage to the secondary windings by the transformer action.
- Based on the movement of the core, three cases arise.

Case 1

When there is no movement of arm

$E_{S1} = E_{S2}$

$E_0 = E_{S1} - E_{S2} = 0$

Case 2

When the core moves towards side S_1

Hence, $E_{S1} > E_{S2}$

$E = E_{S1} - E_{S2} = +\text{ve}$

Output voltage will be in phase to the primary winding voltage.

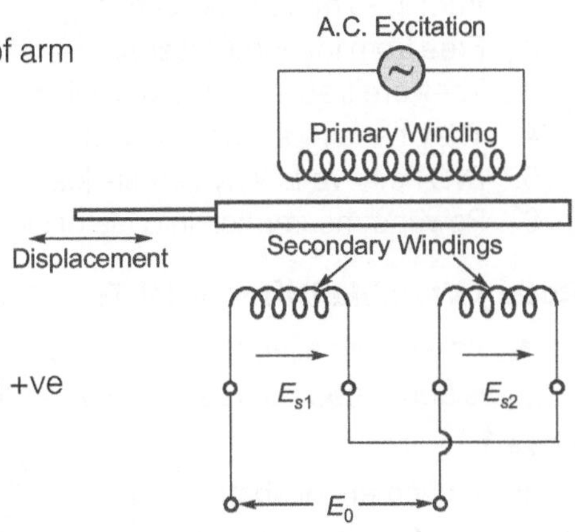

Differential output $(E_0) = (E_{s1} - E_{s2})$

Case 3

When the core moves towards sides S_2

Hence, $E_{S1} < E_{S2}$

$E_0 = E_{S1} - E_{S2} = -\text{ve}$

Output voltage will be 180° out of phase with respect to primary winding voltage.

Note:

The carriers frequency at the source in the LVDT is kept 10 times higher of the highest expected frequency of the input signal.

Output Characteristics of LVDT

- The output is linear for small displacement i.e. upto 5 mm and starts to deviate from linearity after it.
- The curve never starts from zero position, it always starts from some finite value it is on account that at null position there is some residual flux present across the secondary which leads to the residual voltage.

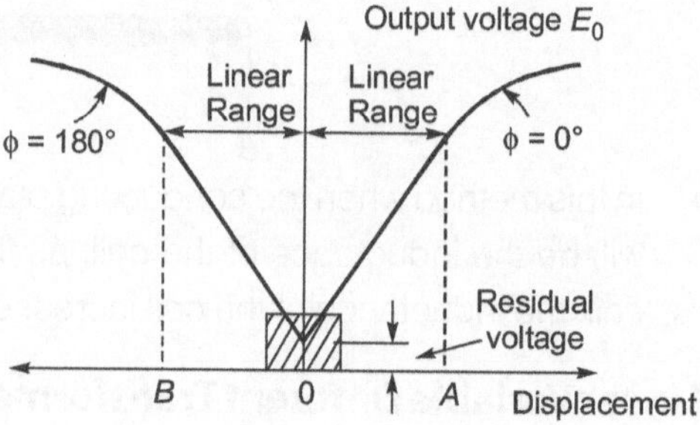

Advantages of LVDT

1. They can be used for the measurement of displacement ranging from 1.25 mm to 250 mm.
2. Free from frictional losses.
3. They are free from all external environmental effects.
4. They have high sensitivity of the order of 40 V/mm.
5. LVDT's have low hysteresis loss.
6. Power consumption in case of LVDT's is low, which is less than 1 W.

Rotary Variable Differential Transformer (RVDT)

- As the core attached to the shaft rotate, the differential output voltage is generated from secondary windings.

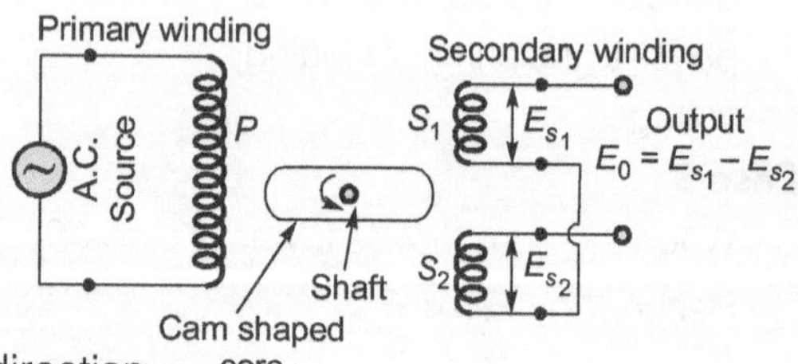

Case 1

No movement of shaft

$$\theta = 0,$$

$$E_{S1} = E_{S2}$$

$$E_0 = E_{S1} - E_{S2} = 0$$

Case 2

Shaft move in clock wise direction

Hence, $E_{S1} > E_{S2}$

$$E_0 = E_{S1} - E_{S2} = +ve$$

Case 3

Shaft move in anti-clockwise direction

Hence, $E_{S1} < E_{S2}$

$$E_0 = E_{S1} - E_{S2} = -ve$$

■■■■

Piezo-Electric Transducers

5

- These are the transducers which work on the principle of piezo-electric effect.
- According to the piezo-electric effect a voltage is induced in a piezo-electric crystal when it is subjected to a mechanical stress.
- The electric field induced is proportional to strain applied.
- The converse of this is also true i.e. the piezo-electric material gets strained when they are subjected to a electric-field.

Example of piezo-electric materials

Quartz, Rochelle Salt, Barium Titanate ($BaTiO_3$), $PbTiO_3$, Lead Zirconoate ($PbZrO_3$), Potassium Dihydrogen Phosphate (KDP), Ammonium Dihydrogen Phosphate (ADP).

Expression For Voltage Induced

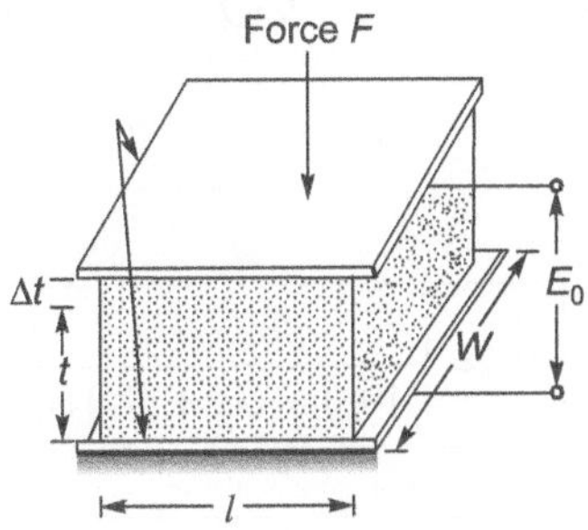

1. Charge, $\boxed{Q = K_q \times F}$

   K_q = Charge sensitivity of crystal (C/N)
   F = Applied force (N)

2. $$\boxed{E_0 = \frac{K_q}{\varepsilon} \times \frac{F}{A} \times t}$$

   (E = O/P voltage, t =Crystal thickness)

3. $\boxed{E_0 = g.P.t}$ (g=Voltage sensitivity, $g = \frac{K_q}{\varepsilon_0 \varepsilon_r}$ V-m/N)

4. $\boxed{E_0 = g.E.\Delta t}$ (Δt = change in thickness of crystal due to applied force F)

Note:

- Quartz and Rochelle Salt are naturally occurring Piezo-electric materials.
- The Piezo-electric effect is direction sensitive, a tensile force produces a voltage of one polarity while a compressive force produces a voltage of opposite polarity

Equivalent Circuit of Piezo-electric Transducers

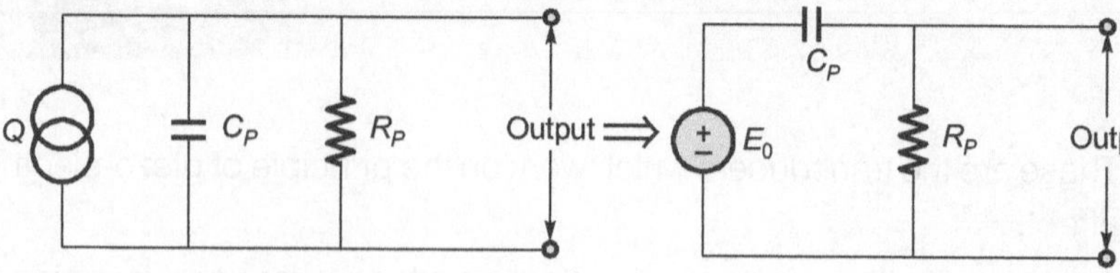

$$Q = K_q x_i \quad \text{Coulomb}$$

$$E_0 = \frac{Q}{C_p} = \frac{K_q \cdot F}{C_p}$$

Where K_q = Charge sensitivity; C/m; x_i = Displacement ; m

Loading Effects of Piezo-electric Transducer

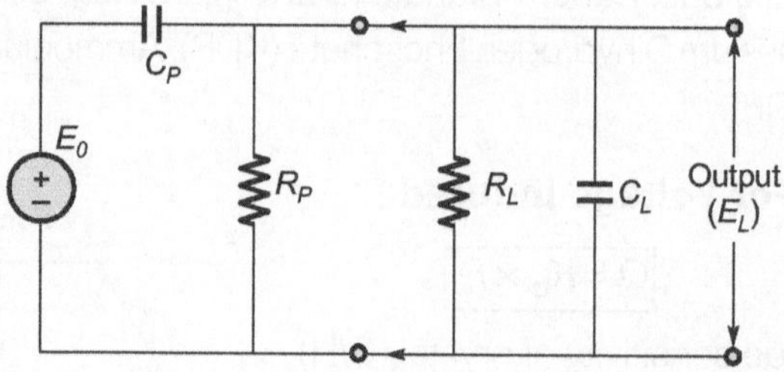

$$E_L = \left[\frac{j\omega C_p R_L}{1 + j\omega(C_p + C_L)R_L}\right] E_0$$

$$|E_L| = \left[\frac{\omega C_p R_L}{\sqrt{1+\omega^2(C_p + C_L)^2 R_L^2}}\right] E_0$$

$$|E_L| = KF\left[\frac{\omega R_L}{\sqrt{1+\omega^2(C_p + C_L)^2 R_L^2}}\right] \quad \left(\because E_0 = \frac{K_q F}{C_p}\right)$$

- At medium and high frequency $\omega^2(C_p + C_L)^2 R_L^2 >> 1$

$$E_L = \frac{E_0 C_p}{(C_p + C_L)}$$

Note: ....

- The piezo-electric transducer responds to the dynamic displacements only do not respond to static displacement.
- The load capacitance is a combination of stray capacitance, capacitance of load and the capacitance of the cable.

....

Frequency Response

Frequency Response of piezo electric transducer is similar to the frequency response of high pass filter.

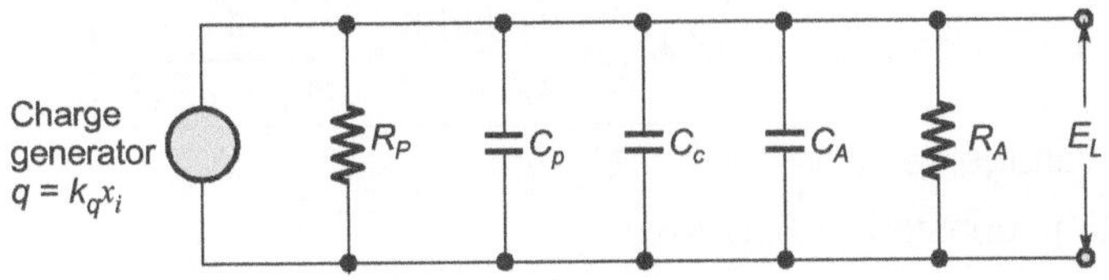

$$\frac{E_L}{X_i(s)} = \frac{K\tau s}{1+\tau s}$$ $$\frac{E_L}{X_i}(j\omega) = \frac{K\tau j\omega}{1+\tau j\omega}$$

$\tau = RC$ sec

where $R = \dfrac{R_A R_P}{R_A + R_P}$, $C = C_p + C_c + C_A$

The amplitude ratio is

$$M = \left|\frac{E_L}{KX_i}(j\omega)\right| = \frac{1}{\sqrt{1+(1/\omega\tau)^2}}$$

The phase shift $\phi = \pi/2 - \tan^{-1} \omega\tau$ rad

At high frequency $\omega >> 1$

$M = 1, \ \phi = 0$

Hence high frequency sensitivity is $K = E_L/x_i$

Charge Amplifier

Charge Amplifier converts charge into voltage directly.

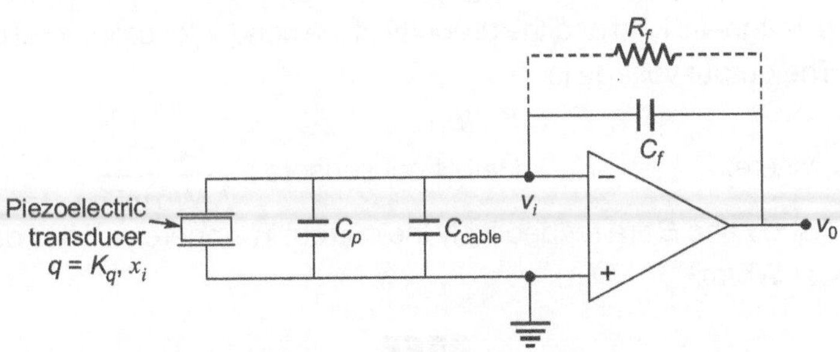

$$\boxed{V_0 = \frac{K_q x_i}{C_f}} \qquad \boxed{\frac{E_0}{X_i(s)} = \frac{K\tau s}{1+\tau s}}$$

The amplitude ratio is $\boxed{M = \frac{1}{K}\left|\left(\frac{V_0(j\omega)}{x_i(j\omega)}\right)\right| = \frac{1}{\sqrt{1+1/\omega^2\tau^2}}}$

K = voltage sensitivity = K_q / C_f V/m

τ = time constant = $R_f C_f$ sec

K_q = charge sensitivity C/m

x_i = input displacement

Advantages of charge amplifier

- The output voltage is proportional to the charge produced by the piezo-electric transducer.
- The amount of charge present is not affected by the cable capacitance.
- The sensitivity, K, as well as the time constant are independent of the capacitance of the crystal and also that of connecting cables.

Hall Effect

- The principle of working of a Hall Effect Sensor is that, if a strip of conducting material carries a current in the presence of a transverse magnetic field as shown in figure, a difference of potential is produced between the opposite edges of the conductor.

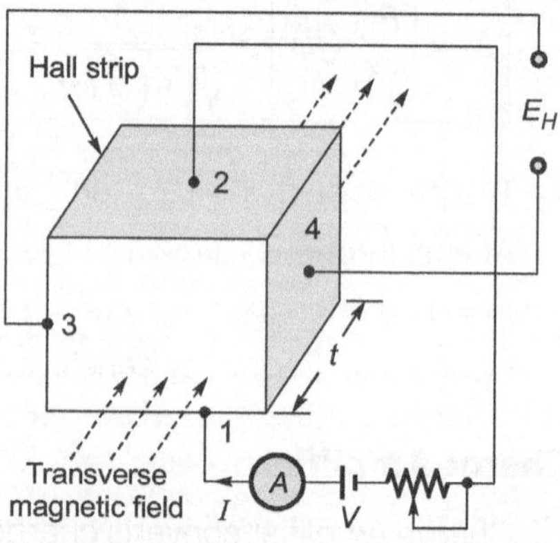

- The magnitude of the voltage depends upon the current, the strength of magnetic field and the property of the conductor called Hall Effect. The output voltage is,

$$E_H = K_H IB/t$$

where, K_H = Hall effect sensors ; $\frac{\text{V-m}}{\text{A-Wb m}^{-2}}$

and I and B are respectively the current in ampere and flux density in Wb/m^2.

■■■■

Measurement of Flow

6

- Flow of material in a process or system can be measured by a variety of methods, depending upon the material and its condition, the type of flow, the volume, the range and the rangeability, the pressure and the temperature range, the accuracy needed and the control required.

Laminar Flow

- When the fluid is flowing through the conduit at very low velocities, the motion of the individual particles are usually along the lines which are parallel to the conduit walls.
- The velocity of the fluid profile thus obtained is greatest at the centre of the conduit and decreases outwards (near the wall).
- This smooth flow of fluid layers one over the above is termed as the **Laminar Flow.**

Turbulant Flow

- When the flow velocity of the fluid is increased more than the critical velocity then the fluid profile no longer remain linear but show a random motion which is called as the **Turbulent Flow.**

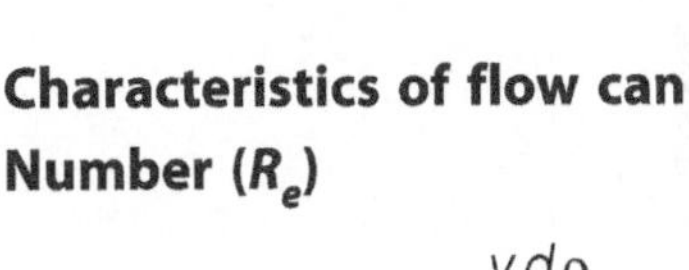

Laminar Flow

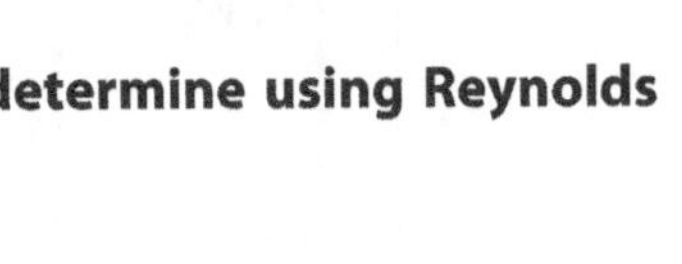

Turbulant Flow

Characteristics of flow can be determine using Reynolds Number (R_e)

$$R_e = \frac{v d \rho}{\mu}$$

where, v = velocity of flow, (m/sec)

d = diameter of pipe, (m)

ρ = density of fluid, (kg/m^3)

μ = viscosity of fluid (Ns/m^2)

Note:

- Fluid flow is considered laminar for Reynolds number upto 2100.
- Fluid flow becomes turbulent for Reynolds number greater than 4000.
- In the transition phase between 2100 and 4000, the fluid flow may be either laminar or turbulent.

Variable Head Flow Meters

- These type of meters are the obstruction type meters as they are placed in between the path of the fluid flow and hence puts restriction in the passage of fluid.
- A pressure differential or head is created across the restriction, and this pressure differential is a measure of the fluid flow.

Venturimeter

- The venturimeter is a device which is used for the measurement of discharge through a pipe.
- Venturimeter is based on the Bernoull's principle, which states that ("for a fluid flow, an increase in speed of the fluid occurs simultaneously with a decrease in pressure or a decrease in the fluids potential energy").

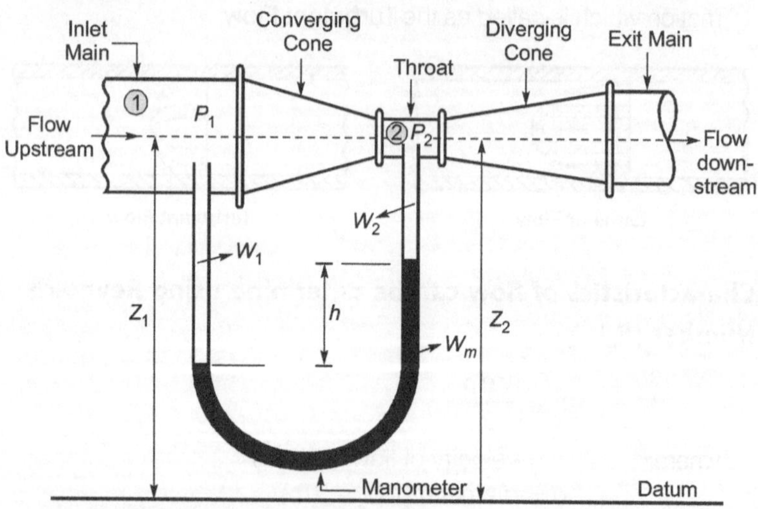

- The venturimeter may be used in any position, horizontal, vertical or inclined.
- The size of the venturimeter is expressed in terms of the inlet and the throat diameter. e.g.: a 300 × 150 mm venturimeter, fits in a pipe of diameter 300 mm and the throat diameter of 150 mm.

Average velocity of fluid flow at the throat of venturimeter is given by

$$V_2 = \frac{C_d}{\sqrt{1-\left(\frac{A_2}{A_1}\right)^2}} \sqrt{\frac{2g}{W}(P_1 - P_2)}$$

Where

P_1 and P_2 = Pressure at inlet and throat respectively (N/m^2).

A_1 and A_2 = Cross-sectional areas of inlet and throat respectively (m^2).

W = Specific weights of fluid. (N/m^3).

In terms of discharge rate

$$Q = \frac{C_d A_2}{\sqrt{1-\left(\frac{A_2}{A_1}\right)^2}} \sqrt{\frac{2g}{W}(P_1 - P_2)}$$

Advantages

- They have low head loss about 10% of the differential pressure head.
- They have a high coefficient of discharge (0.95 to 0.97) on account of low loss, hence they can be used to measure high flow rates in the pipes having diameter of the order of few meters.

Disadvantages

- They have large size which renders them unsuitable for application where space is less.
- Their cost is high due to their large size, also the cost of installation is high.

Note: ....

- The venturimeter may be used in any position, horizontal, vertical or inclined.
- Small size of throat leads to high pressure differential and hence high sensitivity.
- Diameter of throat must not be very small, as it will lead to the cavitation of the fluid in the throat.
- Venturimeter can be used for the measurement of flow of water, gases, liquids, dirty liquids, slurries.

....

Orifice Plates

- The orifice plate is a device used for measuring the volumetric flow rate.
- It uses the same principle as a Venturimeter, i.e. Bernoull's principle.

Types of orifice plates :

Concentric Orifice Plate

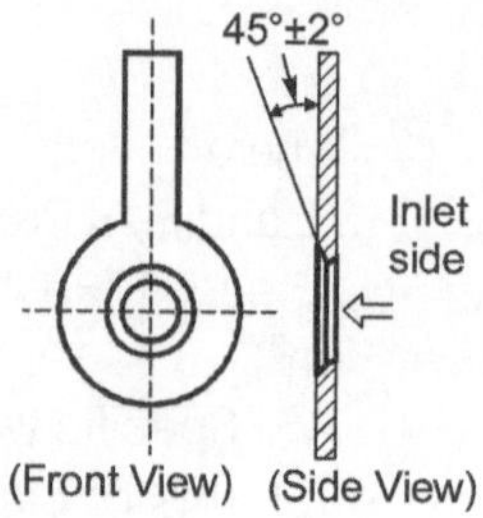

- Concentric orifice plate has a sharp-edged concentric circular hole.
- They are generally made up of stainless steel, but when they are to be protected against corrosion, they are made with nickel and monel.
- The accuracy of the concentric orifice plates is high.

Eccentric Orifice Plate

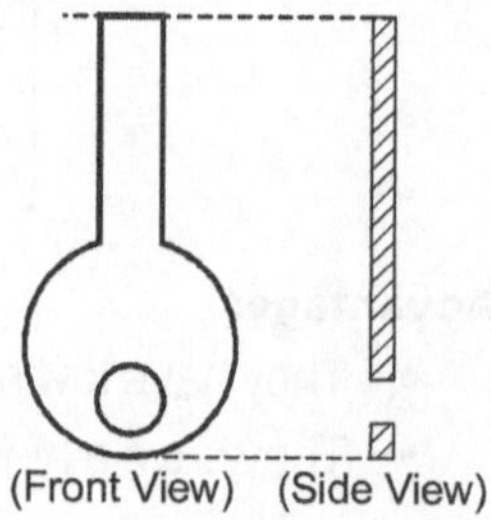

- These plates have bore at the tangent to the plate at the diameter of 0.98 of that of the pipe.
- These plates are useful for the flow measurement of the fluids having suspended particles.

Segmental Orifice Plate

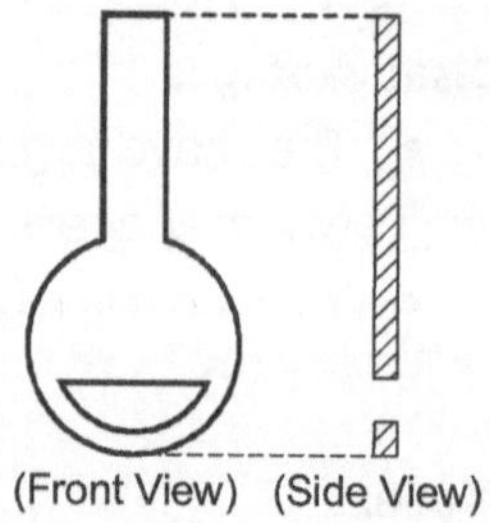

- Segmental Orifice Plate has the boring as shown in the figure.
- They are used for the same application as are the eccentric orifice plates.
- The circular opening has diameter 0.98 to the diameter of the pipe and is placed at the bottom of the pipe.

Quadrant Edge Orifice Plate

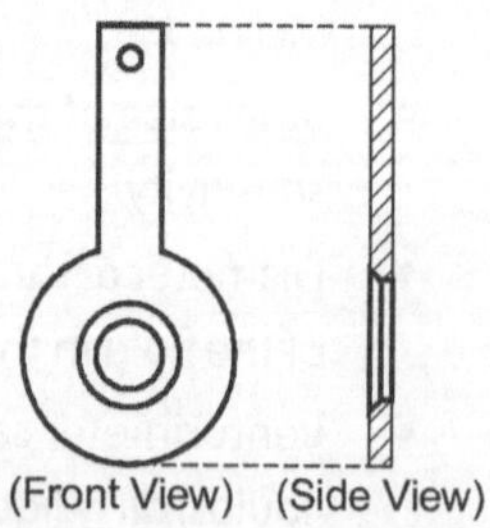

- These plates are so constructed that the edge is rounded so as to form a quarter circle as shown in the figure.
- They are more suited for the reynolds number range from 10,000 or above and down to 3000 to 5000.
- Quadrant plates are used for heavy crudes, syrups and slurries and viscous flows having reynolds number below 10000.

Working of Orifice Plate

- It uses the same principle as a Venturimeter, i.e. Bernoull's principle.
- Vena-contracta is a point where the liquid jet issued from the orifice has the smallest diameter. Hence it is the point where the highest differential pressure obtained.

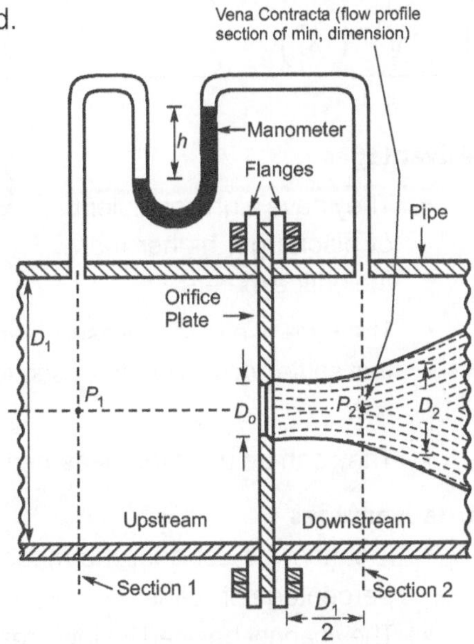

Velocity at Vena contracta is

$$V_{2(actual)} = \frac{C_d}{\sqrt{1-\left(\frac{A_0}{A_1}\right)^2}}\sqrt{\frac{2g}{W}(P_1-P_2)}$$

Discharge rate

$$Q = C_d \frac{A_0}{\sqrt{1-\left(\frac{A_0}{A_1}\right)^2}}\sqrt{\frac{2g}{W}(P_1-P_2)}$$

where, A_0 = Area of the Vena-contracta is, $A_2 = A_0$

All constants have their usual meaning as given in the previous meters.

Advantages

- Low initial cost.
- Smaller in size as compared to the venturi meters.
- Accuracy is high.

Disadvantages

- Lower discharge coefficient (0.6), which is less as compared to venturimeter. **Hence the sensitivity is low as compared to venturimeters.**
- Head loss is high (60 – 70%) is case of orifice plate.
- Inaccuracy may come on account of erosion, corrosion of the plates.

Flow Nozzles

- A flow nozzle is similar to a venturimeter but without a diverging cone.
- Unlike an orifice plate, the jet contraction in case of flow nozzle is less because of its curved shape on the upstream side.

The discharge through a flow nozzle is

$$Q = \frac{C_d}{\sqrt{1-\left(\frac{A_2}{A_1}\right)^2}}\sqrt{\frac{2g}{W}(P_1 - P_2)}$$

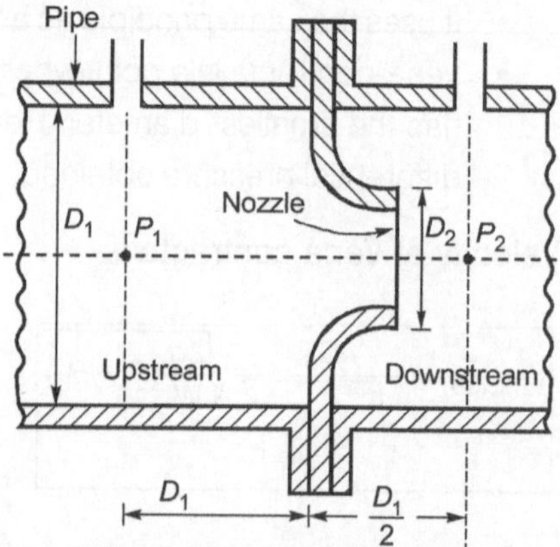

Advantages

- They have high coefficient of discharge, higher than the orifice meters.
- They are useful for measurement of flow of fluids containing solids that settle and also widely used for high pressure/temperature steam flow.
- They can be used for measurement of high flow velocities.

Disadvantages

- It cannot be used for the measurement of flow that contains high percentage of solids.
- They cannot be used for low pressure heads, because they have low pressure recovery.

Pitot Tube

- The principle of working of Pitot Tube is the "transformation of the kinetic energy of the liquid into potential in the form of the static head".
- The tube consists of two tubes which are concentric to each other.

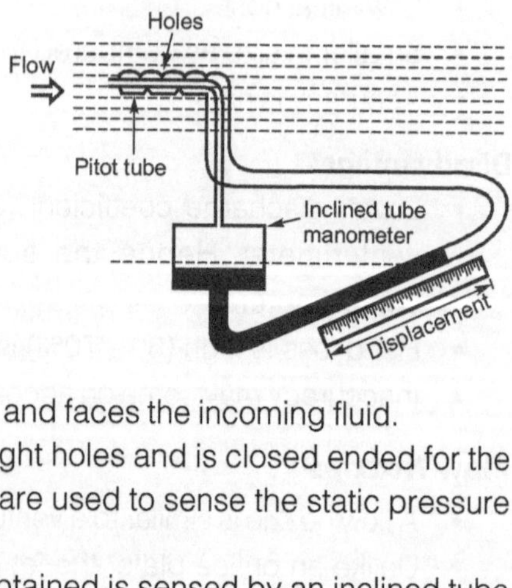

- The inner tube is open ended and faces the incoming fluid.
- The outer tube have four to eight holes and is closed ended for the upstream fluid. But the holes are used to sense the static pressure of the incoming fluid.

  The differential pressure so obtained is sensed by an inclined tube manometer.

$$\text{Velocity } (V) = \sqrt{\frac{2g}{W}(P - P_0)} = \sqrt{\frac{2g}{\rho}(P - P_0)}$$

where, P = pressure at the inlet (stagnation pressure).
P_0 = static pressure
ρ = density of the fluid
v = velocity of fluid flow through the pipe

Advantages

- The loss of head is negligible by the insertion of the pitot tube.
- They are very cheap, and cheaper than the venturimeter, orifice plates and flow nozzles.

Disadvantages

- The differential head produced by them is low, hence their sensitivity is low. They require high flow velocity to produce measurable heads.
- They have small openings which may get clogged if solid particles pass through them.

Variable Area & Variable Head Flowmeters (Weirs & Flumes)

Weirs

- These are the devices which are commonly used for the flow measurement of an open channel or an open stream.
- These type of devices essentially form a dam with a notch type opening at the top to allow the fluid passage through them.
- The weirs can be classified on the basis of the shape of their notches, they may be rectangular, V and trapozoidal notches.

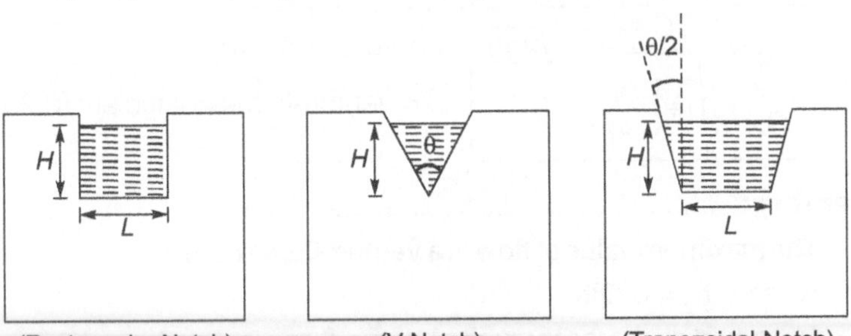

(Rectangular Notch) (V-Notch) (Trapezoidal-Notch)

Discharge through rectangular notch

$$Q = \frac{2}{3} C_d L \sqrt{2g}\, H^{1.5}$$

Discharge through *V*-notch $Q = \frac{8}{15} C_d \sqrt{2g}\, H^{2.5} \tan\frac{\theta}{2}$

Discharge through trapozoidal notch

$$Q = \underbrace{\frac{2}{3} C_d L \sqrt{2g}\, H^{1.5}}_{\text{Rectangular part}} + \underbrace{\frac{8}{15} C_d \sqrt{2g}\, H^{2.5} \tan\frac{\theta}{2}}_{V\text{-Notch part}}$$

where, C_d = Coefficient of discharge

H = height of the liquid in the notch.

L = length of the notch in case of rectangular and trapozoidal notch.

Note: ....

- Weirs are primarily used in water works and waste and sewage systems. They can be used to measure flow ranging from a flow liters per minute to millions of liters per day.

....

Flumes

The figure below shows a Venturi-flume, with the sides arranged to form an inlet and the throat, the throat is thinner in size than the inlet, so as to form a venturi like shape.

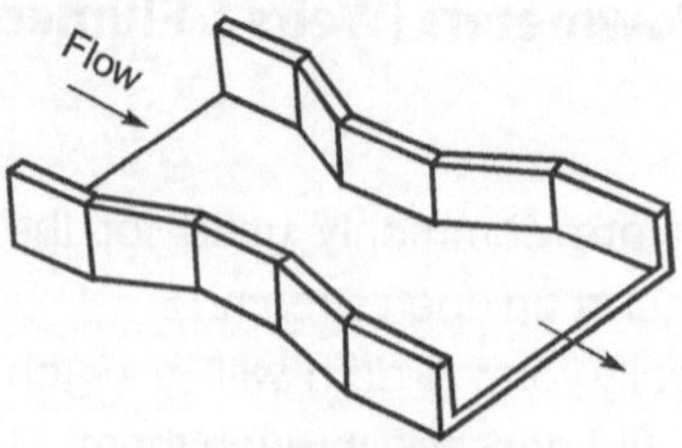

(Venturi Flume)

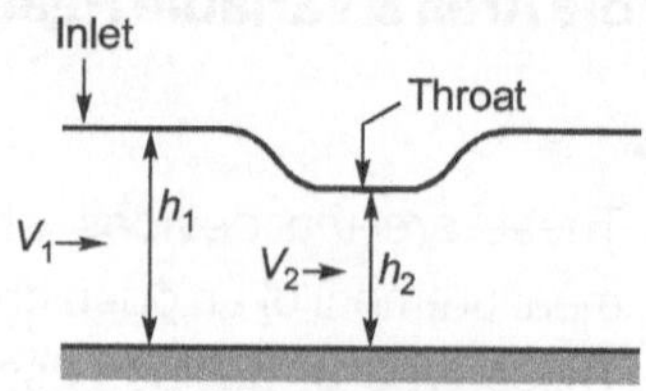

(Venturi Flume)

The actual discharge through a flume can be

$$Q_{(actual)} = \frac{CA_2}{\sqrt{1+\left(\frac{A_2}{A_1}\right)^2}} \sqrt{2gh}$$

where, $h = h_1 - h_2$

C = Venturi-flume coefficient (0.95 to 1)

Remember: ....

- The maximum value of flow in a Venturi-flume occurs when, $h_2 = (2/3)h$
- Maximum discharge through a Venturi-flume is given by $Q_{max} = 1.7\, d_2 H^{1.5}$ where, d_2 = inlet diameter of the Venturi-flume.
- Accuracy of flumes is higher than that of the weirs.

....

Variable Area Type Flowmeters

Rotameter

- Rotameter is a constant pressure drop type variable area flowmeter.
- The fluid flow from bottom to top in a rotameter.
- The density of float is greater than the density of the fluid flowing through it, so that the bouyant force alone is not sufficient to lift the float.
- The increase in the fluid flow at the inlet, rises the float up in the rotameter tube and an decrease in flow brings down the position of the float, hence the float position depends on the inlet fluid flow.
- The float attains the dynamic equilibrium when the upward force (forced due to flowing material, plus the fluid bouyancy) and the downward force (due to weight of the float) on the float balances each other.

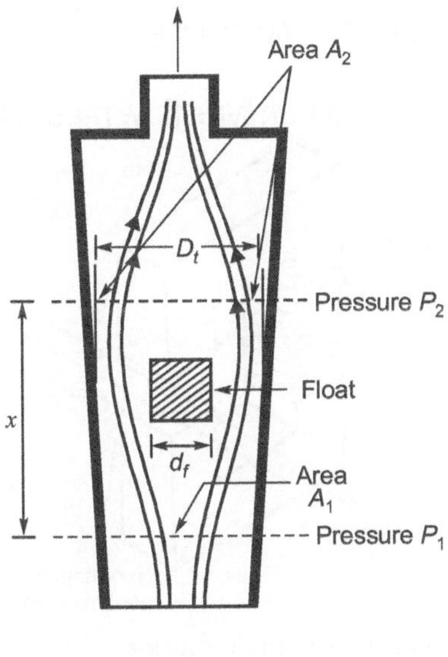

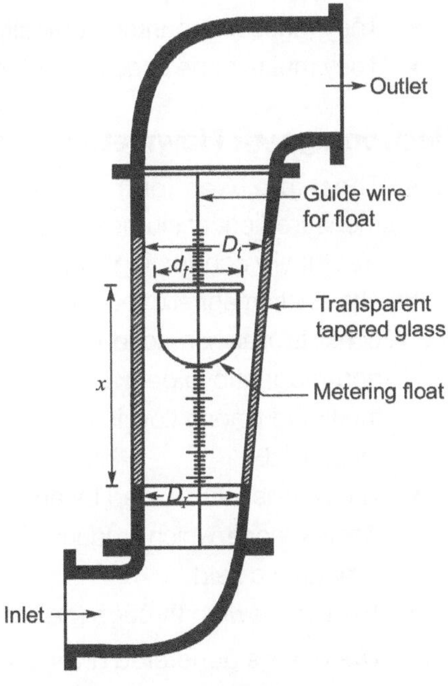

Discharge rate

$$Q = C_c A_2 \sqrt{2g \frac{V_f}{A_f}\left(\frac{\rho_f}{\rho} - 1\right)}$$

A_2 = area between float & tube (m^2)
C_c = coefficient of contraction
ρ = density of fluid (kg/m^3)
ρ_f = density of float (kg/m^3)
V_f = volume of float (m^3)

Note:

- The discharge (volume flow rate) for the rotameter is directly proportional to the displacement of the float (x).

$$\boxed{Q \propto x}$$

- In order to make the volume flow rate independent of the density of the fluid, the density of the float is made twice the density of the flowing fluid.

$$\boxed{\rho_f = 2\rho}$$

Advantages of Rotameter

- They can be used for the measurement of flow of liquid, gases and vapours. They are useful for measurement of low flow rates.
- They are independent from the effects of viscosity upto a certain range.
- The accuracy of the rotameters is higher than the accuracy of the obstruction type meters.

Disadvantages

- They must be mounted vertically.
- They must not be used under high pressure conditions.

Electromagnetic Flowmeters (non-obstruction type flow meters)

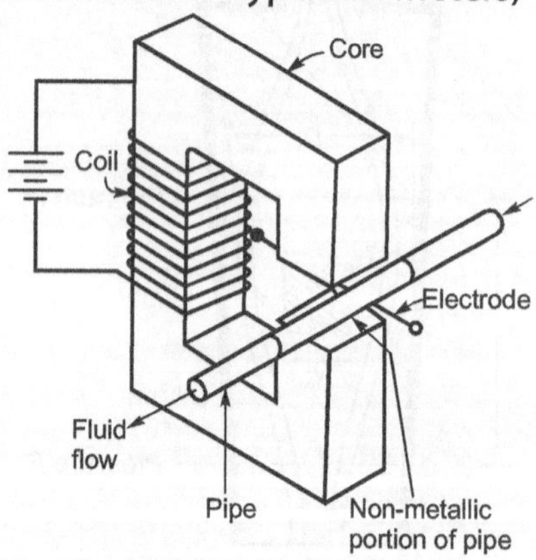

- These are used for the flow measurement of slurries, and electrically conducting liquid.
- The electromagnetic flow meter consists of an non-conducting, non-magnetic pipe carrying the liquid whose flow is to be measured.
- The pipe is surrounded by an electromagnet which produces a magnetic field.
- The fluid flowing through the pipe is conducting in nature.
- The voltage generated given as $\boxed{E = Blv}$

  where, B = flux density, l = length of conductor - diameter of pipe
  v = average velocity of the conductor (fluid)

Remember:

- Non-conducting pipe has to be used, because if we use a conducting pipe, the output will get short circuited.

Advantages

- They do not produce any pressure drop.
- The output voltage just depend on the speed of flow.
- Bidirectional flow can easily be measured by just reversing the connections.

Disadvantages

- Conductivity of the fluids must be high, so that they can act as conductors.
- Trapped bubbles of gas cause error, so the meter needs electrical and mechanical cleaning from time to time.

Turbine Flowmeter (Volumetric Flow Meters.)

- Turbine flowmeter are the devices, in which a small turbine wheel is placed in the pipe along the path of the fluid flow.
- It has a permanent magnet scaled inside the rotor body and is polarized at 90° to the axis of rotation.
- The flowing fluid rotates the blades of the turbine and each blade forms a magnetic circuit with the permanent magnet and coil in the meter housing and hence the output voltage is generated in the turbine.
- The generated output voltage has the form of a sine wave whose frequency is proportional to the angular velocity of the blade.
- The output voltage generated by turbine flow meter is given as

$$E = bm\,\omega_r \sin m\,\omega_r t$$

$$|E| = bm\,\omega_r$$

$$f = \frac{m\omega_r}{2\pi}$$

where m = number of blades

b = amplitude of angular variation in magnetic flux (Wb)

ω_r = angular velocity of wheel or turbine (radian)

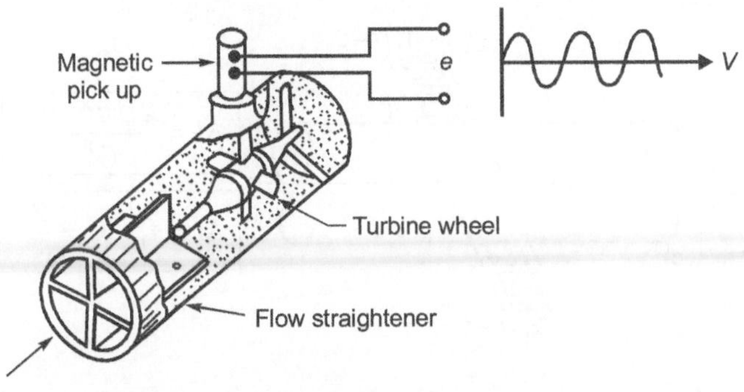

Advantages

- They can be used for the measurement of totalized flow with high accuracy.
- This flow meter has a good dynamic response.

Disadvantages

- Errors may be caused by the excessive frictional torques.
- Errors occur due to the wear and tear and also due to the corrosion of the turbine parts.
- They can not be used to measure low flow rates.

Ultrasonic Flowmeters

- Using the ultrasonic transducers (like piezo-electric crystal) the flowmeter can measure the average velocity along the path of an emitted beam of ultrasound, by averaging the difference in measured transient time between the pulses of ultrasound propagating into and against the direction of flow.
- There are various arrangement in which the crystal are placed to measure the flow.

Case-I: The crystal are placed inside the tube:

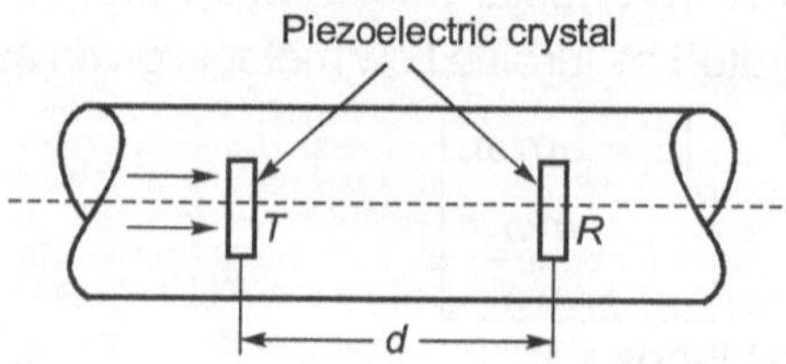

$$\Delta t = \frac{2dv}{C^2 - v^2}$$

Case-II: The crystals (*T* and *R*) are placed outside the tube

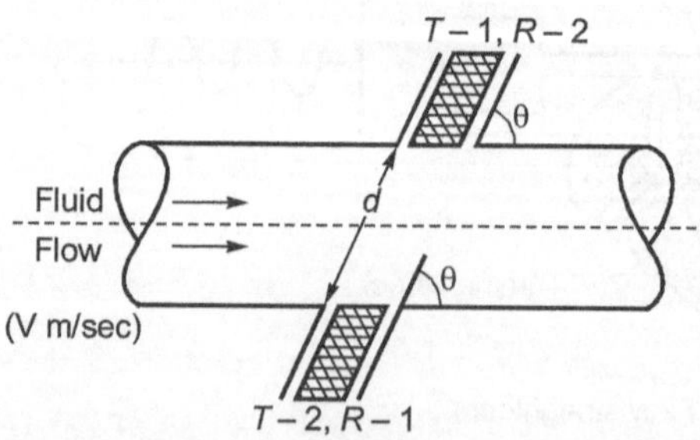

$$\Delta t = \frac{2d\cos\theta}{C^2} v$$

$$\Delta f = f_1 - f_2 = \frac{2v\cos\theta}{d}$$

Case-III: Ultrasonic Dopeller Flowmeter

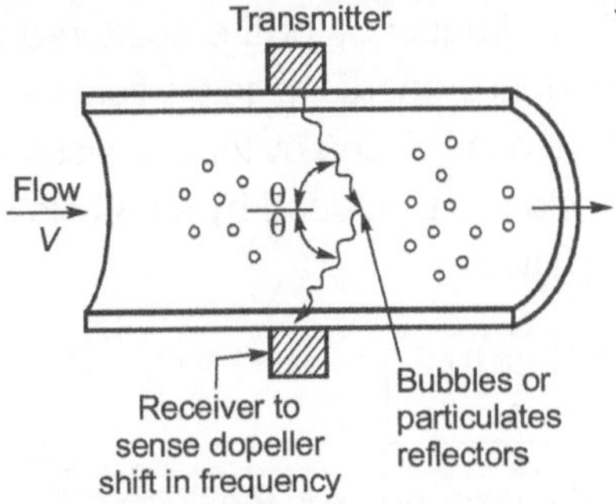

The frequency shift is given by

$$\Delta f = f_t - f_r = \frac{2 f_t \cos\theta \, v}{C}$$

where, f_t = frequency of the transmitter

f_r = frequency of the receiver

Advantages

- They do not offer any obstruction in the path of fluid flow.
- Their output is independent of viscosity, density and temperature of the fluid.
- They have linear relationship between input and the output.
- They can be used for bidirectional flow.
- They have very good dynamic response.
- Their output is in the electrical form so can be fed to any readout (analog/digital) and also can be used for telemetry applications.
- They have very high accuracy of the order of ±2% of the full scale values. They can be used for the biomedical flow measurements.

Disadvantages

- Their cost is high and they are complex in design.
- The acoustic component used in their manufacturing are expensive.

Laser Doppler Aneamometer (LDA)

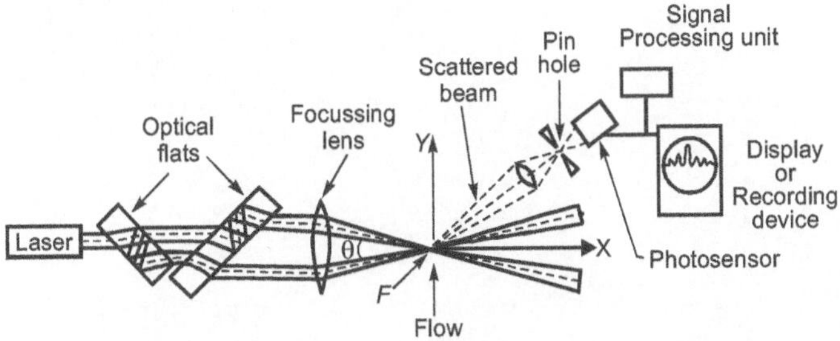

- The operating principle of this instrument involves the focusing of the laser beam at a point where the velocity is to be measured, this light is scattered by the tracer particles present in the fluid, this scattered light is than observed by a photodetector.

- The velocity of these tracer particles which is equal to the fluid velocity, causes a doppler shift of the frequency of the scattered light & produces a photodetector signal proportional to the fluid velocity.
- The velocity of the flow can be easily found by the expression of the frequency (f) of the electric signals produced by a particle moving across dark & bright fringe pattern.

$$\boxed{f = \frac{2V\sin\theta/2}{\lambda}}$$

where, λ = wavelength of the laser beam in the fluid.
V = velocity of fluid flow

Advantages

- High frequency response of the order of MHz.
- Can be used for the measurement of flow of both liquids as well as gases.
- Can be used in the biomedical application for the measurement of blood flows.

Disadvantages

- They involve a transparent channel.
- This method cannot be employed for the clean flow and the insertion of the tracer particles is necessary.
- Error may occur if the velocity of the beam scattering particles (tracer particles) is not equal to the flow velocity of the fluid.

■■■■

Measurement of Pressure

7

Pressure

Pressure of the fluid contained in a container is defined as the force per unit area exerted by the fluid on a container wall.

Static Pressure

- When the fluid is in equilibrium, the pressure at a point is identical in all direction and is independent of orientation. This is called as static pressure.
- Static pressure may be considered as the pressure that is experienced, if moving along the stream.

Dynamic Pressure

- When there is a pressure gradient occur within a continum of pressure, the attempt to restore equilibrium results in fluid flow from regions of higher pressure to regions of lower pressure. In this case the pressures are no longer independent of direction and are called dynamic pressure.

Total or Stangnation Pressure

Pressure exerted by the fluid stream when the stream is brought to rest isentropically

Velocity Pressure

Pressure exerted only due to the motion of the fluid.

Total Pressure = Velocity Pressure + Static Pressure

Absolute Pressure

- It is the absolute value of force per unit area exhibited by the fluid on the walls of the container.

Gauge Pressure

- It represents the difference between the absolute pressure and the local atmospheric pressure.

Vaccum pressure

- The amount by which the atmospheric pressure exceeds the absolute pressure is known as the vaccum pressure.

Relationship between absolute, gauge, atmospheric and Vaccum Pressure

$$P_g = P_a - P_s$$

Where, P_g = Gauge Pressure
P_a = Absolute Pressure
P_s = Atmospheric Pressure
P_v = Vaccum Pressure

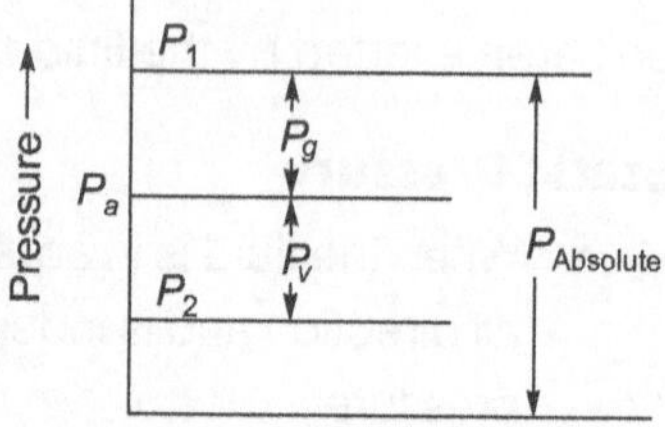

Remember: ....

- Unit of pressure
  1 inch of water = 249.1 N/m^2 = 249.1 Pa (Pascal)
  1 mm of mercury = 1 torr = 133 N/m^2 = 133 Pa
- The atmospheric pressure at sea level is 101.3 kN/m^2 or 760 mm of hg.

....

Moderate Pressure Measurement

Manometers

- Measures the unknown pressure by balancing it against the gravitational force of liquid heads.
- Manometers can be used to measure the dynamic pressures.
- Manometers measure the differential pressure.
- Manometers can be used to measure pressure ranging from as low as vaccum to a high pressure of about 0.1 MN/m^2.

Manometric Fluid Characteristics

- Low viscosity, low thermal coefficient of expansion
- Non-corrosive, non-sticky
- Low surface tension, low vapour pressure

Note: ....

- Mercury is the widely used manometric fluid because of its wide temperature range (from –20°C to 350°C).

....

U-tube manometer

- In the U-tube manometer the height h (i.e the difference in the two levels) is a function of the pressure difference.

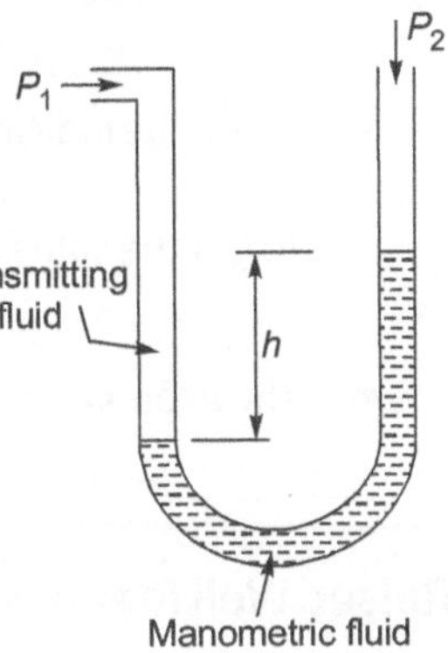

$$(\Delta P) = P_1 - P_2 = gh(\rho_m - \rho_f)$$

where, g = acceleration due to gravity (m/sec^2)

ρ_m = specific gravity of manometric fluid (kg/m^3)

ρ_f = specific gravity of the transmitting fluid (kg/m^3)

$$\text{Pressure sensitivity}, (S) = \frac{1}{g\rho_m}$$

Remember:

- If P_1 is the input pressure then the U tube manometer may be used for measurement of
  (a) Absolute pressure when, $P_2 = 0$.
  (b) Gauge pressure when, P_2 = atmospheric pressure
  (c) Differential pressure when, P_2 as the second pressure

Well Type Manometer

- A well type manometer is a manometer which is obtained when one of the two legs of an ordinary manometer being replaced by a large well, as shown in the figure below.

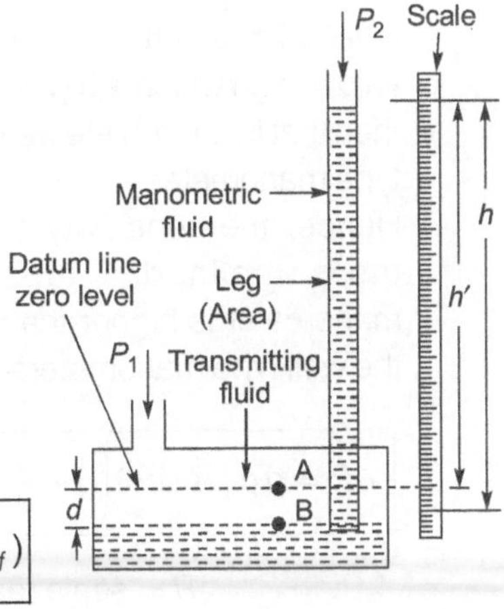

$$\Delta P = P_1 - P_2 = gh'\left(1 + \frac{a}{A}\right)(\rho_m - \rho_f)$$

Remember:

- The value indicated on the scale as h', has to be multiplied with $\left(1+\frac{a}{A}\right)$ to get the value of h.
- The expression $\left(1+\frac{a}{A}\right)$ is known as the area-correction factor.

Raised Well Manometer

- In the raised well type of manometer, the well is located at some raised level, by using this configuration the zero level is made fixed, unlike the case of the well type manometers we don't have to use the correction factor to find the height of the fluid in the column.

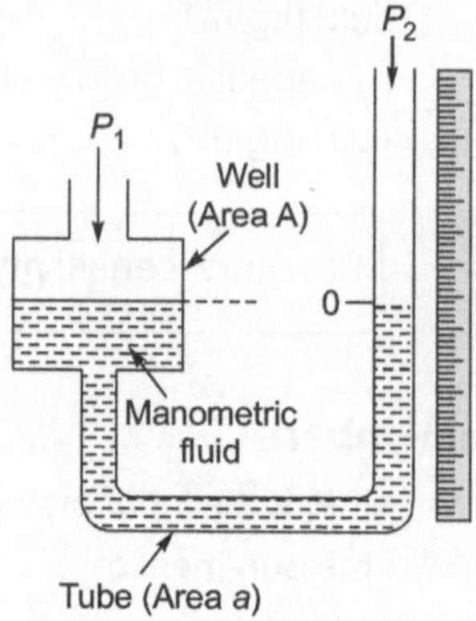

Inclined Tube Manometer

- Due to the inclination of the leg, for a small input pressure P_1, the rise in the level of the manometric fluid is high which is higher then that in case of the well type manometer.
- Hence, the sensitivity of the inclined tube manometer is higher than the well type manometers.

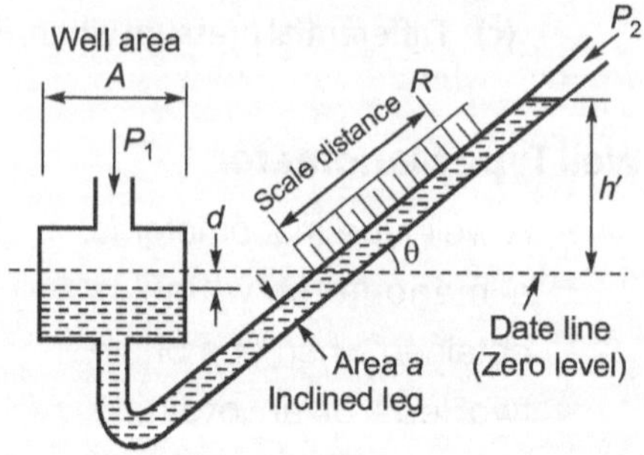

$$\Delta P = g\rho_m R\sin\theta\left(1+\frac{a}{A}\right)$$

R = scale reading in (m)

θ = angle of inclination of inclined leg

Remaining symbols have their usual meanings

Note:

- Order of sensitivity of the manometers

  Inclined tube manometer > Well type manometer > U-tube manometer

Advantages of Manometers

- They are simple in construction.
- They have high accuracy.
- They can be used for the measurement of pressure, temperature, flow and other process variables.

Disadvantages

- They have poor dynamic response.
- They have fragile construction and hence their portability is less.
- They have small range of operation of the order of about 1000 kN/m^2.
- The manometric fluids density depend on the temperature, hence the error may arise on account of the change in temperature.

Dynamic Response of Manometers

Normalized transfer function is

$$G(s) = 2\rho a g \frac{X(s)}{P(s)} = \frac{2ag/V}{s^2 + \frac{B}{\rho V}s + \frac{2ag}{V}}$$

where $X(s)$ = steady state output displacement of manometric fluid (m)

$P(s)$ = input pressure (N/m^2); a = area of tube (m^2)

B = coefficient of viscous friction (N-sec^{-1})

On comparing with second order standard transfer function

for, ω_n (natural frequency) = $\sqrt{2ag/V}$ rad/sec^{-1}

ζ(damping ratio) = $B/2\rho\sqrt{2agV}$

Barometer

- It measures atmospheric pressure directly.
- The rise in the level of the mercury in the leg of the barometer is on the account of the atmospheric pressure (P_s) as shown in the figure.
- The level rise to a height 'h', this height is a measure of the atmospheric pressure (P_s).

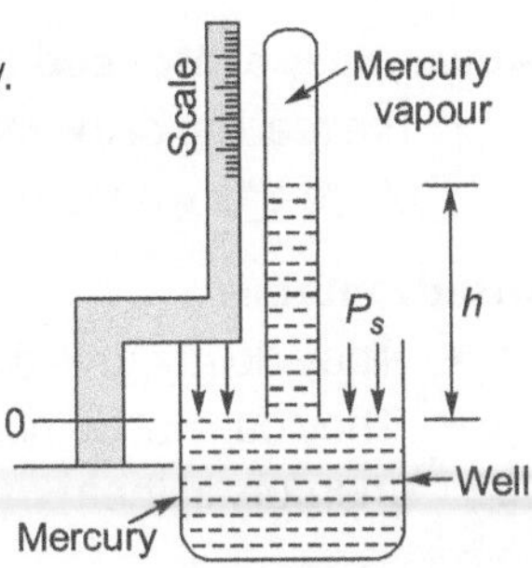

$P_s = \rho_m gh$ g = acceleration due to gravity; ρ_m = density of the mercury; h = height of the liquid in the leg; P_s = atmospheric pressure

Low Pressure (Vaccum) Measurement

- **Vaccum,** the pressure below the atmospheric pressure (760 torr) is regarded as vaccum.
- The very low pressure (vaccum) is expressed in the units like torr, micrometer.

$$\boxed{1\text{ torr} = 1\text{mm of } Hg = 1000\text{ micrometer}} \quad \boxed{\text{atmospheric pressure} = 760\text{ torr}}$$

Mc Leod Gauge

- The principle employed for the working of Mc Leod Gauge is Compression of a sample of a low pressure gas (gas whose pressure is to be measured) to a level which is sufficiently high. Such that it can be read with the use of a manometer.

  The pressure of the gas (P) can be find as:

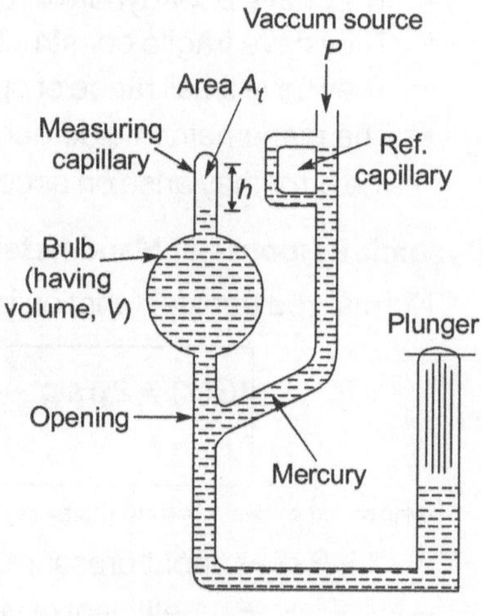

$$\boxed{P = \frac{h^2 A_t}{V - A_t h} \approx \frac{h^2 A_t}{V}}$$

where, h = height of mercury column in the reference capillary (m)

A_t = area of cross-section of the measuring capillary tube (m^2)

V = volume of the bulb

Advantages of Mc Leod Gauge

- The reading of the Mc Leod gauge is not influenced by the composition of the gas whose pressure is to be measured.

Disadvantages

- They do not give the continuous output.
- They can not be used in hazardous conditions because of the use of mercury in them.

Knudsen Gauge

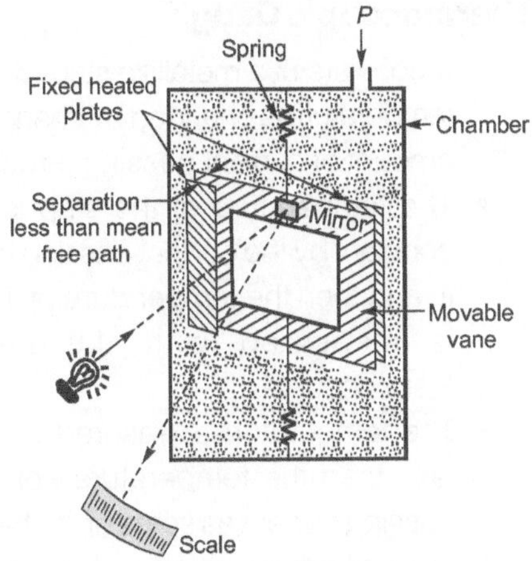

- Due to the higher temperature of the plates, the molecules striking the vane have higher velocity than those which are leaving the vane.
- These striking electrons from the plates to the vane produces a net force (F) and thus leads to the movement of the vane.
- The force produced on the vane by the molecules is directly proportional to the pressure of the gas for a given values of T and T_g.
- The pressure of the gas can be then given as:

$$P = \frac{kF}{\sqrt{T/T_g - 1}}$$

where, k = constant
T = temperature of the plates
T_g = temperature of the gas / movable vane

Advantages of Kundsen Gauge

- Insensitive to the gas composition. Measures pressure which are much lower than the range of Mc Leod gauge.
- This gauge is used to measure low pressure ranging from 10×10^{-9} to 10×10^{-3} torr.

Viscosity Gauge

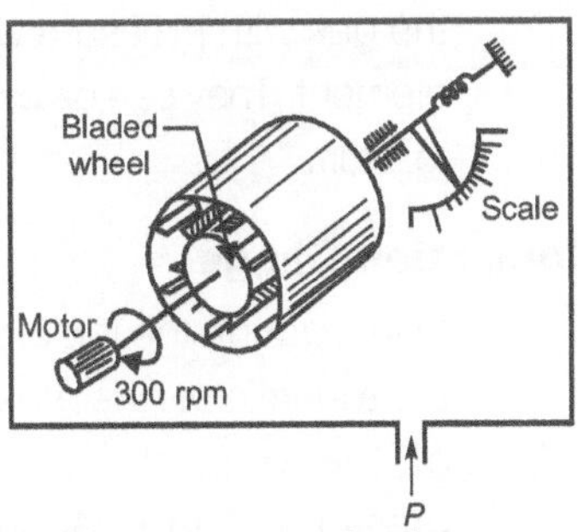

- Since for the low pressure (typically below 10×10^{-3} torr) the viscosity of the gas is proportional to the pressure of the gas, so we have to measure the viscosity of the gas.
- Viscosity can be measured by measuring the torque required to rotate the cylinder at a constant speed within the another cylinder and the space between the two cylinder is filled with the gas whose viscosity is to be measured.
- They can be used to measure the temperature down to about 10×10^{-3} torr. They are sensitive to the composition of the gas whose pressure is to be measured.

Thermocouple Gauge

- It consists of a metallize strip whose temperature is changed by changing the amount of current passing through it.
- The temperature of the strip is made constant by the current passing through it, and then the temperature of the strip is measured with the help of thermocouple.
- The temperature measured is always less than the temperature set by the passing current as some of the heat loss takes place on account of conduction.

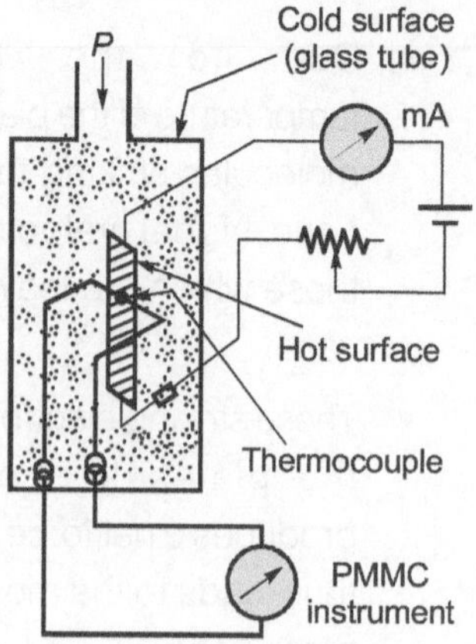

- The amount of heat loss due to conduction is different for different gas pressure. So the measured temperature can then be calibrated in terms of the gas pressure.

Pirani Gauge

- It consists of four coils connected to each other and forming one of the arms of the Wheatstone bridge.

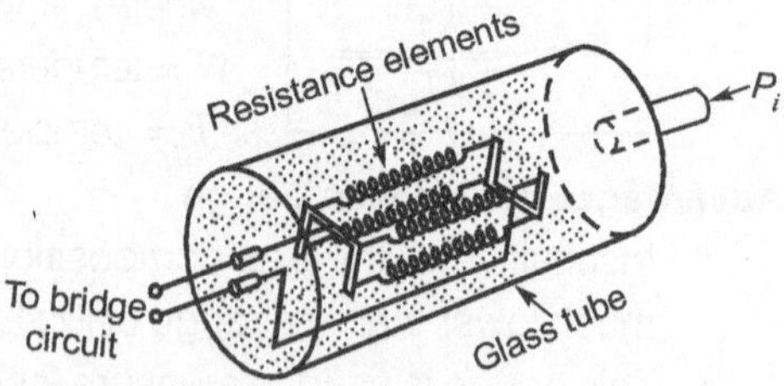

- Pirani gauge is based on the principle of the change in conductivity of the gas with pressure which changes the temperature of the heated element. They can be used to measure pressure ranging from 10×10^{-6} to 1 torr.

Ionization Gauge

- In this gauge the ion current I_i is a function of the pressure of the gas, greater the pressure of the gas in the chamber, the greater is the output (i.e. the ion current, I_i).

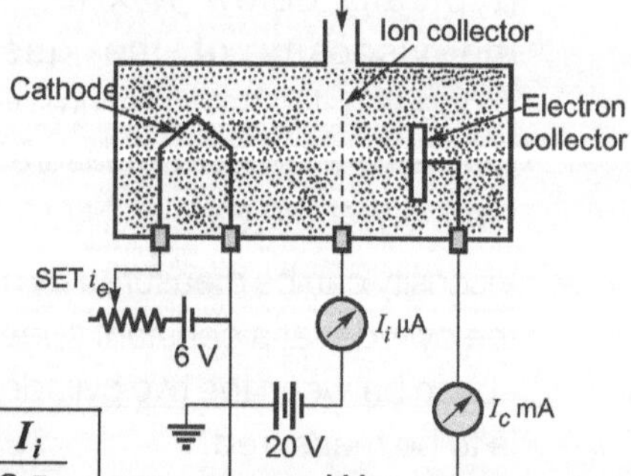

$$S = \frac{I_i}{P I_e}$$

$$P = \frac{I_i}{S I_e}$$

I_i = ion current, gauge output (A); I_e = electron current (A);
P = pressure of the gas in the chamber

Measurement of Pressure using elastic elements (high pressure measurement)

Bourdon Tube

C-Type

- In C-Type bourdon tube one end called the tip of the tube is sealed and is called free end and the other end of the tube is fixed to a socket where the pressure to be measured is applied.
- The internal pressure tends to change the section of the tube.
- The movement of the tip is ideally proportional to the pressure applied. The displacement of tip is

$$\Delta a = 0.05 \frac{aP}{E}\left(\frac{r}{t}\right)^{0.2}\left(\frac{x}{y}\right)^{0.33}\cdot\left(\frac{x}{t}\right)^{3}$$

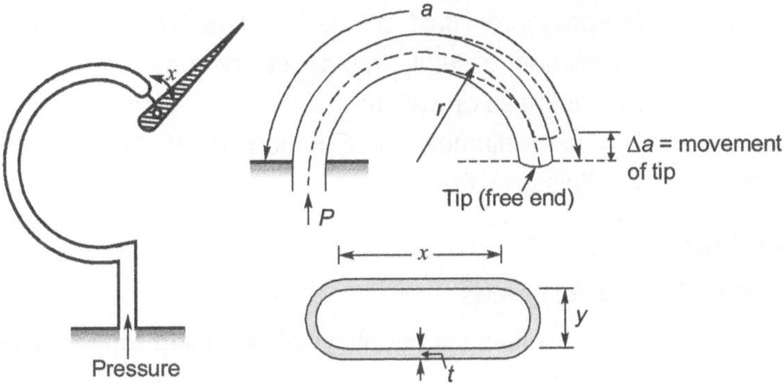

Sprial Type

- These tubes are made by winding several turns of the tube with its flattened cross-section in the form of a spiral.
- When the pressure to be measured is applied to the spiral, it tends to uncoil producing a relatively long movement of the tip (as compare to C-type) whose displacement can be used for indication.

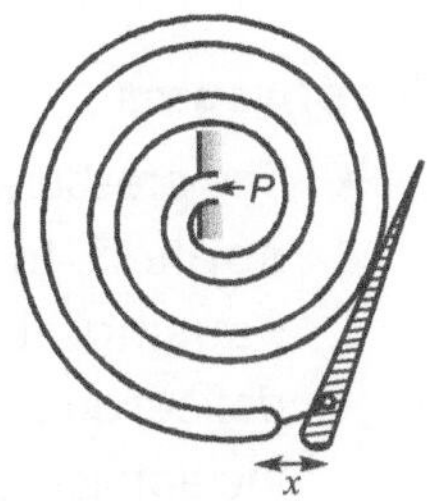

Remember: ....

- Materials used for Bourdon tubes
  Brass, alloy steel, stainless steel, bronze, phosphor, bronze, beryllium copper.
- Pressure gauges using bourdon tube elements are made with ranges from 760 mm of mercury to 700 MPa.

....

Bellows

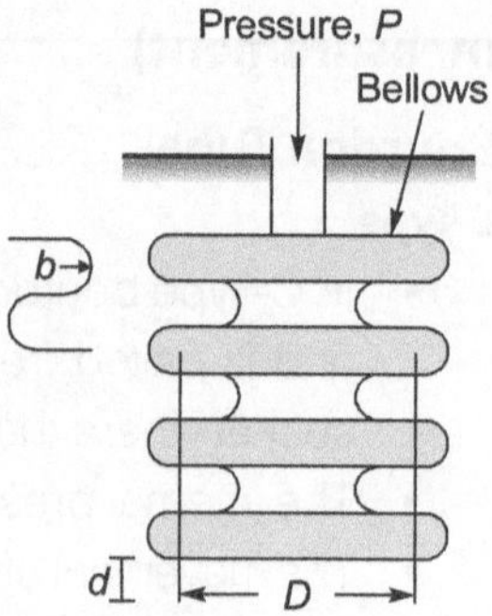

- A metallic bellows is a series of circular parts, resembling the folds in a accordinan as shown in figure.
- These parts are formed or joined in such a manner that they are expanded or contracted axially by changes in pressure.

The displacement of bellows elements is given by

$$d = \frac{0.453\,Pb\,n\,D^2\sqrt{1-\nu^2}}{Et^3}$$

where P = pressure, N/m², b = radius of each corrugation, m
n = number of semi-circular corrugations,
t = thickness of wall, m
D = mean diameter, m, E = modulus of elasticity, N/m²
and ν = Poisson's ratio

Remember:

- **Materials used for Bellows**

  Brass, bronze, beryllium copper, alloys of nickel and copper, steel and monel.

Diaphragms

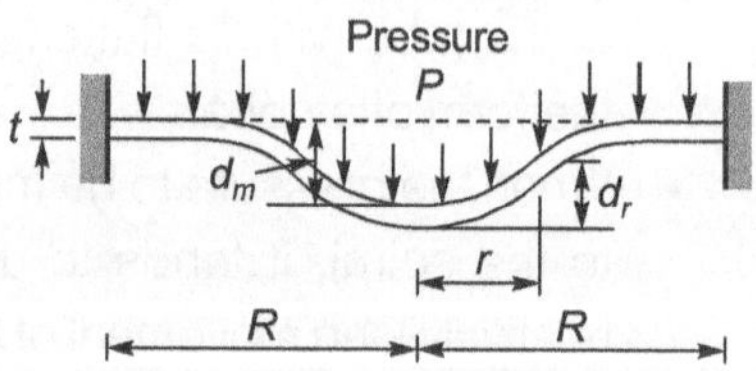

- The pressure to be measured is applied to the diaphragm causing it to deflect, the deflection being proportional to the applied pressure.
- The maximum deflection, d_m and the deflection at any radius, d_r are given by following expressions
- The ranges are 0 - 50 N/m2 to 0 - 200 kN/m².

$$d_m = \frac{3P}{16\,Et^3}R^4\left(1-\nu^2\right) \qquad d_r = \frac{3P\left(1-\nu^2\right)}{16\,Et^3}(R^2 - r^2)^2$$

Measurement of Very High Pressure (> 70000 MPa)

Bridgman Gauge

- The principle of the Bridgman Gauge is the change of the resistance with the change in the pressure.
- The resistance of the coil is then changed according to a linear relationship given by

$$\boxed{R = R_1(1 + b\,\Delta P)}$$

where, R = Resistance in ohms of the coil at pressure of 1 atmosphere (in Ω)

b = pressure coefficient of the resistance

ΔP = gauge pressure

- Materials used for making the coil are Manganin and Gold chrome.

Type of measuring device	Pressure range
Manometer	10 to 10^6 Pa
Bourdon gauge	10^3 to 5×10^8 Pa
Elastic diaphragm with LVDT or capacitance or resistance strain gauge transducer	100 to 10^8 Pa
Piezo-electric transducer	10^4 to 10^8 Pa
Hydrostatic compression gauge	100 to 10^5 atm
McLeod gauge	10^{-5} to 1 Torr
Pirani gauge	10^{-4} to 1 Torr
Ionisation gauge	10^{-12} to 10^{-3} Torr
Knudsen gauge	10^{-8} to 10^{-2} Torr

■■■■

Measurement of Temperature

8

- Temperature is defined as the degree of hotness or coldness of a body an environment measured on a definite scale.

Temperature Scales & their relationship

- Kelvin (K)
- Celsius (°C)
- Fahrenheit (°F)
- Rankine (R)

$$K = {}^\circ C + 273.15$$
$$R = {}^\circ F + 459.69$$
$${}^\circ F = 32 + \frac{9}{5}{}^\circ C$$
$$R = \frac{9}{5}K$$

Non-electrical Methods

Liquid in Glass thermometer

- In this thermometer as the temperature is raised the greater expansion of the liquid compared with that of glass causes it to rise in the capillary or stem of the thermometer and the height of rise is used as a measure of temperature as indicated by the eteched markings.
- Desirable properties of liquid used in glass thermometer
  - The temperature dimensional relationship should be linear.
  - Large coefficient of expansion.
  - Clearly visible when drawn into thread.
  - Should not stick to the capillary walls.
- Lower temperature measuring range of variouis liquids used in thermometer are

  Mercury = – 38.87° C

  Alcohol = – 75° C

  Toluol = – 90°C

  Pentane = – 200°C

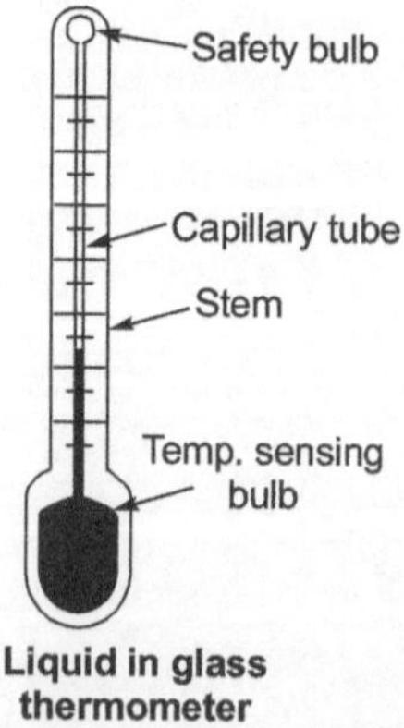

Liquid in glass thermometer

Bimetallic Strip

- A bimetallic thermometer consists of a bimetallic strip which is constructed by bonding together two thin strips of two different metals such that they cannot move relative to each other.
- The differential change of expansion of two metals results in bending of the bimetallic strip with change in temperature.
- The bimetallic strip bends towards the side whose metal has a lower thermal expansion coefficient when there is increase in temperature and reverse happens when there is decrease in temperature.

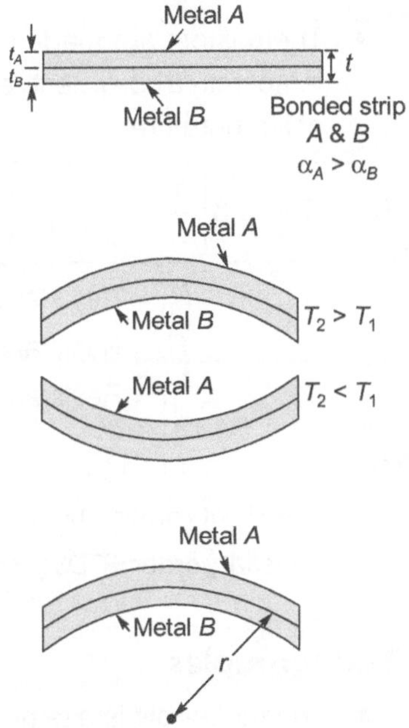

The radius of curvature is given by

$$\boxed{r = \frac{2t}{3(\alpha_A - \alpha_B)(T_2 - T_1)}}$$

$T_2 - T_1$ = Change in temperature

α_A, α_B = Thermal coefficient of expansion of metals A & B respectively.

- Materials used are : Invar, Nickel iron alloys

Electrical Methods

Resistance Thermometer (RTD)

- RTD's have positive temperature coefficient, i.e. their resistance increase with the increase in the temperature.

For Low temperature

$$\boxed{R_T = [R_0(1 + \alpha \Delta T]} \qquad \boxed{\text{Sensitivity, } (S) = \alpha R_o}$$

R_T = Resistance at temperature T°C

R_o = Resistance at temperature O°C

α = Resistance temperature co-efficient; /°C

ΔT = Charge in temperature

For high temperature

For a change temperature from t_1 to t_2

$$\boxed{R_2 = R_1 + R_0\,\alpha\left(t_2 - t_1\right)}$$

Thermistor

- Thermistors are made up of semiconductors. They have negative temperature coefficient i.e. their resistance decreases with the increase of temperature.

$$R = R_0 \exp \beta\left[\frac{1}{T} - \frac{1}{T_0}\right] \qquad S\left(\frac{\Omega}{°C}\right) = \frac{dR}{dT} = R_0 \exp\left[\beta\left(\frac{1}{T} - \frac{1}{T_0}\right)\right]\left(\frac{-\beta}{T^2}\right)$$

R_0 = Resistance at T_0 K

R_r = Resistance at T K

β = Resistance temperature coefficient

Note:

- Sensitivity order : Thermistor > RTDS > Thermocouples
- Linearity order : RTDS > Thermocouple > Thermistor

Thermocouples

- Thermocouple is a temperature measuring device, which is based on the principle of following effects:

  Seeback effect : When the two dissimilar metals are joined together, an emf will exist between the two points A and B which is a function of the junction temperature.

  Peltier effect : If the two different materials are connected in such away that a current is drawn, the e.m.f. generated (on account seeback effect) gets altered.

  Thomson effect: If any temperature gradient exists along either or both of the materials, the junction may undergo an addition small alteration.

  The e.m.f. so produced between the terminals A and B is given by

$$E = a(\Delta\theta) + b(\Delta\theta)^2 \qquad S = \frac{dE}{d\theta}$$

where, E = e.m.f. developed (mV)

a, b = constants

$\Delta\theta$ = temperature difference between the measuring (hot) junction and the reference junction (°C)

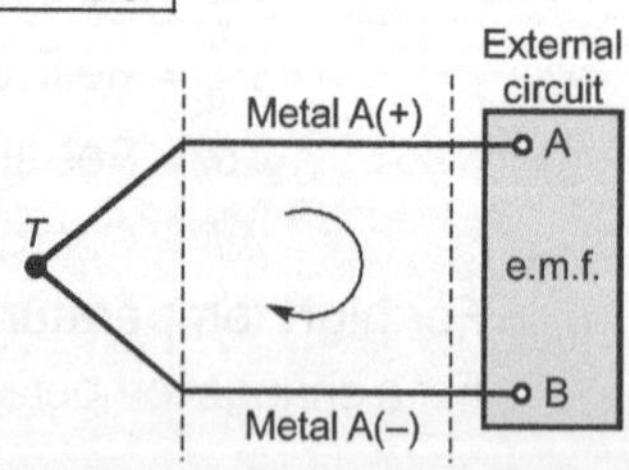

Thermo-Electric Laws

Law of Intermediate Temperature

- This states that the emf generated in a thermocouple with junctions at temperatures T_1 and T_3 is equal to the sum of the emf's generated by similar thermocouples, one acting between temperatures T_1 and T_2.

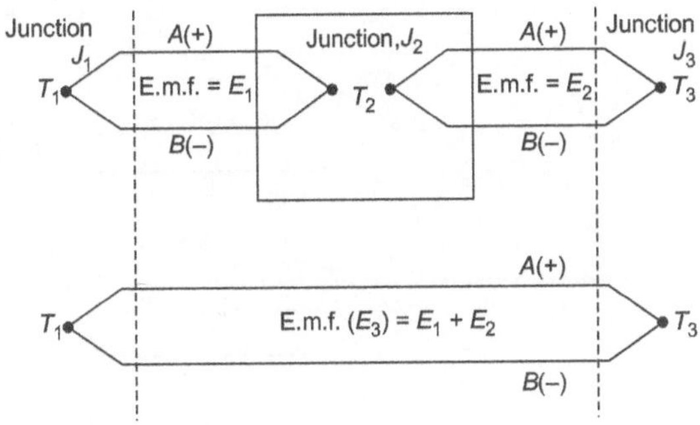

Law of Intermediate Metals

- In a circuit which consists of two dissimilar homogenous metals having the junctions at different temperatures, the e.m.f. so developed will not be affected if a third metal be made a part of the circuit, provided the temperatures of the two junctions are same.

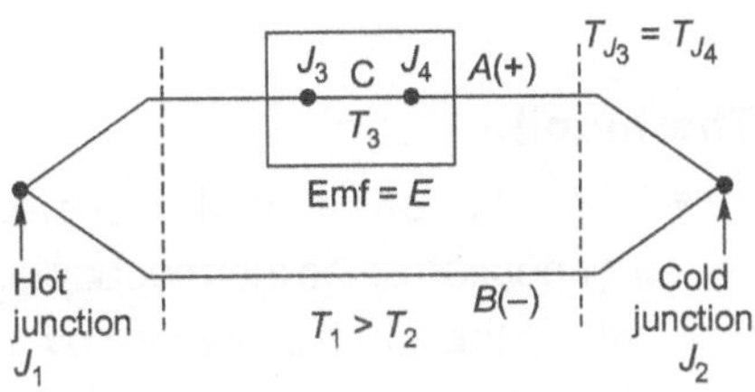

Type	Thermocouple materials*	Approximate sensitivity in (µV/°C)	Useful temperature range (°C)
T	Copper-Constantan	20 - 60	–180 to +400
J	Iron-Constantan	45 - 55	–180 to +850
K	Chromel-Alumel	40 - 55	-200 to +1300
E	Cheromel-Constantan	55 - 80	–180 to +850
S	Platinum-Platinum/10% Rhodium	5 - 12	0 to +1400
R	Platinum-Platinum/13% Rhodium	5 - 12	0 to +1600

Various Connections of Thermocouples

Series connection of thermocouples

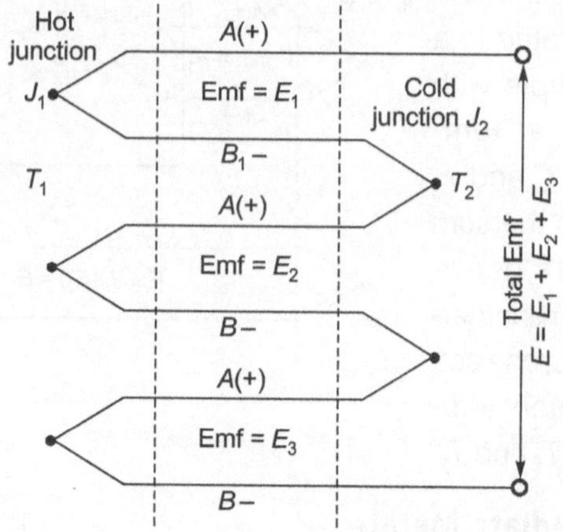

Thermopile

- It is the case of the series connection of the thermocouples where the e.m.f. produced by all the individual thermocouples is the same.

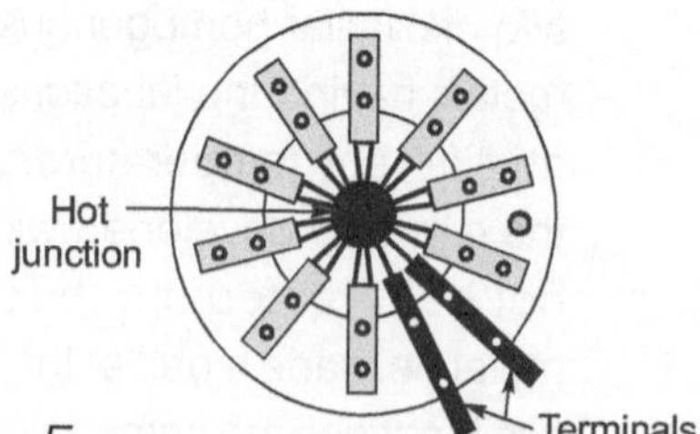

  i.e., $E_1 = E_2 = E_3 .... = E_n$

  Hence, their output comes out to be $E = nE_1$

Parallel Connection

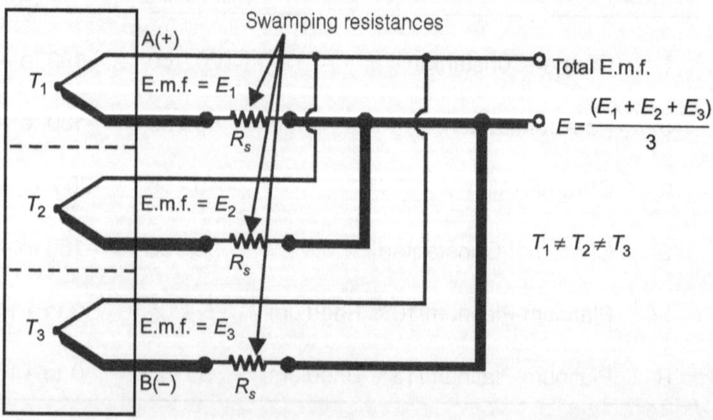

Comparison between Thermocouple, RTDs and Thermistors

Thermocouple	RTD	Thermistors
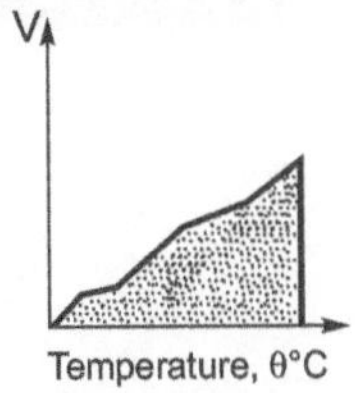	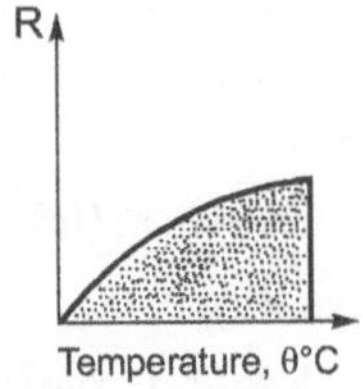	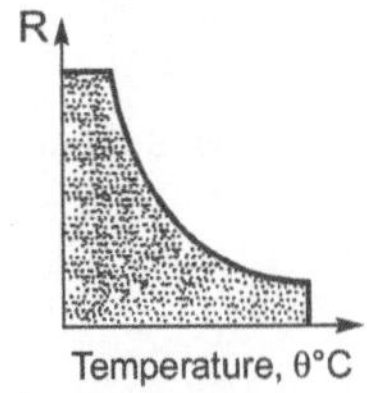
Used for high temperature measurement typically upon 1400°C.	Used in the range of –250 to 1000°C.	Used for low temperature measurement.

Advantages

Thermocouple	RTD	Thermistors
• Self powered (active devices)	• Most stable	• They have high output (hence highly sensitive)
• Simple in construction	• Most accurate	• They have high output (hence highly sensitive)
• Rugged	• Linearity is higher than of thermocouples (Linear)	
• They are inexpensive		
• They have wide temperature range		

Disadvantages

Thermocouple	RTD	Thermistors
• They are non-linear	• Passive	• Non-linear
• Low output voltage (Low sensitivity)	• They are very expensive	• Self-heating
• Require reference	• Self-heating	• Passive
• Less stable	• Small change in resistance with temperature	• They are fragile

Radiation Pyrometer (For high temperature)

- They measure temperature according to Stefen's Boltzman Law. The thermal radiation emitted by a gray body is given by

  $\boxed{q = \epsilon\sigma T^4}$ σ (Stefen's Boltzman constant) = 57.2×10^{-9} w/m^2 h^{-4}

  T = absolute temperature (K); ∈ (emissivity) = q/q_b; q = heat radiated by the gray body W/m^2; q_b =heat radiated by the black body W/m^2

Remember:

- Order of measurement of temperature
  (High temperature measurement to Low temperature measurement)
  Pyrometers > Thermocouples > RTD > Thermistors

■■■■

9

Measurement of Force

Force is defined as,

$$F = m \times a \text{ Newton (N)}$$
$$m = \text{mass (kg)}$$
$$a = \text{acceleration (m/s}^2\text{)}$$

Measurement of Force

Elastic Sensing Elements

- The displacement/deflection of elastic sensing elements varies with the applied force as,

$$\boxed{F = kx}$$

where,
F = applied force in (N)
x = deflection displacement in (m)
k = stiffness in (N/m)

$$S \text{ (Sensitivity)} = \frac{x}{F} = \frac{1}{k} \qquad [\because\ F = kx]$$

Note:

- **For higher sensitivity of the elastic sensing elements the value of the stiffness for the element must be small.**
- **Higher the stiffness lesser will be the settling time and hence faster will be the response, so in order to have a faster response the stiffness of the element must be high.**

Types of elastic sensing elements

Cantilever

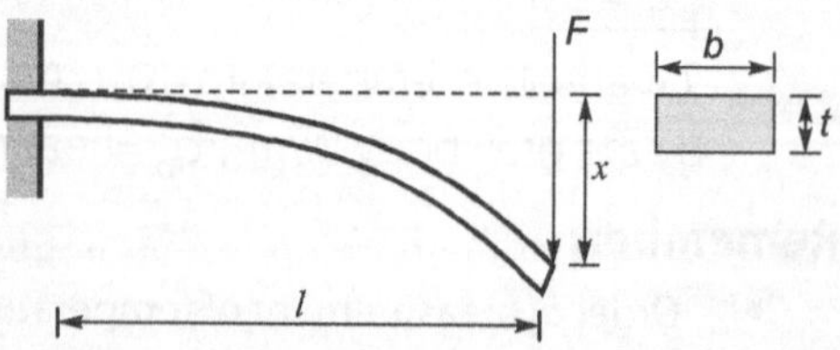

Deflection of the free end is x,

$$x = Fl^3/3\,EI$$

where, F = applied force in (N)
l = length of the cantilever beam in (m)

E = modulus of elasticity

I = moment of inertia $= \left(\frac{1}{12}\right), bt^3$ in(m^4)

b = width of the cantilever in (m)

t = thickness of the cantilever in (m)

$$F = \frac{x(3EI)}{l^3} = \frac{3EI}{l^3}x$$

$$\therefore \quad k = \frac{3EI}{Fl^3} \qquad [\because \ F = kx]$$

Simply supported beam

Deflection for simply supported beam is given as

$$x = \frac{Fl^3}{48EI}$$

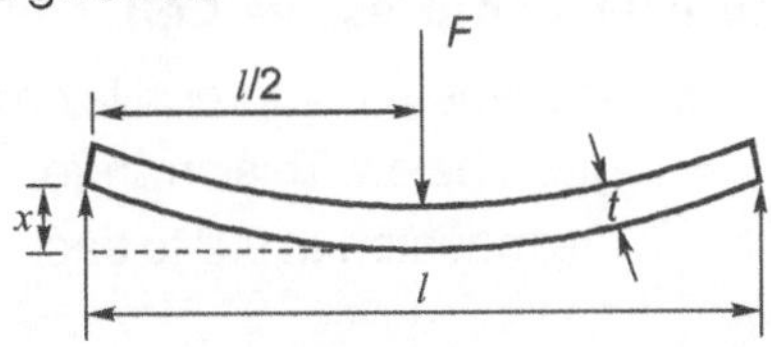

Stiffness, $$k = \frac{F}{x} = \frac{48EI}{l^3}$$

Built in Beam

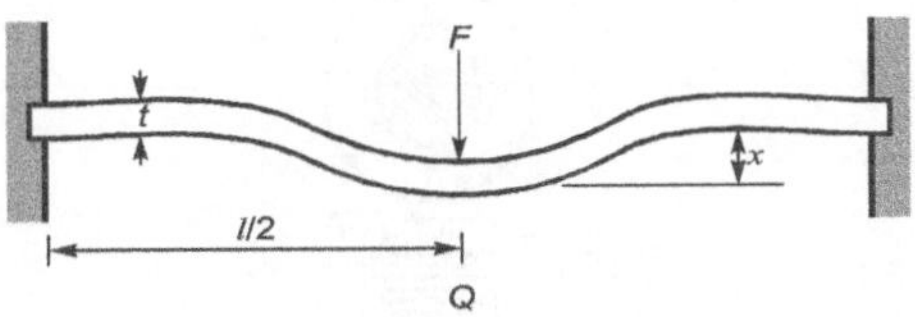

Deflection, $$x = \frac{Fl^3}{192EI}$$

Stiffness, $$k = \frac{F}{x} = \frac{192EI}{l^3}$$

Disc built-in around the perimeter (i.e. force is applied around central circle of radius r).

Deflection produced in this type of beam is

Deflection,

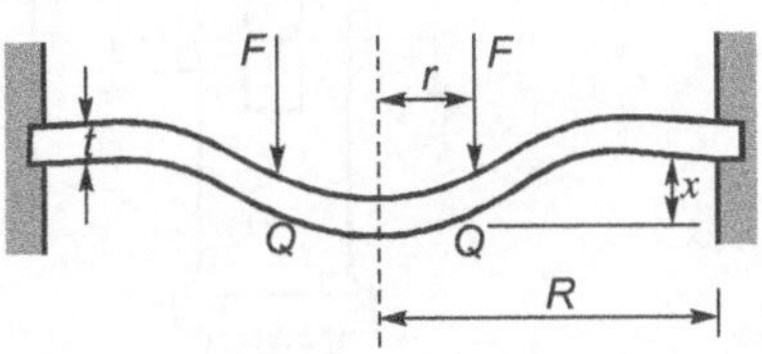

$$x = \frac{3F(1-\nu^2)}{2\pi Et^3}\left\{\frac{1}{2}(R^2 - r^2) - r^2 ln\left(\frac{R}{r}\right)\right\}$$

where, ν = Poissions ratio

and the other symbols have their usual meaning as shown in the above diagram.

Disc built-in around the perimeter and with uniform pressure applied.

Deflection, $x = 3PR^4 \frac{(1-\nu^2)}{16Et^3}$

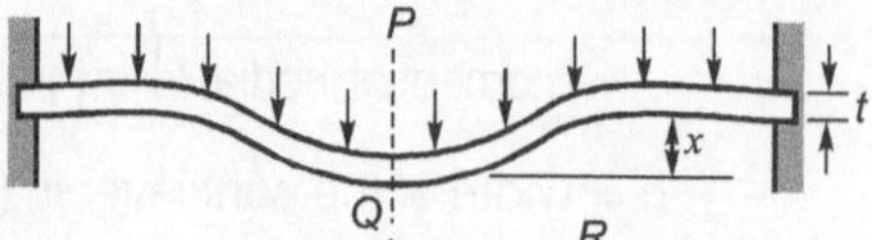

(Disc built in around amplifier)

where, P = uniform pressure

ν = poissions ratio

These type of elastic elements are linear for $x \leq 0.5$t.

Load Cell

- They employ an elastic member as a primary transducer and the strain gauge as a secondary transducer.

Tensile-Compressive Cell

- The tensile-compressive cell which is cylindrical is shown in the figure given below. This arrangement consists of four strain-gauges which are mounted at 90° to each other.

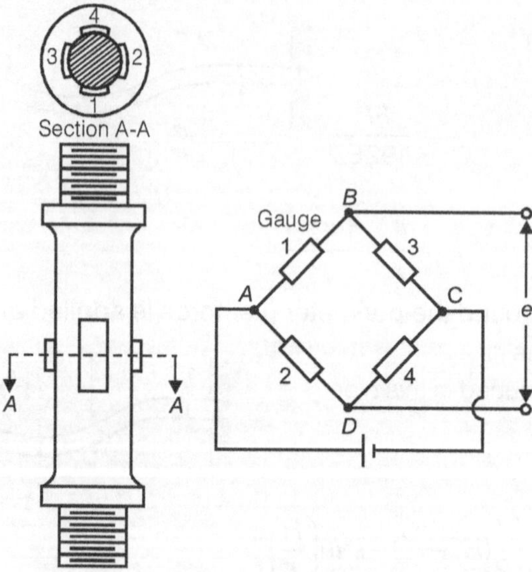

(a) Load cell (b) Strain Gauge Bridge

(Tensile compressive cell using strain gauges)

- Two of the strain gauges experience tensile stress while the other two are subjected to compressive stresses.

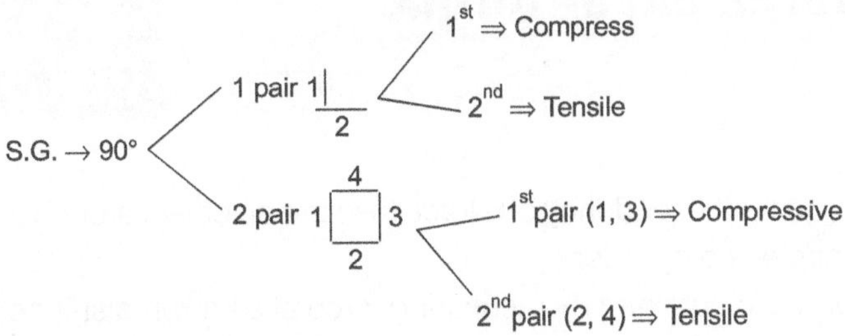

Compressive → $R - \Delta R$

Tensile → $R + \Delta R$

$$V_0 = \frac{E}{4} G_F \varepsilon (1 + \mu) \rightarrow 1 \text{ pair}$$

$$V_0 = \frac{E}{2} G_F \varepsilon (1 + \mu) \rightarrow 2 \text{ pair}$$

Piezo-electric Load Cell

- In case of piezo-electric load cell when the force (F) is applied on the piezo-electric crystal, it leads to the separation of charge on the opposite face of the crystal. The charge is then sent to the signal conditioning stage (i.e. the charge amplifier), the circuit is shown in the below figure.

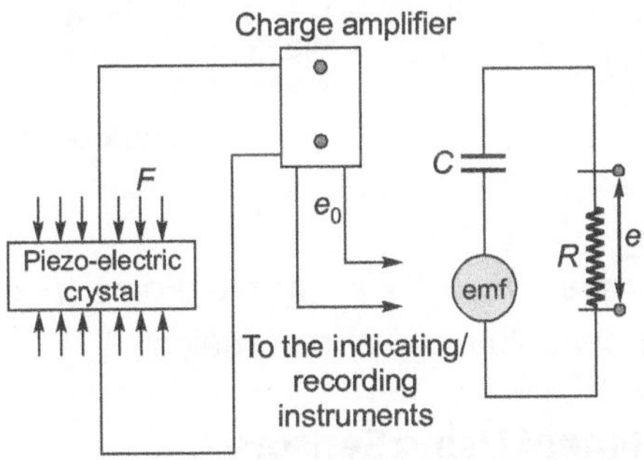

(a) (b)

■■■■

10

Measurement of Torque

- Torque is the tendency of a force to rotate an object around an axis of a circle or a cylinder.
- Torque is obtained by measuring force at a known distance.
  $T = F.r$ in (Nm);
  where, F = force in (N); r = distance perpendicular to the force where force is to be measured in (m).

Measurement of Torque

Torque Reaction Methods

- In this method either the power sources or the sink is mounted in trunnion bearing as shown in the figures below. This concept of mounting the source or the sink in trunnion bearings is called as the cradling.

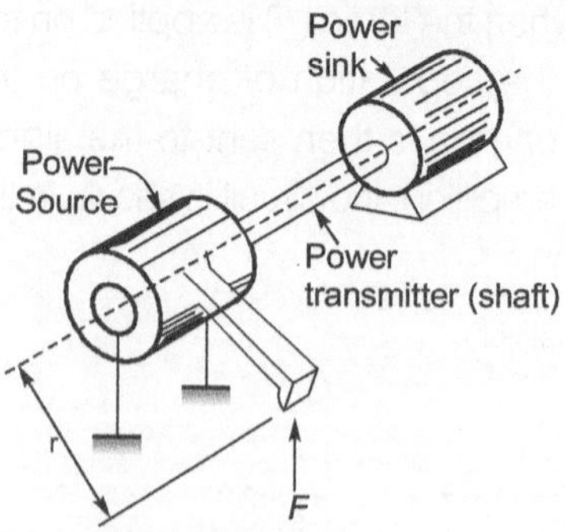

Cradled Source

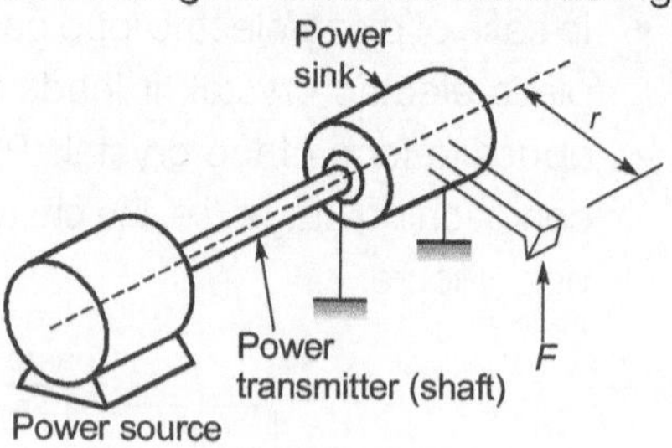

Cradled Sink

- Reaction torque $T'_0 = F \times r$ where, Force (F) can be calculated using either pendulum or platform scales.

Torque Measurement Using Sensors

Torsion Bar

- In this method a torsion bar of a specific length is made a part of the shaft and act as a sensing element, as shown in the figure below.

$$T = \frac{\pi}{16} G d^3 \theta$$

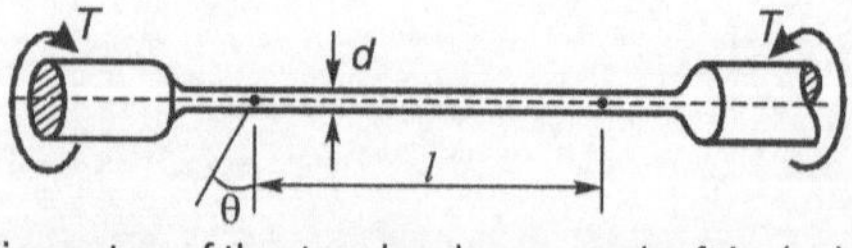

Where,
G = Shear modulus (GN/m³); d= Diameter of the torsion bare or shaft in (m)
θ = Shear strain

Strain Gauge Torque Meters

- In this method of torque measurement, the two strain gauges are mounted on a shaft at an angle of 45° with the axis of the shaft. The torque thus measured is given by

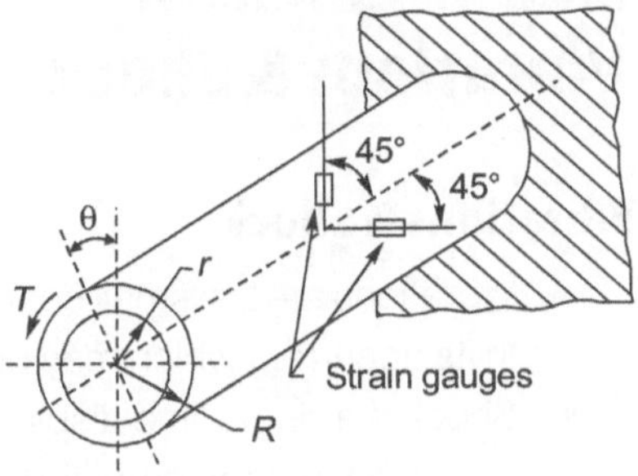

$$T = \frac{\pi G(R^4 - r^4)}{2L}\theta \qquad \varepsilon_{45^\circ} = \pm\frac{TR}{\pi G(R^4 - r^4)}$$

Where, G=Modulus of Rigidity (N/m^2); R = Outer radius of the shaft (m)
r = Inner radius of the shaft (m); L = Length of shaft (m)
θ = Angular deflection of shaft (rad); ε_{45° = Strain at 45°.

- To increase the sensitivity in the torque measurement a full bridge arrangement is used in which all the four gauges are mounted at an angle 45° to the shaft axis, as shown in the figure below.

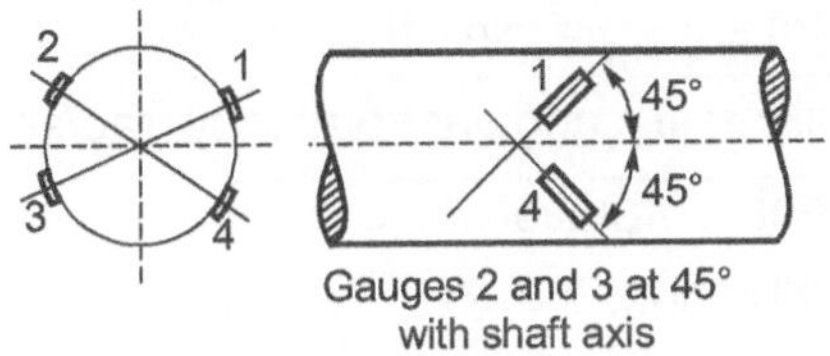

(a) Shaft with strain gauges

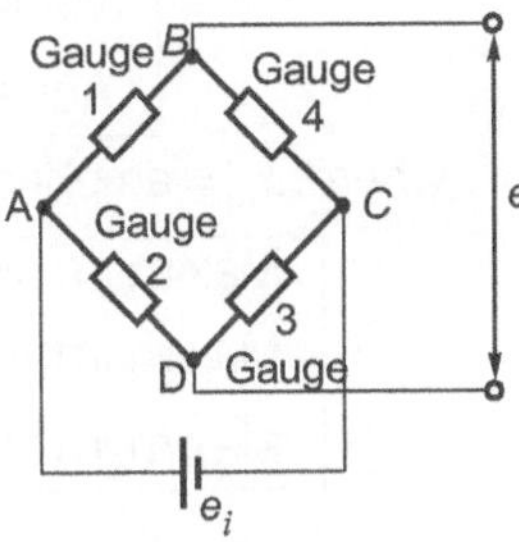

(b) Gauges connected in Bridge circuit

- The gauges are then put in the bridge arrangement, with the gauges 1 and 3 are placed opposite to each other and also 2 and 4 are put in the opposite arms.

Note:

- **For measuring torque using strain the gauges are connected in the bridge circuit such that the gauge on which same strain is applied (tensile or compressive) are opposite arm of the bridge while the gauges on which opposite strain is applied are in the adjacent arm of the bridge.**

■■■■

11

Measurement of Vibrations & Shock

Vibrations & Shock

- Vibrations are the small continuous displacement with repetitiveness in its occurring which may be periodic or non-periodic.
- Shock is a single time displacement which decay as time progresses.
- Shock have large amplitude as compared to the vibrations.
- The frequency of occurrence of displacement is high in case of vibrations, and very low in case of shock.

Nature of Vibrations

- Vibrations are the sinusoidal displacement of the vibrating member about its mean position.
- The nature of vibration depends on its amplitude and frequency of occurrence.

$$\text{Displacement } (x) = x_m \sin\omega t$$
$$\text{Velocity } (v) = x_m\omega \cos \omega t$$
$$\text{Acceleration } (a) = -x_m\omega^2 \sin \omega t$$

where, x_m = amplitude of vibrations; ω = angular frequency (rad/sec)

$$\text{Maximum displacement of vibration} = x_m$$
$$\text{Maximum velocity of vibrations} = x_m\omega$$
$$\text{Maximum acceleration of vibrations} = -x_m\omega^2$$

Vibrating Wedge

- When the amplitude of vibratory motion is greater than 1 mm a simple vibration measuring device named vibrating wedge is used.
- It consists of a simple wedge whose edges are inclined at an angle 'θ'. It is attached to the body whose seismic acceleration is to be measured.

$$X_m = A \tan\frac{\theta}{2}$$

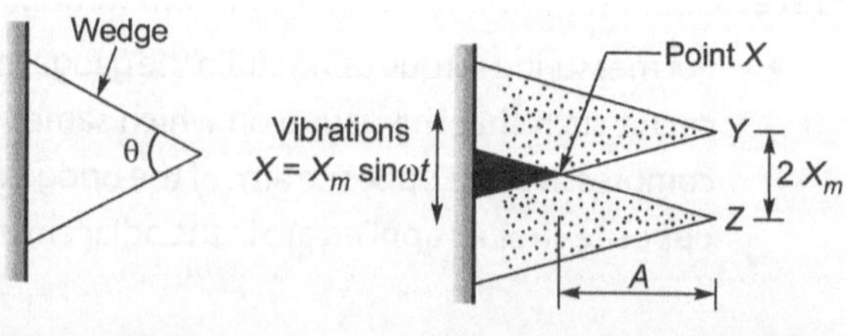

(a) Body is in rest (b) Body is in motion

Seismic Transducers

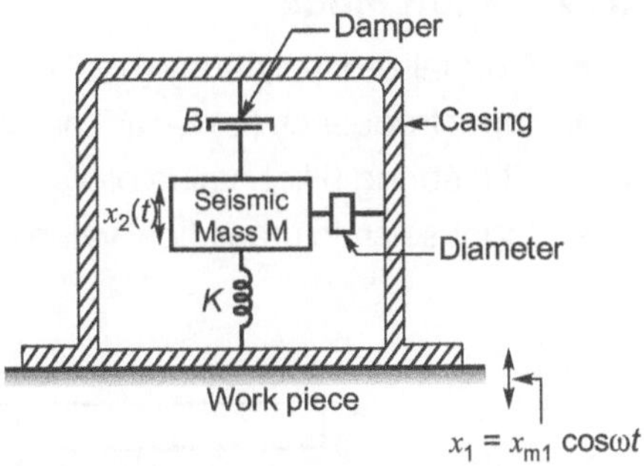

- In this device the seismic mass (M) is connected through a parallel spring and a damper arrangement to the housing frame, the housing frame is then attached to the body whose vibrations are to be measured.
- At steady state, the difference between the amplitude of vibrations of the input and that of the seismic mass i.e. $(x_2 - x_1)$,

$$(x_2 - x_1)_m = \frac{x_{m_1} u^2}{\left[(1-u^2)^2 + (2\zeta u)^2\right]^{1/2}} \quad \text{and} \quad \phi = \tan^{-1}\left(\frac{2\zeta u}{1-u^2}\right) \text{ rad.}$$

where, u = normalized frequency = ω/ω_n

$$\omega_n = \sqrt{k/M} \text{ and } \zeta = B/2\sqrt{kM}$$

- This type of transducers works in two modes.

Displacement mode

- Normalized frequency ($u \geq 2$)
- Input frequency (ω) >> ω_n (natural frequency)
- Weak spring (Low value of k)
- Heavy seismic mass (high value of M)

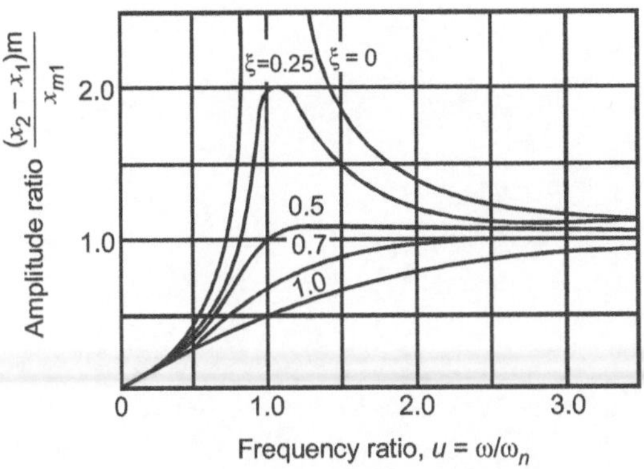

Acceleration mode

- Normalized frequency ($u \leq 0.4$)
- Input frequency (ω) << ω_n (natural frequency)
- Stiff spring (High value of k)
- Light seismic mass (low value of M)

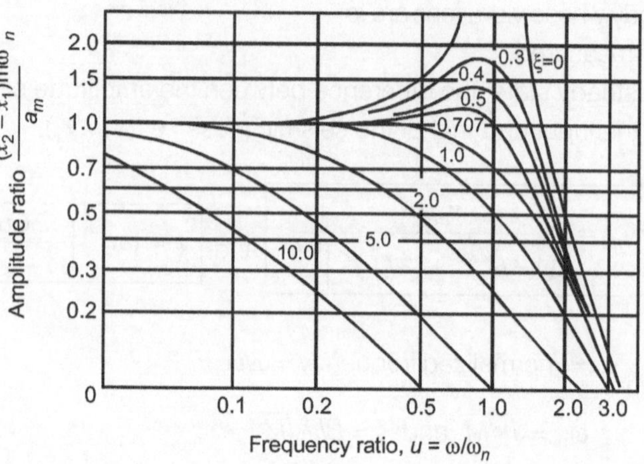

■■■■

12 Measurement of Liquid Level

Sight Glass Method (Direct Methods)

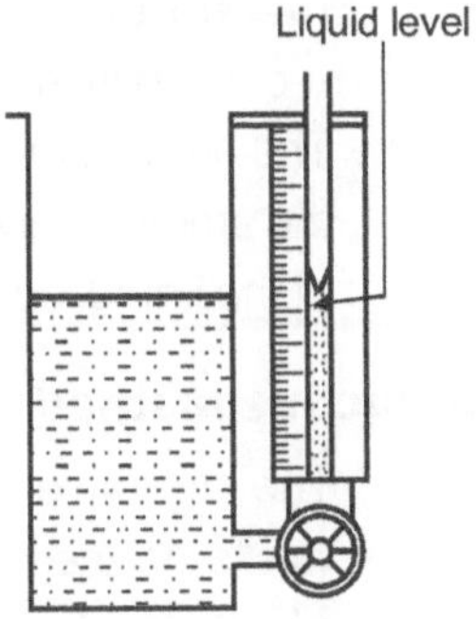

(Open Container)

- This method is used for level measurement of non-corrosive, non-staining and non sticky liquids.
- When the level of liquid increases in the tank, then the level of the liquid in the sight glass also increases linearity.
- The sight glass have the markings on it which gives the value of liquid level directly.
- These methods are used to measure liquid level upto about 1 m.

Inferential Methods

Buoyancy method

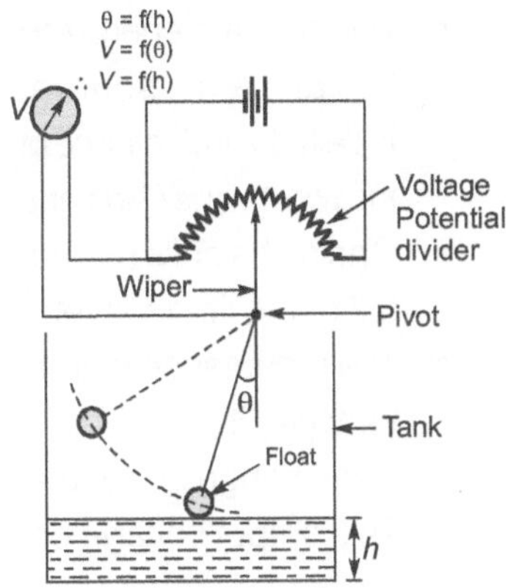

- Buoyancy method for the measurement of liquid level include the measurement using a float.
- The density of float is less as compared to the density of the liquid whose level is to be measured, so that the float do not get immersed in the liquid.
- When the liquid level in the tank increases, then the height 'h' increases and hence the float gets raised, leading to an increase in the angle 'θ'.
- With the variation of 'θ' the wiper movement on the potentiometer will take place which in turn will change the output voltage (V) in the voltmeter.

Resistive Method

- In this method the mercury is used as a conductor, a number of contact rods are present at various liquid level with resistance R in their arms as shown in the figure below.

- When the level of the liquid rises the electrode connected to the tube wall and the highest of the immersed rods gets short and the output current is then recorded in the ammeter.
- The output current recorded by the ammeter gives an indication of the liquid level (h) in the tank.

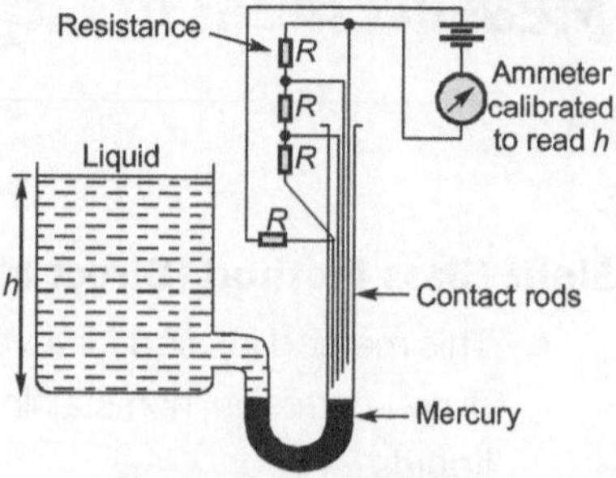

Measurement of liquid level by resistive method

Inductive Method

- The inductive method works on the principle of the change in permeability of the inductor with the change in the liquid level.
- The coils are energized by an a.c. source. Initially the inductance of the two coils is constant.

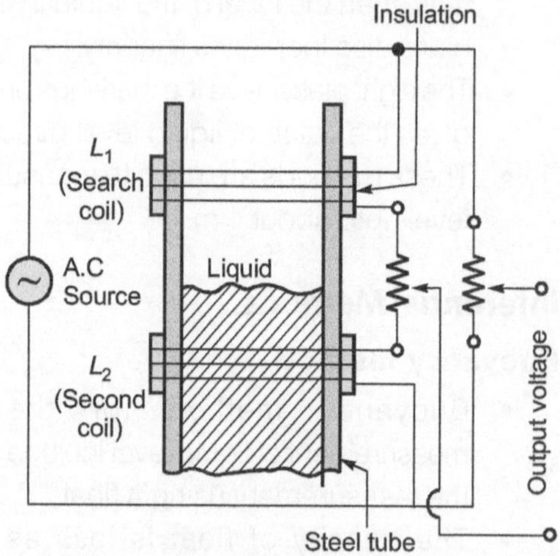

- One of the coils (L_1) acts as a search coil, which search for the rise or fall of the liquid level.
- The inductance of the search coil changes as the liquid rise or fall in between it.
- This change in inductance leads to the change in the output voltage. This output voltage can be calibrated in terms of the liquid level.

Capacitive Methods

Variable Area Method

- In this method one of the connection is taken from the outer container and other from the cylindrical rod completely dipped inside the container.
- The liquid acts as dielectric and the capacitance is varied with the height of the liquid as is given below in the expression for the capacitance.

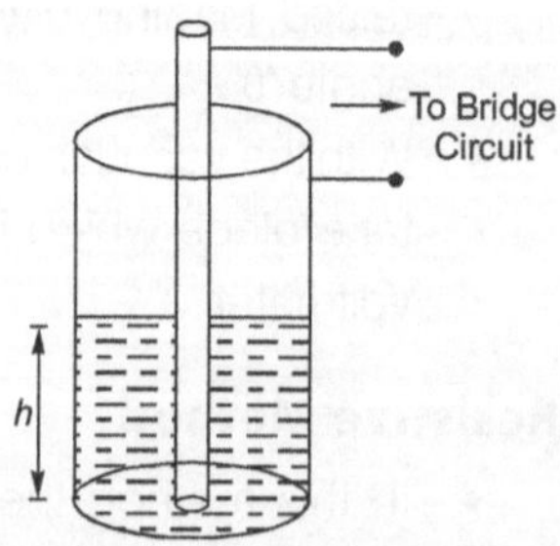

$$C = \frac{2\pi\varepsilon h}{\log_e\left(\frac{d_2}{d_1}\right)} F$$

where, ε = permittivity of the insulator (F/m)
h = height of the material (m)
d_1 = diameter of the metal rod (m)
d_2 = external diameter of the insulator (m)

Capacitive voltage Divider Method

- For the case of a highly conductive liquid, the surface of the liquid can be taken as one of the plates of a parallel plate capacitor.
- As the level of the liquid rise or fall in the container the value of the capacitance C_2 varies, this varied the output voltage E_0.

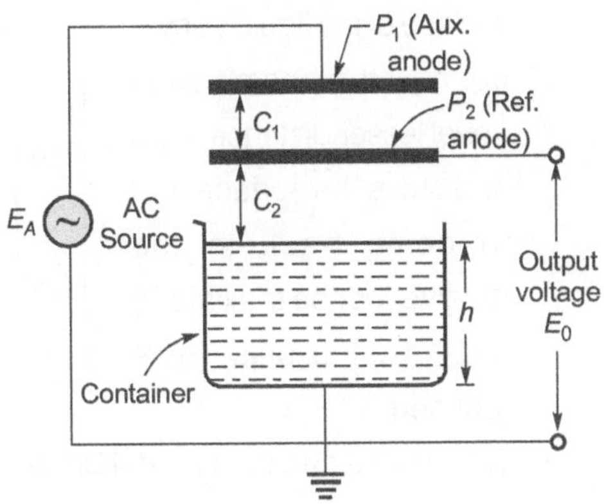

$$E_0 = \frac{C_1}{C_1 + C_2} E_A$$

Measurement of Liquid Level with Gamma Rays

- As the level of the liquid increases in the tank the intensity of the gamma rays reaching the counter decreases as some of the rays are absorbed in the liquid.
- The output of the Geiger-Muller counter is in the form of pulses, the pulses thus obtained are sent to the counter.

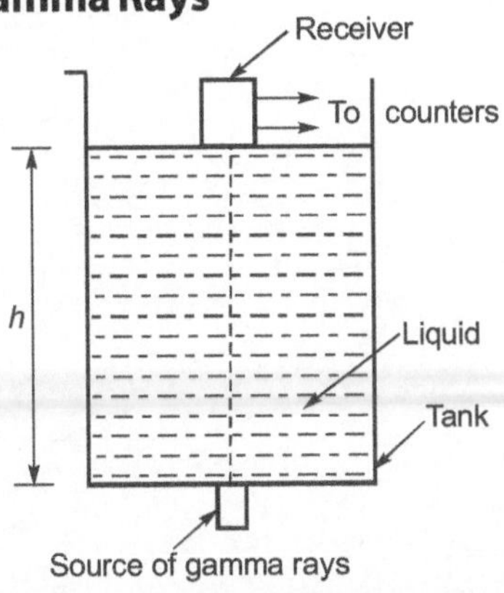

- Output intensity of the gamma rays passing through liquid is given by

$$\boxed{I = I_0 e^{-\mu h}}$$

where I_0 = intensity when the tank is empty
μ = atteneuation coefficient (m^{-1})
h = height of liquid level in the tank (m)

Ultrasonic Method

- As the level of the liquid increases in the tank the surface of the liquid comes up, now the beam has to travel lesser distance to go and strike the surface and to come back to the receiver hence the time (t) taken by the pulse gets reduced.

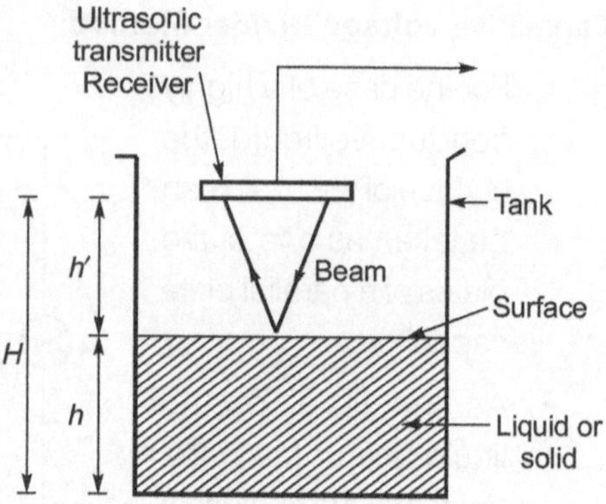

- The time (t) can then be calibrated in terms of the liquid level.

$$\boxed{t \propto h' \propto (H - h)}$$

where, t = time taken by the beam to strike the surface of the liquid and come back

■■■■

Measurement of Velocity

13

Measurement of Angular Velocity

Tachometer

- An instrument used for the measurement of angular velocity of shaft either by registering the total number of revolutions during the period of contact, or by indicating directly the number of revolutions per minute.

DC Tachnometer Generator

- D.C. tachnometer generators consists of a small armature which is coupled to the machine whose speed is to be measured. This armature revolves in the field of a permanent magnet.

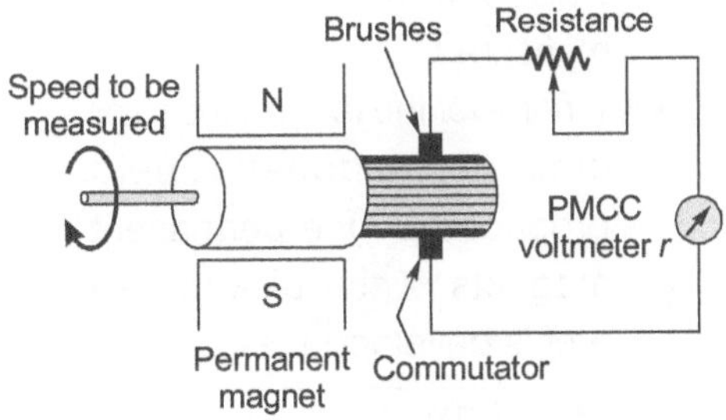

- The emf generated is proportional to the product of flux and speed. Since the flux of the permanent magnet is constant, the voltage generated is proportional to speed.
- The polarity of output voltage indicates the direction of rotation.

AC Tachnometer Generator

- It has rotating magnet which may be either a permanent magnet or an electromagnet.

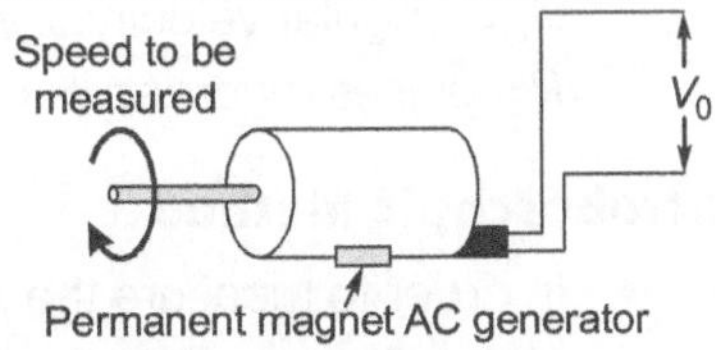

- The coil is wound on the stator and the rotation of the magnet causes an emf to be induce in the stator coil.
- The amplitude and frequency of this emf are both proportional to the speed of rotation.

Photoelectric tachometer

- When the hole comes in between the light source and light sensor the sensor turns on, and

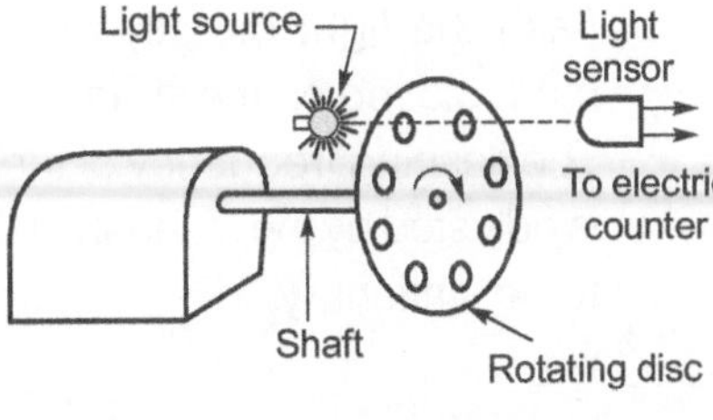

- When the opaque portion of the disc comes in between the light source and sensor no output is observed from the light sensor.

Note:

Gating Period: It is the fixed period of time, for which the number of pulses produced by the sensor are recorded.

Variable Reluctance Tachometer

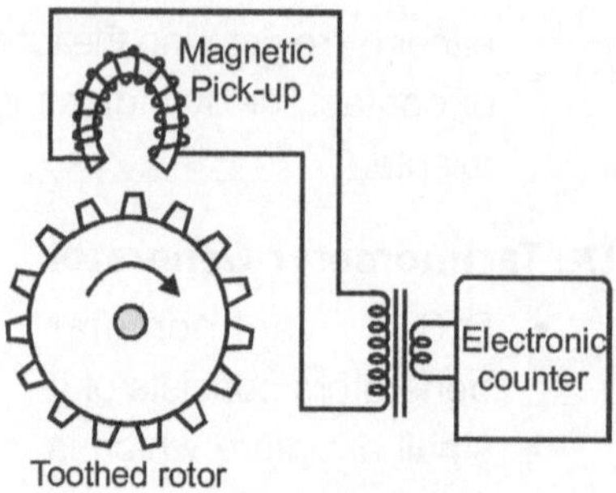

- The tachometer generator consists of a metallic toothed rotor mounted on the shaft whose speed is to be measured.
- Whenever the teeth of the rotor cuts the magnetic field produced by the permanent magnets of the pick-up, an emf is developed in it.

$$\boxed{E = bm\omega_r \sin m\omega_r t}$$

$$\boxed{\text{Speed of the shaft(n)} = \frac{P}{m}\text{ rps} = \frac{P}{m} \times 60\text{ rpm}}$$

where

b = amplitude of the flux variation; m = number of teeth

ω_r = angular velocity of wheel

P = pulses recorded per second

Stroboscopic Method

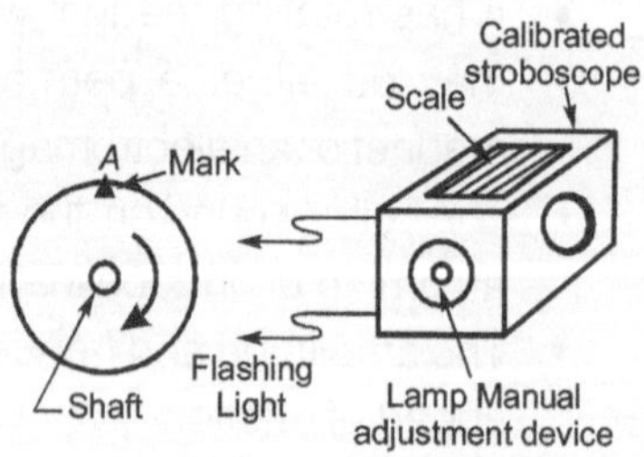

- In order to measure the speed of the shaft, a distinctive mark is made on the shaft or the disc attached to the shaft.
- The stroboscope is made to flash the light directly on the mark placed on the shaft.

The flashing frequency of the light coming through the stroboscope is adjusted by the manual adjustment available until the mark appears to be stationary.

- This frequency of the flashing light is equal to the speed of the shaft. The scale of the stroboscope is calibrated directly in terms of the speed of the shaft.

Case - I : Single line image formed at single flashing frequency due to m marks on the disk

$$\boxed{\text{speed of shaft } (N) = \frac{f}{m} \text{ rps}}$$

Case - II

Single line image formed at multiple frequencies

$$\boxed{\text{speed of shaft } (N) = \frac{f_h f_l (n-1)}{(f_h - f_l)} \text{ rps}}$$

where,

f_h = highest flashing frequency

f_l = lowest flashing frequency

n = no. of different frequencies for which stationary image is obtained

Measurement of Linear Velocity

1. Electro-Magnetic Transducers

This transducer utilizes the voltage produced in a coil on account of change in flux linkages resulting from change in reluctance due to linear velocity.

$$\boxed{e_0 = K\frac{dR}{dt}}$$

where, B = Flux density; wb/m^2

A = Area of coil; m^2

N = Number of turns of coil

V = Relative velocity of magnet with respect to coil

2. Digital Transducers

$$\boxed{\text{Velocity}(v) = \frac{nd}{t}}$$

d = Distance between marks ; m

n = No. of marks passing the given point in t seconds.

3. Doppler Effect

$$\boxed{\Delta f = f_t - f_r = \frac{2f_t \cos\theta}{C}V}$$

where, f_t and f_r = Frequency of the transmitter and receiver respectively.

C = Ultrasound speed

V = Linear velocity

■■■■

14 Measurement of Humidity

Humidity

- The amount of water-vapours present in the air is called as the humidity.

Specific Humidity

- It is the mass of water vapour per unit mass of the dry air.

Dry-Bulb Temperature

- It is the temperature of the air-water mixture as measured by the thermometer exposed to the mixture.

Wet-Bulb Temperature

- It is the temperature indicated by the thermometer covered with a cloth saturated with the liquid.

Relative Humidity (ϕ)

- Relative humidity of a mixture is the ratio of the actual mass of vapour present in the air-water mixture to the mass of vapour required to produce the saturated mixture at the same temperature.

$$\boxed{\phi = \frac{P_v}{P_g}}$$

where P_v = actual partial pressure of vapour

P_g = saturation pressure of the vapour at the temperature of the mixture

Methods of Measuring Relative

(i) Galvonometric Hygrometer

- In this hygrometeric method the air-water mixture is kept in contact with a certain chemical until all the water content of the mixture is absorbed by the chemical.
- The increase in the weight of the chemical is then calibrated in terms of the humidity content of the mixture.

(ii) Electrical Hygrometer

- Resistive Hygrometer
  - Some hygrometric salt like lithium chloride (LiCl) exhibit change in resistivity with humidity.
  - The resistance of the material may change to a wide range of about 10^4 to 10^9 ohm for the humidity change from 100 to 0 percent.
  - The higher the relative humidity, the greater the moisture LiCl will absorb hence, the lower will be the resistance.

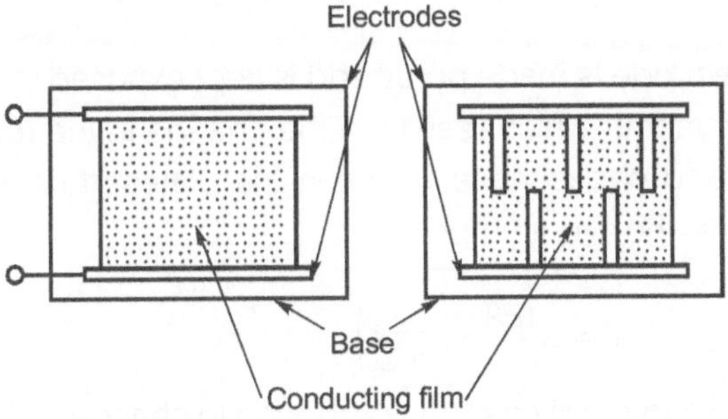

Capacitive Hygrometer

- Some hygroscopic materials change their dielectric constant with the change in humidity.
- This change in dielectric constant can be detected by a capacitive transducer by placing the hygroscopic material in between the two parallel plates of a capacitor.

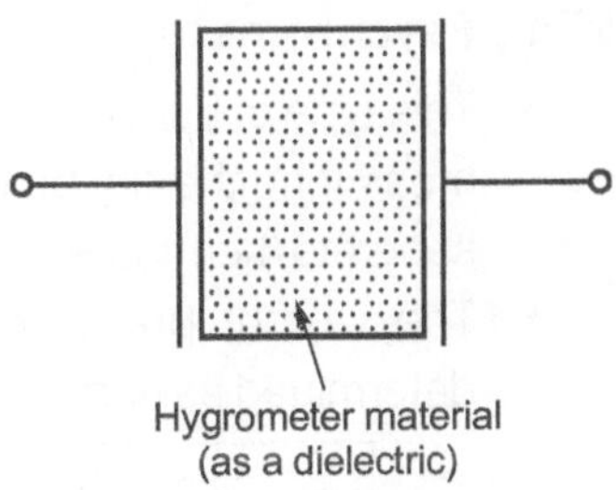

Microwave Refractometer

- In case of microwave refractometer, a system consisting of two cavities each of which is coupled to a klystron is employed.
- One of the two cavities contains the dry air and the other has the sample mixture whose humidity has to be calculated.
- The change in the dielectric constant changes the frequency of one oscillator, and the difference in the two frequencies is then calibrated in terms of the humidity of the mixture.

Aluminium Oxide Hygrometers

- The aluminium-oxide coating on the anodized aluminium exhibits a change in dielectric constant and the change in resistivity with the change in humidity.

- The hygrometer which use this property is made by using the base aluminium as one electrode and depositing a very thin electrode (usually gold) on the opposite side of the oxide as is given in the figure.

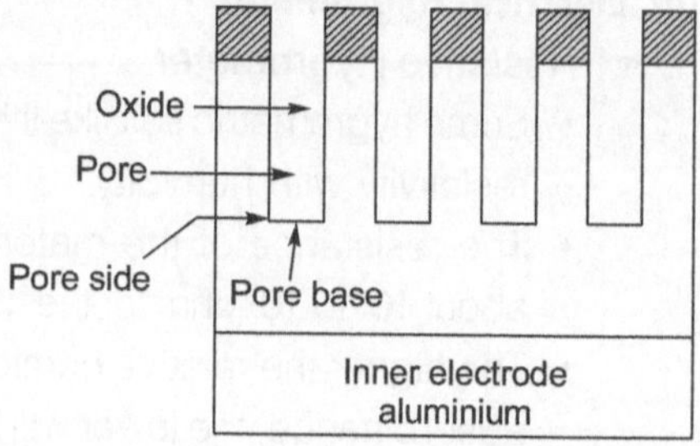

- This electrode is made porus and is kept exposed to the mixture of air and vapour which results to the change in the resistivity of the plates and capacitance between the plates, thus leading to the change in the overall impedance

$$Z = \sqrt{R^2 + \left(\frac{1}{\omega C}\right)^2}$$

when, R and C will change, Z is bound to change

Sling Psychrometer

- This type of humidity measuring scheme measures the humidity by measuring the dry bulb and wet-bulb temperatures with the help of seperate thermometers.
- The vapour pressure of the mixture of air and vapour is than determined as under.

$$\boxed{P_v = P_{gw} - \frac{(P - P_{gw})(T_{DB} - T_{WB})}{1537.8 - T_{WB}}} \quad \text{...(i)}$$

where, P_v = actual vapour pressure (N/m²)

P_{gw} = saturation pressure corresponding to wet-bulb temperature (N/m²)

P = total pressure of the mixture (N/m²)

T_{DB} = Dry bulb temperature (°C)

T_{WB} = Wet bulb temperature (°C)

Finding the relative humidity

Hence for an ideal gas,

$$\boxed{\frac{m_v}{m_{sat.}} = \frac{P_v}{P_g}}$$

Hence by putting the value of P_v to equation (i), the value of m_v (mass of water vapour) can be easily calculated which is a measure of humidity of the mixture.

■■■■

15 Measurement of Viscosity

Viscosity

- Viscosity is the inherent property of the fluid that offers resistance to the flow.
- Viscosity of Liquids decreases with increase in temperature.
- Viscosity of Gases increases with the increase in temperature.
- The coefficient of discharge (C_d) in case of the flow meters varies with the viscosity of the fluid.

$$\boxed{\text{Dynamic viscosity } (\mu) = \frac{\text{Shear stress}}{\text{Velocity gradient}}}$$

SI unit = $Nsec/m^2$

Commonly used unit = Poise

$$\boxed{1 \text{ SI unit of dynamic viscosity} = 10 \text{ Poise}}$$

$$\boxed{\text{Kinematic viscosity } (\nu) = \frac{\text{Dynamic viscosity}}{\text{Density}} = \frac{\mu}{\rho}}$$

Units of kinematic viscosity in SI = m^2/sec.

$$\boxed{1 \text{ SI unit of kinematic viscosity} = 10^4 \text{ stoke (St)}}$$

Viscometers

- The devices/instruments which are used for the measurement of viscosity are called the viscometers.

Falling Sphere Viscometer

- In this type of viscometer, a sphere ball of known diameter (d) is released in the guide tube into the liquid of density (ρ_l).
- The ball continue to accelerate downwards under the force of gravity until it attain the terminal velocity (V_T). (When the force due to gravity is just balanced by the buoyancy and viscous drag force of the liquid)

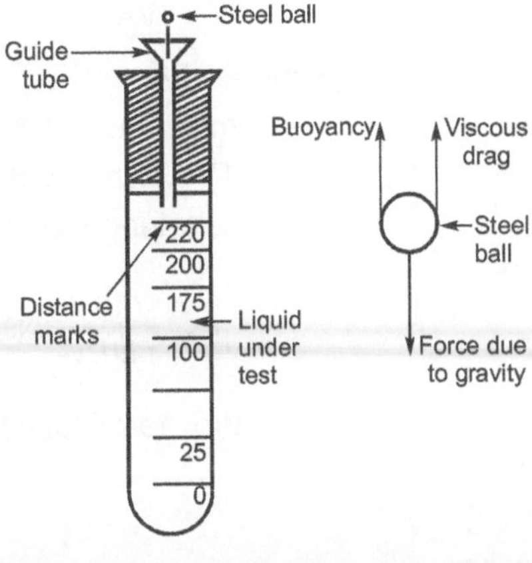

Calculating the Terminal Velocity

- We calculate the time, the sphere takes to fall 150-200 mm of distance, the terminal velocity can then be find by finding the ratio of the distance and time.

  The kinematic viscosity is given as

$$\boxed{\nu = \frac{k\,d^2\, g(\rho_s - \rho_l)}{\rho_l(18\,V_T)}\ \text{m}^2/\text{sec}}$$

where, k = constant (a correction factor)

d = diameter of the sphere (m)

V_T = terminal velocity (m/sec)

ρ_s and ρ_l = densities of the sphere & the liquid respectively (kg/m^3)

Falling Piston Viscometer

- In falling piston viscometer, the piston is allowed to fall in the fluid whose viscosity is to be determined. The terminal velocity of the piston, is then determined.

$$\frac{V}{y} = \frac{dV}{dy} \qquad (\because y \text{ is small})$$

also as, $\tau = F/A$

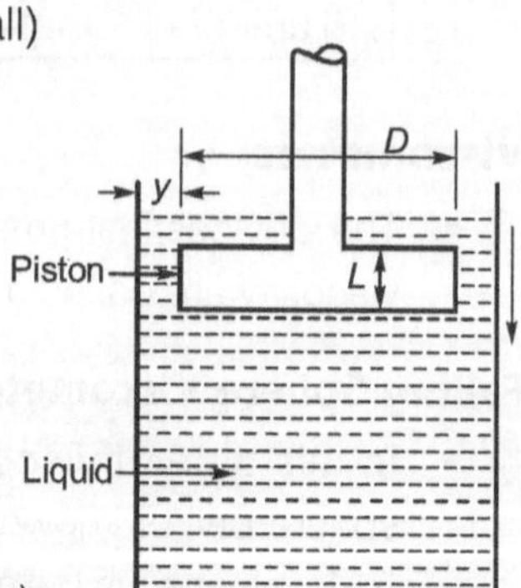

$$\therefore \text{Dynamic viscosity } (\mu) = \frac{F/A}{dV/dy}$$

$$= \frac{Fy}{VA}\ \text{Ns/m}^2$$

Now F = mg and $A = \pi$DL

m = mass of the piston (kg)

D = diameter of the piston (m)

L = length of the piston (m)

$$\therefore \qquad \mu = \frac{mg\,y}{\pi D L V_T}\ \text{Ns/m}^2$$

V_T = terminal velocity of the piston (m/sec)

Capillary Tube Viscometer

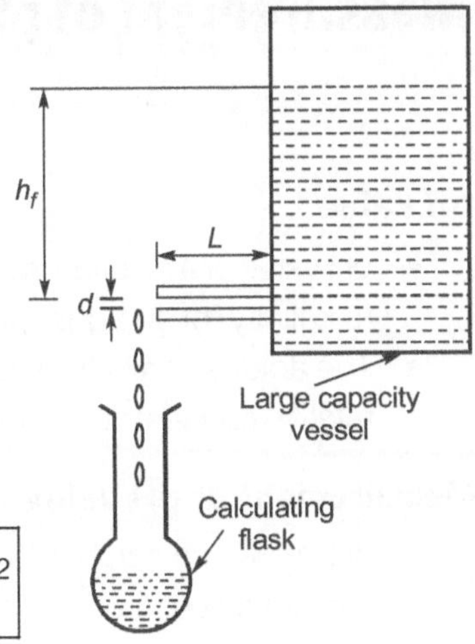

- In capillary tube viscometers, the fluid is allowed to fall through a narrow capillary from the storage tank into the collecting flask.
- The dynamic viscosity for the laminar flow of the fluid can be calculated by the Poiseuill's equation,

$$\text{Dynamic Viscosity}(\mu) = \frac{\pi \rho g h_f d^4}{128\, QL} \text{Ns/m}^2$$

where, ρ = density of liquid (kg/m^3)
h_f = friction head loss of flow (m)
d = diameter of capillary tube (m)
Q = liquid flow rate (m^3/sec.)
L = length of capillary tube (m)

- The flow rate '*Q*' is calculated by noting the time taken by the fluid to fill a specific volume of the flask.

Orifice Type Viscometers

- These meters determines the viscosity by noting the time taken by the fluid to fill a particular fixed volume of the flask at the constant temperature conditions.
- These type of meters are used to find the viscosity of oil and fuel.
- The most common type of orifice viscometers are, Redwood viscometer used in U.K. and Saybolt viscometer used in U.S.A.

Redwood Viscometer

- In a Redwood viscometer the fluid is allowed to flow through the orifice, and the time '*t*' is noted to fill the flask upto a volume of 50 ml. The viscosity is then '*t*-Redwood seconds'.

Saybolt Viscometer

- In a Saybolt viscometer, the time taken by the fluid flowing through the orifice to fill a 60 ml of the flask is calculated, the viscosity is then expressed in terms of Saybolt universal seconds.

■■■■

Measurement of pH Value

16

pH Value

- pH value of a solution is a measure of the degree of acidity or basicity (alkalinity) of an aqueous solution.
- The acidity or alkalinity of an aqueous solution is determined by the relative concentrations of hydrogen and hydroxyl ions in the solution.

Measurement of pH Value

- pH value of a solution is the negative logarithm of the hydrogen ion concentration.

$$pH = -\log_{10}[H^+]$$

- In pure water, the concentration of H^+ ions is 10^{-7} gm/*l* at 25°C. So, the pH value of water is 7.
- The pH value for the acidic solutions is in the range 0 – 7 and for the alkaline solutions in the range of 7 – 14.

Note: ....

- **When the hydrogen ions are present in excess in the solution, the solution becomes acidic, and**
- **When the hydroxyl ions are present in excess, the solution becomes alkaline (basic).**
- **Dissociation constant [k] is a constant value equal to 10^{-14}, it is the product of hydrogen ion concentration and the hydroxyl in concentration of a solution.**

$$[k] = 10^{-14} = [H^+]\,[OH^-]$$

....

Measuring pH Value

- The pH value of a solution is measured by using pH electrode.
- It consists of a pair of electrodes (measuring and reference electrode), both the electrodes are dipped in the solution of unknown pH.
- These two electrodes form two half-cells, the net potential developed is the difference between the individual electric potential developed in each half cell.
- The potential developed in the reference cell is constant, and the measuring cell potential depend on the hydrogen – ion concentration of the solution and is governed by the Nernst's equation.

$$E = E_0 + \frac{RT}{nF} \ln(aC)$$

where, E = e.m.f. of the half cell

E_0 = e.m.f. of the half cell under saturated condition.

R = Gas constant (8.314 J/°C).

T = absolute temperature (K).

n = No. of electrons transferred.

F = Faraday constant = 96493 C mol^{-1}

a = activity coefficient ($0 < a < 1$); (for a very dilute solution, $a \to 1$)

C = molar concentration of ions.

Measuring Electrode

- The measuring electrode is made of thin sodium ion selective glass.
- A potential is developed across the two surfaces of thin glass bulb, when dipped in aqueous solution.
- This potential developed is sensitive to the H^+ ion concentration, having a **Sensitivity of 59.2 mV/pH at 25°C.**
- The Buffer solution inside the glass bulb has a constant H^+ ion concentration and provides electrical connection to the lead wire.

Reference Electrode

- The basic purpose of the reference electrode is to provide continuity to the electrical circuit, since the potential across a single half-cell cannot be measured.

  Net potential between the two electrodes is given by:

$$\Delta v_2 = -2.30 \frac{RT}{\tau} \log \frac{C_H}{C_R}$$

where, R = universal gas constant, 8314J/kg.mol °K

T = absolute temperature °K

J = Faraday's constant, 0.647×10^7 C/kg.mol

C_H = hydrogen ion concentration in solution

C_R = concentration in glass electrode

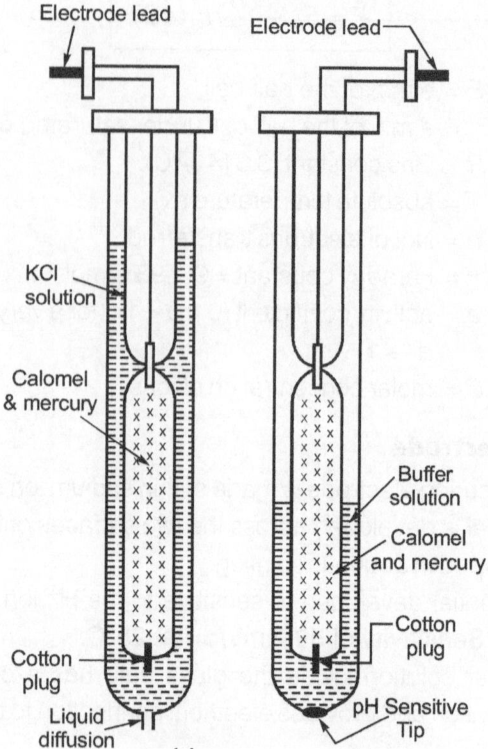

(a) Reference Electrode

(b) Measuring Electrode

■■■■

A Handbook on
Instrumentation Engineering

9

Optical Instrumentation

CONTENTS

BASIC OPTICS

1

Basic Optics

- Light or visible light is the portion of electromagnetic spectrum that is visible to human eye and is responsible for the sense of sight.
- Visible light has a wavelength in the range of about 380 or 400 nm to about 760 or 780 nm, with a frequency of 405 THz To 790 THz.

  Speed of light in vacuum = 3×10^8 m/sec.

$$\text{Refractive index of a material} = \frac{\text{Speed of light in vaccum}}{\text{Speed of light in that medium}}$$

Reflection

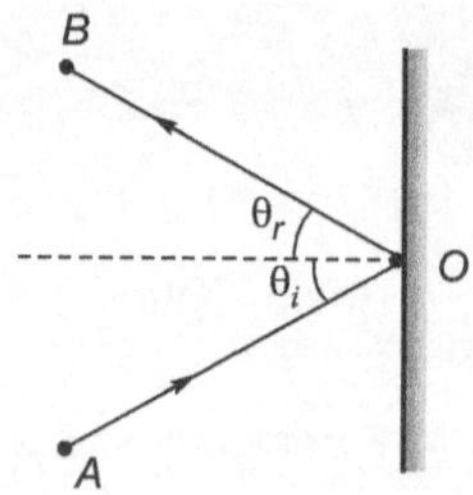

- Bending of light in the same medium in which it is travelling, when it strikes the surface of some other medium is called as the Reflection of light.

  AO = Incident light ray
  OB = reflected light ray
  θ_i = incident angle
  θ_r = reflected angle

Law of reflection

$$\angle\theta_i = \angle\theta_r$$

Refraction

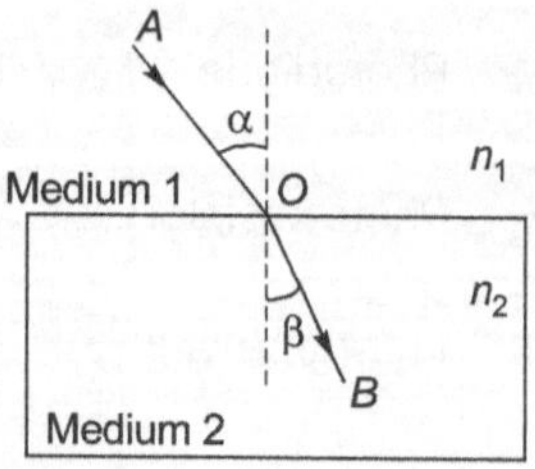

- Bending of light into the medium to which it strikes is called as the Refraction of light.

  AO = incident Ray

  OB = Refracted Ray

  α = angle of incidence; β = angle of Refraction
  n_1 = RI of medium 1; n_2 = RI of medium 2

Snell's Law

According to Snell's law $n_1 \sin\alpha = n_2 \sin\beta$

Conditions of Refraction

(i) Rarer to Denser medium:

$$\boxed{\theta_1 > \theta_2}$$

when a ray of light travels from rarer to a denser medium it bends towards the normal.

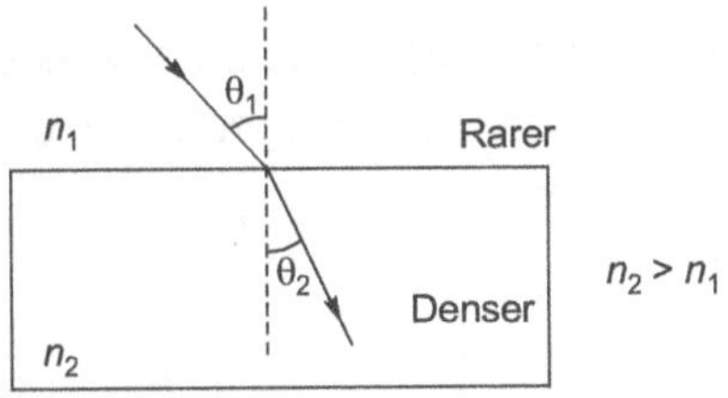

(ii) Denser to Rarer Medium:

$$\boxed{\theta_1 < \theta_2}$$

when a ray of light travels from denser to a rarer medium it bends away from the normal.

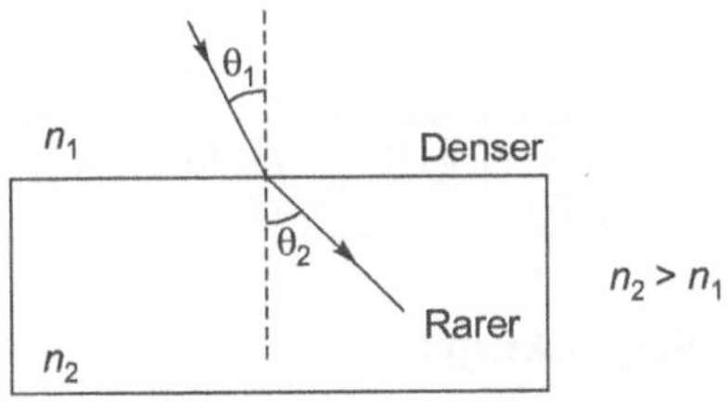

Refractive Index of one Medium with Respect to other

$_1\mu_2$ = RI of 2nd medium with respect to 1st.

$$_1\mu_2 = \frac{\mu_2}{\mu_1}$$

μ_2 = RI of medium 2nd with respect to air

μ_1 = RI of medium 1st with respect to air

Note:

- Light velocity change from one medium to the other.
- Velocity of light in any medium = $\frac{\text{Velocity of light in air}}{\text{RI of medium } (\mu_m)}$

$$\boxed{V_m = \frac{3 \times 10^8}{\mu_m}} \qquad \boxed{_1\mu_2 = \frac{\mu_2}{\mu_1} = \frac{V_{m_1}}{V_{m_2}} = \frac{\lambda_1}{\lambda_2}}$$

Brewster's Law

- This law states that when light is incident at a polarizing angle the reflected and refracted ray are at right angle to each other

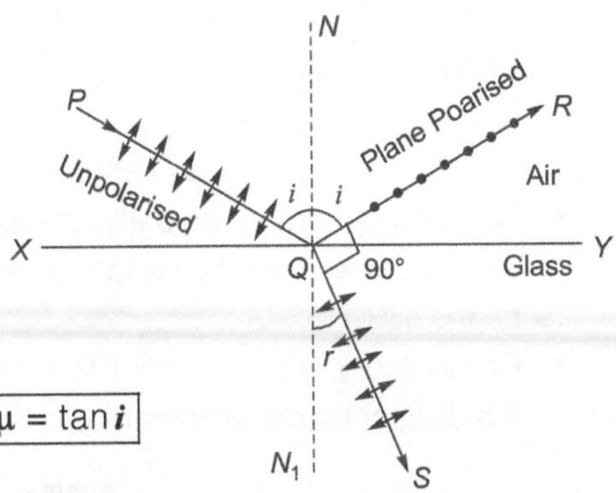

$$\boxed{\mu = \tan i}$$

Reflection Loss

- When light falls on any medium then not all the light is refracted, some part of light is reflected back this is called as the reflection loss.

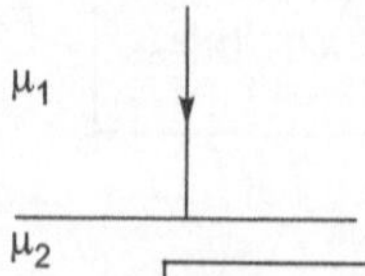

$$\text{Reflection loss (RL)} = \left(\frac{\mu_2 - \mu_1}{\mu_2 + \mu_1}\right)^2$$

$$\boxed{\text{Total \% Reflection loss} = \left(\frac{\mu_2 - \mu_1}{\mu_2 + \mu_1}\right)^2 \times 100\%}$$

Critical Angle

- The critical angle is the angle of incidence at which the angle of refraction become 90° to the normal.

  AO = Incident Ray

  OB = Refracted Ray

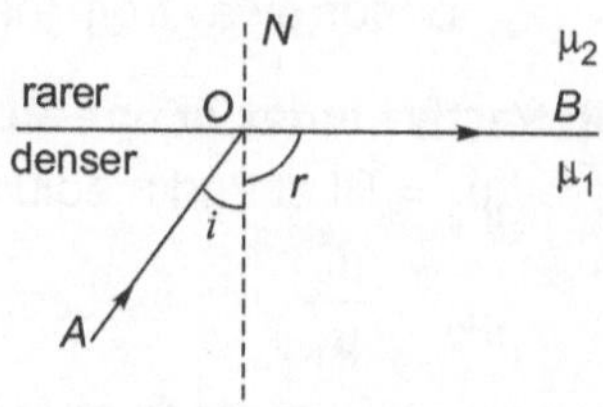

$$\boxed{\sin i_c = \frac{\mu_2}{\mu_1}}$$

Total Internal Reflection (TIR)

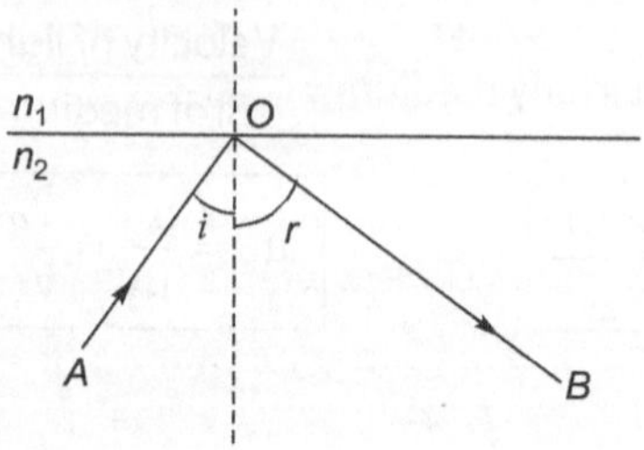

where $n_2 > n_1$

$i > \theta_c$

θ_c = critical angle

- Bending of light in the same medium when it strikes the surface of any other medium at an angle greater then the critical angle is called as the TIR.
- Compulsory condition for TIR to occur in that light must travel from the denser to the rarer medium.

■■■■

2

FIBER OPTICS

Optical Fiber

- Optical fiber is a very thin and flexible medium with a cylindrical shape.

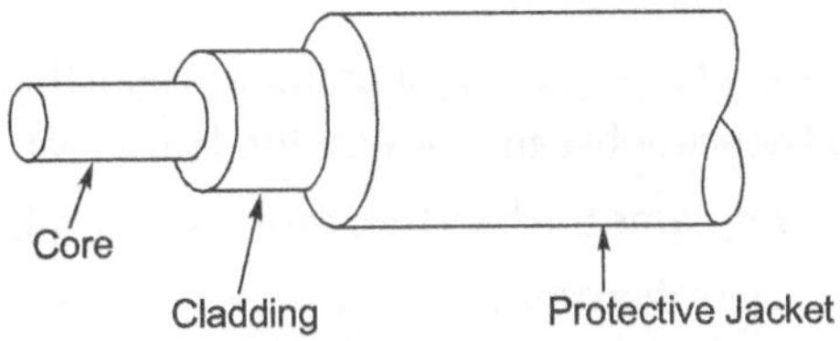

(Optical Fiber Crossection)

It consists of three sections.

(i) Core (ii) Cladding (iii) Protective Jacket

- Principle of working of optical fiber is total internal reflection of light (TIR).
- The refractive index of core is always greater than the refractive index of the cladding, so that TIR can take place.

Refractive Index Sequence

$$\mu_{CORE} > \mu_{CLADDING} > \mu_{JACKET}$$

Acceptance Angle

- The maximum angle of incidence to the fiber for which if light is incident, it gets totally internally reflected and hence travels successfully through the fiber is called as the acceptance angle (A.A.).

In air $$A.A = \sin^{-1}\left(\sqrt{n_1^2 - n_2^2}\right)$$

In medium of RI (μ) $$A.A. = \sin^{-1}\left(\frac{\sqrt{n_1^2 - n_2^2}}{\mu}\right)$$

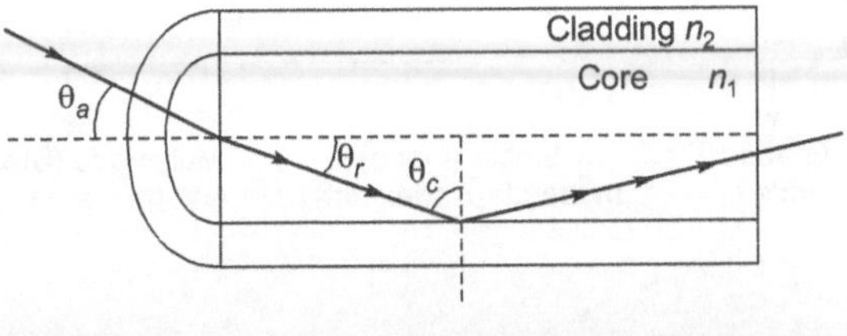

Numerical Aperture

- Numerical aperture is the light gathering ability of the optical fiber. Mathematically is the sine of the acceptance angle (θ_a).

$$\boxed{\text{N.A.} = \sqrt{n_1^2 - n_2^2} = \sin\theta_a}$$

For all light gathering N.A. of the fiber is unity.

Note:

- **The larger the N.A. the greater the amount of light that a optical fiber can accept, and hence light can be transmitted to a greater distance.**
- **If the N.A. is very large the band width of the transmission is decreased.**
- **N.A. is always less than unity.**

Materials used to manufacture optical fibers

1. All glass fibers — Core → glass; Cladding → glass
2. All plastic fibers — Core → plastics; Cladding → plastics
3. Plastic clad silica (PCS) fibers — Core → silica/glass; Cladding → plastics
4. Core → plastic; Cladding → glass } not used

- In (1) and (2) combination doping elements are added to change its refractive index. Thus the original material is SiO_2, we can dope some other materials GeO_2 or P_2O_5,
  To increase (↑) R.I. dope with GeO_2 or P_2O_5
  To decrease (↓) R.I. dope with B_2O_3.

Classification of Optical fiber

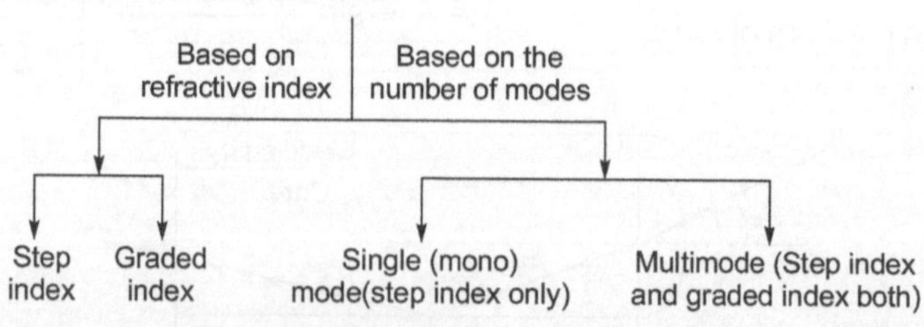

Based on R.I.

1. Step index

   In this R.I. changes suddenly like step function between core and cladding

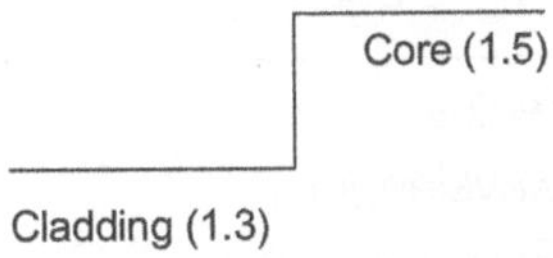

(Step index fiber)

Drawback

- Pulse distortion is more.
- Losses are more.
- Information carrying capacity is less.

2. Graded Index
   - R.I. is not uniform within the core.
   - At the axis of core R.I. is maximum.

   From core to cladding RI decreases

   Light propagation through graded index:

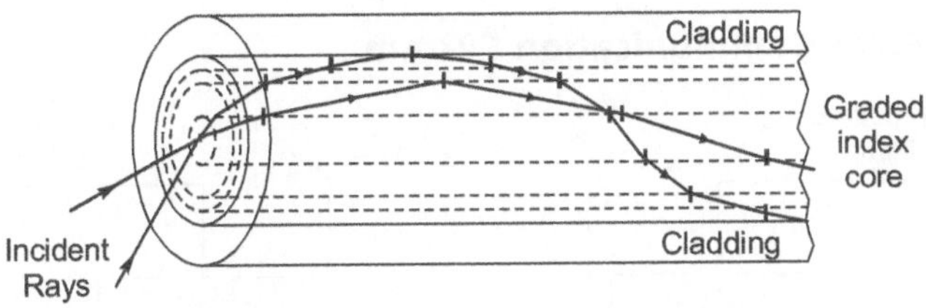

- Ray of light is offered a parabolic or the sinusoidal path when passed through a graded index fiber.

Advantages

- Pulse distortion is reduced
- Losses are reduced
- Information carrying capacity is increased

Disadvantages

- Difficult to manufacture
- costly

Based on Mode

1. Single or Monomode optical fiber
   - Narrow core is used (2 μm to 10 μm).
   - $\mu_{core} - \mu_{cladding}$ = Very small.
   - It has only one path for transmission i.e. only a single mode can travel through it.

- It is always step index as core is narrow which cannot be made graded index.
- V-number decides how many mode or path will be there.

$$V = \frac{2\pi r}{\lambda}\sqrt{n_1^2 - n_2^2} \leq 2.4045$$

r = radius
λ = wavelength
n_1 = RI of the core
n_2 = RI of the cladding

2. Multimode optical fiber
   - Wide core is used (> 50 μm)
   - More than one mode can travel through it

$$\text{No. modes or path} = \frac{V^2}{2}\left(\begin{matrix}\text{for step}\\ \text{index fiber}\end{matrix}\right)$$
$$= \frac{V^2}{4}\left(\begin{matrix}\text{for graded}\\ \text{index fiber}\end{matrix}\right)$$

Optical Fiber Communication System

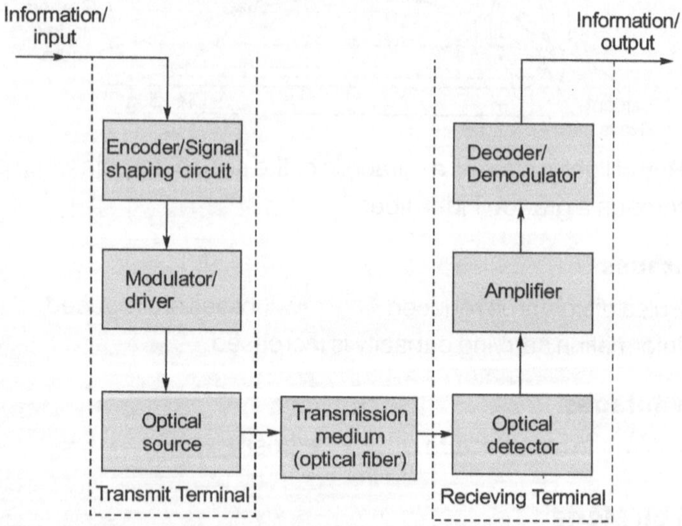

Characteristics of Optical Communications

1. Attenuation loss must be low.
2. Pulse distortion must be low.
3. Bandwidth (information carrying capacity) must be high.

1. **Attenuation Losses**
   - The output at the output end of the fiber is found reduced in power as compared to the input due to absorption, scattering, reflection etc.

$$\text{Attenuation (db/km)} = \frac{10}{L}\log\frac{P_o}{P_i}$$

Where, P_0 = output power; P_i = input power
L = length of the optical fiber

Ideally attenuation losses in the fiber must be 0 dB/km, but it is not practically feasible.

Attenuation Losses Occur Due to

(i) Absorption (ii) Scattering losses (iii) Radiation losses

2. **Pulse Distortion or Dispersion**
   - The broadening of the light pulses at the output of the optical fiber cable is called as the dispersion. Its unit is (n sec/km).

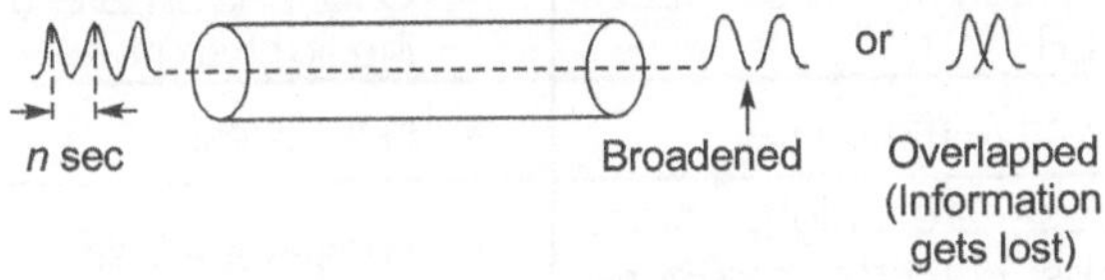

$$\Delta T = \frac{n_1 L}{c}\Delta \qquad \text{(For step index)}$$

$$\Delta T = \frac{n_2 L}{2c}\Delta^2 \qquad \text{(For graded index)}$$

$$\Delta = \frac{n_1}{n_2} - 1$$

where, c = speed of light

ΔT for graded index > ΔT for step index

3. **Bandwidth**
   - Bandwidth of an optical fiber is the maximum number of pulses that can be sent and received without overlapping.
   - The maximum number of pulses are to be limited by the distortion, as we have to keep some time gap between two successive pulses ΔT which reduces the number of pulses and hence the bandwidth.

Preferable Characteristic of Optical Fiber are

- High Bandwidth
- Low losses
- Low core diameter
- Low pulse distortion
- Low numerical aperture (Low. N.A.)

Coupling Losses of the Fiber

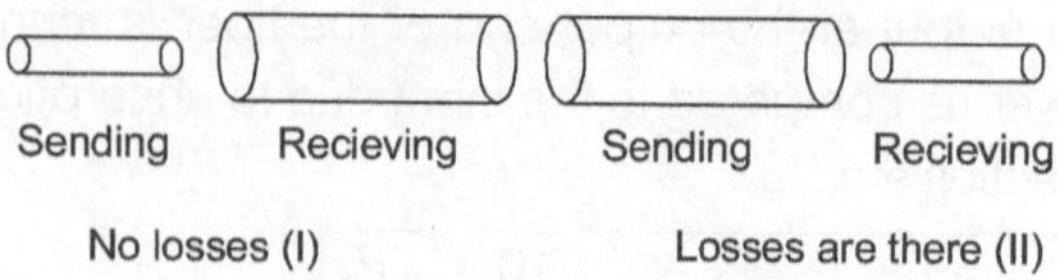

- Losses will occur in case (II) and no losses will be there in case (I)

$$\text{Loses} = 20 \log \frac{(NA)_{sending}}{(NA)_{Reciever}}$$

Comparison of Optical Fibers

Modewise Comparison

Multimode	Singlemode
• Core diameter is large so easy to launch the multiple modes	• Core diameter is less, it carries only a single mode
• Light source used LED, LASER	• Only LASER are used as the light source
• They have more N.A.	• N.A. is less
• Losses are high, since number of modes are more	• Losses are less
• Bandwidth is low	• Bandwidth is high
• Cost is low	• Cost is high

Material wise Comparison

All glass	Plastic silica	All plastic
Losses are less	Medium losses	High losses
Low mechanical strength	Medium mechanical strength	High mechanical strength
Low N.A.	Medium N.A.	Low N.A.
Long distance application (∵ of low losses)	Medium distance application	Short distance application

Applications of Optical Fibers

1. Industrial applications
2. Medical
3. Military operations
4. Telecommunication

■■■■

3

LASERS

Lasers

- LASER is an acronym of Light Amplification By stimulated Emission of Radiation.
- A laser is a device that emits light through a process of optical amplification based on the stimulated emission of photons. The light beam obtained by LASER is highly directional and intense.

Principle of Laser

- Consider the case when an atom or the molecule in an excited state with energy E_2 and makes a transition to lower energy state E_1. It emits electromagnetic radiation of frequency n given by

$$h\nu = E_2 - E_1$$

$$\nu = \frac{E_2 - E_1}{h}$$

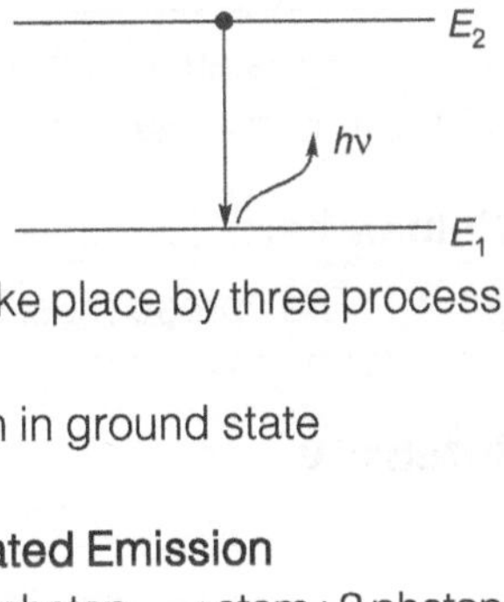

where h = Plank's constant

- The transition of electron in an atom can take place by three process

(i) **Absorption**

atom + photon ⟶ atom* ; atom = atom in ground state

atom* = atom in the excited state

(ii) **Spontaneous Emission**

atom* ⟶ atom + photon

(iii) **Stimulated Emission**

atom*+photon⟶atom+2 photon

Population Inversion

- Let N_1 and N_2 be the number of atoms in the lower and upper levels respectively.
- When the photon of energy $h\nu$ is incident on it may lead to transition from E_2 to E_1 or from E_1 to E_2.
- If $N_2 < N_1$ = Absorption of light is dominant.

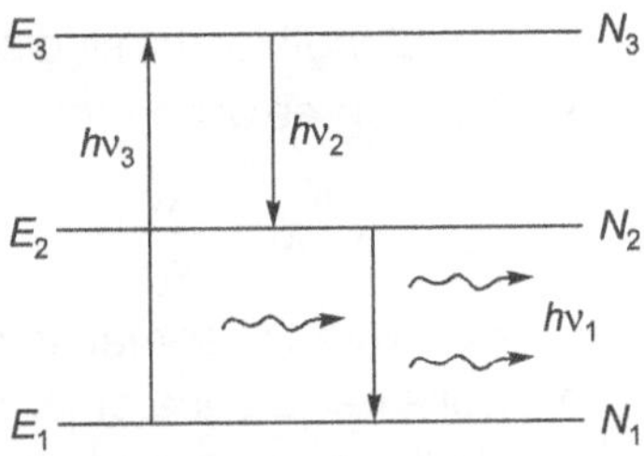

- If $N_2 > N_1$ = Stimulated emission will dominate.
- When this happens, the population inversion, takes place. To start population inversion some starting external energy must be supplied to the atoms of the material. Population inversion constitutes the basic principle of the LASER action.

Characteristics of Laser Beam

(i) Very high directionality (ii) high intensity
(iii) Monochromatic (iv) Coherent

Directionality

(a) **Angle of divergence or angular spread**

$$\Delta\theta = \frac{1.22\,\lambda}{d}$$

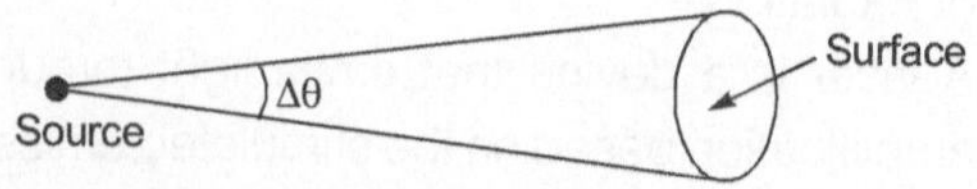

d = aperture diameter of light source from where the light is comming.
λ = wavelength of the light comming from the laser source.
$\Delta\theta$ = angular spread of the laser beam.

(b) **Areal Spread**

$$\text{Areal spread} = \pi(r\theta)^2$$

r = distance of the surface from the laser source (m).

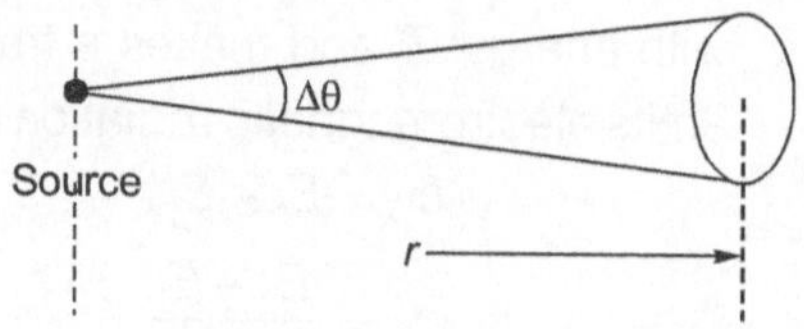

Remember:

- The spread must be low for a better and efficient laser beam.

Intensity

$$\text{Intensity } (I) = \frac{\text{Power}}{\text{Area}}$$

Monochromaticity

- A laser beam is highly monochromatic i.e. it has only a single wavelength or the single colour.
- The degree of non-monochromaticity is defined by (ϵ)

$$\epsilon = \frac{\Delta\nu}{\nu_0} \qquad \nu_0 \Rightarrow \text{Central frequency}$$

$\Delta\nu$ = width of spread in frequency

- Consider a case of large number of modes travelling into a laser cavity of length (L).
  The difference in frequencies between consecutive modes is $\Delta\nu$ and the length of the cavity is L

$$\therefore \quad \Delta\nu = \frac{c}{2L} \qquad \text{No. of modes} = \frac{\text{Bandwidth}}{\Delta\nu}$$

Coherence

- Coherence is a measure of the degree of phase correlation that exists in the radiation field of light source at different locations & at different times.
- Light from a conventional source is a jumble of tiny seperate waves that cancel each other in a random order. The wavefront so produced changes from one point to other and varies from instant to instant. Hence there are two seperate concepts of coherence:
  (i) Spatial coherence (ii) Temporal coherence

Spatial Coherence

- If two fields at two different points on a wavefront of a electromagnetic, wave has a constant phase difference over any time '*t*', they possess spatial coherence.

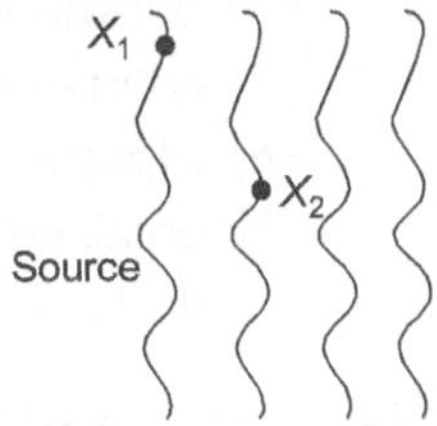

Temporal Coherence

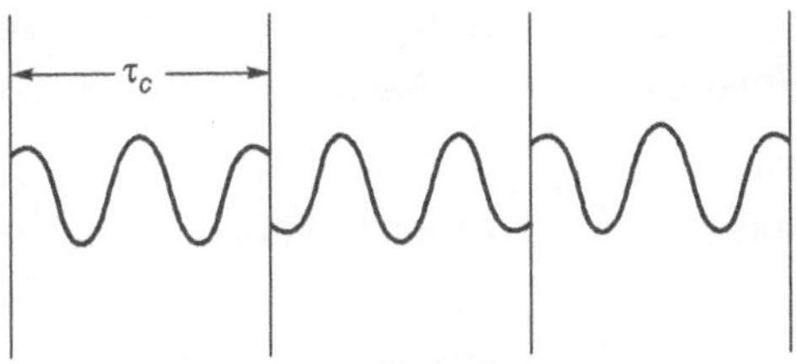

(Electromagnetic wave with coherence time τ_c)

- Temporal coherence refers to the correlation between the field at a point and the field at the same point at a later time.
- The time for which the wave remain sinusoidal and do not break in phase is called as the coherence time (τ_c).
  The distance travelled by the wave in the time interval τ_c is called as the length of coherence (*l*)

$$\boxed{l = c\tau_c}$$

Bandwidth ($\Delta\nu$) is related to τ_c as

$$\boxed{\Delta\nu = \frac{1}{\tau_c}}$$

Basic Parts in Laser

1. Active Material

- It is the material kept inside the cavity.
- It participate in the laser generation process.

2. Excitation/Pumping Mechanism

- The process by which the atoms are raised from the lower level to the upper level so that neccessary population inversion is obtained is called as the pumping.

  It is classified as:

  (a) **Optical Pumping:** In this method the population inversion is attained by means of strong source of light such as gas discharge, flash lamp or arc lamp.

  (b) **Chemical Pumping:** In this method population inversion is obtained by suitable chemical reaction without any need of other source of energy.

  (c) **Electrical Pumping:** In this method population inversion is obtained by means of an intense electric discharge in the medium and is suited to gas medium.

Classification of Lasers

Solid	Gas	Semiconductor
• Ruby laser (made of Ruby crystal)	• He-Ne laser Atomic laser	• Ga As laser
• Nd-YAG laser	• Argon laser (Ion laser)	
	• CO_2 laser (molecular laser)	

Gas Laser

- Highly monochromatic light
- Most pure spectrum
- Continuous and pulsed output
- Power variation can be from low value to high value (i.e. mW to MW)

Semiconductor Diode Laser

- Its output is slightly broad.
- Most widely used laser due to low cost and simple to design and operate.

■■■■

LIGHT EMITTING DIODE (LED)

4

LIGHT EMITTING DIODE (LED)

- An LED is a forward Biased p-n junction which emits visible light when ON current is made to flow through it.

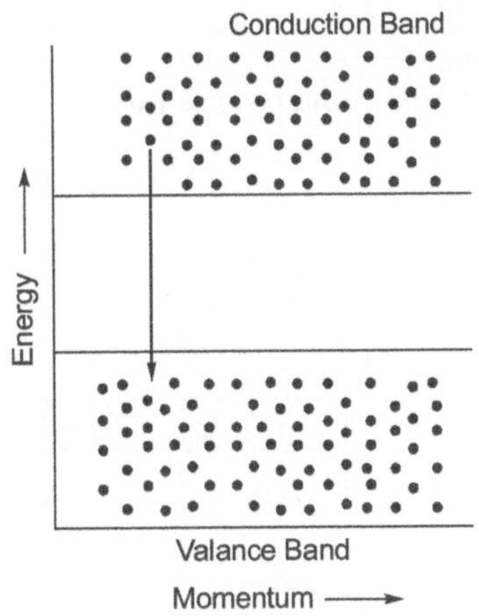

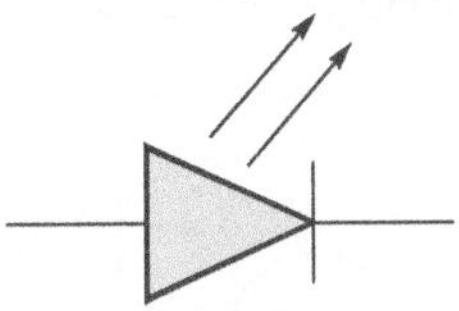

Symbol

- The light is emitted when electrons from the N-side cross the junction and recombine with the holes on the P-side.
- The electrons on the N-side are in heigher level (i.e. conduction band) whereas the holes on the P-side are in lower level (i.e. valence band), whenever the electron falls from conduction to valance band the recombination takes place and the energy is released, this released energy is in the from of either heat or light. The maximum energy released in case of Si, and Ge is in the from of heat whereas in case of Ga As, GaP, Ga, AsP energy released is in the form of light.

Note:

- **GaAs has very high efficiency in terms of photons emitted per carries.**

- Different materials give different energy gap and hence have different colour emission.
  GaAs ⟶ infrared radiation
  GaP ⟶ red or green light
  GaAsP ⟶ red or yellow light

Characteristics of LED

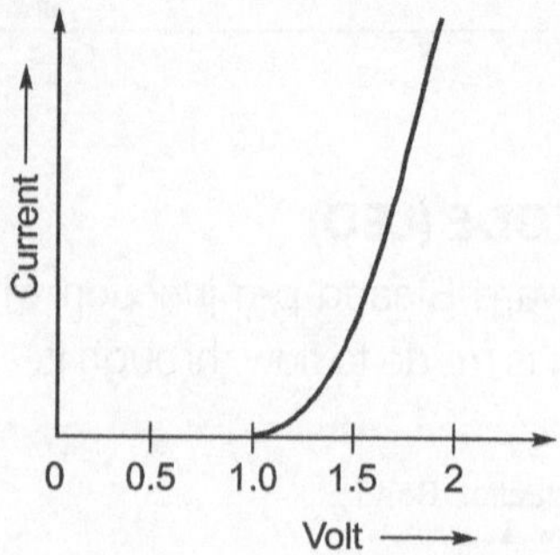

Unlike an ordinary diode, the cut-off voltage for an LED is 1 volt.

Current Through an LED

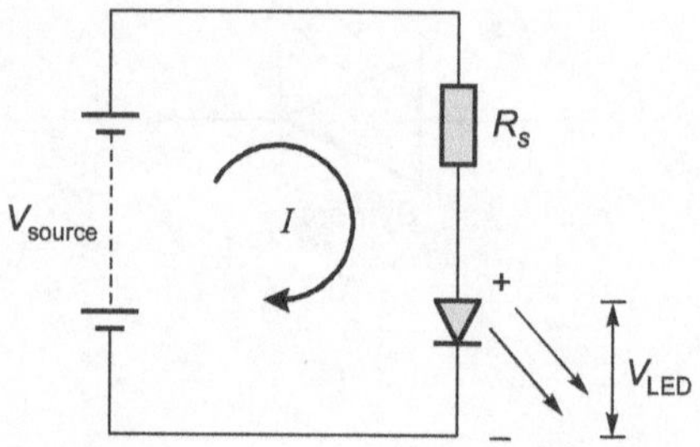

$$I = \frac{V_{source} - V_{LED}}{R_s}$$

Remember: ....

- Never connect an LED with a battery or a power supply directly.
- Led must have a resister in series to limit the current to a safe value.

$$R_s = \frac{(V_{source}) - (V_{LED})}{I}$$

....

Internal Quantum Efficiency

$$\eta = \frac{\text{Light energy emitter}}{\text{Light energy emitter + Heat energy emitter}}$$

$$\eta = \frac{R_r}{R_r + R_{nr}} \times 100\%$$

where

R_e = Radioactive Recombination (light)

R_{nr} = Non-Radioactive Recombination (heat)

■■■■

PHOTO DIODE

5

Photodiode

- A photodiode is a type of photodetector capable of converting light into either current or voltage.

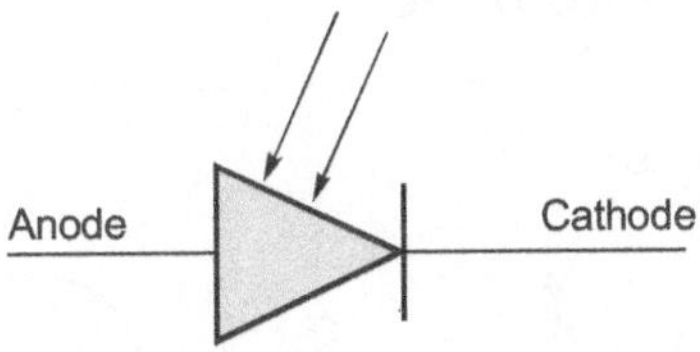

- This device operates in **reverse bias** and the electic field developed across the p-n junction sweeps the mobile charge carriers to their respective majority sides. Hence a deplection region is formed. This barrier stops the flow of majority carrier and supports the flow of only the minority carries and hence the leakage current flows.
- This leakage current is due to both the effects:
    1. In the absence of light falling on the reverse biased photodiode, some leakage current flows called the dark current.
    2. Another leakage current is due to the incidence of light or photons on the diode.
    3. This leakage current is proportional to the current or light faling on the photo diode.

Sensitivity

- Sensitivity of a photodiode is the amount of current flow per unit light irradiance.

$S = \frac{I}{H}$; where = I = Current flowing; H = Irradiance

Note:

- Dark current must be minimised to enhance the sensitivity of the photodiode. The dark current doubles for every 10°C increase in temperature for silicon and this decrease the signal to noise ratio.

Advantages of photodiode over photoconductors

- Better frequency response
- Linear
- Less Noisy

Disadvantages of Photodiode

- Small active area
- Rapid increase in dark current
- Require amplification at low illumination level

Characteristics of Photodiode

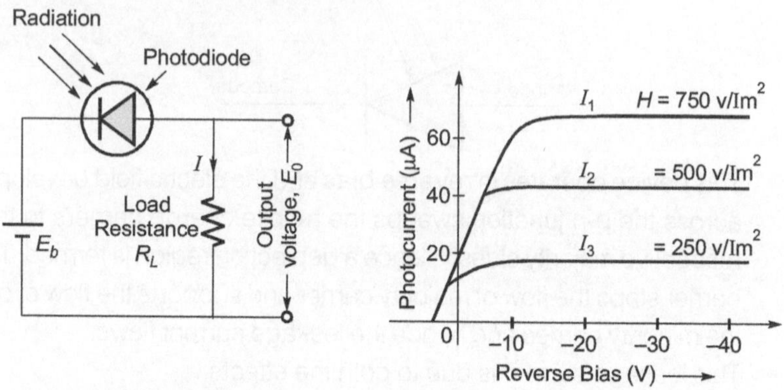

$I_1 > I_2 > I_3$

Photodiode always operates in reverse Bias.

Quantum Efficiency

- The quantum efficiency (η) is defined as the fraction of incident photons which are absorbed by the photoconductor and generate electron which are collected at the detector terminal.

$$\eta = \frac{\text{Number of electrons collected}}{\text{Number of incident photons}} = \frac{R_e}{R_p}$$

where

R_p = incident photon rate (photons/sec)

R_e = electron rate (electron/sec)

Responsivity

$$\boxed{R = \frac{I_p}{P_o}\ (\text{AW}^{-1})}$$

where

R = Responsivity

I_p = Output photocurrent in Amperes

P_o = Incident optical power (W)

$$\boxed{R = \frac{\eta e}{hf} = \frac{\eta e \lambda}{hc}}$$

Current flow through photo dioide (I_p)

$$\boxed{I_p = R \times A \times I}$$

where A = area of photo diode (mm^2)

I = light intensity (mW/mm^2)

Long Wavelength Cut-off

- To built up the leakage current in the photodiode, the energy of the photon must be taken greater than or equal to the energy gap between the valance and the conduction band called band gap (E_g).

$$\therefore \quad h\nu \geq E_g$$

$$\frac{hc}{\lambda} \geq E_g$$

or

$$\lambda \leq \frac{hc}{E_g}$$

So the threshold for the conduction or detection, in commonly called as the long wavelength cut-off point λ_c.

$$\boxed{\lambda_c = \frac{hc}{E_g}}$$

■■■■

PHOTO RESISTOR

6

Photo-Resistor

- Photo-resistor also called as the LDR (Light Dependent Resistor) is a resistor whose resistance decreases with the increasing incident light. It can also be reffered to as the photoconductor.

Symbol

- In case of photo-resistor the current flow increases with the increase in the irradience. When the light falling on its surface increases, its resistance decrease leading to the increase in the current flow in the circuit.

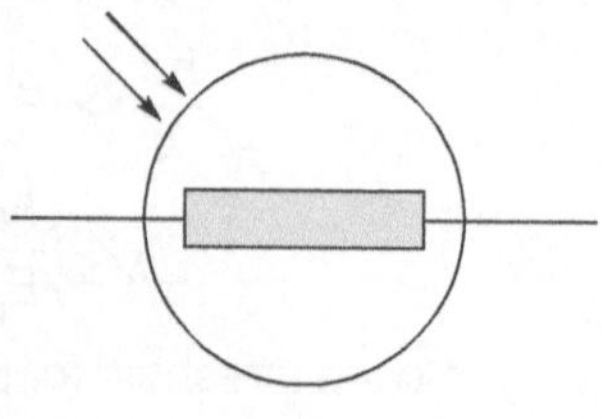

Dark Resistance

- When the cell is kept in the darkness the resistance is called as the Dark Resistance

$$R_t = R_i + (R_f - R_i)\left[1 - \exp\left(-\frac{t}{\tau}\right)\right]$$

where

R_t = resistance at any time 't' (Ω)

R_i = final resistance after the light is incident (Ω).

R_f = dark resistance (Ω)

Remember:

- The resistance of the photoresistor decreases exponentially with the increase in the illumination on its surface.

Materials used to make Photoresistors are

CdS, CdSe, PbS, PbSe

Sensitivity

- The sensitivity of a photoresistor is defined as: $S = \frac{\Delta R}{\Delta H}\ \Omega/\text{W-m}^{-2}$

where

ΔR = Change in resistance (Ω); ΔH = Change in irradiation (W/m^2)

■■■■

INTERFERENCE

7

Interference

- When two or more light waves superimpose on each other, the resultant wave is produced by adding the two waves algebraically keeping their phase difference into consideration.
- This kind of interaction of the waves in the space leads to the phenomenon called as the intereference.
- The phenomenon of interference leads to the formation of the dark and the bright bands.

Constructive Interference

- When the creast of one wave falls on the creast of the other wave and the trough of one wave fall on the trough of the other, the resulting wave well get added up, and hence its brightness will get increased.

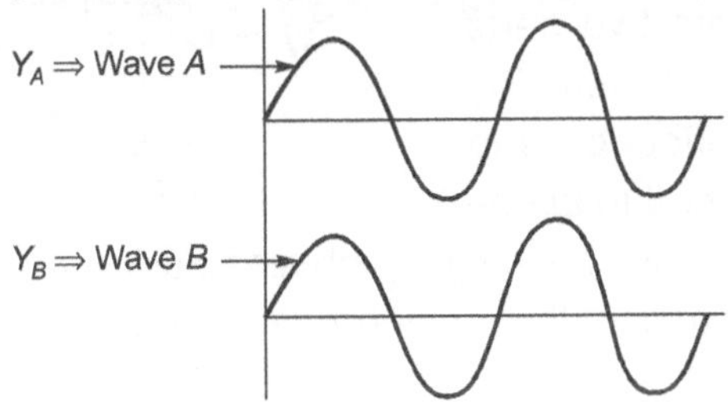

$$Y = Y_A + Y_B$$

Destructive Interference

- When the creast one wave fall on the trough of the other and its trough falls on the creast of the other the resulting wave is obtained by the subtraction of two waves hence the brightness gets reduced.

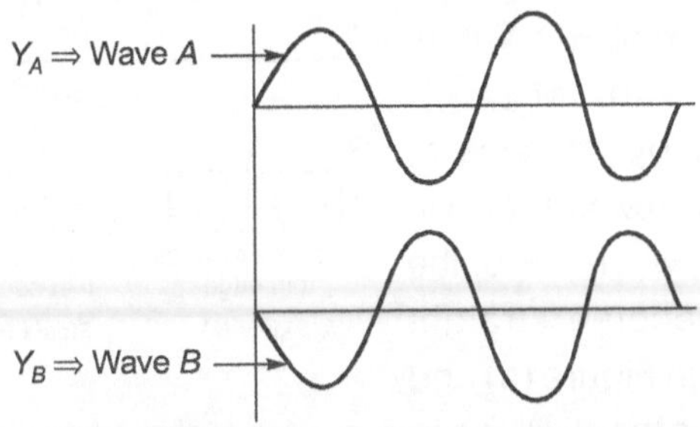

$$Y = Y_A - Y_B$$

Brightness of the wave is reduced in case of destructive interference.

Path Difference

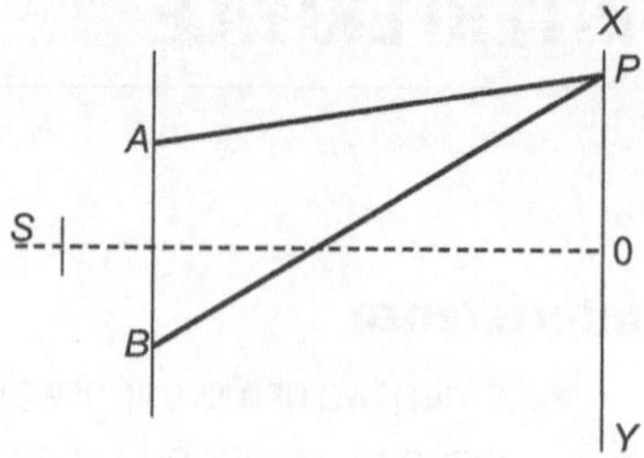

For maximum intensity at point P.

$\Delta = BP - AP = 2n(\lambda/2) = n\lambda$

For minimum intensity at point P

$\Delta = PB - AP = (2n + 1)\,\lambda/2$

where, $n = 0, 1, 2, 3.....$ etc.

Interferometer

- It is a device which is used to produce interference. The device seperate a light beam into two beams and bring them together to produce interference.

Michelson Interferometor

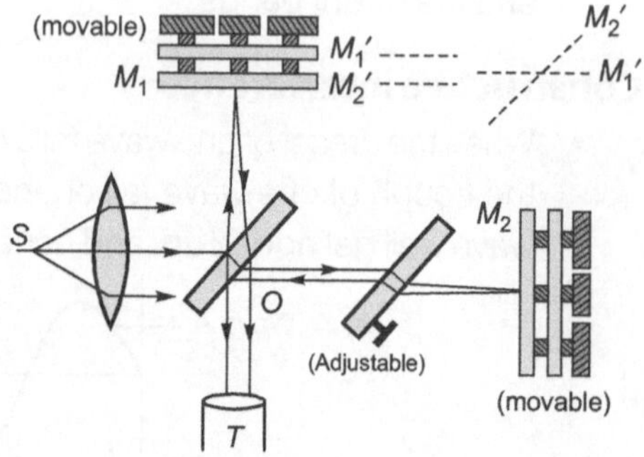

- In michelson interferometer, the amplitude of a light beam comming from an extended source is divided into two parts of equal intensity by partial reflection and transmittance through the mirrors, the two beams then obtained are sent at right angles to each other and are than brough together by reflection through mirrors and focused on a telescope for observation.

Basic Parts:

1. Monochromatic light source 2. Beam Splitter 3. Two Mirrors

Type of Fringes

(i) Circular fringes : For circular fringes to be formed the mirror M_1 and the virtual mirror M'_2 (which is the image of M_2) must be parallel i.e. $M_1 \parallel M'_2$.

(ii) Localised fringes : When the M_1 and M'_2 are seperated by distance d, the fringes are as shown in Figure (a), when $d = 0$, the fringes become straight as shown in Figure (c) and beyond this they will become convex to the other side as is shown in Figure (b) These fringes produced can be viewed by a telescope.

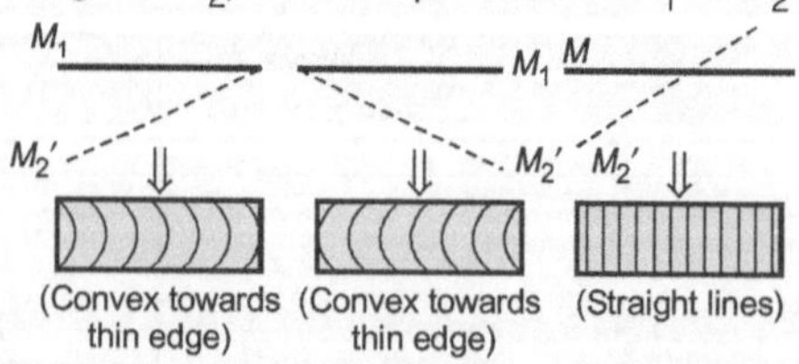

Figure (a) Figure (b) Figure (c)

Application of Michelson Interferometer

(i) Finding the wavelength of monochromatic light

$$n\lambda = 2(\Delta x) = 2(x_2 - x_1)$$

where, n = No. of fringes; λ = Wavelength of monochromatic light
Δx = Micrometer screw movement; x_2 = Final position
x_1 = Initial position

(ii) Calculation of thickness and the refractive index (μ) of any transparent sheet.

(a) 

(b) 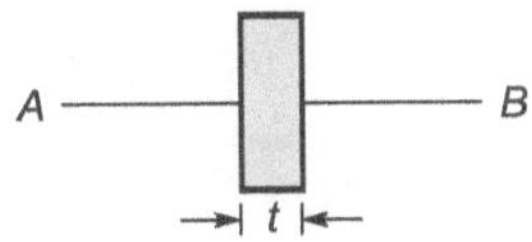

$$n\lambda = 2(\Delta x) = 2t(\mu - 1)$$

(iii) Resolution of Spectral lines

$$\therefore \quad \boxed{\Delta\lambda = \lambda_1 - \lambda_2 = \frac{\lambda_1\lambda_2}{2x} = \frac{(\lambda_{av})^2}{2x}}$$

(iv) **Calculation of thickness of sheet width by other method. (Young's Method)**

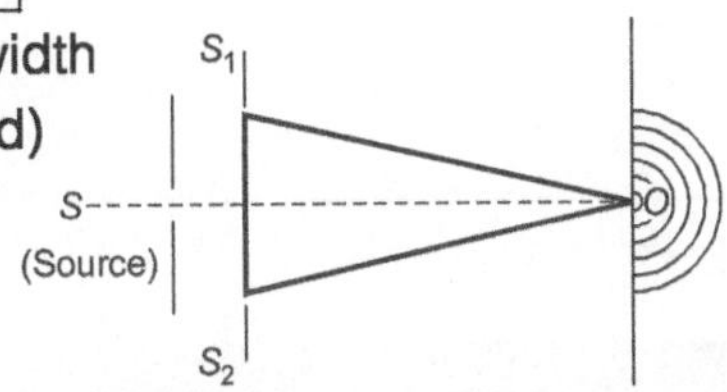

$$n\lambda = t(\mu - 1)$$

$$\boxed{t = \frac{n\lambda}{(\mu - 1)}}$$

Young's Double Slit Experiment

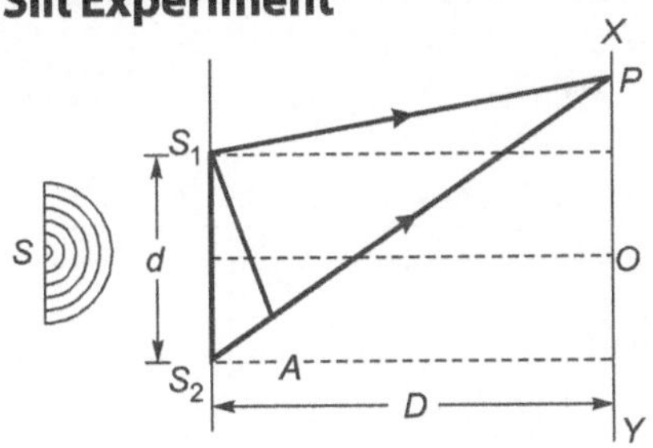

Bright fringe

Distance of nth bright fringe from the contral position '*O*' is

$$\boxed{x_n = \frac{D}{d} n\lambda}$$

Dark fringe

Distance of nth dark fringe from the contral position '*O*' is

$$\boxed{x_n = \frac{D}{d}(2n+1)\frac{\lambda}{2}}$$

Fringe width

- It is denoted by (β). The distance between only two successive bright or dark fringes is called as the fringe width. $\boxed{\beta = \frac{D}{d}\lambda}$

■■■■

A Handbook on

Instrumentation Engineering

10

Communication

CONTENTS

Basics

1

Communication System

A communication system is used to transfer or exchange the information between two points.

Basic Elements of Communication System

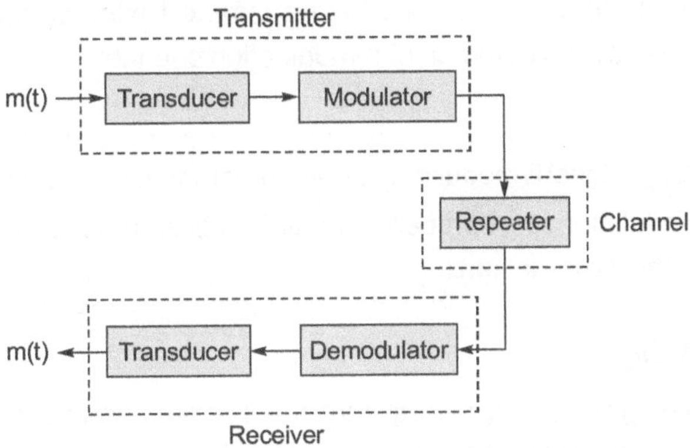

Communication system basically consists transmitter, channel and receiver.

1. **Transducer:** It is a device which converts one form of energy to another form. The input transducer convert the message signal into a time varying electrical signal.
2. **Modulator:** It is used to perform the modulation.
3. **Channel:** It is a physical medium that carries the electrical signal. It is used for connection between transmitter and receiver.
4. **Repeaters:** These produce a fresh copy of transmitted signals.
5. **Demodulator:** It performs reverse operation of modulator i.e. it converts high frequency signal to low frequency signal.

Modulation

Modulation is a process that causes a shift in the range of frequencies in a signal.

Need of Modulation

1. To decrease the length of transmitting and receiving antenna.
2. The low frequencies are attenuated fast and therefore low frequency signal can not be transmitted over a large distance. By translating the low frequency component to high frequency component, long distance communication is then possible.
3. By varying the signal power which is being transmitted, required signal to noise ratio (S/N) can then be obtained.
4. Frequency division multiplexing (FDM) is possible and therefore large number of signals can then be transmitted with different carrier frequencies over a common communication channel.

Note:

- **Message signal is modulating signal and it modulates carrier signal.**
- **In modulation some properties of carrier signal are varied in accordance with the message signal.**

Base Band Signal

The message signal generated from the information source is called base band signal. Base band signal has significant frequency component near to zero or low frequencies.

Band Pass Signal

It is a signal having significant frequencies component for range of frequencies away from zero frequency or low frequency.

Spectrum

It is frequency domain representation of a signal.

Bandwidth

It is defined as band of group of frequencies for which amplitude of signal is not zero.

Note:

- **Carrier frequency is much higher than message signal frequency.**
- **Noise mainly added to signal in the channel.**

■■■■

Fourier Series, Energy and Signals

2

Fourier Series

Fourier Representation of Periodic Function

$$f(t) = a_0 + 2\sum_{n=1}^{\infty}[a_n \cos n\omega_0 t + b_n \sin n\omega_0 t]$$

where $f(t)$ = Periodic function

$\omega_0 = \frac{2\pi}{T_0}$ = Fundamental frequency

T_0 = Fundamental period of function f(t)

$$a_0 = \frac{1}{T_0}\int_{-T_0/2}^{T_0/2} f(t)dt$$ ......mean value of function

$$a_n = \frac{1}{T_0}\int_{-T_0/2}^{T_0/2} f(t)\cdot \cos n\omega_0 t\, dt$$

$$b_n = \frac{1}{T_0}\int_{-T_0/2}^{T_0/2} f(t)\cdot \sin n\omega_0 t\, dt$$

Complex Exponential Fourier Series

$$f(t) = \sum_{n=-\infty}^{\infty} c_n \exp(-jn\omega_0 t)$$

Where,

$$c_n = \frac{1}{T_0}\int_{-T_0/2}^{T_0/2} f(t)\cdot \exp(-jn\omega_0 t)\, dt$$ for n = 0, ± 1, ± 2...

Remember:

Fourier series is used to represent a periodic signal where as Fourier transform is used to represent a non-periodic signal.

- For distortionless transmission

$$\boxed{y(t) = k\,g(t - t_d)}$$

where
$y(t)$ = Output
$g(t)$ = Input
k = Gain
t_d = Time delay

- Transfer function required for distortionless transmission

$$\boxed{H(\omega) = ke^{-j\omega t_d}}$$

Paley-Wiener criterion

The necessary and sufficient condition for amplitude response $|H(\omega)|$ to be realizable

$$\boxed{\int_{-\infty}^{\infty} \frac{ln|H(\omega)|}{1+\omega^2} d\omega < \infty}$$

- **Energy contained in a given signal f(t)**

$$\boxed{E = \int_{-\infty}^{\infty} f^2(t)dt}$$

Parseval's theorems

$$\boxed{E = \frac{1}{2\pi}\int_{-\infty}^{\infty} |F(\omega)|^2\, d\omega}$$

- **Energy spectral density**

$$\boxed{G_E(\omega) = |F(\omega)|^2}\text{ , J/Hz}$$

it is the energy contained in the signal per unit bandwidth

- **Auto correlation function (ACF)**

$$\boxed{R(\tau) = \int_{-\infty}^{\infty} f(t)\cdot f(t-\tau)d\omega}$$

■■■■

Analog Modulation

3

Amplitude Modulation

Amplitude of the carrier signal (high frequency) is varied with the amplitude of modulating signal keeping frequency and the phase of the carrier fixed.

Single Tone Modulation (Sinusoidal Signal)

- Modulating Signal

$$v_m(t) = V_m \cos\omega_m t$$

where, ω_m = Modulating frequency; V_m = Amplitude of modulating signal

- Carrier Signal

$$v_c(t) = V_c \cos\omega_c t$$

where, ω_c = Carrier frequency; v_c = Amplitude of carrier signal

Remember:

- $\omega_c >> \omega_m$
- f_m = 50 Hz - 15 kHz ; f_c = 535 kHz - 1605 Hz
- AM broadcast range is 535 MHz - 1605 MHz.

Amplitude-Modulated (AM) Signal

$$v_{AM}(t) = [V_c + K_a V_m \cos\omega_m t]\cos\omega_c t$$

or

$$v_{AM}(t) = V_c[1 + m_a \cos\omega_m t]\cos\omega_c t$$

$$\boxed{m_a = K_a \frac{V_m}{V_c}}$$

where, m_a = Modulation index = Depth of modulation
= Percentage modulation

K_a = Constant of proportionality, and depends upon circuit from where AM signal has been generated

Note:

- Unless specified take $K_a = 1$ then $m_a = \frac{V_m}{V_c}$.
- Practically $0 < m_a < 1$, generally take value of m_a = 0.45 to 0.6.
- If m_a=0 then unmodulated signal. If m_a=1 then 100% modulated signal.

Representation of Various AM Signal

1. AM-DSB/FC: It is a standard signal

$$v_{AM}(t) = \underbrace{V_c \cdot \cos\omega_c t}_{\text{Free carrier}} + \frac{1}{2}m_a V_c \left[\underbrace{\cos(\omega_c + \omega_m)t}_{\text{Upper side band}} + \underbrace{\cos(\omega_c - \omega_m)t}_{\text{Lower side band}}\right]$$

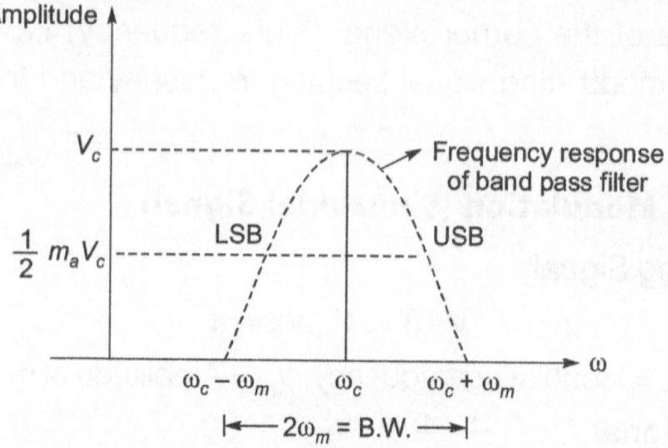

Spectrum of AM-DSB/FC sigal

Important Points:

- AM-DSB/FC signal represents standard AM signal which requires maximum amount of power and maximum bandwidth for its transmission.
- Such system is used for broadcast purpose since its circuit configuration is simplest.
- Carrier carries no information. USB and LSB carry equal information since both have same frequency component with reference to the centre frequency ω_c.
- For the transmission of such signal, a band pass filter is used, having centre frequency of ω_c and a bandwidth of 2 ω_m.
- For the transmission of such signal, a band pass filter is use, having centre frequency of ω_c and a bandwidth of 2 ω_m.
- **Total power required for transmission of AM-DSB/FC**

$$P_{Total} = P_{carrier} + P_{LSB} + P_{USB} \qquad \boxed{P_t = P_c\left(1 + \frac{m_a^2}{2}\right)}$$

where, P_c = Carrier power; P_{USB} = Upper side band power

P_{LSB} = Lower side band power

Note: ....

If carrier is modulated by several modulating signals then over all modulation index:

$$\boxed{m_a = \sqrt{m_{a1}^2 + m_{a2}^2 + \cdots m_{an}^2}}$$

2. AM-DSB/SC

$$v_{AM}(t) = \frac{1}{2} m_a V_c \left[\cos(\omega_c + \omega_m)t + \cos(\omega_c - \omega_m)t\right]$$

3. AM-SSB/FC

$$v_{AM}(t) = V_c \cdot \cos\omega_c t + \frac{1}{2} m_a V_c \left[\cos(\omega_c \pm \omega_m)t\right]$$

(+ → USB, − → LSB)

4. AM-SSB/SC

$$v_{AM}(t) = \frac{1}{2} m_a V_c \left[\cos(\omega_c \pm \omega_m)t\right]$$

Different AM System: At $K_a = 1$

System	P_t	P_s (saved power)	BW	% power saving	Complexity	SNR
AM-DSB/FC	$P_c\left(1+\frac{m_a^2}{2}\right)$	–	$2\omega_m$	–	Min.	Max.
AM-DSB/SC	$P_c \times \frac{m_a^2}{2}$	P_C	$2\omega_m$	67%	↓	Low
AM-SSB/FC	$P_c\left(1+\frac{m_a^2}{4}\right)$	$P_c + \frac{m_a^2}{4}$	ω_m	16%	↓	High
AM-SSB/SC	$P_c \times \frac{m_a^2}{4}$	$P_c\left(1+\frac{m_a^2}{4}\right)$	ω_m	83%	Max.	Min.

Note:

- AM-DSB/FC system is the best from the circuit configuration point of view. It has least complexity in the circuit but requires maximum bandwidth as well as maximum power.
- AM-SSB/SC system is the worst from the circuit complexity point of view. It has maximum complexity in the circuit but power and bandwidth requirement are minimum.
- AM-DSB/FC system are used for broadcast purpose whereas AM-SSB/SC system is used for point to point communication.

- Transmission Efficiency

$$\eta = \frac{\text{Useful power}}{\text{Total power}}$$

$$\boxed{\eta = \frac{P_s}{P_T} \times 100\%}$$

Case-1: AM-DSB/FC

$$\eta = \frac{m_a^2}{2 + m_a^2} \times 100\%$$

If $m_a = 1$ then $\eta = 33\%$.

Case-2: AM-DSB/SC

$\eta = 100\%$

- Percentage power savings

$$\% \text{ Power saving} = \frac{\text{Power saved}}{\text{Total power}}$$

Case-1 : AM-DSB/SC

$$\% \text{ Power saving} = (1 - \eta) = \frac{2}{2 + ma^2}$$

Case-2 : AM-SSB/SC

$$\% \text{ Power saving} = \frac{4 + ma^2}{2(2 + ma^2)}$$

Note: ....

- For full carrier system the transmission efficiency depends upon modulation index (m_a).
- For full carrier system the maximum efficiency is only 33% for maximum modulation index of unity.
- For suppressed carrier system the transmission efficiency is always independent of modulation index and is always 100%.

....

Multi Tone Modulation (Non Sinusoidal)

Representation of Various AM-Signal

1. AM-DSB/FC

Standard signal

In time domain

$$v(t) = (A + f(t))\left[\frac{1}{2}\left(e^{j\omega_c t} + e^{-j\omega_c t}\right)\right]$$

In frequency domain

$$V(\omega) = \frac{1}{2}\left[\{F(\omega - \omega_c) + F(\omega + \omega_c)\} + \pi A\{\delta(\omega - \omega_c) + \delta(\omega + \omega_c)\}\right]$$

2. AM-DSB/SC

Standard signal

In time domain

$$v(t) = f(t)\left[\frac{1}{2}\left(e^{j\omega_c t} + e^{-j\omega_c t}\right)\right]$$

In frequency domain

$$V(\omega) = \frac{1}{2}[F(\omega - \omega_c) + F(\omega + \omega_c)]$$

Important Points:

- In suppressed carrier signal, the free carrier is absent.
- Modulation represents frequency translation. Therefore original spectrum centered about zero frequency if translated to frequency spectrum centered at ω_c.
- Using modulation any low pass signal is modulated to band pass signal.
- Negative frequencies in the original spectrum have no physical meaning. These are represented since
  - (a) The frequency spectrum of any signal is always an even function of ω and therefore negative and positive frequencies occurs in pair.
  - (b) Any negative frequency can be made a positive frequency by frequency translation or using modulation.

AM Modulators and Demodulators

- Product modulation is used for generation of AM waves using non linear devices.
- Switching modulator is also used for generation of AM waves.
- Balanced modulator and ring modulators are used for generation of DSB-SC waves.
- SSB wave is generated using analog multiplier and band pass filter.
- Filter method and phase shift methods are also used for generation of SSB wave.

- **Enveloped Detector**

  It is used for detection of AM wave.

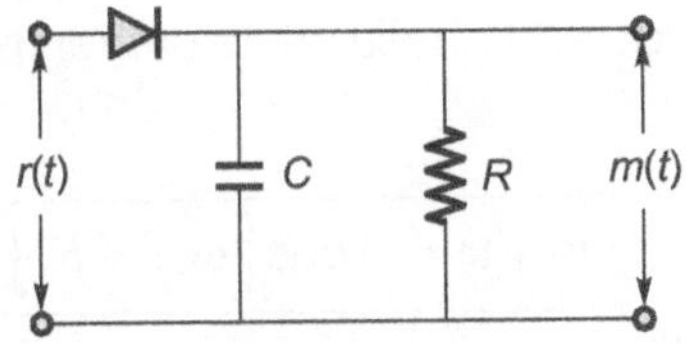

$r(t)$ is received signal and $m(t)$ is message signal and for better reception RC must be selected such as

$$\frac{1}{f_c} << RC << \frac{1}{\omega}$$

where, ω is bandwidth of message signal (in Hz).

Note: ....

- Demodulation of DSB-SC and SSB waves is done using coherent carrier signal at receiver.
- Envelope detector can also be used to recover message by passing received VSB signal through it.

....

Angle Modulation

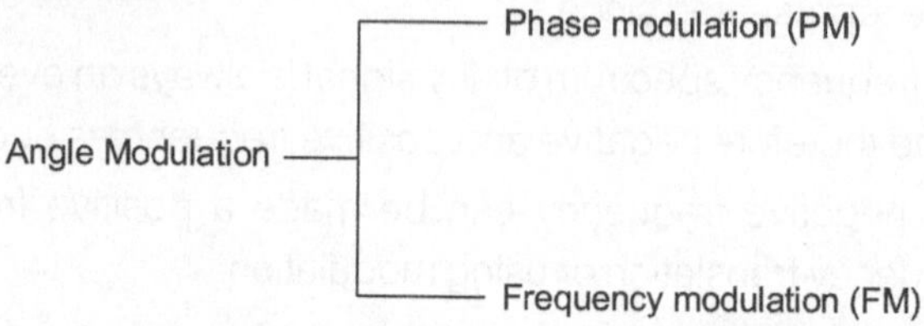

Phase Modulation (PM)

In phase modulation the phase of the carrier is varied in accordance with instantaneous value of the amplitude of the modulating signal keeping the amplitude and frequency of the carrier fixed.

General Expression

$$v_{PM}(t) = A\cos[\omega_c t + K_p f(t)]$$

where, $f(t)$ = Modulating signal

K_p = Phase sensitivity or phase deviation

Frequency Modulation (FM)

In frequency modulation the frequency of the carrier is varied in accordance with the instantaneous value of the amplitude of modulating signal keeping amplitude and phase of the carrier fixed.

General Expression

$$\boxed{V_{FM}(t) = A\cos\left[\omega_c t + K_f \int f(t)dt\right]}$$

where, $f(t)$ = Modulating signal

K_f = Frequency sensitivity

If $f(t) = V_m \cos\omega t$

then $$\boxed{V_{FM}(t) = A\cos[\omega_c t + m_f \sin\omega_{mt}]}$$

where, $m_f = K_f \dfrac{V_m}{\omega_m}$ = Modulation index

Note:

- A pure PM signal or a pure FM signal cannot be obtained since the phase change or the frequency change are inter-related.
- In PM the modulation index is directly proportional to the amplitude of the modulating signal where as the modulating index in the FM depends upon the amplitude as well as the frequency of modulating signal.

- Total Power Required for FM Signal

$$v_{FM}(t) = A\cos\left[\omega_c t + K_f \int f(t)dt\right]$$

$$\boxed{P = \frac{A^2}{2}}$$

- Frequency Deviation

$$\boxed{\delta = m_f\ \omega_m} \text{ rad/sec.}$$

or

$$\boxed{\delta = m_f\ f_m} \text{ Hz}$$

δ represents total bandwidth require for the transmission of FM signal.

Note:

- Bandwidth require for FM signal is always more than the bandwidth require for the AM signal.
- ω_m control the spacing between two successive sideband.
- The modulation index (m_f) controls the number of significant sidebands.

FM Standards

Remember:

- Carrier frequency = 88 MHz – 108 MHZ
- Maximum frequency deviation = ±75 kHz
- Intermediate frequency (I.F.) fi = 10.7 MHz

- Bandwidth

Practical bandwidth required is given by Carson's rule

$$\boxed{\text{B.W.} \cong 2\omega_m(m_f + 1)} \text{ rad/sec.}$$

or

$$\boxed{\text{B.W.} \cong 2(\delta + \omega_m)} \text{ rad/sec.}$$

Case-1 : Narrow band FM (NBFM)

$$m_f << 1 \qquad \text{B.W.} \cong 2W_m$$

Case-2 : Wide band FM (WBFM)

$$m_f >> 1 \qquad \text{B.W.} \cong 2\delta$$

It is used in FM radio broadcast.

Note:

- Same bandwidth required for AM-DSB system.
- NBFM used in mobile radio.

Remember:

- The modulation index determines the number of sidebands. Components have significant amplitudes.
- In FM, the total transmitted power always remains constant, but with increased depth of modulation, the required bandwidth, is increased.
- Deviation ratio is the shift in carrier frequency from its resting point compared to the amplitude of the modulating voltage.
- A deviation ratio of 5 is the maximum allowed in commercially broadcast FM.

Angle Modulators and Demodulators

- FM wave can be generated using VCO caused direct method.
- FM wave can be generated using varactor diode.
- FM and PM both can be generated using reactance tubes.
- Another method is first generate narrow band FM then pass through it frequency multiplier.
- Demodulation is normally done using PLL.

Remember:

The FM system requires larger bandwidth and therefore the information is transmitted through larger number of frequency components or larger number of sidebands. Therefore the quality of recovered signal using FM system is much better as compare to that using AM system. Therefore FM system is Hi-Fi system and therefore has the ability to reproduce same quality of signal which was actually transmitted.

Superheterodyne receiver

(i) **Sensitivity:** It is the minimum signal that should be present at input of a receiver to get standard output. It depends on gain of the receiver.

(ii) **Selectivity:** It is the ability of a receiver to reject the unwanted frequency component.

(iii) **Fidelity:** It is the ability of the receiver to reproduce all the frequency component at the output of the receiver.

- Intermediate frequency

$$\boxed{f_i = f_l - f_s}$$

where, f_l = Local oscillator frequency

f_s = desired signal frequency

- Image frequency

It is undesired frequency and to be rejected.

$$\boxed{f_{si} = 2f_l - f_s}$$

$$\boxed{f_{si} = f_s + 2f_i}$$

where, f_{si} = image signal frequency

- Image Signal Rejection

$$\boxed{\alpha = \sqrt{1 + Q^2 \rho^2}}$$

where, $\alpha_{dB} = 20 \log \alpha$

$$\rho = \left(\frac{f_{si}}{f_s} - \frac{f_s}{f_{si}}\right)$$

Q = Quality factor

Note:

- If two tuned circuit are connected in cascade then $\alpha = \alpha_1 \cdot \alpha_2$.
- The image signal rejection represents the ratio of desired signal amplitude to the amplitude of undesired signal.
- For any good radio receiver the parameter α must have high value.

■■■■

Pulse Modulation

4

Pulse Analog Modulation

1. **Pulse amplitude modulation (PAM):** The amplitude of pulse is varied in accordance with the height of the modulating signal keeping the width and the position of the pulse constant.
2. **Pulse width modulation (PWM):** The width of the pulses is varied in accordance with the height of modulating signal keeping amplitude and position of pulse constant.
3. **Pulse position modulation (PPM):** The position or the location of the pulses is varied in accordance with the height of the modulating signal keeping the amplitude and the width of the pulses constant.

Note:

- Speed control of DC motor is done using PL/m.
- PWM is generated using 555 timer in monostable multivibrator mode.
- PPM is generated using 555 timer by using PWM as a trigger signal in monostable multivibrator mode.

Comparison of various pulse modulation signals

	Circuit Complexity	BW	SNR
PAM	Minimum	$\sim 2\omega_m$	Minimum
PWM	↓	$\sim 10\omega_m$	↓
PPM	Maximum	$\sim 1/t_r$	Maximum

where, t_r = Rise time of each pulse

Note:

- PPM is most effective pulse analog modulator scheme in forms of SNR.
- SNR should high, it is an advantage.
- The aperture effect occurs only when flat topped sampling is used. This error voltage is acceptable since the generation of flat top sampling is least complex.

Time Division Multiplexing (TDM)

1. In TDM entire time interval into smaller time slots and corresponds to each time slot, the sample from a specified signal is transmitted over a common communication channel.
2. A common sampling frequency is use for transmission of various signal.
3. Simplex circuit and less costly.
4. Used for discrete signals such as pulse modulated signals or pulse code modulated signals (PCM).

Frequency Division Multiplexing (FDM)

1. In FDM entire frequency band available is divided into smaller frequency bands and corresponding to each frequency band the signals are transmitted with different carrier frequencies at all times.
2. A separate transmitter is require for the transmission of each signal.
3. Complex circuit and therefore more costly.
4. FDM system is used for continuous signals such as AM and FM signals.

Pulse Digital Modulation

Sampling Theorem

In order to recover the original modulating signal from its sampled version the signal must be having a sampling frequency of greater than or equal to twice of highest modulating frequency component contain in the given signal i.e.

$$\boxed{f_s \geq 2f_m} \quad ; \quad \boxed{\omega_s \geq 2\omega_m}$$

where, f_s = Sampling rate

The Nyquist rate of sampling represents the minimum rate sampling so that the original signal can be recovered from its sampled version.

$$\boxed{f_{s(\text{min})} = f_{s(\text{Nyquist})} = 2f_m} \text{ Samples/sec}$$

where, f_m = Maximum modulating frequency component

Pulse Code Modulation (PCM)

1. The signal to noise ratio of PCM signal is very high.
2. Only two level signal and therefore the noise can always be minimise by passing the PCM signal through a limiter circuit.
3. For detecting the PCM signal a threshold level is fixed and incoming voltage is detected on the basis of whether the incoming voltage is

more than or less than this threshold level. The detection does not depend upon the absolute value of the voltage receive but depends only upon the relative voltage receive with reference to the threshold value.

4. Very low probability of error. Error in the detection takes place only when
   (a) The noise exceeds the threshold value.
   (b) The signal is detected at the time instant where the nose exceeds the threshold value.

Note:

The PCM signal cannot be transmitted through antenna since the sampling is done at a rate comparable to highest modulating frequency component. Therefore the PCM signal is transmitted through a transmission line or co-axial cable and therefore the range transmission is limited.

- **Quantization levels**

$$L = 2^n$$

where, n = Number of bits/sample
L = Number of quantization levels

- **Step size**

$$\delta = \frac{2V_m}{L} = \frac{2V_m}{2^n}$$

where, V_m = Maximum amplitude of sinusoidal signal

- **Quantization Noise**

$$N_q = \frac{\delta^2}{12}$$

- **Signal to Noise Ratio (SNR)**

$$\frac{S_0}{N_q} = \frac{3}{2}L^2 = \frac{3}{2}2^{2n} \quad ; \quad \left(\frac{S_0}{N_q}\right)_{dB} \cong 1.76 + 6n$$

- **Bit rate**

$$R_b = nf_s$$

where f_s = Number of samples per sec.

- **Min-bandwidth**

$$BW_{min} = \frac{1}{2}R_{b(min)} = nf_m$$

Note: ....

The bit coading parameter 'n' is adjusted depending upon.

- Availability of B.W. of transmission.
- Required value of the signal to noise ratio.
- The complexity of circuit which we can afford for the transmission of given signal.

....

Delta Modulation

As the bit coding parameter increases the number of quantization level of the quantizer will increase. Therefore design of the quantizer becomes more complex.

The use of quantizer is avoided in the delta modulation system. It has much simpler circuit and SNR of the delta modulation system is comparable to the PCM system since both are digital modulation system.

Error Signal

$$\boxed{\Delta(t) = f(t) = \tilde{f}(t)}$$

where, $f(t)$ = Present sample value of the input signal

$\tilde{f}(t)$ = Latest approximation to $f(t)$

Remember: ....

$$\boxed{f_s\big|_{\text{DM system}} > f_s\big|_{\text{PCM system}}}$$

f_s = Sampling Frequency

$$\boxed{\text{BW}\big|_{\text{DM system}} > \text{BW}\big|_{\text{PCM system}}}$$

....

Noise in DM system

1. Granular noise
2. Slope overload noise

(a) To minimize granular noise:

(i) Step size 'δ' should be decreased to limit the range of transmission.

(ii) Sampling frequency increased, f_s ~ 3 to 4 times of f_s.

(b) To avoid slope overload noise

$$\boxed{\delta \cdot f_s \geq 2\pi f_m A} \; ; \; \boxed{A_{max} = \frac{\delta f_s}{2\pi f_m}}$$

To avoid or minimize slope overload noise, the above mention condition must be satisfied and it is controlled by solving 4 parameters.

(i) Maximum amplitude of the modulating signal.

(ii) Frequency of the modulating signal.

(iii) Step size 'δ'.

(iv) The sampling frequency f_s.

Note: ....

If these conditions are specified, still the slope overload noise will always occur. To minimise the slope overload noise under this condition a delta modulator with variable step size is used. Such delta modulator is then called, adaptive delta modulator (ADM).

....

Adaptive Delta Modulation (ADM)

- In adaptive delta modulation, step size is chosen in accordance with message signal sampled value to overcome slope overload error and hunting.
- If message is varying at a high rate then step size is high and if message is varying slowly the step size is small.

Note: ....

- In case of ADM and DM bandwidth required is almost same.
- The PCM and DM signal cannot be transmitted through antenna, since the sampling is done at the rate f_s or Δf_s, therefore such signals are then transmitted through a transmission line so that the range of transmission is limited.
- To transmit such signals through antenna, a high frequency carrier is modulated using the PCM or DM signals as the modulating signals, therefore a low pass signal is converted to a band pass signal and is then transmitted through antenna for long distance communication.

....

■■■■

Digital Carrier Modulation

5

- In digital carrier modulation system, the modulating signal is digital which is the output of PCM or DM system and carrier signal is high frequency sinusoidal carrier.
- In digital carrier modulation system one of the properties of the carrier, namely amplitude, frequency or the phase is varied at a time with the binary modulating signal.
    1. **Amplitude shift keying:** The generation of ASK signal is simplest, is amplitude dependent, maximum noise is introduced and therefore ASK system has lowest signal to noise ratio.
        - ASK signal uses ON-OFF signaling.
        - In ASK probability of error (P_e) is high.
        - The ASK system is used for telegraphy.
    2. **Frequency shift keying:** The circuit configuration of FSK system is most complex, requires large bandwidth for its transmission but has relatively high S/N ratio.
        - FSK signal uses NRZ signaling.
        - In case of FSK, P_e is less.
        - Multiplexing is difficult.
        - Used in MODEM.
    3. **Phase shift keying:** The PSK system has relatively high SNR, relatively complex circuit, require lesser bandwidth as compare to FSK system.
        - PSK signal uses NRZ signaling.
        - This system has lowest probability of error
        - The PSK in its modified form is used for satellite communication or for the mobile communication.

- **Comparison between ASK, PSK, FSK**

Parameter	ASK	PSK	FSK
E	$\frac{1}{2}A^2T$	$\frac{1}{2}A^2T$	$E' \equiv 2E$
P_s	$\frac{1}{2}\left(\frac{E}{T}\right)$	$\left(\frac{E}{T}\right)$	$\left(\frac{E}{T}\right)$
$P_{e(min)}$	$\text{erfc}\sqrt{\frac{P_sT}{\eta}}$	$\text{erfc}\sqrt{\frac{2P_sT}{\eta}}$	$\text{erfc}\sqrt{\frac{P_sT}{\eta}}$
BW	$\frac{2}{T_b}$	$\frac{2}{T_b}$	$(f_2 - f_1) + \frac{2}{T_b}$

Binary Phase Shift Keying (BPSK)

- Waveform b(t) is a NRZ (non-return-to-zero) binary waveform
- Transmitted Signal

$$V_{BPSK}(t) = b(t)\sqrt{2P_s}\cos\omega_0 t$$

where, P_s = Signal power

$b(t)$ = 1 V for Logic level 1

= −1 V for Logic level 0

- Received signal

$$V_{BPSK}(t) = b(t)\sqrt{2P_s}\cos(\omega_0 t + \theta)$$

where, θ = Phase shift corresponding to time delay θ/ω_0

Phase shift depends on the length of the path from transmitter to receiver

- Power spectral density of the BPSK signal

$$G_{BPSK}(f) = \frac{P_s T_b}{2}\left\{\left[\frac{\sin\pi(f-f_0)T_b}{\pi(f-f_0)T_b}\right]^2 + \left[\frac{\sin\pi(f+f_0)T_b}{\pi(f+f_0)T_b}\right]^2\right\}$$

where, T_b = Bit duration

- Energy contained in a bit duration

$$E_b = P_s T_b$$

- Bandwidth

$$BW = \frac{2}{nT_b}$$

where, n = Number of input

Remember: ....

- Differential phase-shift keying (DPSK) and differential encoded PSK (DEPSK) are modifications of BPSK.
- DPSK avoids the need to provide the synchronous carrier required at the demodulator for detecting a BPSK signal.

....

Quadrature Phase Shift Keying (QPSK)

- Four quadrature signals

$$V_m(t) = \sqrt{2P_s}\cos\left[\omega_0 t + (2m+1)\frac{\pi}{4}\right] \quad .... m = 0, 1, 2, 3$$

- Bandwidth

$$\boxed{\text{BW} = \frac{2}{2T_b} = R_b}$$

M-ARY PSK

$$\boxed{V_m(t) = \sqrt{2P_s}\cos(\omega_0 t + \phi_m)} \quad ....m = 0, 1, 2, ...(M-1)$$

- Phase Angle

$$\boxed{\phi_m = (2\,m + 1)\frac{\pi}{M}}$$

Binary Frequency Shift Keying

$$\boxed{V_{\text{BFSK}}(t) = \sqrt{2P_s}\cos[\omega_0 t + d(t)\Omega t]}$$

where, $d(t)$ = 1 for logic levels 1 of data waveform
= –1 for logic levels 0 of data waveform
Ω = Constant offset from nominal carrier frequency

Note: ....

$$\boxed{\text{BW}_{(\text{BFSK})} = 2 \times \text{BW}_{(\text{BPSK})}}$$

....

Minimum Shift Keying

- Transmitted signal

$$\boxed{V_{\text{MSK}}(t) = \sqrt{2P_s}\left[\left\{b_e(t)\sin 2\pi\left(\frac{1}{4T_b}\right)\right\}\cos\omega_0 t + \left\{b_0(t)\cos 2\pi\left(\frac{1}{4T_b}\right)\right\}\sin\omega_0 t\right]}$$

- Most important and useful feature of MSK is its phase continuity

$$\boxed{\int_0^{T_b} \sin\omega_H t \cdot \sin\omega_L t\, dt = 0}$$

- Probability of error in the detection of any signal.

$$\boxed{P_{e(\min)} = \text{erfc}\left(\frac{a}{\sqrt{\eta E/2}}\right)}$$

where, erfc(u) = Complementary error function
a = Threshold level

■■■■

6 Random Variables and Noise

Theory of Random Signals and Process

1. If p(x) is the probability density function

$$\int_{-\infty}^{\infty} p(x)dx = 1$$

2. Cumulative distribution function (CDF)

$$P(X) = \begin{cases} \int_{-\infty}^{x} p(x)dx & , X \le x \\ 1 & , X < \infty \end{cases}$$

3. Avg. or dc value of signal/mean

$$\bar{X} = E(X) \equiv m = \int_{-\infty}^{\infty} x\,p(x)dx$$

4. DC power contained in the given signal

$$P_{dc} = m^2$$

5. Total power contained in the given signal/mean square value

$$\bar{X}^2 = E(X^2) \equiv \int_{-\infty}^{\infty} x^2 p(x)dx$$

6. AC power contained in the signal/variance

$$P_{AC} = \sigma_x^2 = \left[E(X^2) - m^2\right]$$

where, σ_x = Standard deivation

7. RMS voltage

$$V_{rms} = \sqrt{\bar{x}^2} = \sqrt{E(x^2)}$$

Noise

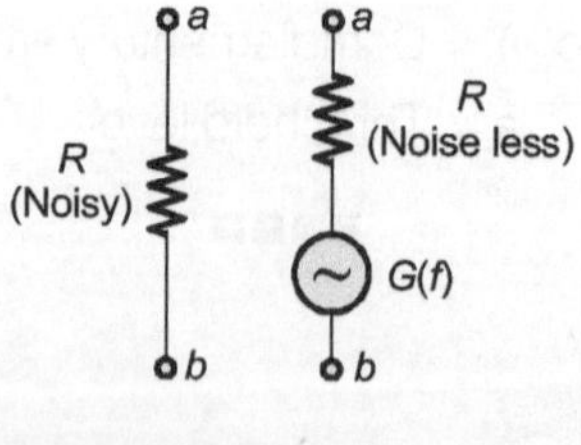

- **Power spectral density of thermal noise**

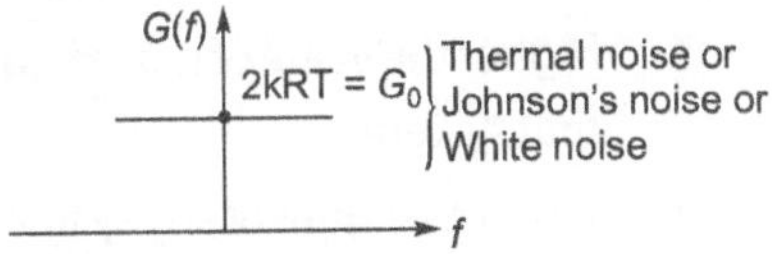

$$\boxed{G(f) = 2kRT}$$

where, k = Boltzmann constant
$= 1.38 \times 10^{-23}$ J/°K

- **Noise power** $$\boxed{N_0 = \int_{-\infty}^{\infty} G(f)df = \int_{-\infty}^{\infty} 2kRT\,df = \infty}$$

- The thermal noise requires infinite amount of power for its generation and infinite amount of BW for its transmission.
- Band limited thermal noise.

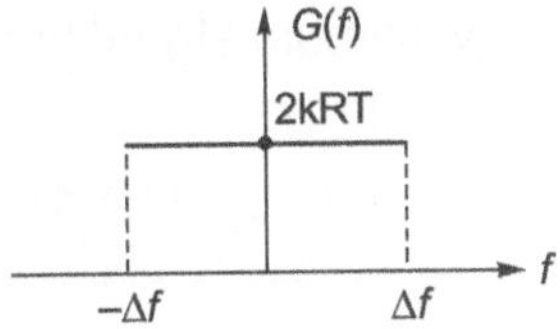

$$\boxed{G(f) = 2kRT} \text{ W/Hz } \quad \ldots\ldots -\Delta f < f < \Delta f$$

$$\boxed{N_0 = \int_{-\Delta f}^{+\Delta f} G(f)df = 4kRT\Delta f} \ W$$

$$\boxed{V_{rms} = \sqrt{N_0} = \sqrt{4kRT\Delta f}} \ V$$

Noise Figure of a Communication System

- The noise figure of any communication system is a measure of total noise power contributed by that communication system.
- For an ideal communication system or for a noise-less system the noise figure has a minimum value of unit.

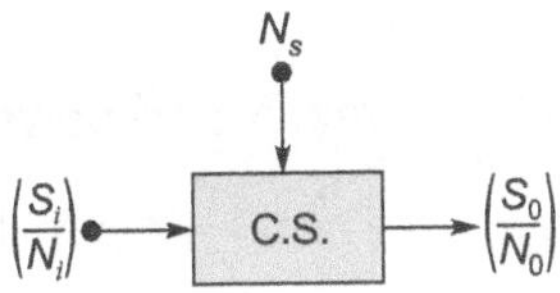

$$\boxed{F = \frac{\text{Input SNR}}{\text{Output SNR}}}$$

$$\boxed{F = 1 + \frac{N_S}{AN_i} = 1 + \frac{T_{eq}}{T_0} = 1 + \frac{R_{eq}}{R_s}}$$

where, A = Power gain; N_s = System noise power; N_i = Input noise power; T_{eq} = Equivalent noise temperature; T_0 = Room temperature R_{eq} = Equivalent input noise resistance; R_s = Source resistance

- Noise figure of cascaded Communication Systems

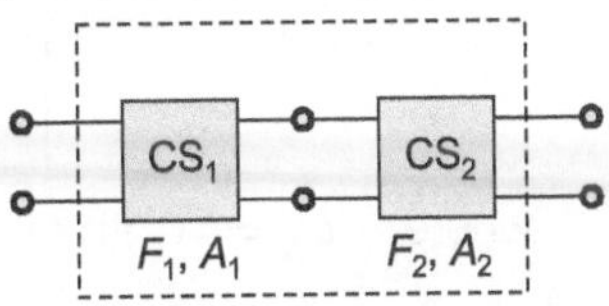

$$\boxed{F = F_1 + \frac{F_2 - 1}{A_1} + \frac{F_3 - 1}{A_1 A_2} + \ldots}$$

Note:

- Higher the value of noise figure, higher is the contribution of the noise by that system.
- The noise figure of any system does not depend upon signal power at the input. Therefore the noise figure remain same irrespective of the variation in the input signal power.
- By cascading the power gain of the overall system increases to a very large value but contribution of the noise due to cascading is only marginal. Therefore cascading of the communication system is always preferred.

- **Maximum signal to noise ratio**

$$\rho_{max} = (SNR)_{max} = \int_{-\infty}^{\infty} |S(\omega)|^2 \, d\omega = \frac{1}{\pi\eta} \int_{-\infty}^{\infty} |S(\omega)|^2 \, d\omega$$

$$\rho_{max} = \frac{2E}{\eta}$$

where, $S(\omega)$ = Fourier transform of message signal $s(t)$

Note:

Maximum SNR at output depends upon the spectrum and energy contained in the input signal.

- **Power sectral density of Thermal Noise**

$$G_V(f) = 2\,kRT \quad (V^2/Hz)$$

- **Thermal Noise power**

$$T_N = \frac{G_V(f)}{4R} = \frac{kT}{2} \quad (W/Hz)$$

- **Power spectral density of white noise**

$$G_{WN}(f) = \frac{\eta}{2}$$

where, η = constant

- **Noise equivalent Band-width**

$$B_N = \frac{\pi}{2} B_{eff.}$$

where, B_N = Width of white noise

■■■■